THE ENGLISH LANGUAGE

An Introduction

THE ENGLISH LANGUAGE

An Introduction

W. NELSON FRANCIS

THE ENGLISH UNIVERSITIES PRESS LIMITED
ST. PAUL'S HOUSE WARWICK LANE
LONDON EC4

FIRST PRINTED IN GREAT BRITAIN 1967
REPRINTED 1969, 1970

SBN 340 04987 1

PRINTED IN GREAT BRITAIN FOR THE ENGLISH UNIVERSITIES PRESS LTD.,
BY LOWE AND BRYDONE (PRINTERS) LTD.

CONTENTS

Contents

inadequacies of the equipment used has meant the late appearance of Volume 7, but I hope the reader will feel that the contents are of sufficient merit and topicality that this volume will be as useful as earlier ones appear to have been. My purpose in attempting the move to cheaper printing methods was two-fold. Firstly, to speed up the editing and production, and, secondly, to bring about a significant reduction in price through lower printing costs. It is a matter of chagrin for me that the contributors to Volume 7 suffered inadvertently, knowing they had played their part by presenting scripts on time but then having had to sit through a long interval when nothing appeared to happen. I hope they and our readers will forgive me.

On a pleasanter note, I am pleased to report that my successor as Senior Reporter will be Malcolm Rose. As most afficionados of mass spectrometry will know, Malcolm co-authored with me a text-book on mass spectrometry designed as a teaching aid. He is a contributor to Volume 7 and I wish him well in his efforts in the future. I can only trust reviewers of this series will be as kind and critically helpful to him as they have been to me. In fact, one of the better aspects of being Senior Reporter has been the interest taken in the series by reviewers and others who have suggested improvements and topics for inclusion. My grateful thanks go to them for their kindness.

Volume 7 sees the welcome arrival of four new regular contributors in David Harvey, Trevor Kemp, Ivan Powis, and Malcolm Rose, together with 'one-off' contributors Professor C. J. Lorquet and Professor S. A. Safron.

Last, but not least, in my list of special mentions, I should like to commend the staff of The Royal Society of Chemistry, especially Philip Gardam, for friendly, willing help and guidance in producing this series.

Signing off from this rambling foreword is about as dismal as writing one's own obituary, so without further ado I shall bid you all "Farewell!" and "Go to it, Malcolm!"

October 1984 R. A. W. Johnstone

Foreword

In previous volumes of this series I have taken the opportunity of welcoming new contributors and thanking past contributors for their efforts. It is with mixed feelings that I report my own departure as Senior Reporter. For twelve years I have had responsibility for determining the contents of the volumes, asking (or sometimes twisting arms!) people to contribute, editing and liaising with the staff of The Royal Society of Chemistry. By and large it has been a pleasurable experience, occasionally fraught with panic when manuscripts were late or contributors dropped out at the eleventh hour to publication. The days spent reading the various manuscripts reinforced my own literature surveys of mass spectrometry and made me realize how widespread a technique it has become. At one extreme, there are purely analytical applications of mass spectrometry utilizing advanced instrumentation, ionization methods, and data systems. At the other extreme, basic research into ion chemistry has progressed at an astonishing pace. In between those extremes, mass spectrometry is used either centrally or peripherally in many other areas of research, so much so that one imagines the attractions of the technique would be more fully realized if there could be much cheaper instruments available. With prices of quality, routine mass spectrometers hovering around £250,000, requests for funds to purchase them need to be very well presented and/or shown to be highly cost-effective for there to be more than a sporting chance of the money being made available.

On a personal note, I am more than displeased with my abortive efforts to produce Volume 7 through modern printing methods, utilizing the word processor to provide the input for modern type-setting and printing machines. This failure due to technical

ISBN 0-85186-318-3
ISSN 0305-9987

Printed in Great Britain by
Whitstable Litho Ltd., Whitstable, Kent

A Specialist Periodical Report

Mass Spectrometry

Volume 7

A Review of the Recent Literature Published between July 1980 and June 1982

Senior Reporter

R. A. W. Johnstone, *Department of Organic Chemistry, University of Liverpool*

Reporters

J. H. Bowie, *Department of Organic Chemistry, University of Adelaide, Adelaide, South Australia, Australia*
R. H. Cragg, *Department of Chemistry, University of Kent, Canterbury*
D. E. Games, *Department of Chemistry, University College of Wales, Cardiff*
D. J. Harvey, *Department of Pharmacology, University of Oxford*
I. Howe, *Churchill College, University of Cambridge*
T. Kemp, *V. G. Analytical Ltd., Altrincham, Cheshire*
J. C. Lorquet, *Département de Chimie Générale et de Chimie Physique, Université de Liège, Liège, Belgium*
I. Powis, *Physical Chemistry Laboratory, University of Oxford*
M. E. Rose, *Department of Biochemistry, University of Liverpool*
S. A. Safron, *Department of Chemistry, Florida State University, Tallahassee, Florida, U.S.A.*

The Royal Society of Chemistry
Burlington House, London W1V 0BN

1

Ionization Processes and Ion Dynamics

BY I. POWIS

1 Introduction

Compared with previous volumes in the series this first chapter will be found to exclude coverage of simple determinations of thermochemical quantities. Nevertheless the field to be surveyed remains broad with many new developments, and the necessity for selective reporting remains. In choosing material, I have been guided by the desire to present work that advances our understanding of the fundamental processes experienced by ionized molecules.

2 Ionization Processes

Photoionization.- Methods for the calculation of molecular photoionization cross-sections are being actively developed and evaluated. The partial-channel cross-section for ionization to a given final state is given by equation (1) where $\hat{\mu}$ is the

$$\sigma = \frac{4}{3}\pi^2\omega \mid \langle\Psi^i \mid\hat{\mu}\mid \Psi^f\rangle \mid^2 \qquad (1)$$

dipole-moment operator, ω the photon frequency, and Ψ^i and Ψ^f the initial-state and final-state (ion plus electron) wavefunctions. The function Ψ^i may be evaluated with the Born-Oppenheimer approximation by means of standard computational methodology; usually *ab initio* single-configuration functions are used, although semi-empirical[1] and correlated[2] initial-state functions have also been employed. The principal difficulties lie in constructing the final-state wavefunction, Ψ^f, which can be treated as the product of a bound ionic core and a continuum orbital for the ejected electron. A 'frozen-core' approximation may be invoked and the initial-state orbitals used to generate the ionic hole-state. In view of Koopman's theorem this is acceptable, at least for the outer-valence-shell ionizations. It is the treatment of the continuum function that serves to characterize most usefully the various calculational schemes. Early approaches made use of plane waves (with or without orthogonalization to the remaining orbitals), coulomb waves, and single-centre pseudo-potential methods.

The continuum orbital may in principle be obtained as a solution to the Schrödinger equation (2) where ε is the electron

$$\left[-\frac{\hbar^2}{2m}\nabla^2 + V_{N-1} - \varepsilon\right]\phi = 0 \tag{2}$$

kinetic energy and V_{N-1} is the static-exchange potential of the N-1 orbitals in the ion core. In general, the non-central nature of the static-exchange potential makes the solution of equation (2) difficult, but single-centre expansion techniques applicable to diatomic and linear polyatomic species such as HCl,[3] HF,[4] N_2,[2,5,6] O_2,[7] and CO_2[8] have been described. The main problem is to ensure convergence in the calculations, but the use of the Schwinger variational method in particular seems to be well behaved and to provide solutions of Hartree-Fock accuracy.[2,8] Single-centre expansions with a simpler static potential approximation have also been employed.[9]

A more tractable and generally applicable method involves replacing the non-local static-exchange potential with localized model potentials; typically the X_α local-exchange scheme is used in which a parameterized, statistically averaged exchange term, based upon the electron density, is used in the Hamiltonian. Several local-exchange schemes have been compared against static-exchange calculations for atomic photionization and found to be fairly accurate,[10] a result that is also said to be relevant to molecular systems. In the multiple-scattering method (MSM) for molecules, the non-central potential is partitioned into spherical regions and localized model potentials, the so-called muffin tin potential, employed for each region. Photoionization cross-sections are then readily calculated. Although less accurate than single-centre expansion techniques, MSM calculations may be applied routinely to polyatomics such as BF_3,[11] H_2O, and H_2S.[12] The general reliability of such calculations appears to be fairly good, although in a comparison with experimental results for CS_2 and COS it has been noted that the results for the polar COS molecule and the inner-valence orbitals of CS_2 are less good than for the outer-valence orbitals.[13]

In solving equation (1), either a length or velocity form of the dipole operator may be used; when Ψ^i and Ψ^f are exact eigenfunctions of the electronic Hamiltonian these are entirely equivalent, but when only approximate wavefunctions are available different results are obtained. It has been proposed that the discrepancy between the two forms of the calculation may be used as an indication of the quality of the approximations being used.[4] Thiel has compared the use of both forms with MSM calculations for a number of diatomics[1] and concludes that the dipole-length form is to be preferred. This result has since been adopted for MSM calculations of HCN photoionization.[14]

A third major category of calculations is available that avoids the use of model potentials and yet circumvents the problems associated with obtaining a continuum wave in a non-central non-local molecular-ion potential. In this approach[15] the system Hamiltonian is diagonalized over a very large basis set of both compact and diffuse functions. The bound-state, so-called improved virtual orbitals (IVO) that result are used to generate a pseudo-spectrum of discrete transitions. The photoionization cross-section may then be obtained by smoothing the pseudo-spectrum oscillator strengths using Stieltjes-Tchebycheff moment theory (STMT). The virtual orbitals in this STMT approach are usually computed at the Hartree-Fock level by the use of a static-exchange Hamiltonian. As such, the calculations are potentially more accurate than the MSM method but have the drawback that, since the continuum wavefunction is never generated, information on angular distributions and so on can not be obtained. Like MSM, this method can be applied to non-linear polyatomics such as H_2O.[16,17]

Explicit comparisons of the Schwinger single-centre expansion method with both MSM and STMT calculations have been made by Lucchese *et al.*[2,8] Their results suggest that STMT calculations may depend critically on there being an appropriate distribution of bound virtual orbitals in order to avoid oversmoothing of the oscillator-strength distribution with a consequent loss or distortion of detailed structure in the photoionization cross-section. Implicit comparisons of these methods are also contained in the many results discussed below. To summarize the present position, it seems that for the most accurate calculations single-centre expansion techniques are preferred, where such methods are applicable. Otherwise, STMT calculations can provide detailed quantitative cross-sections. MSM calculations generate semi-quantitative cross-sections, are widely applicable, and, unlike STMT, can be used to investigate angular distributions and so on.

Much interest focuses on the ability of calculations to predict shape resonances in photoionization. A shape resonance arises when the outgoing electron is excited to a quasi-bound electronic state supported by a centrifugal barrier generated by the electron's angular momentum. It is a purely one-electron phenomenon and should be reproducible by the above calculations. Methods for the graphical representation of the continuum electron waves have been developed.[18,19] Figure 1 shows the f ($l = 3$) partial-wave component of the σ_u channel in the $X\ ^2\Sigma_g^+\ N_2^+$ ionization at the resonant energy. The f-like nature is evident from the nodal structure. Significantly, there is a dramatic increase in electronic amplitude near the nuclei, which gives the

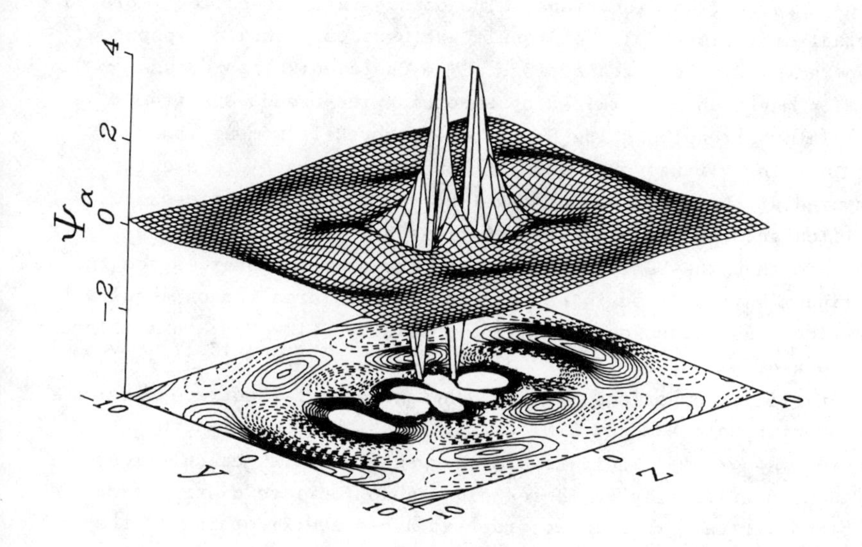

Figure 1 $N_2\sigma_u$ continuum f-wave eigenchannel wavefunction at shape resonance. The nucleii lie in the y-z plane along the z-axis

(Reproduced with permission from *J. Chem. Phys.*, 1981, 75, 4546)

enhanced ionization probability. This also means that the resonance is very sensitive to the internuclear separation, R, and can result in non-Franck-Condon vibrational distributions. Anisotropic effects arise because the resonances occur in specific channels, *i.e.* for a continuum orbital of given symmetry. Consequently, when a given symmetry channel is preferentially enhanced by resonance, one expects to see this reflected in the photoelectron angular distribution. Recently, Poliakoff *et al.*[20] demonstrated that the degree of polarization of the photoionized $B\ ^2\Sigma_u^+\ N_2^+$ fluorescence also reflects contributions from different ionization channels and so will be sensitive to shape resonances. Finally, shape resonances may be of value in achieving specific molecular-orientation effects in photoionization processes, *e.g.* of absorbed molecules.

Experimental photoelectron asymmetry parameters, β, have been measured for a range of small molecules of interest in the theory of photoionization,[21] including H_2O[22] and H_2S.[23] Possibly the most dramatic example of a shape resonance is found in the ionization of $C\ ^2\Sigma_g^+\ CO_2^+$,[24-26] where the experimentally determined β parameter dips sharply at around 42 eV. STMT calculations predict such a resonance in the σ_u channel, but at a photon energy of around 35 eV,[27] whereas Schwinger variational calculations of the photoionization cross-section do place the resonance at ~42 eV.[8] MSM-X_α calculations are also found to be broadly in agreement with experiment[25] and to predict resonances for the first four ionic states. However, the C-state β dip is predicted to be too narrow and too deep. This may be a vibrational effect as the discrepancy with R-averaged calculations is less marked. Vibrationally resolved measurements of the $A\ ^2\Pi_u$ ionization show that β decreases with increasing vibrational quantum number, again in accordance with calculations.[25,28]

Vibrationally resolved photoelectron angular distributions have been reported for $X\ ^2\Sigma_g^+\ N_2^+$.[29] In those regions that are free of autoionization $\beta(v = 1)$ is found to be 0.4 - 0.8 units greater than $\beta(v = 0)$, a phenomenon that can be interpreted in terms of shape resonance. Although Grimm[30] has suggested that, for the $v = 2$ level, autoionization may contribute to the variations of β, the shape resonances have been widely studied. Raseev *et al.*[6] and Lucchese[5] have performed single-centre, frozen-core, static-exchange calculations for σ_g^{-1} ionization *via* σ_u and π_u channels that indicate a σ_u-channel shape resonance. Using the Schwinger variational method, Lucchese[2] has further improved and extended these calculations and claims considerably better results than are achieved with MSM or STMT methods. One feature of the Schwinger calculation is that correlation was considered in the initial-state wavefunction. The STMT method has been extended to allow for coupling of the continuum states[31] by replacing the separated-channel, static-exchange calculation of virtual orbitals with the time-dependent Hartree-Fock (RPAE) approximation. The theoretical treatment of inner-shell ionizations has been tackled for N_2 (and CO) by the use of c.i. calculations of ionic hole-states in the static-exchange approximation.[32] Very similar behaviour, displaying a σ-channel resonance, is found for the ionization of $X\ ^2\Sigma^+$ CO^+.[12,15,33]

Gustafsson[34] has reported cross-sections for O_2 ionization in the 20 - 45 eV region. The $b,B\ ^{4,2}\Sigma_g^-$ states in particular show evidence of resonances, the former at around 21 eV. However, a higher-resolution study claims that the resonance structure lies at ~19 eV,[35] and vibrational branching ratios also indicate some sort of resonance at this energy.[36] The 19 eV structure is only 0.4 eV wide, unusually narrow for a shape resonance but rather broad for an autoionization resonance. Various calculations predict a shape resonance in the σ_u channel,[7,37] but at around 21 eV photon energies, and the exact nature of the σ_g^{-1} ionization remains uncertain. The ionization of both $X\ ^3\Sigma_g^-$ and a $^1\Delta_g$ neutral O_2 to the $^2\Pi_g$ ionic state has been treated in different calculations using an orthogonalized coulomb-wave continuum function.[38]

Like O_2, NO is an open-shell molecule whose ionization is perhaps less well understood than that of N_2 and CO. Both MSM[39] and STMT[40] calculations have been performed for these ionizations with only moderate agreement with experimental data;[41-43] in particular, the exact position of the σ-channel resonance for the ionization to $c\ ^3\Pi$ is uncertain.

Of prime importance in the development of our understanding of ionization processes is the system represented in equation (3).

$$^1\Sigma_g^+\ H_2\ (v'', J'') + h\nu \rightarrow {}^2\Sigma_g^+\ H_2^+\ (v', N') + e^-\ (l) \qquad (3)$$

Measurements of the overall photoelectron angular distribution over a wide range of photon energies[44] produced β values of around 1.7, somewhat lower than predicted by many previous calculations.

However, it is known that asymmetry parameters for those ionizations in which the rotational angular momentum, N', changes by 2 units are very different from those transitions in which $\Delta N = 0$ and are much more sensitive to the details of calculation. Pollard *et al.*[45] have measured $\beta_{\Delta N = 2}$ values for $v' = 0$ from rotationally resolved photoelectron spectra, and they compared their results with the theoretical values of Itikawa[46] (see the Table). Unlike most previous calculations, these allow for the

Table Photoelectron asymmetry parameters for H_2^+ ($v' = 0$) $\Delta N = 2$ ionizations

λ/Å	*Expt.*[a]	*Theoretical*		
		p *wave*[b]	p + f *wave*[b]	p *wave* + *autoionization*[c]
584	0.87 ± 0.19	0.234	0.749	-
736	0.08 ± 0.15	0.207	0.348	0.2

[a] J.E. Pollard, D.J. Trevor, J.E. Reutt, Y.T. Lee, and D.A. Shirley, *Chem. Phys. Lett.*, 1982, 88, 434. [b] Y. Itikawa, *Chem. Phys.*, 1979, 37, 401. [c] M. Raoult, Ch. Jungen, and D. Dill, *J. Chim. Phys. Phys.-Chim. Biol.*, 1980, 77, 599

inclusion of $l = 1$ and $l = 3$ waves for the outgoing electron, and this clearly produces better agreement at 584 Å. The converse may seem to be true at 736 Å, although the discrepancies might be attributable to autoionization effects at this wavelength. Jungen and co-workers[47] have performed calculations that allow for autoionization with the results included in the Table. So far these calculations only treat a pure *p*-electron wave ($l = 1$). Clearly more work is needed to understand results of this type.

The work of Jungen is part of a general development of multi-channel quantum defect theory (MQDT) to treat the phenomenon of vibrational and/or rotational autoionization in the hydrogen molecule. The initial application was to autoionization of rotationally cold para-hydrogen ($J'' = 0$) just above the $v' = 0, N' = 0$ ionization threshold, and excellent agreement with experimental data was claimed.[48] Subsequently, vibrational autoionization in the 790 - 760 Å region has been investigated.[49]

Other developments of MQDT have been reported,[50,51] but our general level of understanding of autoionization phenomena in molecules other than hydrogen remains on a much less quantitative basis. Eland[52] has discussed the position regarding electronic autoionization in a review-type article. A number of papers have reported and assigned Rydberg series in the autoionization structure of SO_2,[53,54] NO,[55] and acetonitrile,[56] and examples of autoionization in the production of various electronic states of O_2^+ have appeared.[35,57,58]

The question of how autoionization affects vibrational-level populations in ions has been tackled by determining branching ratios at photon wavelengths corresponding to autoionizing resonances in N_2[59] and O_2.[60,61] The Fano configuration interaction theory of the Rydberg-continuum coupling can be extended to provide detailed branching ratios in the region of a resonance, with information on the coupling being derived from the autoionization peak shapes, although at the peak of a resonance a simplified equation is applicable.[61] The full theory has yet to be evaluated fully. In studies of N_2O and CS_2[62] Eland notes that the influence of autoionization on the vibrational branching may be observed well outside the resonance peak.

The effects of autoionization are often clearly seen as non-Franck-Condon distributions of vibrational bands in threshold photoelectron spectra of molecules such as N_2O.[63] A close examination of threshold (*i.e.* zero kinetic energy) electron production in the *A*-*X*-band Franck-Condon gap of COS has been described.[64] In this region, Rydberg series converging on the *B* ionic state clearly contribute to the ionization despite highly unfavourable vibrational overlap integrals. The mechanism by which electronic energy is converted to vibrational energy is not at all clear, although such 'resonant' autoionization seems to be a widespread phenomenon. In connection with a previous study of N_2O,[65] it was suggested that resonant autoionization may be enhanced for Rydberg levels coupled to a neutral dissociation continuum and that incipient predissociation may precede electron ejection. It is interesting in this regard to note an observation made by Stockbauer[66] that there is a marked increase in threshold electron production from H_2O just as the onset for OH^+ fragment is reached. This is a region in which autoionization processes are dominant and neutral predissociation is known to compete with ionization. One may speculate that threshold electron emission occurs favourably from the predissociation channel {equation (4)} once the thermodynamic threshold is reached.

$$H_2O + h\nu \rightarrow H\cdots OH^+ + e^- \rightarrow H + OH^+ + e^- \quad (4)$$

Considerable interest has been expressed in the ionization of acetylene. Autoionizing Rydberg levels have been identified just above the $X\ ^2\Pi_u\ C_2H_2^+$ threshold,[67] but it is the ionization of this state in the 13 - 16 eV photon range that has proved most controversial. Strong, broad structure is evident in the cross-section in this region and is variously described as a minimum at ~14 eV or two maxima at ~13.3 and ~15.3 eV.[68,69] Non-Franck-Condon vibrational distributions and changes in angular distributions are also evident in photoelectron spectra recorded at these wavelengths.[70-72] The broadness of the resonance structure was initially interpreted as being indicative of a possible shape resonance at 14 eV, but MSM-X_α and STMT calculations provide no confirmation of this.[69,71] Rather, Hayaishi *et al.*[68] have performed *ab initio* calculations indicating that strong valence transitions, $3\ \sigma_g \rightarrow 3\ \sigma_u\ (\sigma^*)$ and $2\ \sigma_u \rightarrow 1\ \pi_g\ (\pi^*)$, lie at the positions of the cross-section maxima. These valence states converge to the $^2\Sigma_g$- and $^2\Sigma_u$-state ionization thresholds, respectively, and it is suggested that they autoionize. The width of the resonances is explained by the considerable geometric changes involved. A slightly different interpretation emerges from static-exchange calculations.[69,73] The $3\ \sigma_g$ excitations are found to be of a Rydberg rather than valence nature, and strong coupling effects between the autoionizing channels and the continuum are indicated. Additional contributory effects are considered, but it is evident that no definitive explanation can yet be given.

Many of the experimental techniques for studying photoionization make use of synchrotron radiation as a tunable source of vacuum u.v. Of particular note is the use of time-of-flight electron-energy analysers that exploit the pulsed nature of the synchrotron source.[36,41] Such an analyser has excellent electron-transmission properties, especially for low-energy electrons. An alternative laboratory source of photoionization data uses the continuum of 'pseudo-photons', or virtual photon field, generated by fast electron impact. This technique and the use of the Bethe-Born theory to relate electron-scattering data to optical oscillator strengths have been reviewed.[74] Brion's group have recently studied photoionization of NO[42] and COS and

and CS_2[75-77] by these methods. Such experiments also serve to illustrate the close relationship between electron- and photon-impact ionization.

Molecular-beam sources are now commonly used. In addition to offering experimental advantages of improved signal intensity, rotational cooling, *etc.* they permit the study of a range of van der Waal's molecules. Clusters of acetone,[78] COS,[79] CO_2 and N_2O,[80] and Ar[81] have all been studied by photoionization techniques. In a study of CS_2 clusters the photoionization spectra were observed to show autoionization structure characteristic of, but slightly red-shifted from, the monomer.[82] This suggests that to a good approximation the CS_2 molecule can be regarded as retaining its chemical identity within the cluster. In the case of the dimer, the species $CS_2^+\cdots CS_2$ and $CS_2^*\cdots CS_2$ are initially formed by photon absorption. Subsequently, the ion or excited Rydberg state undergoes an essentially zero-kinetic-energy collision with its partner, which can be observed as an ion/molecule reaction[83] or chemi-ionization.[84] Because the collision time is short, very short-lived excited states may be observed in this manner.

In contrast, the dimers of CO, N_2, NO,[85] and O_2[86] show none of the autoionization structure of the corresponding monomers,[85] suggesting that the Rydberg states formed are highly vibrationally excited and rapidly dissociate before ionization takes place. The H_2 dimer provides a fascinating demonstration of the competition between predissociation and autoionization. Much, though not all, of the Rydberg structure in the monomer photoionization spectrum is present in the dimer.[87] It has been deduced that predissociation competes effectively with vibrational autoionization *via* $\Delta v = -2$ transitions ($\tau \sim 10^{-8}$ s) but not with $\Delta v = -1$ transitions ($\tau \sim 10^{-11}$ s) so that only $\Delta v = -2$ autoionizing lines are absent from the spectrum.

Near to their ionization thresholds, the dimers Ar_2[88,89] and Kr_2[90] have been found to display extensive Rydberg structure that does not resemble the corresponding atomic spectrum. Between the $^2P_{3/2}$ and $^2P_{1/2}$ atomic thresholds some of the structure for these dimers and for Xe_2 can be correlated with atomic Rydberg levels,[91] but interaction between the atoms in these dimers is clearly significant in determining the autoionizing levels. In contrast, the photoionization spectra of the mixed dimers formed between Ne and the heavier rare gases do correlate closely with the heavier atom's atomic Rydberg spectra, so that in this instance it is again possible to regard the molecular absorptions as slightly perturbed atomic lines.

Electron-impact Ionization.- Autoionization seems likely to be as prevalent a phenomenon in electron-impact ionization as in photoionization. It has been clearly established in studies of the rare gases Xe and Kr,[92,93] where structure in the ionization-efficiency curves may be assigned to a series of Rydberg states converging to the $^2P_{1/2}$ ionization limit. Likewise, many features found in the ionization curves of COS and CS_2 have been related to Rydberg levels of the neutral, although others are thought to be due to negative-ion formation.[94] A new study of the dissociative ionization of methane used mono-energetic electrons in the range 19 - 22 eV.[95] Here, too, it has been claimed that changes in the slopes of the ion-yield curves may be correlated with Rydberg states produced by excitation of the $2a_1$ electron. These are most accurately identified in the curves for the H^+ fragment.[96] Five other thresholds for H^+ formation have been observed at electron energies above 25 eV,[97] but it is believed that these are due to dissociation of the excited $CH_4{}^+$ states that are produced.

It is conventional to assume that, for electron energies of 70 eV or above, the 'optical limit' is approached and that the electron-molecule interaction in some ways resembles photon impact. Thus, a fairly simple integral relationship between electron- and photon-impact ionization curves, including their resonance features, can be expected. A corollary is that electric-dipole selection rules should apply to electron-impact ionization, though this has not been extensively tested. Recently, the rotational distribution following electron-impact ionization of supercooled N_2 (most likely frozen into $J'' = 0$ or 1 levels) to $B\ ^2\Sigma_u^+\ N_2{}^+$ ($v' = 0$) has been investigated as a function of impacting electron energy by monitoring of the fluorescence.[98] Considerable rotational excitation was apparent, particularly at low electron energies, which gave 'hotter' ions than higher electron energies. Even at electron energies of several hundred electron volts, the authors conclude that the dipole selection rules are being violated. Such a breakdown of the Born approximation is not unexpected for slow electrons but is certainly surprising at 500 eV. A qualitative explanation offered is that higher-order multipole interactions occur between the ion and the ejected electron, which may of course have low kinetic energy even for high-energy impacting electrons;

it is these exit-channel effects that cause the rotational angular momentum to change by several units rather than the postulated initial transition to an excited (possibly virtual) autoionizing state.

The angular distributions of H^+ ions produced by 30 - 70 eV electron ionization of H_2 have been measured.[99] The distributions show peaks parallel to the electron beam, but marked dependences on the electron and fragment-ion kinetic energy were observed. In part, these results may be explained in terms of dissociation from the two lowest states of H_2^+, but other unknown mechanisms also contribute. Below the ground state of H_2^+, electron impact generates H^+ by ion-pair formation.[100] Other molecules in which dissociative ionization has been investigated include SO_2, SO_3,[101] and CS_2.[102]

Multi-photon Ionization.- Molecular multi-photon ionization (m.p.i.) must be one of the most rapidly developing areas of chemical research at the present time, with more than an order of magnitude increase in the annual number of papers published in just a couple of years. This is stimulated not just by an interest in m.p.i. *per se* but also by the many potential applications of the technique that are relevant to a wide cross-section of the chemical community, though not least to mass spectroscopists. Perhaps partly in consequence of this fact, workers active in the m.p.i. field have diverse research backgrounds, itself a healthy development since it should contribute to the breakdown of the false dichotomy that has long existed between studies of ion and neutral reaction dynamics.

The term m.p.i. is quite generally applied to any system in which a number of photons are absorbed to take the system above its first ionization potential with resulting ion formation. Typically, visible- and near-u.v.-laser sources are used, but i.r.-laser m.p.i. of, for example, N_2 has been reported.[103] The photon uptake may be a true multi-photon, coherent absorption or it may be a sequential (strictly a multiple-photon) process. Often a mixture of both is involved.

Several groups have measured effective cross-sections for a variety of molecules, using high-powered excimer lasers.[104-106] In these cases the cross-sections are occasionally found to be several orders of magnitude greater than average, a fact often attributable to the presence of an intermediate state of the molecule that is resonant with an integral number of incident photons. This is resonance-enhanced ionization (r.e.m.p.i.). Because the increasing state densities above the excited resonant level make subsequent transitions more probable, it is the initial resonant step that is usually rate limiting in r.e.m.p.i., and, to a first approximation, the variation of resulting ion current gives the spectrum of the intermediate state. This is of course the basis of m.p.i. spectroscopy, the most mature application of m.p.i. It has been pointed out, however, that even non-resonant ionization can display structure when the ionizing ground-state population is depleted by a resonant transition to a non-ionizing (*e.g.* predissociating) level.[107]

M.p.i. spectroscopy has been well reviewed and the conditions required for observing m.p.i. phenomena have been discussed.[108-111] In keeping with the general theme of this chapter, purely spectroscopic applications of m.p.i. that yield information on only the neutral resonant state are excluded from the following discussion.

It is found that not only the total ion current but also the ion-fragmentation patterns carry a wavelength-dependent signature. Striking examples of this are found for hexa-2,4-diyne[112] and for triethylene diamine (DABCO)[113] where, as different resonant transitions are excited, the ionization dynamics clearly change too. This gives rise to the suggested development of a two-dimensional mass spectrometry in which a sample is characterized by its wavelength *versus* mass-abundance behaviour. Further, the resonant nature of the initial step makes it feasible to ionize selectively one component of a mixture. This has been demonstrated in the case of *cis*- and *trans*-1,2-dichloroethene[114] where better than 10:1 discrimination against the unwanted isomer was achieved. Isotope-selective ionization of NO,[115,116] of K_2,[117] and of aniline containing ^{13}C at natural abundance[118] has also been reported. Additionally, schemes to ionize and hence detect selected isotopes of Ar, Kr, or Xe in the presence of 10^{15} atoms of other isotopes of the same element[119] and to achieve even greater selectivity[120] for the identification of $^{14}C^{16}O$ have been considered.

A common shorthand notation to describe r.e.m.p.i. processes indicates n, the number of photons required for the initial resonant step, and m, the number subsequently absorbed to ionize and fragment the molecule, as $(n + m)$. Many molecules are 'colourless' in the wavelength regions considered so that $n > 1$, and this first absorption is a concerted process. Therefore the resonant step is of low intrinsic probability, and intense lasers are required for m.p.i. The transition is, however, readily characterized, whereas the determination of m, and more particularly the modes by which these photons are taken up, is one of the major preoccupations of m.p.i. fragmentation studies.[121] These post-resonance (ionizing) steps are often kinetically saturated by the high laser fluxes employed, and consequently n-photon-absorption cross-sections may be used in estimating the sensitivity of $(n + m)$ ionization. It can then be shown that 100% ionization efficiencies are feasible. Using a moderate-power (1 kW) laser, up to 25% efficiencies have been noted for a collection of organic molecules.[122] In this case about 10^6 ions were produced per laser pulse and a detection limit of 2 in 10^9 molecules was estimated for such a set-up. A consequence of this extreme sensitivity is that mass spectra may be recorded for a wide range of laser intensities and, when this is done, very marked variations in the fragmentation pattern can appear, giving a 'third dimension' to m.p.i. mass spectra.

Boesl *et al.*[123] report that ^{13}C benzene molecules may be selectively ionized by 10^7 W cm^{-2} and 10^9 W cm^{-2} irradiation, but only in the latter case does extensive fragmentation result. It is known that m.p.i. can be effected so as to produce other aromatic ions with minimal fragmentation,[124,125] and this illustrates a major potential of m.p.i. for use as a controllable, high-intensity, 'soft' ionization source. This has been most effectively demonstrated for van der Waals species;[126-128] in each case parent ions are produced in abundance. At the other extreme, with high laser intensities, extensive skeletal fragmentation is well known to occur, and even doubly charged ions have been produced by m.p.i. of benzene[129] and UF_6.[130]

M.p.i. techniques have been used to obtain accurate ionizing potentials in experiments that use two exciting lasers. The first laser generates a high concentration of excited-state neutrals, which are then ionized by the second laser. Scanning the second laser wavelength enables the ionization onset to be identified. Alkali-metal dimers[128,131] have been studied by this method, as have various aromatics.[132-134] A number of caveats apply, as in single-photon ionization-potential determinations, in particular concerning the accurate identification of the adiabatic transition. M.p.i. holds a number of advantages here. The purely experimental ones arise from the use of a high-intensity, high-resolution, tunable source of visible radiation to generate high-quality data.

In addition, there is time for the nuclear co-ordinates to relax in the intermediate state, giving rise to Franck-Condon factors that are different from those in a single-photon ionization. By choosing different vibronic levels of the intermediate, the vibrational structure of the ion-yield curves may be extensively explored to identify the truly adiabatic transition. Another possible source of error has been identified by Duncan *et al.*,[134] who comment on the field ionization of highly excited Rydberg states by the source field in their apparatus ($\sim$175 V cm^{-1}).

Suggestions of more specialized applications of m.p.i. sources have arisen, for example as a combined laser description/ionization of condensed species[135] or as a clean, pulsed, low-temperature, high-intensity source of metal-atom ions generated from organometallics.[136] There are some possibly undesirable effects accompanying the use of high-powered lasers that may be largely avoided by the choice of suitable operating conditions. These include secondary ionization by the emitted photoelectrons,[137] collisional effects enhanced by the laser field,[138] competing third-harmonic generation,[139] and gain saturation effects.[140]

The high sensitivity and detailed spectroscopic information obtained by m.p.i. cause it to be a powerful tool for the detection and analysis of reaction products. NO_2 from the i.r. multi-photon dissociation of $MeNO_2$ has been monitored this way,[141] as has NO produced in an atmospheric flame[142] and methyl radicals produced by pyrolysis.[143,144] A rotational-state analysis of the NO fragment from NO_2 photodissociation has been described.[145] Indeed, m.p.i. can be compared very favourably with laser-induced fluorescence (l.i.f.) as a means of state-selective detection of very low concentrations of products.

Both techniques depend on the population of an *n*-photon resonant level, which in l.i.f. is allowed to fluoresce but in m.p.i. is removed by further excitation and ionization. The intrinsic sensitivity of both methods is therefore the same; with l.i.f., however, only photons emitted into a restricted solid angle can be collected, whereas it may be arranged to collect all the charged species produced by m.p.i. Furthermore, only 50% of the initial state can be excited in l.i.f., whereas potentially all species may be ionized in m.p.i. since the subsequent ionization step can compete effectively with stimulated emission from the intermediate state. Nevertheless, both techniques are usually limited by the quantum efficiencies of various relaxation processes.

For example, predissociation competes with ionization in the m.p.i. of ammonia as indicated by the intensity depletion and lifetime broadening of successive rotational lines.[146] Even so, ionization steps in m.p.i. can compete with relaxation steps more successfully than does fluorescence, so that even non-fluorescing species may be detected by m.p.i.

Smalley and co-workers have investigated the lifetimes of intermediate states in m.p.i. Bromobenzene was ionized in a (1 + 1) process, going through its $^{1}B_2$ state.[147] From the optical linewidths a lower lifetime of about 10^{-12} s may be inferred. The state is efficiently converted to the low-lying triplet that was not ionized at the wavelengths used. By using two delayed lasers, one for initial excitation and one for $^{1}B_2$ ionization, this intersystem crossing was monitored as a loss of ion current. In this instance a lifetime of $<10^{-8}$ s was inferred, *i.e.* comparable with the laser pulse widths. A single-laser experiment would therefore ionize bromobenzene efficiently even though the lifetime is too short for fluorescence detection. When a higher-frequency second laser was employed in very similar experiments on benzene[148] and toluene[149] the triplet state was ionized and its lifetime probed by varying the relative delay between the lasers.

Attempts to gain a mechanistic understanding of the r.e.m.p.i. process have often centred around modelling of the variation of observed fragmentation patterns with laser power.[121] One of the earliest and currently most extensively studied systems in this context is benzene. It is now accepted that r.e.m.p.i. fragmentation of this molecule occurs not by photon absorption up a neutral 'ladder' of excited states that subsequently (dissociatively) autoionize but rather by absorption by the ions. Switching from the neutral to ion ladders happens as soon as energetically possible. Further evidence to this effect has been forthcoming. Boesl *et al.*[150] have performed two colour (visible and u.v. photons) delayed-pulse experiments. They have observed that, while u.v. photons alone can produce parent ions, the extent of fragmentation is greatly increased by a second pulse of visible light and that this effect is largely independent of whether or not the visible pulse is delayed relative to the u.v. pulse. Hence, absorption of the 3 - 5 visible photons that are deduced to be necessary to produce the observed degree of fragmentation occurs through a relatively long-lived state and not short-lived neutral Rydberg levels.

More direct evidence is obtained in a variation of this experiment. A variable ion-repelling voltage can selectively eject charged species produced in the first pulse from the focal volume of the visible laser before it is fired.[151] As this is done, the effects of the second laser are minimized, thus convincingly demonstrating that it is the ionic and not neutral species that take up the extra photons required for fragmentation.

In a rather different experiment, Reilly and co-workers[152] measured the photoelectron-energy distributions accompanying benzene ionization by ArF and KrF lasers. The electron energies are consistent with ionization taking place when only two or three photons have been absorbed. Again it must be that the further photons that are energetically required for fragmentation are taken up after ionization has occurred. Differences between the total ion currents produced by these two lasers can be explained in terms of a simple kinetic model that assumes rapid ionization of the resonant intermediate.[153] Similar photoelectron energies have been recorded in a dye-laser r.e.m.p.i. experiment.[154] Ion kinetic energies were also measured, and different fragments were found to have similar, near-thermal distributions. The mechanistic implications of this result have yet to be explored fully.

The above experiments do not unambiguously determine whether it is solely the parent ion or possibly also fragment ions that absorb the extra photons. Rebentrost *et al.*[155,156] have proposed a statistical rate-equations model to account for the details of the benzene fragmentation in which it is assumed that the parent ion sequentially absorbs multiple photons, always converting back to a vibrationally excited ground electronic state after each step and prior to any dissociation taking place. When dissociations do occur they take place *via* parallel and, fragment internal energy permitting, consecutive reaction paths without any photons being absorbed by the fragment species. Quite good agreement with experimental intensity-dependence data may be demonstrated, although no wavelength dependence is predicted since it is always ground-electronic-state parent ions that are assumed to be dissociating. Nevertheless, there are grounds for doubting the physical content of this model. First, it would require approximately 60 eV to be absorbed by the parent ion to make C^+ the most abundant fragment.[156] Before such high excitation energies are achieved, dissociation must become very rapid and liable to occur in preference to further photon absorption.[157] Second, experiments with picosecond laser pulses result in fragmentation that stops at the C_4^+/C_3^+ stage;[129] given that dissociation lifetimes are probably comparable to the laser pulse widths, this suggests that the further fragmentation normally observed results from secondary- and tertiary-ion photodissociation. Third, in a refinement of the two colour laser experiments, the second laser pulse has been spatially as well as temporally separated from the first.[158] A time-of-flight arrangement can then be used to cause ions of selected mass to pass in front of the visible laser as it fires; in this manner photodissociation of the fragment ions has been demonstrated.

Dietz *et al.*[157] have therefore considered a statistical rate-equations model in which competition between unimolecular reaction and photon absorption by every species entering the scheme is permitted. Individual rates are obtained from RRKM-type calculations, and it is estimated that no more than two photons can be absorbed before dissociation rates exceed photon-absorption rates. The model works well, and the process of successive fragment-ion dissociation it describes seems to be appropriate for benzene.

A completely different mode of behaviour has been encountered in m.p.i. studies of organometallics such as nickelocene and ferrocene,[136,159-161] which are fully photodissociated prior to ionization of the metal atom. Only bare metal ions are formed, and atomic transitions are evident in the m.p.i. spectrum. Metal carbonyls show very similar behaviour,[162,163] although photo-fragmentation is incomplete at some wavelengths, enabling singly co-ordinated ions to be detected as well.[164,165] Dimethylmercury proves to be an exception in that it is believed to produce Hg^+ ions through fragmentation of the parent ion;[166] no atomic energy lines are observed in the m.p.i. spectrum, and at low laser intensities the parent ion can be detected.

Other examples of neutral predissociation preceding the ionization step are found with NH_3[167] and benzaldehyde.[168,169] Below 500 nm, photofragmentation of NO_2 becomes more likely and the m.p.i. spectrum takes on the features of the spectrum of the NO molecule.[170] Even more extensive predissociation has been reported for m.p.i. of CCl_2F_2.[171] The major fragment ion observed is CF^+, and an impressive high-resolution p.e.s. shows diatomic-like features. The authors conclude that the reaction sequence is as shown in equation (5).

$$CCl_2F_2 \rightarrow CClF_2 \rightarrow X\ CF \rightarrow B\ CF \rightarrow CF^+ + e^- \qquad (5)$$

Silberstein and Levine[172] have described a statistical theory of m.p.i. fragmentation patterns in which no reference is made to the specific dynamics and reaction channels of the system; rather, it is the available phase space for all conceivable products (including ion pairs[173]) that is used to determine the product branching ratios. A maximal-entropy formalism is employed, subject to constraints that include the average energy uptake by the system. A simple consequence of this theory is that the fragmentation of a molecule should be a function of the mean energy absorbed but not of the photon wavelength nor of the isomeric structure. This gives rise to a simple evaluation procedure for the theory: can identical fragmentation patterns be observed for (i) different laser wavelengths and (ii) different isomers merely by altering the laser intensity (and presumably therefore finding the point at which the same mean energy is absorbed)? The answer to (i) is, for a variety of molecules, yes[174] as it is also to (ii) when the isomers azulene and naphthalene are compared.[175] Silberstein and Levine have themselves published a more quantitative evaluation of the theory[176] using the same experimental data as above.[174] It should be stressed that such a theory does not purport to have didactic value as regards the ionization and dissociation dynamics; rather, it provides a yardstick of extreme statistical behaviour against which real systems may be compared.

In a markedly different vein, quantum-mechanical methods, such as are employed for the calculations of single-photon ionization cross-sections, have been applied to the theoretical description of multi-photon ionization of diatomic molecules. Two-photon, non-resonant ionization of H_2 has been treated by formulating the problem as a one-photon ionization of a perturbed orbital.[177] Resonant ionization, on the other hand, has been treated by Cremaschi, whose examination of the m.p.i. of NO has recently been extended by consideration of the three-photon (2 + 1) r.e.m.p.i. process[178] and five-photon ionizations.[179] In the latter case, both (3 + 2) and (4 + 1) cross-sections were calculated from *ab initio* bound-state wavefunctions and an orthogonalized coulomb-wave continuum function.

The NO molecule has also been the subject of experimental investigations of Franck-Condon factors in multi-photon ionization.

Photoelectron-energy distributions were measured for the (2 + 2) process[180] shown in equation (6).

$$NO\ X\ {}^2\Pi\ (v'' = 0) \xrightarrow{2h\nu} NO\ A\ {}^2\Sigma^+\ (v') \xrightarrow{2h\nu} NO^+\ X\ {}^1\Sigma^+\ (v) + e^- \qquad (6)$$

Ionization from each of the intermediate vibrational levels v' = 0, 1, 2, or 3 in turn gave photoelectron spectra peaking around the Δv = 0 ionizing transition, in accord with the Franck-Condon factors applying to the ionizing step. There is thus an implication that, by tuning the laser to a specific vibronic level of the intermediate state, a degree of control over the ion vibrational distribution may be exerted, possibly exciting vibrational levels not reached in one-photon ionization. A re-examination[181] of the m.p.i. p.e.s. of NO at much higher electron-energy resolution (~15 meV) revealed clearly that a distribution of vibrational levels was in fact populated (see Figure 2).

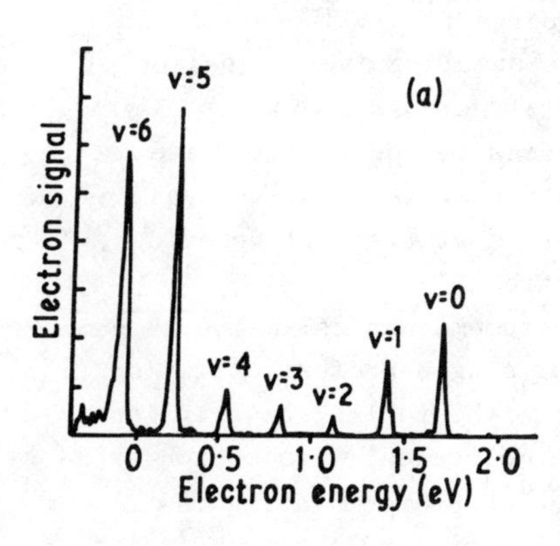

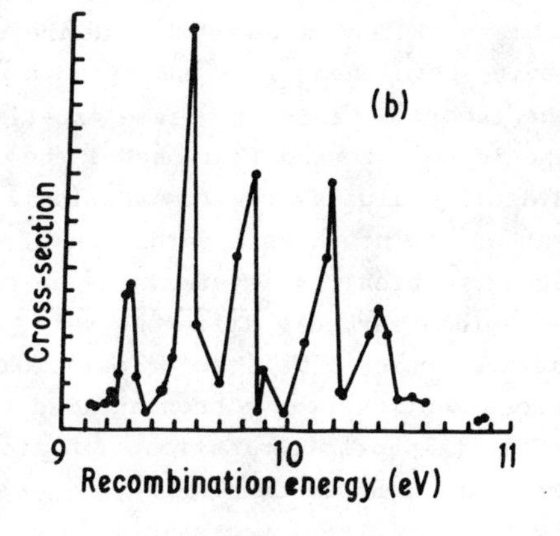

Figure 2 (a) Four-photon m.p.i. electron-energy spectrum of NO^+ produced *via* the intermediate $A\ {}^2\Sigma^+$ (v = 0) state
(Reproduced with permission from *Chem. Phys. Lett.*, 1982, 88, 576)

(b) 'Pseudo-photoelectron spectrum' of NO^+ showing variation of charge-exchange cross-section with primary-ion recombination energy
(Reproduced with permission from *Chem. Phys. Lett.*, 1980, 76, 380)

However, both studies also found plentiful generation of threshold electrons (which correspond to the highest energetically accessible vibrational levels being populated), in clear violation of the Franck-Condon principle. This seems to be an example of the resonant autoionization phenomenon encountered in single-photon ionizations.

A very detailed study of m.p.i. mechanisms in I_2 has been published by de Vries *et al.*[182] Angle-resolved and kinetic-energy-resolved measurements of the fragment species were obtained by these authors. Experiments with a 192 nm laser, in which two photons are absorbed, were found to produce signals showing distortions caused by space-charge effects, yet small unaffected signals were also present. It was deduced that these result from long-lived states that dissociatively autoionize only after the initial charge cloud has dispersed. The kinetic-energy distributions for I^+ were bimodal with high- and low-energy ions having different angular distributions. It was concluded that two repulsive curves of I_2 must be involved. When a 580 - 600 nm pump laser and a 248 nm ionizing laser were used, bimodal kinetic-energy distributions were again observed, both for I^+ and I^-. Lack of either laser-induced electron detachment from I^- or space-charge effects was interpreted as evidence that I^- is not formed until after the laser pulse and arises from slow predissociation of an excited state of I_2. The bimodal distribution could be ascribed to (1 + 2) and (2 + 1) r.e.m.p.i. processes occurring concurrently, and, by varying the delay between the two lasers, intermediate-state lifetimes of about 3 μs and 1.1 μs, respectively, were determined.

A rate-equations model was used in the work on I_2 as well as for benzene, as discussed previously. Another molecule whose r.e.m.p.i. mass spectrum has been discussed using the rate-equations approach is acetaldehyde.[183,184] The various fragment-ion intensities show strong but different laser-flux-dependent behaviour. Although exact quantitative agreement with experiment is not achieved, the use of relatively simple kinetic models does prove to be an impressively powerful technique for unravelling mechanistic details of r.e.m.p.i. processes. For the chemist, this approach is familiar and conceptually appealing. Certain conditions do have to be fulfilled for rate-equations models to be valid;[185] essentially these require that the coherent steps are dominated by incoherent processes. This is usually the case for multi-photon-ionization experiments and the lasers typically used have in any case short coherence times. Nevertheless, coherence effects are sometimes apparent,[164] in which case the simple rate-equations approach would be inapplicable.

Other molecular systems whose multi-photon ionization has been investigated include alkyl iodides,[186,187] tertiary amines (particularly DABCO),[113,188,189] *p*-xylene,[190] and 1,1,1-trichloroethane.[191] This last molecule was shown to have an enhanced m.p.i. cross-section when the ground state was initially excited by an infrared-laser pulse. The first reported r.e.m.p.i. metastable dissociation was found for aniline {equation (7)}.[192]

$$C_6H_5N^+ \rightarrow C_5H_6^+ + HCN \quad (k \sim 2 \times 10^6 \ s^{-1}) \qquad (7)$$

Between about 266 and 285 nm this was observed as strongly distorted time-of-flight mass peak shapes. The ionization is believed to be a (1 + 2) process, hence the 285 nm onset corresponds to a total energy uptake of 13.1 eV. Baer and Carney[193] have determined that the ground-state ion produced by single-photon ionization in the 12.6 - 13.6 eV range has a lifetime ranging between 100 and 0.3 μs. They found no evidence of metastable ions produced by m.p.i. with visible photons, indicating that the behaviour is dependent upon the exact states populated. As these authors remark, the observation of metastable parent ions in r.e.m.p.i. is significant because it indicates the presence of species that are stable for the duration of the laser pulse. A study of H_2S r.e.m.p.i. by the same authors[194] is of some interest because the chemistry of specific electronic states of the molecular ion has been well studied by p.e.p.i.c.o. experiments and theoretical calculations.

Charge Transfer and Penning Ionization.- Three factors are generally held to influence charge-transfer cross-sections: the energy defect, the Franck-Condon overlap factors between initial and final vibrational states, and the translational energy of the collision partners {equation (8)}.

$$A^+ + M \rightarrow M^+ + A \qquad (8)$$

The energy defect, ΔE, is defined as the difference between the recombination energy of $A^+ + e^-$ and the ionization energy of M^+. When A^+ is in an excited state, the recombination energy may take a variety of values depending on the final state of A. Similarly, M^+ may be produced, energy permitting, in a range of states. If it is understood that momentum transfer is negligible, then the cross-section will be enhanced at resonance, *i.e.* when $\Delta E \approx 0$. Furthermore, if the recombination energy of A^+ is well defined, energy-selected M^+ ions will be produced at resonance. Charge transfer has often been used for just this purpose, although the actual ionization mechanism is not well understood.

Tedder and Vidaud[195-197] have systematically recorded charge-exchange mass spectra (c.e.m.s.) of molecules using up to 49 primary ions, A^+, whose recombination energies span a range of 7.8 - 21.6 eV. A plot of total ionization cross-section against recombination energy resembles the photoelectron spectrum of these molecules.[195] Their results[196] for NO^+ ($X\ ^1\Sigma^+$) are of particular interest since they clearly demonstrate both resonance enhancement, when a vibrational level of NO^+ coincides with the primary-ion recombination energy, and the role of favourable $NO \rightarrow NO^+$ Franck-Condon factors in the ionization (see Figure 2b). However, the c.e.m.s. for individual fragment ions, from ethyl acetate, *etc.*,[197] reveal onsets at nominal recombination energies well below the ions' known thresholds, indicating deviations from a simple description of the process.

In part, this may be due to the presence of electronically excited A^+ states in the experiment. Such a situation may exert a strong influence on the observed behaviour since excited-state ions can have different, and often larger, cross-sections than the ground-state ions. This has been observed for electronic states of N^+ and O^+,[198,199] NO^+,[200] and Cl^+,[201] with various neutral collision partners. Vibrational excitation of polyatomic ions may also be important. Vibration-dependent cross-sections have been observed for the charge transfer of N_2^+ with N_2, though not with Ar.[202,203] Orth *et al.*[204] used collisional relaxation to control and examine internal-energy effects for the system $N_2^+ + SO_2$. Simple charge transfer is strongly exothermic, but the dissociative channel shown in equation (9) only becomes exothermic for $v \geqslant 2$. This is

$$N_2^+(v) + SO_2 \rightarrow SO^+ + O + N_2 \qquad (9)$$

reflected in the changing branching ratios for the two channels as the internal energy is increased.

Modern experimental techniques, particularly photoelectron-photoion coincidence methods, permit charge transfer to be studied with vibrational states resolved. For instance, Baer and Murray[205] have obtained cross-sections for the process shown in equation (10).

$$NH_3^+ (v) + NH_3 (v = 0) \rightarrow NH_3 (v') + NH_3^+ (v'') \tag{10}$$

The resonance requirement is easily met in symmetric charge transfers so that Franck-Condon factors should be of main importance. In ammonia these vary by a factor of more than 35, but the authors found that for $v = 0 - 10$ the charge-transfer cross-sections for unresolved final states v' and v'' varied by a factor of only 2, peaking at $v = 9$.

Vibrational effects have been more closely studied for reaction (11). Campbell *et al.*[206] report cross-sections that peak

$$H_2^+ (v) + Ar \rightarrow H_2 + Ar^+ \tag{11}$$

at $v = 1$, although both Tanaka *et al.*[207] and Houle *et al.*[208] agree that the maximum is at $v = 2$. The reaction is endoergic for $v = 0$ or 1, and the peak at $v = 2$ corresponds both to a near-resonant condition and to the most favourable Franck-Condon factors between H_2 and H_2^+. In contrast to the role of vibrational energy, translational energy between 0 and 20 eV has only a limited effect on the reaction, the barely endoergic $v = 1$ cross-section being the most markedly enhanced by the additional source of energy.[207,208]

Product-state distributions may be determined if they are luminescent. Tsuji *et al.*[209] have observed emission from SO^+ in the dissociative charge-transfer system He^+/SO_2. Neuschafer *et al.*[210] have studied emission from A CO^+ resulting from He^+ charge transfer with CO, over a range of collision energies up to 1 keV. Production of the excited A state is 2 eV endoergic whereas the ground state is 0.52 eV exoergic, yet up to 25% of the total ionization resulted in the excited state. Individual vibrational levels, $v \leqslant 8$, were resolved, and rotational distributions indicating high rotational temperatures were deduced.

The role of relative translational energy cannot be neglected when endoergic reactions are observed. Lindinger *et al.*[211] have studied the variation of cross-section for $Ar^+ + O_2$ with collisional energy. The cross-section initially falls slightly but rises very dramatically at 0.4 eV. At this point production of a $^4\Pi_u$ O_2^+ becomes energetically possible and is indeed strongly favoured by the Franck-Condon factors. This behaviour was also evident in a study of charge transfer of X and an O_2^+ with various neutral molecules,[212] where the excited O_2^+ generally has the larger cross-section. Reaction with H_2 was fairly typical. At translational energies of a few eV or less the excited-state cross-section was found to approach the Langevin limit whereas the endoergic charge transfer from the ground state was not observed. As the collisional energy was increased to several keV, the ground-state (endoergic) cross-section slightly increased whereas the excited-state (exoergic) cross-section decreased more markedly.

If Langevin cross-sections are assumed, no dependence on an ion's spin-orbit state would be predicted. Nevertheless, Campbell *et al.*[213] have measured different branching ratios for the $^2P_{1/2}$ and $^2P_{3/2}$ states of rare-gas ions undergoing symmetric charge transfer. In crossed-beam studies of $Ar^+ + H_2O$[214] as well as $H_2^+ + Ar$[215] at low energies there is evidence for two distinct mechanisms: (i) an electron 'jump' at large impact parameter and (ii) a more complex adiabatic process involving an intimate collision complex. Seely[216] has considered the possibility of a charge-transfer enhancement in such complexes by intense radiation fields.

Charge transfer is not necessarily restricted to singly ionized species. Single-electron transfer has been observed for various doubly charged diatomic ions with N_2, yielding N_2^+.[217] Ar^{2+}, on the other hand, is believed to form N_2^{2+}, which very rapidly dissociates to $N^+ + N^+$.[218]

Penning ionization is analogous to charge transfer in that ionization energy is acquired from an excited collision partner, in this case a metastable atom. Niehaus has recently reviewed this topic.[219] The translational-energy dependence of the Penning ionization of water with He ^1S,^3S atoms has been investigated.[220] The He^*-H_2O surface is strongly attractive, and the Penning ionization can be well represented as spontaneous transition into the ionized continuum of a lower-lying state. Penning ionization of rare gases by Ne $^3P_{0,2}$ is, in contrast, typical of a weakly attractive van der Waals collision pair.[221] The Penning electron spectrum shows clear bands corresponding to the $^3P_{3/2,1/2}$ states of the resulting ion.

Niehaus, Morgner, and associated workers[222-224] have developed an electron-ion coincidence apparatus especially suited to the study of Penning ionization. The first system to be studied was that of He ^3S and H_2.[222] Electron-energy distributions were recorded in coincidence with HeH_2^+, HeH^+, and H_2^+ ions. Vibrational structure in the electron spectrum resembles an only slightly perturbed H_2^+ photoelectron spectrum so that a two-step mechanism {equation (12)} was proposed: (a) ionization at long range followed by (b) an 'ion-molecule' reaction on the final-state surface.

$$He^* + H_2 \rightarrow He\cdots H_2^+ (v) + e^- \quad (12a)$$

$$He\cdots H_2^+ (v) \rightarrow He + H_2^+ \rightarrow HeH^+ + H \quad (12b)$$

Only for $v = 0$ was HeH_2^+ observed. HeH^+ was found for $v \geqslant 4$, becoming increasingly more probable as v increases. This behaviour is in accordance with there being an effective translational energy of about 20 meV for step (b), as was also deduced from a model of the initial Penning-ionization step. The $He^* + H_2$ and $He^* + H$[225] systems are, of course, of particular importance in studies of Penning ionization because they are amenable to detailed theoretical calculations.

The same coincidence apparatus was employed for studies of Penning ionization in NO_2 and SO_2.[223,224] These systems were found to be somewhat more complicated in that two entrance channels are available. One is a repulsive quartet, the other a strongly attractive ionic surface.

Ionization Potentials.- Although it is not intended to review routine determinations of ionization potentials in this section, the timely appearance of an excellent compendium of u.v. photoelectron spectra, together with assignments, deserves to be mentioned.[226]

The most familiar methods of calculating ionization energies, by Koopman's theorem and ΔSCF calculations, suffer from well known drawbacks, particularly in their treatment of correlation energy. Nevertheless, systematic tests of semiempirical molecular-orbital programs such as MNDO[227] suggest that such methods will continue to find applications where simple, cheap calculations are desired. It is possible to calculate ionization potentials, accurate to a few tenths of an electron volt, for small molecules by using the many-body Green's function and the related equations-of-motion approach.[228-231] A simple one-particle model, which this method implies, is inappropriate for inner-valence-shell ionizations where concomitant electron excitation occurs. In such regions the so-called two-particle-hole Tamm-Dancoff approximation (2ph-TDA) may be used with Green's function calculations.[232-235] Since ground-state correlation effects are then only partially included, the results are somewhat less accurate than for outer-valence-shell ionizations.

An alternative, though formally related, approach to the calculation of ionization potentials utilizes third-order Rayleigh-Schrödinger perturbation theory (RSPT). Though less accurate than the best Green's function calculations with large basis sets, this method is found to be of comparable accuracy when using similar, smaller basis sets.[236] Similarly, when attention is paid to the choice of basis set the RSPT method compares favourably with CI calculations.[237-239]

A configuration-interaction approach applicable to the calculation of open-shell ionization potentials has been described by Schirmer *et al.*[240] The two-hole-one-particle method of these authors is related to the (2ph-TDA) approximation used with Green's function calculations.

Different methods need to be adopted when large molecules are considered. Muller *et al.*[241] have suggested a frozen-orbital approximation that uses a pair-potential approach to treat correlation effects. X_α calculations have most often been used for transition-metal complexes[242] but are finding wider application.[243-245] A study by de Alti *et al.*[246] claims that the LCAO-X_α method may be at least as accurate as Koopman's theorem/Hartree-Fock calculations for small molecules.

Doubly Charged and Negative Ions.- Tsai and Eland[247] have investigated the formation of doubly charged ions in photoionization with He-I and He-II lamps. Initial double ionization comprised a substantial fraction of the total ion yield at 30.4 nm, perhaps more so than electron-impact ionization at comparable energies. An empirical formula for estimation of the appearance energy of doubly charged ions was proposed, based upon a range of existing data, which these authors tabulated. The M^{2+} ions detected in this work have implied lifetimes of at least 20 μs but many more such ions must fragment in less time under the influence of the coulomb repulsion, to be observed as singly charged species.

The fragmentation of doubly charged hydrocarbon ions, produced by 75 eV electron impact, was identified by Brehm and de Frenes,[248] who used a coincidence arrangement to detect the two fragment ions, A^+ and B^+ {equation (13)}.

$$AB^{2+} \rightarrow A^+ + B^+ \quad (13)$$

They estimated that up to 20% of the ions were initially formed doubly charged. Hence, double ionization seems to be a more frequent occurrence than is generally recognized.[249] Coincidence detection has also been employed to study the dissociation of N_2^{2+} into N^+ fragments.[250]

Griffiths *et al.*[251] report the photodissociation of doubly charged ions of benzene and benzonitrile. When excited to a predissociating state, the kinetic energies of fragment ions were found to be independent of photon wavelength. It was suggested that the energy release was predominantly coulombic repulsion, $T = e^2/r$, and that the value of r, which could thus be deduced, might carry some structural information concerning the fragmenting species.

The doubly charged ion of methane has aroused some interest. Using a charge-stripping technique Ast *et al.*[252] deduced an ionization potential of 30.6 eV, substantially less than the value of 40.7 eV from earlier Auger spectroscopy work. Furthermore, whereas a lifetime of $<10^{-15}$ s had been predicted for CH_4^{2+}, in this experiment the ion was clearly stable for the instrumental flight time (~3 μs). A number of calculations have since indicated that CH_4^{2+} has a planar D_{4h} equilibrium structure[253-255] and that the Auger work must relate to a highly distorted ion structure. The poor Franck-Condon factors for vertical ionization would explain the non-appearance of this ion in electron-impact mass spectra. Further, the planar configuration has a relatively large barrier (~1 eV) to dissociation.[254,255] In contrast, the distorted CH_4^{2+} ion must rapidly dissociate to $CH_3^+ + H^+$.

Negative-ion formation by dissociative electron attachment has been studied by Illenberger and co-workers. F^- ions were observed from fluoroethylene[256] and other fluorocarbons.[257] In the latter case, kinetic-energy releases were obtained which, in some instances, were in line with simple statistical expectations but were sometimes much greater. Wang and Franklin[258] have measured ionization efficiency curves and kinetic-energy releases for dissociative electron attachment in sulphuryl halides. Several of the cross-sections for halogen fragment ions show two resonances, the second of which can only be tentatively ascribed to the formation of an excited state of SO_2.

The inverse process of electron detachment from anions has been induced with infrared radiation.[259] The mechanism is one of vibrational photodetachment, which is suggested to be analogous to vibrational autoionization. Beynon and co-workers[260,261] have observed collisionally induced double charge stripping from negative ions produced in a mass spectrometer. Excitation of the positively charged products was evident.

3 Unimolecular Decay

Theoretical Advances.- There can be no doubt that RRKM/QET theory is still the predominant tool for framing theoretical descriptions of unimolecular rates of polyatomic ions. The approach it embodies can be variously characterized as a transition-state theory and a statistical theory. Both these aspects of RRKM theory continue to receive attention as our understanding of the underlying dynamics develops.[262]

The concept of a transition state is intellectually appealing and yet there are problems associated with the use of naïve, intuitive ideas about the transition state in actual calculations. The current status of transition-state theory has been succinctly reviewed by Pechukas.[263] Criteria for identifying the classical transition state are now well established. Although techniques for locating saddle points on potential-energy surfaces continue to be discussed,[264] there is, in the case of many ionic systems, a clear pragmatic objection to this method, namely the absence of any such potential maximum or 'hump' in the reaction co-ordinate. In any case there are strong dynamical arguments in favour of characterizing the transition state as a dividing surface in system phase space carrying minimum flux. Microcanonical transition-state theory would be exact if each reactive trajectory crossed the transition-state surface once only; any other surface over-estimates the reactive flux and hence the rate. A variational criterion therefore exists that places the transition state at the point of minimum flux. Truhlar and Garrett[265] have formulated several versions of variational transition-state theory (VTST). In the microcanonical version (μVT) one locates the transition state at the point along the minimum-energy reaction path where the sum of states is a minimum. These procedures can be compared favourably with accurate quantal calculations and other varieties of transition-state theory for simple model systems.[266]

Chesnavich *et al.*[267] have applied related, though less rigorous, procedures to calculations for the $C_4H_8^+$ system. In common with many ion dissociations there are grounds for suspecting at least two dynamical bottlenecks in this system; one a tight, inner transition state, perhaps primarily affecting the lifetime distributions, the other a loose orbiting transition state situated on the centrifugal barrier that principally influences product-energy distributions. In the transition-state switching model proposed by these authors control of the reaction switches between the two transition states according to local flux criteria.

Another area of conceptual difficulty surrounds the reaction coordinate. Lorquet[262] has stressed the multi-dimensional nature of the reaction path, particularly when non-adiabatic interactions are involved. In any case, it has long been qualitatively recognized how general topological features of a reaction surface can influence product-energy distributions. More quantitative treatments of the effects of curvature in the minimum-energy reaction path have recently appeared.[268,269]

Questions regarding the rate and extent of intramolecular energy transfer are quite clearly of central importance to an understanding of both the power and limitations of statistically based unimolecular reaction theories. Trajectory calculations make a major contribution in this area.[270] Although classical or quasi-classical systems (in which initial conditions are restricted to discrete energy levels but are thereafter classically treated) are best understood, there is currently much interest in extending the techniques to quantal calculations.[271]

Classical systems show two distinct categories of behaviour. At sufficiently low energies, the trajectories are variously described as quasi-periodic, regular, stable, *etc.* and are in accord with the KAM theorem. In the language of phase space the trajectories for a system of n coupled oscillators are restricted to an n-dimensional hypersurface or torus. Consequently not all phase space is explored, the system is non-ergodic, and a statistical treatment is inapplicable. At higher energies the trajectories become stochastic, chaotic, irregular, *etc.*, though still not necessarily fully ergodic. Model potentials for two bound oscillators have been most extensively investigated, and the transition between these two behavioural regimes was found to be very distinct, occurring at a well defined critical energy.[271]

Mayne and Wolf[272] have examined the situation with a potential more relevant to unimolecular dissociation. This was a collinear, two-oscillator DIM surface for the triatomic HeH_2^+ with a plausible dissociation asymptote. A number of interesting observations resulted. First, unlike the more abstract model systems there was no clearly defined transition to chaotic behaviour, and quasi-periodic motion persisted above the dissociation threshold. Second, even above the dissociation threshold, not all the non-quasi-periodic trajectories were ergodic; in particular the region of phase space occupied by the regular trajectories was avoided. Third, not all the non-quasi-periodic trajectories above the dissociation threshold led to dissociation in a reasonable time and only the dissociative trajectories (and then only well above the threshold) became ergodic in character. Removal of the restriction to a collinear molecule would be of interest.

Farantos and Murrell[273] have reported trajectory calculations for the triatomic molecules SO_2 and O_3, using highly realistic six-dimensional surfaces. The transition from quasi-periodic to stochastic motion was found to be somewhat blurred, particularly for O_3. This molecular system did appear to follow statistical behaviour in its dissociation lifetimes.

Extensive studies of the C_2H radical system have been conducted by Wolf and Hase. In their initial investigations, ten plausible model potentials were employed and lifetime distributions calculated from classical trajectories.[274] Selecting two of these multi-dimensional, anharmonic surfaces that displayed intrinsic non-RRKM behaviour (*i.e.* non-RRKM lifetime distributions) they examined the nature of the trajectories in closer detail with the aim of identifying reasons for the failure of RRKM theory in these cases.[275] At 6.5 kcal mol^{-1} above the unimolecular threshold they found the trajectories were quasi-periodic to a very high and surprising degree. In further studies with just one of these surfaces, the effects of the manner of initial excitation were investigated.[276] When randomly excited, it was known that dissociation rates less than RRKM predictions would be found because of the significant proportion of quasi-periodic trajectories. Simulated chemical activation on the other hand was found not to form these regular, non-dissociating trajectories, and the rate actually exceeded the RRKM prediction. Despite these differences, and rather significantly, both modes of excitation resulted in similar, essentially RRKM product-energy distributions at the dissociation barrier. Non-statistical distributions amongst the products must therefore reflect exit-channel interactions. In these calculations it was evident that orbital angular momentum was, on average, conserved on passing from the tight critical configuration at the barrier to products; indeed, at the higher energies studied the barrier and final angular momentum states were highly correlated.[276,277] Consequently, the orbital energy in the exit channel is not available for statistical redistribution and passes instead into product translation. As the degree of correlation increases, this dynamical effect is apparently manifested earlier along the reaction path and could account for the slight deviations from RRKM distributions that were noted at the higher activation energies.

The observation in classical studies[272,275] of isolated regions of phase space that support quasi-periodic motion above dissociation thresholds raises inevitable questions about the use and reliability of classical trajectory calculations. The implication is that, although energetically possible, some dissociations may be classically forbidden. Davis and Heller[278,279] have considered the possibility of quantum-mechanically allowed transitions between regions of classically trapped quasi-periodic motion (referred to as dynamical tunnelling), and clearly it is a process that might prove to be highly significant in the dissociation of such systems. Semiclassical approaches to the study of intramolecular dynamics of polyatomic systems are presently available,[280,281] but full quantal treatments are restricted to simple model systems.

In a rather general way the correspondence between classical and quantal treatments can be seen if classical trajectories and wavefunctions are similarly distributed in co-ordinate space. A primary objective of current studies is the identification of quantum analogues of quasi-periodic and chaotic motion.[271,282] Various criteria for recognizing quantum chaos are being proposed and *inter alia* they help to clarify the relationship between classical ergodic behaviour and vibrational-energy redistribution as it affects unimolecular decay rates.

Using essentially exact quantum-mechanical treatments, Waite and Miller[281] have studied lifetime distributions for quasi-bound model systems consisting of two coupled oscillators, one harmonic, the other capable of dissociation by tunnelling through a potential barrier. Not surprisingly, for weak couplings and disparate frequencies, mode-specific behaviour was encountered (*i.e.* the rate was dependent upon the mode of excitation). What was unexpected was that, for a given coupling, the degree of specificity was approximately constant and showed no indications of the energy threshold behaviour that delimits classical quasi-periodic and stochastic behaviour. Similar observations were made for metastable states of a Henon-Heiles model potential,[283] posing questions as to the relationship between classical stochastic motion of a system and its real observed behaviour. As the authors point out, the ergodic hypothesis assumes that microcanonical phase space and time averages are equivalent over a possibly infinite period of time, whereas what is required for the observation of statistical behaviour is the rapid redistribution of energy on the dissociation time-scale.

This problem is directly confronted in the work of Moiseyev and Certain,[284] who have derived quantum-mechanical expressions for the time-dependent energy in a vibrational mode of the Henon-Heiles potential. They suggest that rapid vibrational relaxation of an excited mode be regarded as an indication of quantal chaos. In this manner, the transition from quasi-periodic behaviour when only one oscillator had been excited was identified near the classical threshold; when both oscillators were excited quasi-periodic behaviour persisted to higher energies. For a rather different model, Kay[285] has found that for strong excitation of just one oscillator the classical system was more ergodic than the quantal one, although otherwise there was close correspondence between the two. Further clarification of the wider significance to be drawn from these findings is needed.

The decay of vibrational energy is but one indication of the more general behaviour found for time-correlation functions of dynamic variables. Essentially, the autocorrelation indicates for how long a property exists before the dynamic process causes it to be averaged out. Koszykowski *et al.*[286] computed a number of correlation functions for the Henon-Heiles system that were observed to decay away in the presence of chaotic motion. The rate of decay of a mode-energy correlation function indicates the rate of relaxation of that mode. For a quantum-mechanical system Weissman and Jortner[287] have considered the rapid dephasing of the initially coherent gaussian wave packets to be an indication of quantal chaos. As the wave packet spreads over the surface, its autocorrelation function decays, just as for classical variables.

It is known that if the Fourier transform of one of these correlation functions is generated it gives a frequency spectrum. In the quasi-periodic regime a line spectrum showing fundamental and overtone frequencies results. In the chaotic regime broad series of lines become evident. For classical trajectories this presents a method for distinguishing the two regimes.[271,273] Further, correspondence with a quantum-mechanical system may be demonstrated by comparing the known eigenstates with the spectrum. Heller[288,289] has shown how the vibrational structure in an electric-dipole spectrum may be calculated by Fourier transforming the correlation function of the wave packet prepared in the Franck-Condon transition. In this method the propagation of the wave packet is established by following a small ensemble of appropriately chosen classical trajectories.

In a paper of considerable interest Lorquet *et al.*[290] proposed inverting this procedure to study the short-time dynamics of an excited system by using spectroscopic data. Specifically they take the Fourier transform of vibrational bands of high-resolution photoelectron spectra, which they then interpret as a correlation function of the wave packet prepared on the upper surface, *i.e.* equation (14) holds:

$$|\mathcal{F}\ I(E)| = |\ \langle\phi(0)|\phi(t)\rangle\ | = C(t) \qquad (14)$$

where ϕ is not in general a stationary state but a solution of the time-dependent Schrödinger equation. For a photoelectron-energy resolution of 10 meV the correlation function, C(t), spanning a time interval of 10^{-15} - 10^{-13} s following the transition, is obtained and a qualitative interpretation of these dynamics of very short time intervals becomes possible. Figure 3 presents a number of examples. In the case of the *B* state of N_2^+, which is deep and bound, the correlation function is a periodic function representing the slightly oscillating overlap as the wave packet moves in and out of phase with its initial position. For the *A* state the vibrational amplitude is greater and the minima of the oscillations are much deeper. Very different behaviour is found for the *B* state of HCN^+, where the function C(t) quickly decays but then persists, showing slight undulations. Lorquet *et al.*[290] have interpreted this behaviour, with the aid of a potential surface for the *B* state, as indicating a rapid spreading of the wave packet due to energy transfer within a few vibrational periods, followed by a period of more stable behaviour.

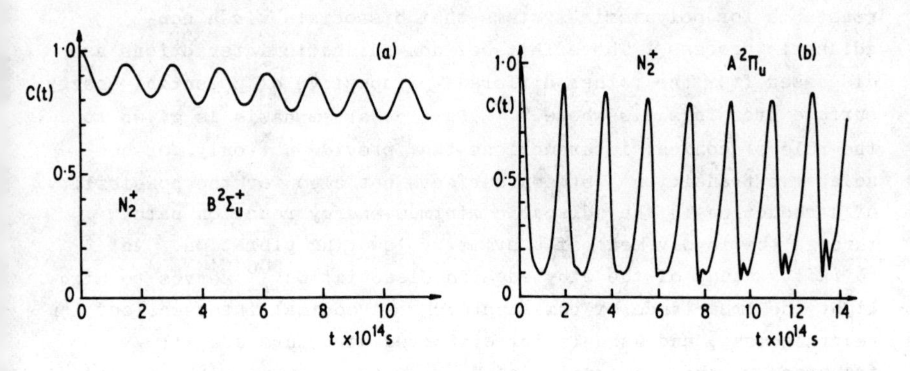

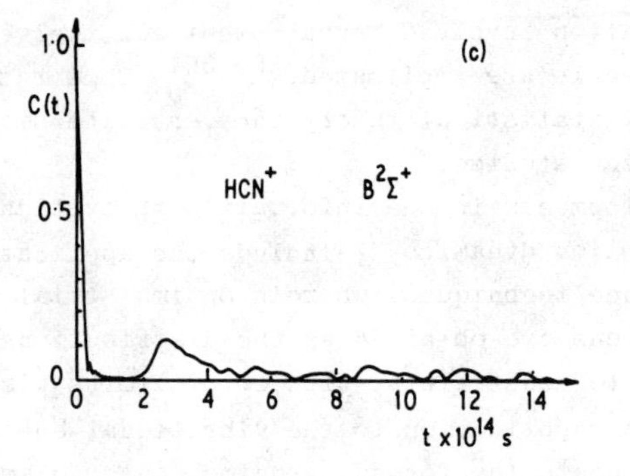

Figure 3 Autocorrelation functions of *(a)* $B\ ^2\Sigma_u^+\ N_2^+$, *(b)* $A\ ^2\Pi_u\ N_2^+$, and *(c)* $B\ ^2\Sigma^+\ HCN^+$ obtained from Fourier transforms of the experimental photoelectron spectra
(Reproduced with permission from *J. Chem. Phys.*, 1982, 76, 4692)

Doll[291] has outlined a unified theory of dissociation, which is a generalization of Slater's theory, avoiding the use of normal mode analysis. When it is desired to apply anharmonic corrections within the RRKM formalism, he has pointed out the use of Monte Carlo sampling techniques for the rapid evaluation of the state density functions.[292,293]

The foregoing theories are implicity concerned with dissociation over a single electronic potential-energy surface, although a strong expression of QET presupposes rapid internal conversion to the ground electronic state. Leach and co-workers[294,295] have summarized the theoretical treatment of radiationless transitions as applied to molecular ions, and Lorquet *et al.*[296-298] have developed a method for calculating rate constants for polyatomic systems that dissociate *via* a non-adiabatic process. The effects of non-adiabatic interactions are discussed from the rather different perspective of potential-energy surface crossings elsewhere.[299] Particular emphasis is given to the role of conical intersections that provide not only for non-adiabatic transitions between surfaces but also for the possibility of a reduction in the adiabatic minimum-energy reaction path through the involvement of a symmetry-lowering vibration. An *ab initio* study of the ethylene-ion dissociation[300] serves to highlight the importance of this approach. A conical intersection, ion rearrangement, and Jahn-Teller distorted structure are all traversed in the elimination of H_2 from the A state. The reaction path is clearly complicated, and a consideration of the changes in nuclear configuration involved reveals that all twelve vibrational modes of the molecule are implicated.[300,301] Some rationalization of the success of statistical theory in even quite small molecules is obtained by such studies.

Recent developments in the information theoretical approach to molecular reaction dynamics[302] include the application of surprisal inference techniques, wherein optimal trial product-energy distributions are obtained as the distributions of maximal entropy, subject to constraints, with error limits also determined.[303] An application to the vibrational-energy distributions of diatomics formed by unimolecular elimination reactions found the only constraint to be the mean vibrational energy, $\langle E_v \rangle$, of the diatomic.[304] This varied with total system energy, E, in a manner that could be interpreted simply as representing a contribution from the fixed reaction-barrier energy, E_b, and the excess energy at the barrier, *i.e.* $\langle E_v \rangle = a(E - E_b) + b(E_b)$, where a and b are constants.

Ion Photodissociations.- Ion photofragment spectroscopy utilizes fast ion beams, coaxial with a laser beam, to obtain spectroscopic detail of dissociative and predissociated states, often with exceedingly high resolution. The experiments may be combined with measurements of fragment-ion angular distributions or kinetic-energy releases and are capable of yielding a wealth of information about ion-dissociation dynamics.[305,306]

The most studied predissociation is that of the $b\ ^4\Sigma_g^-$ state of O_2^+, produced from the metastable $a\ ^4\Pi_u$ state of the ion {see equation (15)}.[306]

$$O_2^+\ (a\ ^4\Pi_u,\ v'') \xrightarrow{h\nu} O_2^+\ (b\ ^4\Sigma_g^-,\ v') \rightarrow O^+\ (^4S^0) + O\ (^3P) \qquad (15)$$

Several hundred resolved transitions to $v' = 4$ or 5 levels have been observed[307] and predissociation lifetimes deduced from the linewidths. Both the $^4\Sigma_g^+$ and $f\ ^4\Pi_g$ states could potentially predissociate the $b\ ^4\Pi_g^-$ state, but the indicated lifetimes suggest that only the former is involved. This is actually in mild disagreement with earlier measurements of the $v' = 3 - 5$ levels,[308] where it was concluded that 22 ± 10% of the ions were branching to the $f\ ^4\Pi_g$ state. A number of calculations have predicted a small barrier in the $f\ ^4\Pi_g$ surface, although there is disagreement as to its exact height, and this might hinder predissociation *via* this state. The calculations of Roche[307] indicated that it would indeed inhibit the predissociation of the $v' = 4$ level. The barrier has been experimentally identified in direct transitions to the $f\ ^4\Pi_g$ surface.[309] Two quasi-bound levels were assigned as $v' = 2$ or 3, and the barrier height was estimated as 48.8 meV for $\Omega' = 3/2$ and 44.7 meV for $\Omega' = 5/2$. The linewidths are consistent with the $v' = 3$ level dissociating mainly by tunnelling, but predissociation by $\alpha\ ^6\Pi_g$ is probably responsible for the dissociation of $v' = 2$ levels. No predissociation of the $v' = 1$ level takes place, but this line has nevertheless been observed by an optical double-resonance technique,[310] confirming the assignment of $v' = 2$ or 3. The dissociative regions of the $f\ ^4\Pi_g$ surface can also be populated by a photon-induced transition from the $\alpha\ ^4\Pi_u$ metastable state. By measuring the kinetic-energy spectrum of the O^+ fragments the *f*-state curve has been precisely mapped out in the region where it crosses the *b*-state curve, found to be near the $v' = 7$ level.[311] This is higher than previously thought. Very recent *ab initio* CI calculations[312] support this finding. The $b\ ^4\Sigma_g^-$ curve was crossed by the $^4\Sigma_g^+$ curve at $v' = 4 - 5$ and by $f\ ^4\Pi_g$ at $v' = 6 - 7$ in these calculations. The spin-orbit couplings between these states were calculated (though not the rotational coupling between the *b* and *f* states) and were found to account well for the experimental data.[307] Although the computed *b*-*f*-state couplings are different from those found by Roche,[307] it was again concluded that $v' = 5$ is little affected by the *f* state. It is apparent that the v' = 7 level should be very strongly predissociated, mainly by the $f\ ^4\Pi_g$ state. Predissociation of the v' = 5 - 9 levels has been observed experimentally,[313] though not with sufficient resolution to test fully the theoretical predictions. An alternative theoretical approach has been outlined by Durup,[314] who considers the spin-orbit interactions in O_2^+ in terms of a diabatic basis set with long-range coupling effects.

Photofragment spectra have been obtained for the He^{2+} transition $X\ ^2\Pi_u^+ \rightarrow A\ ^2\Pi_g^+$. The A state is dissociative and the kinetic-energy distributions of the fragment, He^+, showed discrete vibrational structure in excellent agreement with the ground-state parent-ion potential-energy curve.[315] Similar measurements for excitation from the higher rotational levels of $X\ ^1\Sigma_g^+$ CH^+ show evidence of centrifugally bound levels[316] for the dissociation shown in equation (16).

$$CH^+\ (A\ ^1\Pi) \rightarrow C^+\ (^2P) + H\ (^2S) \qquad (16)$$

Quite detailed assignment of these rotationally quasi-bound levels has been achieved. This effect can be predicted from calculations,[317] and it is suggested that it might also result in an enhanced rate for the atom/ion recombination. Three vibrational levels of $X\ ^2A_1$ O_3^+ have been identified by photofragment kinetic-energy analysis.[318] An additional complication here is the possibility of excess energy being partitioned into product rotation and vibration, and both these effects were observed. At lower energies, rotational energy of the O_2 fragment was favoured to a greater extent than predicted, indicating substantial geometry changes in the transition.

Well resolved rotational lines with no indication of an underlying continuum are observed in the photofragment spectrum of N_2O^+.[319] This apparent lack of direct dissociation suggests that the $^2\Pi$, $^2\Delta$, and $^2\Sigma^-$ dissociation continua are not responsible for predissociating the $A\ ^2\Sigma^+$ state, since they are all dipole-allowed transitions from the $X\ ^2\Pi$ ground state; a $^4\Sigma^-$ state is most likely responsible for the predissociation. Cosby and Helm[320] have observed predissociation of $c\ ^3\Pi$ NO^+ from the v' = 0 - 19 levels. Lifetimes estimated from the linewidths were all greater than 1.6 ns.

Photodissociation is not restricted to di- and tri-atomic species and, for example, has been applied to buta-1,3-diene, with fragment kinetic-energy analysis.[321] A most interesting series of observations concerns the infrared photodissociation of CF_3X^+

(X = I, Br, or Cl). In the experiments of Coggiola *et al.*,[322] absorption of a single 10 μ photon induces dissociation to CF_3^+. Hence the ions must be highly vibrationally excited to near the dissociation limit in the beam. Any photon energy in the 1000 cm^{-1} region should, therefore, be energetically capable of bringing about dissociation and yet the photodissociation spectrum is strongly peaked at 950 cm^{-1}. In the statistical limit, an individual vibrational mode would hold very few quanta of energy even near the reaction threshold, and it is deduced that the 950 cm^{-1} absorption is a $0 \rightarrow 1$ or $1 \rightarrow 2$ transition of the ν_1 mode. This is then a clear example of vibrational predissociation.

In ion cyclotron resonance photodissociation experiments[323] the CF_3I^+ ions are presumably somewhat more relaxed since here multi-photon dissociation (m.p.d.) is observed, again peaking at about 950 cm^{-1}. Only 60% of the CF_3I^+ dissociates, but that which does undergoes a very facile reaction. Various arguments suggest that it is the $X\ ^2E_{3/2}$ state of CF_3I^+ which is stable and the $^2E_{1/2}$ component which dissociates. Photoelectron-photoion coincidence experiments[324] support this analysis since approximately half the $^2E_{1/2}$ ions produced in photoionization lie below the dissociation threshold, giving a 1:1 ratio of unreacted $^2E_{1/2}$:$^2E_{3/2}$ ions. It is less clear why this photodissociation should prove to be so facile. There is no significant population of metastable ions and no evidence for any barrier in the dissociation channel.

Infrared enhancement of visible photodissociation has been observed. The dissociation of cyanobenzene[325] at 568 nm is strongly enhanced by i.r. radiation, the effect peaking at 970 cm^{-1}. It is thought that i.r. m.p.d. of a vibrationally excited state of the ion takes place following internal conversion (i.c.) of an electronically excited intermediate, as shown in reaction (17).

$$C_6H_5CN^+ \xrightarrow{\lambda\ =\ 568\ nm} C_6H_5CN^{+*} \xrightarrow{i.c.} C_6H_5CN^+\ (vib.) \xrightarrow[\text{m.p.d.}]{\lambda\ =\ 568\ nm} C_6H_4^+ + HCN \qquad (17)$$

Similarly, iodobenzene[326] shows infrared enhancement of a two-photon photodissociation at 610 nm, but in this case it was proposed that i.r. pumping of the vibrationally excited intermediate counteracted relaxation effects, so making the second-photon absorption more probable.

The two-photon dissociations of cyanobenzene, *etc.*,[327] and of naphthalene[328] have themselves been studied. Naphthalene may also be four-photon dissociated at 600 nm, although the observed intensity dependence was only $I^{1.5} - I^{2.5}$. A kinetic model was discussed to account for this.[328] Such studies are of course relevant to the m.p.i. of naphthalene.[329]

Beynon and co-workers have utilized four photon energies of an argon-ion laser in attempts to study internal-energy-dependent branching ratios[330,331] and fragment-ion kinetic-energy releases.[332-335] Because of lack of information about the composition of the initial ion beams, the simplest possible assumption is made - that the observed behaviour occurs as a function of change in the mean internal energy, which is the same as the change in photon energy. Clearly this can only be an approximation because of the known wavelength selectivity of laser-induced transitions.

In several instances the translational-energy *versus* photon-energy plots unexpectedly show negative slopes.[332,334,335] For the dissociation of MeI^+ to Me^+ + I, this was ascribed[335] to an increasing contribution from a high-energy channel leading to excited I ($^2P_{1/2}$). The metastable peaks for this dissociation were reported in the same paper and support this interpretation, there being a bimodal kinetic-energy release with the low-energy component having the higher appearance potential. Similar photodissociation results were obtained for I^+ formation and were likewise interpreted as indicating excited I^+ ($^3P_{0.1}$) reaction channels. An alternative explanation based on studies of the full photodissociation spectrum in this wavelength region can be proposed.[336,337] As the photon energy is scanned upwards from 2 eV a 'repetition' of the spectral detail is observed starting at about 2.5 eV. This is due to photodissociation first of the MeI^+ $X\ ^2E_{1/2}$ state and then, at higher energies, of the ground $X\ ^2E_{3/2}$ state. Consequently, more dissociation occurs as the photon energy increases, but the *mean* energy in excess of the dissociation threshold may actually decrease as more of the $^2E_{3/2}$ ions from the beam are caused to react. This is not necessarily to deny that the MeI^+ system is an example of state-specific behaviour, since collisionally induced dissociations also reveal bimodal energy distributions.[335,338]

Ion-fluorescence Studies.- An expanding range of molecular ions has been studied by fluorescence spectroscopy.[339] Depending upon whether the emission wavelength or, in the case of laser-induced fluorescence, the exciting wavelength is scanned, vibronic structure of the ground or excited state is obtained. A knowledge of vibrational frequencies and of interactions such as the Jahn-Teller effect (*e.g.* see ref. 340) greatly assists the investigation of unimolecular decay processes. Direct evidence of ionic-state lifetimes comes from monitoring the fluorescence-decay rates following pulsed ionization and/or excitation.[341,342] An alternative method accelerates ions away from the optical detection system as they are formed by continuous u.v. irradiation.[343] Control of the acceleration field enables fluorescence-decay times of the order 10^{-7} - 10^{-5} s to be measured.

Leach and co-workers[344,345] have used a photoion-fluorescent-photon coincidence technique (p.i.f.c.o.) to measure decay rates and quantum yields. The molecular ions of mono- and di-chloroacetylene[345] display multi-exponential decay curves for the $X\ ^2\Pi_{(u)} \leftarrow A\ ^2\Pi_{(g)}$ transitions, behaviour that is characteristic of coupling with an 'intermediate' final-state density. When dissociative ionization takes place, the p.i.f.c.o. technique enables fluorescence decays for each of the ion fragments to be identified, as has been demonstrated with SO_2.[346] Low fluorescence quantum yields from S_2^+ (which must emit from either the $C\ ^2B_2$ or $D\ ^2A_1$ states) indicate the extent of competition with dissociation. Electronically excited SO^+ and SO are also produced, but pre-dissociation to S^+ results only in ground-state products and so does not fluoresce.

Other ion-fluorescence results are discussed in the next section.

Coincidence Methods of State Selection.- Methods for the detection of ions or fluorescence photons in coincidence with energy-selected photoelectrons provide much valuable and detailed information on ion dynamics as a function of initial internal energy, electronic state, and, in favourable curcumstances, initial vibronic state. These are considered according to the type of information acquired.

Competing Predissociation and Radiative Processes. Two competing dissociation channels have been observed for $B\ ^2\Sigma_g^-\ O_2^+\ (v')$ by photoelectron-photoion coincidence (p.e.p.i.c.o.) experiments.[347,348] The alternative sets of products, O (^3P) + O^+ ($^4S^0$) and O (^1D) + O^+ ($^4S^0$), are clearly distinguished by the accompanying kinetic-energy release. Below v' = 3, only ground-state O (^3P) is produced (on energetic grounds) but, for $v' = 3$, 4, or 5, excited O (^1D) is also produced. The branching ratio for this second channel was found[347] to peak at $v' = 4$, and it was suggested that the dissociating state, tentatively identified as the second $^4\Pi_g$, crossed the $B\ ^2\Sigma_g^-$ curve at around the $v' = 4$ level. Recent calculations[312] indicate that the crossing point is more likely to be $v' = 10$. An alternative indirect predissociation mechanism is therefore proposed, involving coupling to the second $^2\Pi_g$ state, which is in turn predissociated by the $^4\Delta_g$ state.

Maier and co-workers have developed photoelectron-fluorescent-photon coincidence techniques (p.e.f.c.o.) to investigate fluorescence from selected ion states.[349] In this manner, radiative-decay curves and quantum yields have been determined for individual vibronic levels of triatomics[349] and diacetylenes.[350] Some of the triatomics have non-exponential decay curves, indicative of 'intermediate' coupling behaviour, but the diacetylenes decay in a manner characteristic of the 'statistical' limit. The *A* state of the octadiyne cation provides a further example of 'statistical' behaviour.[351]

Comparisons of the fluorescent decay of different electronic states of an ion prove to be interesting. The *C*-state emissions of several fluorobenzenes are apparently at similar wavelengths to the *B*-state emissions.[352] Possibly there are radiationless *C*-to-*B* transitions preceding the fluorescence, which is actually from the *B* state. Excited-electronic-state fluorescence has also been investigated in dicyanoacetylene[353] and fluorocarbon[354] cations.

Non-radiative-decay rates may be inferred from a knowledge of the fluorescence quantum yields and lifetimes. When p.e.f.c.o. studies are complemented by p.e.p.i.c.o. investigations of the various non-radiative (fragmentation) decay paths, a very complete picture of the unimolecular behaviour of an ionic system emerges. The ion N_2O^+ is a case in point. The *A* state is known to be predissociated by the $a\ ^4A''$ ($^4\Sigma^-$) state,[319] which correlates with the lowest-energy products of O^+ and NO^+. A threshold-photo-electron-photoion coincidence (t.p.e.p.i.c.o.) experiment[355] showed that the (0,0,0) level is undissociated whereas the (1,0,0), (2,0,0), and (0,0,1) levels give parent and NO^+ ions, but only traces of O^+. The p.e.f.c.o. experiments of Maier[349,356] indicate radiative decay of these same levels (hence stable parent ions) with quantum yields of unity for the (0,0,0) level and less than unity for the others. Beswick and Horani[357] have performed model calculations for this system, invoking spin-orbit coupling of the states. These calculations accurately explain the observed behaviour. Most impressively, a radiative lifetime for the (2,0,0) level of 35 ns was predicted and subsequently measured as 39 ns, very much shorter than for the other levels.[356] The same $a\ {}^4A''$ $({}^4\Sigma^-)$ state is believed[358] to participate in dissociations that have been observed by the t.p.e.p.i.c.o. experiment[355] in the non-Franck-Condon region between N_2O^+ *X* and *A* states. In this instance, O^+ not NO^+ is the predominant fragment, presumably reflecting features of the system dynamics that are not yet fully understood.

Another ion studied by both p.e.f.c.o. and p.e.p.i.c.o. is *cis*-1,2-difluoroethylene. Although the lowest vibrational levels of the *A* state lie 1 eV above the threshold for HF loss, parent ions were found to be largely stable in p.e.p.i.c.o. investigations of these levels.[359] A direct p.e.f.c.o. measurement[360] of the fluorescence quantum yield is much smaller than suggested by the p.e.p.i.c.o. parent/fragment branching ratio. Consequently, a competing non-radiative, non-fragmenting decay mechanism must be operative and from the experimental data can be assigned a rate constant of about $2.9 \times 10^6\ s^{-1}$, compared to approximately $2.1 \times 10^5\ s^{-1}$ for radiative decay and $4.7 \times 10^4\ s^{-1}$ for HF elimination.[360] Competing radiative- and non-radiative-decay mechanisms for electronically excited hexadiynes,[361-363] chloropropyne,[364] and hexatrienes[365] have been elucidated in related experiments that combine p.e.p.i.c.o. and ion-fluorescence data.

Product Distributions and State-specific Behaviour. Baer *et al.*[366] have measured kinetic-energy-release distributions by t.p.e.p.i.c.o. for the dissociation of ethyl iodide ion and find good agreement, on the whole, with statistical-dynamical-model calculations that conserve angular momentum at the centrifugal barrier. This system thus joins the growing list of instances where such a theoretical model seems to be apt. It has been proposed by Johnson *et al.*[367] that in many of these cases the distribution functions will obey a simple scaling law; that is, the distributions can be reduced to a common form, independent of the internal energy of the system, by a simple transformation of variables. The specific example given was dissociation of acetone ion, for which scaling of both experimental p.e.p.i.c.o. data and calculated statistical-dynamical-model distributions was demonstrated. Where scaling applies, it offers a means for conveniently compacting experimental data, for interpolating the data, and may further reveal changes in the underlying dissociation channel(s) as a deviation from the scaling above an associated threshold.

Powis[368] has found a striking example of the influence of angular momentum and the centrifugal barrier upon product-kinetic-energy distributions in the H-loss dissociation of NH_3^+ (^2E). Above the 0 K reaction threshold the mean energy release rises monotonically with system internal energy, as expected. At and below this threshold the energy release does not tend to zero but rather reaches a minimum and even rises slightly with falling energy. In this region it is of course only the rotationally excited ions that are energetically capable of dissociation, and this observation indicates a correlation between high rotational quantum number and a partitioning of energy favouring product translation. Such behaviour can be understood by considering the need to conserve angular momentum at the centrifugal barrier and is predicted by the usual statistical-dynamical-model calculations;[368] it is particularly marked in the case of the system NH_2^+ + H because of the abnormally large rotational constants that restrict the availability of product rotational states on purely energetic grounds.

Marked state-specific behaviour is evident in the measured kinetic energies of dissociations $NF_3^+ \rightarrow NF_2^+ + F$.[369] The ground-state distributions are described by the statistical-dynamical model but are very different in character and magnitude for the *A* and *B* states of NF_3^+. Here, an impulsive dissociation model is found to be appropriate, implying a much more rapid dissociation. The *C*-state dissociations release much less energy and it seems that a statistical-type dissociation occurs. A detailed consideration suggests that the product ions may exist in an excited state of NF_2^+ lying about 1.5 eV above the ground state. There is some uncertainty about this identification because, whilst it agrees with MO calculations and mass-spectrometry results, it conflicts with the lowest, 2.43 eV, electronic state found by photoelectron spectroscopy.

Isolated-state behaviour has been identified tentatively by means of state-specific energy releases in the excited states of formic acid[370,371] and by deviations from QET predictions that occur at higher excitation energies for the ethanol ion.[372] These same authors[373] find distinct initial-state-dependent branching ratios for the dissociation of $MeNO_2^+$. The fragments NO_2^+ and Me^+ have similar thermodynamic thresholds, yet, while the former is found close to its threshold, Me^+ is not observed in p.e.p.i.c.o. measurements until 1.2 eV above its calculated onset. This onset coincides with the threshold for formation of excited neutral NO_2, and the measured energy release supports the conclusion that excited products are formed. Meisels *et al.*[374] have investigated the isomeric $MeONO^+$ ion by t.p.e.p.i.c.o. experiments. At one point they observed concurrent formation of MeO^+ and NO^+. Dissociation to MeO^+ was clearly metastable ($\tau = 3 \times 10^{-6}$ s), unlike the more rapidly produced NO^+. Consequently, the two channels cannot be competing in the normal sense, and it was suggested that different electronic states were involved.

Kinetic Rates, Mechanisms, and Fragment Branching Ratios. Baer and co-workers have reported t.p.e.p.i.c.o. experiments on the dissociation of three $C_4H_8O_2^+$ isomers. The most stable, butanoic acid,[375] was not found to decompose through metastable channels, in contrast to electron-impact mass spectra of this ion. On the other hand, 1,4-dioxane did produce metastable ions, $C_2H_4O^+$, $C_2H_5O^+$, and $C_3H_6O^+$, whose energy-dependent rates were measured from the asymmetrically broadened time-of-flight (TOF) mass peaks.[376] Their behaviour was modelled by RRKM/QET calculations with an assumed 'tight' transition state. No evidence of isomerization prior to dissociation was found, unlike for the third isomer, ethyl acetate, for which comparison with QET calculations indicated the involvement of an, as yet, unidentified isomer.[377] The question of whether rearrangement steps precede fragmentation has been studied in various other isomeric systems. Isomers of $C_4H_4O^+$, for example, do not appear to rearrange to a common precursor,[378] whereas those of $C_4H_4N^+$ [379] and $C_4H_8^+$ [380] generally do.

More specific information regarding the nature of dissociation transition states has been sought by careful modelling of the energy-dependent fragmentation behaviour of molecular ions of thiophene by RRKM/QET calculations.[381] Although these ions fragment extensively, below 12.9 eV only $C_2H_2S^+$ fragment ions are produced; dissociation rates have been determined for this species from the t.p.e.p.i.c.o. TOF peak shape. Optimal agreement with RRKM calculations results if linear $C_4H_4S^+$ and $C_2H_2S^+$ structures are assumed, though QET calculations of the translational-energy releases are at least 50% in error.

Rosenstock *et al.*[382] utilize a delayed-ion-extraction t.p.e.p.i.c.o. experiment to obtain energy-selected, time-dependent fragmentation patterns. They, too, find that 'tight' transition states are required in the construction of RRKM models for the fragmentation patterns of the ions benzonitrile[382] and pyridine.[383] Decay rates determined by this technique for bromobenzene ion[384] show considerable discrepancies with earlier results derived from metastable TOF peak analysis but are in good agreement with a recent photoionization study.[385] In this case it is not so clear that a 'tight' transition state is adequate to explain the data.

Dannacher *et al.* have been able to deduce absolute rate data from the slopes of high-quality, deconvoluted ion-yield curves measured by p.e.p.i.c.o. experiments.[324,386] For the $A\ ^2A_u$ state of buta-1,3-diene cation[386] total-fragmentation rate constants ranging from 7×10^3 to $10^5\ s^{-1}$ were obtained. These complement data from the alternative technique of TOF peak-shape analysis, which is more suited to the $10^5 - 10^6\ s^{-1}$ rate-constant regime.

The same group has measured breakdown diagrams for acetaldehyde[387] and formaldehyde[388] molecular-ion dissociations by p.e.p.i.c.o. The CO^+ fragment from formaldehyde is of some interest since it is first observed from the A ($v' = 5$) parent-ion state, some 0.72 eV above its thermodynamic threshold. Its formation is accompanied by a substantial energy release[389] indicating that there is a barrier to this dissociation. The reaction to CHO^+ occurs throughout the A state but, from the deuteriated species, CDO^+ appears as a large metastable peak.[390] Since this process has an appearance energy that coincides with the adiabatic A-state ionization potential, and which is about 2 eV above the thermodynamic dissociation threshold, it appears that the rate-determining step must be internal conversion from the A to a vibrationally excited X state of the formaldehyde ion.

Mass-spectrometer Studies and Metastable Ions.- The determination of dissociation mechanisms as such is reviewed in Chapter 2 of this volume. This section commences with a consideration of experimental techniques. A review paper by Beynon[391] discusses the wide range of experiments that can be attempted with current designs of mass spectrometer. One such possibility is collision-induced dissociation (c.i.d.) of selected ions. For collision energies exceeding 1 keV it is found that the collision times are considerably less than a vibrational period, and the energy deposition in the ions somewhat resembles electronic excitation.[392] Stace and Shukla[393] have developed procedures for extracting data on kinetic-energy release from the c.i.d. peaks, for which there are obvious applications.

It can be demonstrated[394] that, for a beam of aligned Na_2^+ ions, the c.i.d. cross-section peaks when the rotational angular momentum is parallel to the relative velocity of ions and collision gas, *i.e.* when the ions are 'broadside on'. While LAB scattering angles are small, further studies[395,396] indicate that there is a correlation between scattering angle and internal-energy deposition and, though this is not yet well enough understood, it seems to rival seriously other techniques for performing state-selected chemistry.

Other results[397] indicate that c.i.d. excitation is comparable to field ionization for the study of short lifetimes. The energy deposition of field ionization itself has been investigated, and the proposition that it should predominantly reflect the thermal distribution was tested.[398,399] Variation of ion-source temperature can provide a well characterized control of the ion energies.

Lifshitz has described a time-dependent method of measuring mass-spectral fragmentation patterns by trapping ions for periods ranging from microseconds to milliseconds. Results for pyridine[400] were in good agreement with QET and delayed-resident-time t.p.e.p.i.c.o. experiments,[383] while time-dependent appearance potentials were noted for hexa-1,5-diyne.[401] For short delays, the parent ion had the lower appearance potential whereas $C_6H_5^+$ appeared at a lower energy after millisecond delays. This was attributed to a slow isomerization to the more stable benzene isomer followed by H loss.

Holmes and Terlouw[402] have documented the scope of metastable peak experiments, drawing attention to the lack of uniformity in the reporting of kinetic-energy data derived from peak-shape measurements. Some standard procedures are proposed for future use. Beynon *et al.*[403,404] have considered how, by selecting ions from different regions of a metastable peak for further study, a degree of control over the internal energy of those ions can be exercised. This follows from two facts: first, the selection of a narrow range of precursor lifetimes ($\sim 10^{-6}$ s) implies a narrow range of precursor energies, and, second, the release of large amounts of kinetic energy in the dissociation will both broaden the peak shape and be associated with a lesser partitioning of energy into internal modes of the fragment ion.

Jarrold *et al.*[405] have investigated the metastable dissociation of H_2S^+ and measured the temperature dependence of the S^+ fragment kinetic-energy distributions. These display structure, identifiable with the vibronic levels of $A\ ^2A_1\ H_2S^+$ which, it is suggested, convert slowly to the $X\ ^2B$ state *via* a Renner-Teller interaction at the linear configuration. Subsequently, the ground state crosses to the dissociative $^4A_2$ surface with which it is spin-orbit coupled. This contrasts with a correlation-diagram approach,[406] where it was suggested that the A state slowly crosssed to a $^2B_2$ state *via* a conical intersection, which was then predissociated by the $^4A_2$ state. The identification of $^4A_2$ as the dissociative state seems well founded.[407]

Considerable interest has been aroused by the finding that the metastable dissociation of the enol ion of acetone displays evidence of non-ergodic behaviour. Isotopic labelling of the α and β carbon atoms {equation (18)} indicates that the loss of the β methyl group

$$\left[C^{\alpha}H_3{-}C(OH){=}C^{\beta}H_2\right]^{+\cdot} \rightarrow \left[C^{\alpha}H_3{-}CO{-}C^{\beta}H_3\right]^{+\cdot} \rightarrow CH_3CO^+ + CH_3^{\cdot} \qquad (18)$$

is preferred by a factor of about 3:2 and is also accompanied by a rather larger energy release[408,409] whereas in a fully randomized acetone intermediate no preferences are to be expected. Measurements of the kinetic-energy-release distributions for this reaction showed a distinct bimodal distribution.[410] The barrier height for isomerization was estimated from appearance-potential data and an estimate of the energy of the metastable precursors was obtained. When compared with acetone ions prepared directly with similar energy, the kinetic-energy release from the enol was very different and so this, too, was taken to indicate non-ergodic behaviour in the enol-ion dissociation. The barrier height is difficult to estimate accurately from appearance potentials. Very recently, the original estimate[410] has been revised upwards from 0.76 to 1.4 eV above the threshold for acetyl-ion formation and a surprisal analysis was performed on the data,[411] but the essential conclusion about non-ergodic behaviour remains the same. However, Johnson *et al.*[367] have pointed out that the bimodal distribution can be decomposed into two maxwellian-shaped curves, one of which is fully consistent with p.e.p.i.c.o. measurements on the (presumed ergodic) dissociation of acetone made at an excess of energy of ?.1 eV, *i.e.* just 0.7 eV above the current best estimate of the barrier height. The smaller energy release was suggested to be due to formation of the CH_2CHO^+ isomer, a feature that is also incorporated in the revised analysis of the metastable data.[411]

It has been suggested that the dissociation of the enol of $EtCOEt^+$ is similarly an example of non-randomization of vibrational energy,[412] but this disagrees with ^{13}C-labelling experiments that find no preference for the loss of either ethyl group.[413] More work on these systems seems desirable. For example, p.e.p.i.c.o. results from acetone are themselves in rather worse agreement with statistical-dynamical-model predictions than many other molecules that have been studied, and the c.i.d. peak profiles of acetone appear to be composite in nature,[393] neither fact being yet understood.

4 Bimolecular Ion/Molecule Collisions

Only a brief consideration of this topic is presented here, selectively emphasizing those aspects that are most pertinent to the study of unimolecular reactions. The use of beam techniques in this field is discussed in Chapter 4 of this volume.

Collision Theory.- The simple Langevin theory of capture cross-sections plays an important role, serving as a reference point against which real behaviour and other theories may be judged. It is a model for two particles interacting under the influence of a central potential in which the existence of orbiting trajectories at a critical impact parameter delimits collision-complex formation. Pechukas[414] has shown that orbiting trajectories exist also for non-central potentials. The most often used extensions to Langevin theory attempt to make allowance for the orientation-dependent ion-dipole interaction of polar molecules. Hsieh and Castleman[415] have rederived a number of these treatments (Langevin, locked-dipole, Barker-Ridge, ADO, and AADO) using a common formalism that helps to demonstrate their relationship with one another. They find shortcomings in these approaches and are led to propose an alternative 'total-energy and angular-momentum-conserved average charge-dipole interaction model(s)' (TEAMS) with the same formalism. Bates[416] has likewise identified problems with the treatment of energy conservation in ADO theory and has proposed a slight reformulation to accommodate his suggested improvements.

Internal Energy Effects.- Vibrational-state selection of the ionic partners in ion/molecule collisions has been achieved in a number of instances using t.p.e.p.i.c.o. techniques. Koyano and Tanaka[417] have investigated the system H_2^+ (v = 0 - 3) + H_2 in this manner. The predominant outcome is proton transfer, the extent of either charge transfer or v-v transfer being negligible. At low, average collision energy (~0.11 eV) the reaction cross-section decreases with increasing vibrational quantum number, though such dependence is much less in evidence at 0.93 eV collision energies. The v = 0 state cross-sections actually follow the Langevin-model predictions quite closely, but this is not true of higher vibrational levels. The reasons for this behaviour are not clear, nor indeed is the exact collision/reaction mechanism.

Rather better understood are the systems H_2^+ + Ar and Ar^+ + H_2, which are asymptotic limits of the same set of $(ArH_2)^+$ collision-complex potential surfaces. Only the system H_2^+ + Ar correlates with the products HAr^+ + H, and the cross-section for this reaction shows strong vibrational enhancement, particularly for collision energies exceeding 6 eV.[418] This seemingly contradicts an earlier experiment where little dependence on vibrational energy was noted,[419] but this is explicable since the v = 0 - 2 levels that show the most marked variations were never fully resolved in that work. The observed vibrational enhancement is very similar to that seen for charge transfer in this collision,[418] suggesting some common dynamical features.

Tanaka *et al.*[420] have used the t.p.e.p.i.c.o. technique to study the other entrance channel, Ar^+ + H_2, with selected spin-orbit states ($^2P_{3/2}$, $^2P_{1/2}$) of Ar^+. These reactants do not correlate with the product HAr^+ and so surface crossings are necessarily involved. Experimentally, it was found that the $^2P_{1/2}$ state has a cross-section 1.5 times greater than $^2P_{3/2}$ (1.3 times in collision with D_2). The collision-energy dependence of the cross-section was further found to be proportional to $E^{-1/2.3}$ for H_2 and to $E^{-1/2.6}$ for D_2, compared with the Langevin-model $E^{-1/2}$ dependence. These observations were explained by considering the electronic-state couplings for the spin-orbit perturbed system.

References

1 W. Thiel, *Chem. Phys.*, 1981, 57, 227.
2 R.R. Lucchese, G. Raseev, and V. McKoy, *Phys. Rev.*, 1982, 25A, 2572.
3 K. Faegri and H.P. Kelly, *Chem. Phys. Lett.*, 1982, 85, 472.
4 K. Faegri and H.P. Kelly, *Phys. Rev.*, 1981, 23A, 52.
5 R.R. Lucchese and V. McKoy, *J. Phys. B*, 1981, 14, L629.
6 G. Raseev, H. Le Rouzo, and H. Lefebvre-Brion, *J. Chem. Phys.*, 1980, 72, 5701.
7 G. Raseev, H. Lefebvre-Brion, H. Le Rouzo, and A.L. Roche, *J. Chem. Phys.*, 1981, 74, 6686.
8 R.R. Lucchese and V. McKoy, *J. Phys. Chem.*, 1981, 85, 2166.
9 F. Hirota, *Chem. Phys. Lett.*, 1980, 74, 67.
10 B. Ritchie, M.S. Pindzola, and W.R. Garrett, *Phys. Rev.*, 1981, 23A, 2905.
11 J.R. Swanson, D. Dill, and J.L. Dehmer, *J. Chem. Phys.*, 1981, 75, 619.
12 M. Roche, D.R. Salahub, and R.P. Messmer, *J. Electron Spectrosc. Relat. Phenom.*, 1980, 19, 273.
13 T.A. Carlson, M.O. Krause, F.A. Grimm, J.D. Allen, D. Mehaffy, P.R. Keller, and J.W. Taylor, *J. Chem. Phys.*, 1981, 75, 3288.
14 J. Kreile, A. Schweig, and W. Thiel, *Chem. Phys. Lett.*, 1982, 87, 473.
15 P.W. Langhoff, T.N. Rescigno, N. Padial, G. Csanak, and B.V. McKoy, *J. Chim. Phys. Phys.-Chim. Biol.*, 1980, 77, 589.
16 J.J. Delaney, V.R. Saunders, and I.H. Hillier, *J. Phys. B*, 1981, 14, 819.
17 G.H.F. Diercksen, W.P. Kraemer, T.N. Rescigno, C.F. Bender, B.V. McKoy, S.R. Langhoff, and P.W. Langhoff, *J. Chem. Phys.*, 1982, 76, 1043.
18 M.R. Hermann and P.W. Langhoff, *Chem. Phys. Lett.*, 1981, 82, 242.
19 D. Loomba, S. Wallace, D. Dill, and J.L. Dehmer, *J. Chem. Phys.*, 1981, 75, 4546.
20 E.D. Poliakoff, J.L. Dehmer, D. Dill, A.C. Parr, K.H. Jackson, and R.N. Zare, *Phys. Rev. Lett.*, 1981, 46, 907.
21 J. Kreile and A. Schweig, *J. Electron Spectrosc. Relat. Phenom.*, 1980, 20, 191.
22 C.M. Truesdale, S. Southworth, P.H. Kobrin, D.W. Lindle, G. Thornton, and D.A. Shirley, *J. Chem. Phys.*, 1982, 76, 860.
23 S. Katsumata, K. Mitani, and H. Shiromaru, *Chem. Phys. Lett.*, 1980, 75, 196.
24 T.A. Carlson, M.O. Krause, F.A. Grimm, J.D. Allen, jun., D. Mehaffy, P.R. Keller, and J.W. Taylor, *Phys. Rev.*, 1981, 23A, 3316.
25 F.A. Grimm, J.D. Allen, T.A. Carlson, M.O. Krause, D. Mehaffy, P.R. Keller, and J.W. Taylor, *J. Chem. Phys.*, 1981, 75, 92.
26 W. Thiel, *Chem. Phys. Lett.*, 1982, 87, 249.
27 N. Padial, G. Csanak, B.V. McKoy, and P.W. Langhoff, *Phys. Rev.*, 1981, 23A, 218.
28 J.R. Swanson, D. Dill, and J.L. Dehmer, *J. Phys. B*, 1981, 14, L207.
29 T.A. Carlson, M.A. Krause, D. Mehaffy, J.W. Taylor, F.A. Grimm, and J.D. Allen, *J. Chem. Phys.*, 1980, 73, 6057.
30 F.A. Grimm, *J. Electron Spectrosc. Relat. Phenom.*, 1980, 20, 245.
31 M.G. Payne, C.H. Chen, G.S. Hurst, and G.W. Foltz, *Adv. At. Mol. Phys.*, 1981, 17, 229.
32 P.W. Langhoff, S.R. Langhoff, T.N. Rescigno, J. Schirmer, L.S. Cederbaum, and W. Domcke, *Chem. Phys.*, 1981, 58, 71.

33 B.E. Cole, D.L. Ederer, R. Stockbauer, K. Codling, A.C. Parr, J.B. West, E.D. Poliakoff, and J.L. Dehmer, *J. Chem. Phys.*, 1980, 72, 6308.
34 T. Gustafsson, *Chem. Phys. Lett.*, 1980, 75, 505.
35 A. Tabche-Fouhaile, I. Nenner, P.-M. Guyon, and J. Delwiche, *J. Chem. Phys.*, 1981, 75, 1129.
36 A. Morin, I. Nenner, P.M. Guyon, O. Dutruit, and K. Ito, *J. Chim. Phys. Phys.-Chim. Biol.*, 1980, 77, 605.
37 J.J. Delaney, I.H. Hillier, and V.R. Saunders, *J. Phys. B*, 1982, 15, L37.
38 A.-L. Roche, K. Kirby, S.L. Guberman, and A. Dalgarno, *J. Electron Spectrosc. Relat. Phenom.*, 1981, 22, 223.
39 S. Wallace, D. Dill, and J.L. Dehmer, *J. Chem. Phys.*, 1982, 76, 1217.
40 J.J. Delaney, I.H. Hillier, and V.R. Saunders, *J. Phys. B*, 1982, 15, 1477.
41 S. Southworth, C.M. Truesdale, P.H. Kobrin, D.W. Lindle, W.D. Brewer, and D.A. Shirley, *J. Chem. Phys.*, 1982, 76, 143.
42 C.E. Brion and K.H. Tan, *J. Electron Spectrosc. Relat. Phenom.*, 1981, 23, 1.
43 T. Gustafsson and H.J. Levinson, *Chem. Phys. Lett.*, 1981, 78, 28.
44 S. Southworth, W.D. Brewer, C.M. Truesdale, P.H. Kobrin, D.W. Lindle, and D.A. Shirley, *J. Electron Spectrosc. Relat. Phenom.*, 1982, 26, 43.
45 J.E. Pollard, D.J. Trevor, J.E. Reutt, Y.T. Lee, and D.A. Shirley, *Chem. Phys. Lett.*, 1982, 88, 434.
46 Y. Itikawa, *Chem. Phys.*, 1979, 37, 401.
47 M. Raoult, Ch. Jungen, and D. Dill, *J. Chim. Phys. Phys.-Chim. Biol.*, 1980, 77, 599.
48 Ch. Jungen and D. Dill, *J. Chem. Phys.*, 1980, 73, 3338.
49 M. Raoult and Ch. Jungen, *J. Chem. Phys.*, 1981, 74, 3388.
50 A. Giuoti-Suzor and H. Lefebvre-Brion, *Chem. Phys. Lett.*, 1980, 76, 132.
51 H. Takagi and H. Nakamura, *J. Chem. Phys.*, 1981, 74, 5808.
52 J.H.D. Eland, *J. Chim. Phys. Phys.-Chim. Biol.*, 1980, 77, 613.
53 J. Erickson and C.Y. Ng, *J. Chem. Phys.*, 1981, 75, 1650.
54 S.P. Goss, J.D. Morrison, and D.L. Smith, *J. Chem. Phys.*, 1981, 75, 757.
55 Y. Ono, S.H. Linn, H.F. Prest, C.Y. Ng, and E. Miescher, *J. Chem. Phys.*, 1980, 73, 4855.
56 D.M. Rider, G.W. Ray, E.J. Darland, and G.E. Leroi, *J. Chem. Phys.*, 1981, 74, 1652.
57 V. Kumar and E. Krishnakumer, *J. Electron Spectrosc. Relat. Phenom.*, 1981, 22, 109.
58 A. Chutjian and J.M. Ajello, *Chem. Phys. Lett.*, 1980, 72, 504.
59 J.B. West, K. Codling, A.C. Parr, D.L. Ederer, B.E. Cole, R. Stockbauer, and J.L. Dehmer, *J. Phys. B*, 1981, 14, 1791.
60 K. Codling. A.C. Parr, D.L. Ederer, R. Stockbauer, J.B. West, B.E. Cole, and J.L. Dehmer, *J. Phys. B*, 1981, 14, 657.
61 J.H.D. Eland, *J. Chem. Phys.*, 1980, 72, 6015.
62 J.H.D. Eland, *Mol. Phys.*, 1980, 40, 917.
63 P.M. Dehmer, J.L. Dehmer, and W.A. Chupka, *J. Chem. Phys.*, 1980, 73, 126.
64 J. Delwiche, M.-J. Hubin-Franskin, P.-M. Guyon, and I. Nenner, *J. Chem. Phys.*, 1981, 74, 4219.
65 T. Baer, P.-M. Guyon, I. Nenner, A. Tabche-Fouhaille, R. Botter, L.F.A. Ferreira, and T.R. Govers, *J. Chem. Phys.*, 1979, 70, 1585.
66 R. Stockbauer, *J. Chem. Phys.*, 1980, 72, 5277.

67 Y. Ono, E.A. Osuch, and C.Y. Ng, *J. Chem. Phys.*, 1982, 76, 3905.
68 T. Hayaishi, S. Iwata, M. Sasanuma, E. Ishiguro, Y. Morioka, Y. Iida, and M. Nakamura, *J. Phys. B*, 1982, 15, 79.
69 P.W. Langhoff, B.V. McKoy, R. Unwin, and A.M. Bradshaw, *Chem. Phys. Lett.*, 1981, 83, 270.
70 R. Unwin, I. Khan, N.V. Richardson, A.M. Bradshaw, L.S. Cederbaum, and W. Domcke, *Chem. Phys. Lett.*, 1981, 77, 242.
71 J. Kreile, A. Schweig, and W. Thiel, *Chem. Phys. Lett.*, 1981, 79, 547.
72 A.C. Parr, D.L. Ederer, J.B. West, D.M.P. Holland, and J.L. Dehmer, *J. Chem. Phys.*, 1982, 76, 4349.
73 L.E. Machado, E.P. Leal, G. Csanak, B.V. McKoy, and P.W. Langhoff, *J. Electron Spectrosc. Relat. Phenom.*, 1982, 25, 1.
74 C.E. Brion and A. Hamnett, *Adv. Chem. Phys.*, 1980, 45, 1.
75 F. Carnovale, A.P. Hitchcock, J.P.D. Cook, and C.E. Brion, *Chem. Phys.*, 1982, 66, 249.
76 M.G. White, K.T. Leung, and C.E. Brion, *J. Electron Spectrosc. Relat. Phenom.*, 1981, 23, 127.
77 F. Carnovale, M.G. White, and C.E. Brion, *J. Electron Spectrosc. Relat. Phenom.*, 1981, 24, 63.
78 A.J. Stace and A.K. Shukla, *J. Phys. Chem.*, 1982, 86, 865.
79 Y. Ono, E.A. Osuch, and C.Y. Ng, *J. Chem. Phys.*, 1981, 74, 1645.
80 S.H. Linn and C.Y. Ng, *J. Chem. Phys.*, 1981, 75, 4921.
81 P.M. Dehmer and S.T. Pratt, *J. Chem. Phys.*, 1982, 76, 843.
82 Y. Ono, S.H. Linn, H.F. Prest, M.E. Gress, and C.Y. Ng, *J. Chem. Phys.*, 1980, 73, 2523.
83 Y. Ono, S.H. Linn, H.F. Prest, M.E. Gress, and C.Y. Ng, *J. Chem. Phys.*, 1981, 74, 1125.
84 M.E. Gress, S.H. Linn, Y. Ono, H.F. Prest, and C.Y. Ng, *J. Chem. Phys.*, 1980, 72, 4242.
85 S.H. Linn, Y. Ono, and C.Y. Ng, *J. Chem. Phys.*, 1981, 74, 3342.
86 S.H. Linn, Y. Ono, and C.Y. Ng, *J. Chem. Phys.*, 1981, 74, 3348.
87 S.L. Anderson, T. Hirooka, P.W. Tiedemann, B.H. Mahan, and Y.T. Lee, *J. Chem. Phys.*, 1980, 73, 4779.
88 P.M. Dehmer and E.D. Poliakoff, *Chem. Phys. Lett.*, 1981, 77, 326.
89 P.M. Dehmer, *J. Chem. Phys.*, 1982, 76, 1263.
90 S.T. Pratt and P.M. Dehmer, *Chem. Phys. Lett.*, 1982, 87, 533.
91 P.M. Dehmer and S.T. Pratt, *J. Chem. Phys.*, 1981, 75, 5265.
92 D. Mathur and D.C. Frost, *J. Chem. Phys.*, 1981, 75, 5381.
93 A. Hashizume and N. Wasada, *Int. J. Mass Spectrom. Ion Phys.*, 1980, 36, 291.
94 M.-J. Hubin-Franskin, P. Marmet, and P. Huard, *Int. J. Mass Spectrom. Ion Phys.*, 1980, 33, 311.
95 D. Mathur, *J. Phys. B*, 1980, 13, 4703.
96 D. Mathur, *Chem. Phys. Lett.*, 1981, 81, 115.
97 R. Locht and J. Momigny, *Chem. Phys.*, 1980, 49, 173.
98 B.M. De Koven, D.H. Levy, H.H. Harris, B.R. Zegarski, and T.A. Miller, *J. Chem. Phys.*, 1981, 74, 5659.
99 J.P. Johnson and J.L. Franklin, *Int. J. Mass Spectrom. Ion Phys.*, 1980, 33, 393.
100 D. Mathur, *Int. J. Mass Spectrom. Ion Phys.*, 1981, 40, 235.
101 O.I. Smith and J.S. Stevenson, *J. Chem. Phys.*, 1981, 74, 6777.
102 M. Miletic, D. Eres, M. Veljkovic, and K.F. Zmbov, *Int. J. Mass Spectrom. Ion Phys.*, 1980, 35, 231.
103 G. Baravian, J. Godart, and G. Sultan, *Phys. Rev.*, 1982, 25A, 1483.

104 R.V. Hodges, L.C. Lee, and J.T. Moseley, *Int. J. Mass Spectrom. Ion Phys.*, 1981, 39, 133.
105 J.J. DeCorpo, J.W. Hudgens, M.C. Lin, F.E. Saalfeld, M.E. Seaver, and J.R. Wyatt, *Adv. Mass Spectrom.*, 1980, 8A, 133.
106 M. Seaver, J.W. Hudgens, and J.J. DeCorpo, *Int. J. Mass Spectrom. Ion Phys.*, 1980, 34, 159.
107 L.J. Rothberg, D.P. Gerrity, and V. Vaida, *J. Chem. Phys.*, 1981, 75, 4403.
108 P.M. Johnson, *Acc. Chem. Res.*, 1980, 13, 20.
109 P.M. Johnson and C.E. Otis, *Annu. Rev. Phys. Chem.*, 1981, 32, 139.
110 V.S. Antonov and V.S. Letokhov, *Appl. Phys.*, 1981, 24, 89.
111 M.J. van der Wiel, *J. Chim. Phys. Phys.-Chim. Biol.*, 1980, 77, 647.
112 T. Carney and T. Baer, *J. Chem. Phys.*, 1981, 75, 477.
113 D.A. Lichtin, S. Datta-Ghosh, K.R. Newton, and R.B. Bernstein, *Chem. Phys. Lett.*, 1980, 75, 214.
114 J.W. Hudgens, M. Seaver, and J.H. DeCorpo, *J. Phys. Chem.*, 1981, 85, 761.
115 H. Zacharias, R. Schmiedl, and K.H. Welge, *Appl. Phys.*, 1980, 21, 127.
116 H. Zacharias, H. Rottke, and K.H. Welge, *Appl. Phys.*, 1981, 24, 23.
117 S. Leutwyler, A. Herrmann, L. Woste, and E. Schumacher, *Chem. Phys.*, 1980, 48, 253.
118 S. Leutwyler and U. Even, *Chem. Phys. Lett.*, 1981, 81, 579.
119 M.G. Payne, C.H. Chen, G.S. Hurst, S.D. Kramer, W.R. Garrett, and M. Pindzola, *Chem. Phys. Lett.*, 1981, 79, 142.
120 M.G. Payne, C.H. Chen, G.S. Hurst, and G.W. Foltz, *Adv. At. Mol. Phys.*, 1981, 17, 229.
121 R.B. Bernstein, *J. Phys. Chem.*, 1982, 86, 1178.
122 U. Boesl, H.J. Neusser, and E.W. Schlag, *Chem. Phys.*, 1981, 55, 193.
123 U. Boesl, H.J. Neusser, and E.W. Schlag, *J. Am. Chem. Soc.*, 1981, 103, 5058.
124 C.T. Rettner and J.H. Brophy, *Chem. Phys.*, 1981, 56, 53.
125 M. Stuke, D. Sumida, and C. Wittig, *J. Phys. Chem.*, 1982, 86, 438.
126 H. Shinohara and N. Nishi, *Chem. Phys. Lett.*, 1982, 87, 561.
127 J.B. Hopkins, D.E. Powers, and R.E. Smalley, *J. Phys. Chem.*, 1981, 85, 3739.
128 D. Eisel and W. Demtroder, *Chem. Phys. Lett.*, 1982, 88, 481.
129 P. Hering, A.G.M. Maaswintel, and K.L. Kompa, *Chem. Phys. Lett.*, 1981, 83, 222.
130 M. Stuke and C. Wittig, *Chem. Phys. Lett.*, 1981, 81, 168.
131 S. Leutwyler, M. Hofmann, H.-P. Harri, and E. Schumacher, *Chem. Phys. Lett.*, 1981, 77, 257.
132 C.D. Cooper, A.D. Williamson, J.C. Miller, and R.N. Compton, *J. Chem. Phys.*, 1980, 73, 1527.
133 K.H. Fung, W.E. Henke, T.R. Hays, H.L. Selzle, and E.W. Schlag, *J. Phys. Chem.*, 1981, 85, 3560.
134 M.A. Duncan, T.G. Dietz, and R.E. Smalley, *J. Chem. Phys.*, 1981, 75, 2118.
135 M. Mashni and P. Hess, *Chem. Phys. Lett.*, 1981, 77, 541.
136 S. Leutwyler, U. Even, and J. Jortner, *J. Phys. Chem.*, 1981, 85, 3026.
137 E.R. Sirkin and Y. Haas, *Appl. Phys.*, 1981, 25, 253.
138 R.E. Demaray, C. Otis, K. Aron, and P. Johnson, *J. Chem. Phys.*, 1980, 72, 5772.
139 J.C. Miller and R.N. Compton, *Phys. Rev.*, 1982, 25A, 2056.
140 K. Bergmann and E. Gottwald, *Chem. Phys. Lett.*, 1981, 78, 515.

141 B.H. Rockney and E.R. Grant, *Chem. Phys. Lett.*, 1981, 79, 15.
142 W.G. Mallard, J.H. Miller, and K.C. Smyth, *J. Chem. Phys.*, 1982, 76, 3483.
143 T.G. DiGiuseppe, J.W. Hudgens, and M.C. Lin, *Chem. Phys. Lett.*, 1982, 82, 267.
144 T.G. DiGiuseppe, J.W. Hudgens, and M.C. Lin, *J. Phys. Chem.*, 1981, 86, 36.
145 G. Radhakrishnan, D. Ng, and R.C. Estler, *Chem. Phys. Lett.*, 1981, 84, 260.
146 J.H. Glownia, S.J. Riley, S.D. Colson, and G.C. Nieman, *J. Chem. Phys.*, 1980, 72, 5999.
147 T.G. Dietz, M.A. Duncan, M.G. Liverman, and R.E. Smalley, *J. Chem. Phys.*, 1980, 73, 4816.
148 M.A. Duncan, T.G. Dietz, M.G. Liverman, and R.E. Smalley, *J. Phys. Chem.*, 1981, 85, 7.
149 T.G. Dietz, M.A. Duncan, and R.E. Smalley, *J. Chem. Phys.*, 1982, 76, 1227.
150 U. Boesl, H.J. Neusser, and E.W. Schlag, *J. Chem. Phys.*, 1980, 72, 4327.
151 R.S. Pandolfi, D.A. Gobeli, and M.A. El-Sayed, *J. Phys. Chem.*, 1981, 85, 1779.
152 J.T. Meek, R.K. Jones, and J.P. Reilly, *J. Chem. Phys.*, 1980, 73, 3503.
153 J.P. Reilly and K.L. Kompa, *J. Chem. Phys.*, 1980, 73, 5468.
154 J.C. Miller and R.N. Compton, *J. Chem. Phys.*, 1981, 75, 2020.
155 F. Rebentrost, K.L. Kompa, and A. Ben-Shaul, *Chem. Phys. Lett.*, 1981, 77, 384.
156 F. Rebentrost and A. Ben-Shaul, *J. Chem. Phys.*, 1981, 74, 3255.
157 W. Dietz, H.J. Neusser, U. Boesl, E.W. Schlag, and S.H. Lin, *Chem. Phys.* 1982, 66, 105.
158 U. Boesl, H.J. Neusser, and E.W. Schlag, *Chem. Phys. Lett.*, 1982, 87, 1.
159 S. Leutwyler, U. Even, and J. Jortner, *Chem. Phys. Lett.*, 1980, 74, 11.
160 S. Leutwyler, U. Even, and J. Jortner, *Chem. Phys.*, 1981, 58, 409.
161 P.C. Engelking, *Chem. Phys. Lett.*, 1980, 74, 207.
162 S. Leutwyler and U. Even, *Chem. Phys. Lett.*, 1981, 84, 188.
163 D.P. Gerrity, L.J. Rothberg, and V. Vaida, *Chem. Phys. Lett.*, 1980, 74, 1.
164 G.J. Fisanick, A. Gedanken, T.S. Eichelberger, N.A. Kuebler, and M.B. Robin, *J. Chem. Phys.*, 1981, 75, 5215.
165 D.A. Lichtin, R.B. Bernstein, and V. Vaida, *J. Am. Chem. Soc.*, 1982, 104, 1830.
166 A. Gedanken, M.B. Robin, and N.A. Kuebler, *Inorg. Chem.*, 1981, 20, 3340.
167 J.H. Glownia, S.J. Riley, S.D. Colson, and G.C. Nieman, *J. Chem. Phys.*, 1980, 73, 4296.
168 V.S. Antonov, V.S. Letokhov, and A.N. Shibanov, *Appl. Phys.*, 1980, 22, 293.
169 V.S. Antonov, V.S. Letokhov, and A.N. Shibanov, *Sov. Phys. - JETP (Engl. Transl.)*, 1980, 51, 1113.
170 R.J.S. Morrison, B.H. Rockney, and E.R. Grant, *J. Chem. Phys.*, 1981, 75, 2643.
171 J.W. Hepburn, D.J. Trevor, J.E. Pollard, D.A. Shirley, and Y.T. Lee, *J. Chem. Phys.*, 1982, 76, 4287.
172 J. Silberstein and R.D. Levine, *Chem. Phys. Lett.*, 1980, 74, 6.
173 N. Ohmichi, J. Silberstein, and R.D. Levine, *J. Phys. Chem.*, 1981, 85, 3369.

174 D.A. Lichton, R.B. Bernstein, and K.R. Newton, *J. Chem. Phys.* 1981, 75, 5728.
175 D.M. Lubman, *J. Phys. Chem.*, 1981, 85, 3752.
176 J. Silberstein and R.D. Levine, *J. Chem. Phys.*, 1981, 75, 5735.
177 B. Ritchie and E.J. McGuire, *Phys. Rev.*, 1981, 24A, 2532.
178 P. Cremaschi, *Chem. Phys. Lett.*, 1981, 83, 106.
179 P. Cremaschi, *Chem. Phys.*, 1981, 63, 85.
180 J.C. Miller and R.N. Compton, *J. Chem. Phys.*, 1981, 75, 22.
181 J. Kimman, P. Kruit, and M.J. van der Wiel, *Chem. Phys. Lett.*, 1982, 88, 576.
182 M.S. de Vries, N.J.A. van Veen, T. Baller, and A.E. de Vries, *Chem. Phys.*, 1981, 56, 157.
183 G.J. Fisanick, T.S. Eichelberger, B.A. Heath, and M.B. Robin, *J. Chem. Phys.*, 1980, 72, 5571.
184 G.J. Fisanick and T.S. Eichelberger, *J. Chem. Phys.*, 1981, 74, 6692.
135 J.R. Ackerhalt and J.H. Eberly, *Phys. Rev.*, 1976, 14A, 1705.
186 D.H. Parker and R.B. Bernstein, *J. Phys. Chem.*, 1982, 86, 60.
187 J. Danon, H. Zacharias, H. Rottke, and K.H. Welge, *J. Chem. Phys.*, 1982, 76, 2399.
188 D.H. Parker, R.B. Bernstein, and D.A. Lichtin, *J. Chem. Phys.*, 1981, 75, 2577.
189 K.R. Newton, D.A. Lichtin, and R.B. Bernstein, *J. Phys. Chem.*, 1981, 85, 15.
190 Y. Takenoshita, H. Shinohara, M. Umemoto, and N. Nishi, *Chem. Phys. Lett.*, 1982, 87, 566.
191 J.H. Glownia, R.J. Romero, and R.K. Sander, *Chem. Phys. Lett.*, 1982, 88, 292.
192 D. Proch, D.M. Rider, and R.N. Zare, *Chem. Phys. Lett.*, 1981, 81, 430.
193 T. Baer and T.E. Carney, *J. Chem. Phys.*, 1982, 76, 1304.
194 T.E. Carney and T. Baer, *J. Chem. Phys.*, 1981, 75, 4422.
195 J.M. Tedder and P.H. Vidaud, *J. Chem. Soc., Faraday Trans. 2*, 1980, 76, 1516.
196 J.M. Tedder and P.H. Vidaud, *Chem. Phys. Lett.*, 1980, 76, 380.
197 J. Jalonen, J.M. Tedder, and P.H. Vidaud, *J. Chem. Soc., Faraday Trans. 2*, 1980, 76, 1450.
198 C.R. Szmanda, K.B. McAfee, and R.S. Hozack, *J. Phys. Chem.*, 1982, 86, 1217.
199 M. Matic, V. Sidis, M. Vujovic, and B. Cobic, *J. Phys. B*, 1980, 13, 3665.
200 J.B. Wilcox, K.L. Harbol, and T.F. Moran, *J. Phys. Chem.*, 1981, 85, 3415.
201 A.B. Rakshit, *Chem. Phys. Lett.*, 1980, 75, 283.
202 B.H. Mahan and A. O'Keefe, *J. Chem. Phys.*, 1981, 74, 5606.
203 B.H. Mahan, C. Martner, and A. O'Keefe, *J. Chem. Phys.*, 1982, 76, 4433.
204 R.G. Orth, J.H. Futrell, and Y. Nishimura, *J. Chem. Phys.*, 1981, 75, 3345.
205 T. Baer and P.T. Murray, *J. Chem. Phys.*, 1981, 75, 4477.
206 F.M. Campbell, R. Browning, and C.J. Latimer, *J. Phys. B*, 1980, 13, 4257.
207 K. Tanaka, T. Kato, and I. Koyano, *J. Chem. Phys.*, 1981, 75, 4941.
208 F.A. Houle, S.L. Anderson, D. Gerlich, T. Turner, and Y.T. Lee, *Chem. Phys. Lett.*, 1981, 82, 392.
209 M. Tsuji, C. Yamagiwa, M. Endoh, and Y. Nishimura, *Chem. Phys. Lett.*, 1980, 73, 407.
210 D. Neuschafer, Ch. Ottinger, and S. Zimmermann, *Chem. Phys.*, 1981, 55, 313.

211 W. Lindinger, H. Villinger, and F. Howorka, *Int. J. Mass Spectrom. Ion Phys.*, 1981, 41, 89.
212 J.B. Wilcox and T.F. Moran, *J. Phys. Chem.*, 1981, 85, 989.
213 F.M. Campbell, R. Browning, and C.J. Latimer, *J. Phys. B*, 1981, 14, 1183.
214 J. Glosik, B. Friedrich, and Z. Herman, *Chem. Phys.*, 1981, 60, 369.
215 R.M. Bilotta, F.N. Preuninger, and J.M. Farrar, *Chem. Phys. Lett.*, 1980, 74, 95.
216 J.F. Seely, *J. Chem. Phys.*, 1981, 75, 3321.
217 J.H. Agee, J.B. Wilcox, L.E. Abbey, and T.F. Moran, *Chem. Phys.*, 1981, 61, 171.
218 H.M. Holzscheiter and D.A. Church, *J. Chem. Phys.*, 1981, 74, 2313.
219 A. Niehaus, *Adv. Chem. Phys.*, 1981, 45, 399.
220 W. Allison and E.E. Muschlitz, jun., *J. Electron Spectrosc. Relat. Phenom.*, 1981, 23, 339.
221 M. Hotop, J. Lorenzen, and A. Zastrow, *J. Electron Spectrosc. Relat. Phenom.*, 1981, 23, 347.
222 A. Munzer and A. Niehaus, *J. Electron Spectrosc. Relat. Phenom.*, 1981, 23, 367.
223 W. Goy, V. Kohls, and H. Morgner, *J. Electron Spectrosc. Relat. Phenom.*, 1981, 23, 383.
224 W. Goy, H. Morgner, and A.J. Yencha, *J. Electron Spectrosc. Relat. Phenom.*, 1981, 24, 77.
225 A. Khan, H.R. Siddiqui, D.W. Martin, and P.E. Siska, *Chem. Phys. Lett.*, 1981, 84, 280.
226 K. Kimura, S. Katsumata, Y. Achiba, T. Yamazaki, and S. Iwata, 'Handbook of HeI photoelectron spectra of fundamental organic molecules', Japan Scientific Societies Press, Tokyo, 1981.
227 H.J. Chiang and S.D. Worley, *J. Electron Spectrosc. Relat. Phenom.*, 1980, 21, 121.
228 M.F. Herman, K.F. Freed, and D.L. Yeager, *Adv. Chem. Phys.*, 1981, 48, 1.
229 I. Cacelli, R. Moccia, and V. Carravetta, *Theor. Chim. Acta*, 1981, 59, 461.
230 J. Baker and B.T. Pickup, *Chem. Phys. Lett.*, 1980, 76, 537.
231 R. Cambi, G. Ciullo, A. Scamellotti, F. Tarantelli, R. Fantoni, A. Giardini-Guidoni, and A. Sergio, *Chem. Phys. Lett.*, 1981, 80, 295.
232 W. von Niessen, W.P. Kraemer, and J. Schirmer, *J. Chem. Soc., Faraday Trans. 2*, 1981, 77, 1461.
233 W. von Niessen and G.H.F. Diercksen, *J. Electron Spectrosc. Relat. Phenom.*, 1980, 20, 95.
234 W. von Niessen, L.S. Cederbaum, W. Domcke, and G.H.F. Diercksen, *Chem. Phys.*. 1981, 56, 43.
235 W. von Niessen, G. Bieri, J. Schirmer, and L.S. Cederbaum, *Chem. Phys.*, 1982, 65, 157.
236 D.P. Chong and S.R. Langhoff, *Chem. Phys.*, 1982, 67, 153.
237 C.B. Backsay, *Chem. Phys.*, 1980, 48, 21.
238 S.R. Langhoff and D.P. Chong, *Chem. Phys. Lett.*, 1982, 86, 487.
239 S.R. Langhoff and D.P. Chong, *Chem. Phys.*, 1981, 55, 355.
240 J. Schirmer, L.S. Cederbaum, and W. von Niessen, *Chem. Phys.*, 1981, 56, 285.
241 W. Muller, C. Nager, and P. Rosmus, *Chem. Phys.*, 1980, 51, 43.
242 H. Sambe and R.H. Felton, *Chem. Phys.*, 1981, 59, 329.
243 G.L. Gutsev and A.E. Smoljar, *Chem. Phys.*, 1981, 56, 189.
244 G.L. Gutsev and S.I. Boldyrev, *Chem. Phys.*, 1981, 56, 277.
245 L. Noodleman, D. Post, and E.J. Baerends, *Chem. Phys.*, 1982, 64, 159.
246 G. de Alti, P. Decleva, and A. Lisini, *Chem. Phys.*, 1982, 66, 425.

247 B.P. Tsai and J.H.D. Eland, *Int. J. Mass Spectrom. Ion Phys.*, 1980, 36, 143.
248 B. Brehm and G. de Frenes, *Adv. Mass Spectrom.*, 1980, 8A, 138.
249 T. Ast, *Adv. Mass Spectrom.*, 1980, 8A, 555.
250 A.K. Edwards and R.M. Wood, *J. Chem. Phys.*, 1982, 76, 2938.
251 I.W. Griffiths, E.S. Mukhtar, F.M. Harris, and J.H. Beynon, *Int. J. Mass Spectrom. Ion Phys.*, 1981, 39, 257.
252 T. Ast, C.J. Porter, C.J. Proctor, and J.H. Beynon, *Chem. Phys. Lett.*, 1981, 78, 439.
253 A.W. Hanner and T.F. Moran, *Org. Mass Spectrom.*, 1981, 16, 512.
254 J.A. Pople, B. Tidor, and P. von R. Schleyer, *Chem. Phys. Lett.*, 1982, 88, 533.
255 P.E.M. Siegbahn, *Chem. Phys.*, 1982, 66, 443.
256 M. Heni, E. Illenberger, H. Baumgartel, and S. Suzer, *Chem. Phys. Lett.*, 1982, 87, 244.
257 E. Illenberger, *Chem. Phys. Lett.*, 1981, 80, 153.
258 J.-S. Wang and J.L. Franklin, *Int. J. Mass Spectrom. Ion Phys.*, 1980, 36, 233.
259 F.K. Meyer, J.M. Jasinski, R.N. Rosenfeld, and J.I. Brauman, *J. Am. Chem. Soc.*, 1982, 104, 663.
260 J.E. Szulejko, J.H. Bowie, I. Howe, and J.H. Beynon, *Int. J. Mass Spectrom. Ion Phys.*, 1980, 34, 99.
261 J.E. Szulejko, I. Howe, and J.H. Beynon, *Int. J. Mass Spectrom. Ion Phys.*, 1981, 37, 27.
262 J.C. Lorquet, *Org. Mass Spectrom.*, 1981, 16, 469.
263 P. Pechukas, *Annu. Rev. Phys. Chem.*, 1981, 32, 159.
264 C.J. Cerjan and W.H. Miller, *J. Chem. Phys.*, 1981, 75, 2800.
265 D.G. Truhlar and B.C. Garrett, *Acc. Chem. Res.*, 1980, 13, 440.
266 B.C. Garrett, D.G. Truhlar, R.S. Grev, and R.B. Walker, *J. Chem. Phys.*, 1980, 73, 235.
267 W.J. Chesnavich, L. Bass, T. Su, and M.T. Bowers, *J. Chem. Phys.*, 1981, 74, 2228.
268 M. Sizum and S. Goursaud, *Chem. Phys. Lett.*, 1981, 79, 269.
269 R.T. Skodje, D.G. Truhlar, and B.C. Garrett, *J. Phys. Chem.*, 1981, 85, 3019.
270 W.L. Hase in 'Potential Energy Surfaces and Dynamics Calculations', ed. D.G. Truhlar, Plenum, New York, 1981, p.1.
271 D.W. Noid, M.L. Koszykowski, and R.A. Marcus, *Annu. Rev. Phys. Chem.*, 1981, 32, 267.
272 H.R. Mayne and R.J. Wolf, *Chem. Phys. Lett.*, 1981, 81, 508.
273 S.C. Farantos and J.N. Murrell, *Chem. Phys.*, 1981, 55, 205.
274 W.L. Hase and R.J. Wolf in 'Potential Energy Surfaces and Dynamics Calculations', ed. D.G. Truhlar, Plenum, New York, 1981, p. 37.
275 R.J. Wolf and W.L. Hase, *J. Chem. Phys.*, 1980, 73, 3774.
276 W.L. Hase and R.J. Wolf, *J. Chem. Phys.*, 1981, 75, 3809.
277 R.J. Wolf and W.L. Hase, *J. Chem. Phys.*, 1980, 73, 3010.
278 M.J. Davis and E.J. Heller, *J. Chem. Phys.*, 1981, 75, 246.
279 E.J. Heller and M.J. Davis, *J. Phys. Chem.*, 1981, 85, 307.
280 B.A. Waite and W.H. Miller, *J. Chem. Phys.*, 1982, 76, 2412.
281 B.A. Waite and W.H. Miller, *J. Chem. Phys.*, 1980, 73, 3713.
282 S.A. Rice, *Adv. Chem. Phys.*, 1981, 47, part 1, 117.
283 B.A. Waite and W.H. Miller, *J. Chem. Phys.*, 1981, 74, 3910.
284 N. Moiseyev and P.R. Certain, *J. Phys. Chem.*, 1982, 86, 1149.
285 K.G. Kay, *J. Chem. Phys.*, 1980, 72, 5955.
286 M.L. Koszykowski, D.W. Noid, M. Tabor, and R.A. Marcus, *J. Chem. Phys.*, 1981, 74, 2530.
287 Y. Weissman and J. Jortner, *Chem. Phys. Lett.*, 1981, 78, 224.
288 E.J. Heller, *Acc. Chem. Res.*, 1981, 14, 368.
289 E.J. Heller in 'Potential Energy Surfaces and Dynamics Calculations', ed. D.G. Truhlar, Plenum, New York, 1981, p.103.

290 A.J. Lorquet, J.C. Lorquet, J. Delwiche, and M.J. Hubin-Franskin, *J. Chem. Phys.*, 1982, 76, 4692.
291 J.D. Doll, *J. Chem. Phys.*, 1980, 73, 2760.
292 J.D. Doll, *Chem. Phys. Lett.*, 1980, 72, 139.
293 J.D. Doll, *J. Chem. Phys.*, 1981, 74, 1074.
294 S. Leach, G. Dujardin, and G. Taieb, *J. Chim. Phys. Phys.-Chim. Biol.*, 1980, 77, 705.
295 G. Dujardin, S. Leach, G. Taieb, J.P. Maier, and W.M. Gelbart, *J. Chem. Phys.*, 1980, 73, 4987.
296 A.J. Lorquet, J.C. Lorquet, and W. Forst, *Chem. Phys.*, 1980, 51, 241.
297 A.J. Lorquet and J.C. Lorquet, *Chem. Phys.*, 1980, 51, 253.
298 A.J. Lorquet, J.C. Lorquet, and W. Forst, *Chem. Phys.*, 1980, 51, 261.
299 J.C. Lorquet, D. Dehareng, C. Sannen, and G. Raseev, *J. Chim. Phys. Phys.-Chim. Biol.*, 1980, 77, 719.
300 C. Sannen, G. Raseev, C. Galloy, G. Fauville, and J.C. Lorquet, *J. Chem. Phys.*, 1981, 74, 2402.
301 J.C. Lorquet, C. Sannen, and G. Raseev, *J. Am. Chem. Soc.*, 1980, 102, 7976.
302 R.D. Levine, *Adv. Chem. Phys.*, 1981, 47, part 1, 239.
303 E. Zamir, R.D. Levine, and R.B. Bernstein, *Chem. Phys.*, 1981, 55, 57.
304 E. Zamir and R.D. Levine, *Chem. Phys.*, 1980, 52, 253.
305 J. Moseley and J. Durup, *J. Chim. Phys. Phys.-Chim. Biol.*, 1980, 77, 673.
306 J. Moseley and J. Durup, *Annu. Rev. Phys. Chem.*, 1981, 32, 53.
307 M. Carre, M. Druetta, M.L. Gaillard, H.H. Bukow, M. Horani, A.-L. Roche, and M. Velghe, *Mol. Phys.*, 1980, 40, 1453.
308 J.T. Moseley, P.C. Cosby, J.-B. Ozenne, and J. Durup, *J. Chem. Phys.*, 1979, 70, 1474.
309 H. Helm, P.C. Cosby, and D.L. Huestis, *J. Chem. Phys.*, 1980, 73, 2629.
310 P.C. Cosby and H. Helm, *J. Chem. Phys.*, 1982, 76, 4720.
311 F.J. Gneman, J.T. Moseley, R.P. Saxon, and P.C. Cosby, *Chem. Phys.*, 1980, 51, 169.
312 C.M. Marian, R. Marian, S.D. Peyerimhoff, B.A. Hess, R.J. Buenker, and G. Seger, *Mol. Phys.*, 1982, 46, 779.
313 J.C. Hansen, M.M. Graff, J.T. Moseley, and P.C. Cosby, *J. Chem. Phys.*, 1981, 74, 2195.
314 J. Durup, *Chem. Phys.*, 1981, 59, 351.
315 J.P. Flamme, T. Mark, and J. Los, *Chem. Phys. Lett.*, 1980, 75, 419.
316 H. Helm, P.C. Cosby, M.M. Graff, and J.T. Moseley, *Phys. Rev.*, 1982, 25A, 304.
317 M.M. Graff and J.T. Moseley, *Chem. Phys. Lett.*, 1981, 83, 97.
318 J.T. Moseley, J.-B. Ozenne, and P.C. Cosby, *J. Chem. Phys.*, 1981, 74, 337.
319 M. Larzilliere, M. Carre, M.L. Gaillard, J. Rostas, M. Horani, and M. Velghe, *J. Chim. Phys. Phys.-Chim. Biol.*, 1980, 77, 689.
320 P.C. Cosby and H. Helm, *J. Chem. Phys.*, 1981, 75, 3882.
321 F.N. Preuninger and J.M. Farrar, *J. Chem. Phys.*, 1981, 74, 5330.
322 M.J. Coggiola, P.C. Cosby, and J.R. Peterson, *J. Chem. Phys.*, 1980, 72, 6507.
323 L.R. Thorne and J.L. Beauchamp, *J. Chem. Phys.*, 1981, 74, 5100.
324 R. Bombach, J. Dannacher, J.-P. Stadelmann, J. Vogt, L.R. Thorne, and J.L. Beauchamp, *Chem. Phys.*, 1982, 66, 403.
325 C.A. Wight and J.L. Beauchamp, *Chem. Phys. Lett.*, 1981, 77, 30.

326 R.C. Dunbar, J.D. Hays, J.P. Honovich, and N.B. Lev, *J. Am. Chem. Soc.*, 1980, 102, 3950.
327 P.N.T. van Velzen and W.J. van der Hart, *Chem. Phys.*, 1981, 61, 325.
328 M.S. Kim and R.C. Dunbar, *J. Chem. Phys.*, 1980, 72, 4405.
329 D.M. Lubman, R. Naaman, and R.N. Zare, *J. Chem. Phys.*, 1980, 72, 3034.
330 E.S. Mukhtar, I.W. Griffiths, F.M. Harris, and J.H. Beynon, *Int. J. Mass Spectrom. Ion Phys.*, 1981, 37, 159.
331 I.W. Griffiths, F.M. Harris, E.S. Mukhtar, and J.H. Beynon, *Int. J. Mass Spectrom. Ion Phys.*, 1981, 38, 127.
332 I.W. Griffiths, E.S. Mukhtar, F.M. Harris, and J.H. Beynon, *Int. J. Mass Spectrom. Ion Phys.*, 1981, 38, 333.
333 E.S. Mukhtar, I.W. Griffiths, R.E. March, F.M. Harris, and J.H. Beynon, *Int. J. Mass Spectrom. Ion Phys.*, 1981, 41, 61.
334 E.S. Mukhtar, I.W. Griffiths, F.M. Harris, and J.H. Beynon, *Org. Mass Spectrom.*, 1981, 16, 51.
335 E.S. Mukhtar, I.W. Griffiths, F.M. Harris, and J.H. Beynon, *Int. J. Mass Spectrom. Ion Phys.*, 1982, 42, 77.
336 S.P. Goss, J.D. Morrison, and D.L. Smith, *J. Chem. Phys.*, 1981, 75, 757.
337 S.P. Goss, D.C. McGilvery, J.D. Morrison, and D.L. Smith, *J. Chem. Phys.*, 1981, 75, 1820.
338 R. Weber, K. Levsen, P.C. Burgers, and J.K. Terlouw, *Org. Mass Spectrom.*, 1981, 16, 514.
339 J.P. Maier, *Acc. Chem. Res.*, 1982, 15, 18.
340 T. Sears, T.A. Miller, and V.E. Bondybey, *J. Chem. Phys.*, 1980, 72, 6749.
341 G. Bieri, E. Kloster-Jensen, S. Kvisle, J.P. Maier, and O. Marthaler, *J. Chem. Soc., Faraday Trans. 2*, 1980, 76, 676.
342 F. Arqueros and J. Campos, *J. Chem. Phys.*, 1981, 74, 6092.
343 R.C. Dunbar and D.W. Turner, *Chem. Phys.*, 1981, 57, 377.
344 S. Leach, G. Dujardin, and G. Taieb, *J. Chim. Phys. Phys.-Chim. Biol.*, 1980, 77, 705.
345 G. Dujardin, S. Leach, G. Taieb, J.P. Maier, and W.M. Gelbart, *J. Chem. Phys.*, 1980, 73, 4987.
346 G. Dujardin and S. Leach, *J. Chem. Phys.*, 1981, 75, 2521.
347 R.G.C. Blyth, I. Powis, and C.J. Danby, *Chem. Phys. Lett.*, 1981, 84, 272.
348 R. Bombach, A. Schmelzer, and J.-P. Stadelmann, *Chem. Phys.*, 1981, 61, 215.
349 J.P. Maier and F. Thommen, *Chem. Phys.*, 1980, 51, 319.
350 J.P. Maier and F. Thommen, *J. Chem. Phys.*, 1980, 73, 5616.
351 J.P. Maier, L. Misev, and F. Thommen, *Helv. Chim. Acta*, 1981, 64, 1985.
352 J.P. Maier and F. Thommen, *Chem. Phys.*, 1981, 57, 319.
353 J.P. Maier, L. Misev, and F. Thommen, *J. Phys. Chem.*, 1982, 86, 514.
354 J.P. Maier and F. Thommen, *Chem. Phys. Lett.*, 1981, 78, 54.
355 I. Nenner, P.-M. Guyon, T. Baer, and T.R. Govers, *J. Chem. Phys.*, 1980, 72, 6587.
356 D. Klapstein and J.P. Maier, *Chem. Phys. Lett.*, 1981, 83, 590.
357 J.A. Beswick and M. Horani, *Chem. Phys. Lett.*, 1981, 78, 4.
358 D.G. Hopper, *J. Chem. Phys.*, 1982, 76, 1068.
359 J.-P. Stadelmann and J. Vogt, *Int. J. Mass Spectrom. Ion Phys.*, 1980, 35, 83.
360 J.P. Maier and F. Thommen, *J. Chem. Soc., Faraday Trans. 2*, 1981, 77, 845.
361 J. Dannacher, J.-P. Stadelmann, and J. Vogt, *J. Chem. Phys.*, 1981, 74, 2094.

362 J. Dannacher, J.-P. Stadelmann, and J. Vogt, *Int. J. Mass Spectrom. Ion Phys.*, 1981, 38, 69.
363 P. Forster, J.P. Maier, and F. Thommen, *Chem. Phys.*, 1981, 59, 85.
364 J. Dannacher and J.-P. Stadelmann, *Chem. Phys.*, 1980, 48, 79.
365 M. Allan, J. Dannacher, and J.P. Maier, *J. Chem. Phys.*, 1980, 73, 3114.
366 T. Baer, U. Buchler, and C.E. Klots, *J. Chim. Phys. Phys.-Chim. Biol.*, 1980, 77, 739.
367 K. Johnson, I. Powis, and C.J. Danby, *Chem. Phys.*, 1981, 63, 1.
368 I. Powis, *J. Chem. Soc., Faraday Trans. 2*, 1981, 77, 1433.
369 P.I. Mansell, C.J. Danby, and I. Powis, *J. Chem. Soc., Faraday Trans. 2*, 1981, 77, 1449.
370 A.V. Golovin, M.E. Akopyan, and Yu.L. Sergeev, *Khim. Vys. Energ.*, 1981, 15, 387.
371 Y. Niwa, T. Nishimura, F. Isogai, and T. Tsuchiya, *Chem. Phys. Lett.*, 1980, 74, 40.
372 Y. Niwa, T. Nishimura, and T. Tsuchiya, *Int. J. Mass Spectrom. Ion Phys.*, 1982, 42, 91.
373 Y. Niwa, S. Tajima, and T. Tsuchiya, *Int. J. Mass Spectrom. Ion Phys.*, 1981, 40, 287.
374 G.G. Meisels, T. Hsieh, and J.P. Gilman, *J. Chem. Phys.*, 1980, 73, 4126.
375 J.J. Butler, M.L. Fraser-Monteiro, L. Fraser-Monteiro, T. Baer, and J.R. Hass, *J. Phys. Chem.*, 1982, 86, 747.
376 M.L. Fraser-Monteiro, L. Fraser-Monteiro, J.J. Butler, T. Baer, and J.R. Hass, *J. Phys. Chem.*, 1982, 86, 739.
377 L. Fraser-Monteiro, M.L. Fraser-Monteiro, J.J. Butler, and T. Baer, *J. Phys. Chem.*, 1982, 86, 752.
378 G.D. Willett and T. Baer, *J. Am. Chem. Soc.*, 1980, 102, 5769.
379 G.D. Willett and T. Baer, *J. Am. Chem. Soc.*, 1980, 102, 6774.
380 T. Hsieh, J.P. Gillman, M.J. Weiss, and G.G. Meisels, *J. Phys. Chem.*, 1981, 85, 2722.
381 J.J. Butler and T. Baer, *J. Am. Chem. Soc.*, 1980, 102, 6764.
382 H.M. Rosenstock, R. Stockbauer, and A.C. Parr, *J. Chim. Phys.-Chim. Biol.*, 1980, 77, 745.
383 H.M. Rosenstock, R. Stockbauer, and A.C. Parr, *Int. J. Mass Spectrom. Ion Phys.*, 1981, 38, 323.
384 H.M. Rosenstock, R. Stockbauer, and A.C. Parr, *J. Chem. Phys.*, 1980, 73, 773.
385 S.T. Pratt and W.A. Chupka, *Chem. Phys.*, 1981, 62, 153.
386 J. Dannacher, J.-P. Flamme, J.-P. Stadelmann, and J. Vogt, *Chem. Phys.*, 1980, 51, 189.
387 R. Bombach, J.-P. Stadelmann, and J. Vogt, *Chem. Phys.*, 1981, 60, 293.
388 R. Bombach, J. Dannacher, J.-P. Stadelmann, and J. Vogt, *Int. J. Mass Spectrom. Ion Phys.*, 1981, 40, 275.
389 R. Bombach, J. Dannacher, J.-P. Stadelmann, and J. Vogt, *Chem. Phys. Lett.*, 1980, 76, 429.
390 R. Bombach, J. Dannacher, J.-P. Stadelmann, and J. Vogt, *Chem. Phys. Lett.*, 1981, 77, 399.
391 C.J. Porter, J.H. Beynon, and T. Ast, *Org. Mass Spectrom.*, 1981, 16, 101.
392 D.J. Douglas, *J. Phys. Chem.*, 1982, 86, 185.
393 A.J. Stace and A.K. Shukla, *Int. J. Mass Spectrom. Ion Phys.*, 1981, 37, 35.
394 E.W. Rothe, F. Ranjbar, and D. Sinha, *Chem. Phys. Lett.*, 1981, 81, 175.
395 A.R. Hubik, P.H. Hemberger, J.A. Laramee, and R.G. Cooks, *J. Am. Chem. Soc.*, 1980, 102, 3997.

396 P.H. Hemberger, J.A. Laramee, A.R. Hubik, and R.G. Cooks, *J. Phys. Chem.*, 1981, 85, 2335.
397 D.J. Burinsky, G.L. Glish, R.G. Cooks, J.J. Zwinselman, and N.M.M. Nibbering, *J. Am. Chem. Soc.*, 1981, 103, 465.
398 W. Brand and K. Levsen, *Int. J. Mass Spectrom. Ion Phys.*, 1980, 35, 1.
399 W. Brand, H.D. Beckey, B. Fassbender, A. Heindricks, and K. Levsen, *Int. J. Mass Spectrom. Ion Phys.*, 1980, 35, 11.
400 C. Lifshitz, *J. Phys. Chem.*, 1982, 86, 606.
401 C. Lifshitz and S. Gefen, *Int. J. Mass Spectrom. Ion Phys.*, 1980, 35, 31.
402 J.L. Holmes and J.K. Terlouw, *Org. Mass Spectrom.*, 1980, 15, 383.
403 C.J. Proctor, A.G. Brenton, J.H. Beynon, B. Kralj, and J. Marsel, *Int. J. Mass Spectrom. Ion Phys.*, 1980, 35, 393.
404 C.J. Proctor, B. Kralj, A.G. Brenton, and J.H. Beynon, *Org. Mass Spectrom.*, 1980, 15, 619.
405 M.F. Jarrold, A.J. Illies, and M.T. Bowers, *Chem. Phys.*, 1982, 65, 19.
406 G. Hirsch and P.J. Bruna, *Int. J. Mass Spectrom. Ion Phys.*, 1980, 36, 37.
407 C.P. Edwards, C.S. Maclean, and P.J. Sarre, *Chem. Phys. Lett.*, 1982, 87, 11.
408 R.C. Heyer and M.E. Russell, *Org. Mass Spectrom.*, 1981, 16, 236.
409 G. Depke, C. Lifshitz, H. Schwarz, and E. Tzidony, *Angew. Chem., Int. Ed. Engl.*, 1981, 20, 792.
410 C. Lifshitz and E. Tzidony, *Int. J. Mass Spectrom. Ion Phys.*, 1981, 39, 181.
411 C. Lifshitz, *Int. J. Mass Spectrom. Ion Phys.*, 1982, 43, 179.
412 D.J. McAdoo, W. Farr, and C.E. Hudson, *J. Am. Chem. Soc.*, 1980, 102, 5165.
413 G. Depke and H. Schwarz, *Org. Mass Spectrom.*, 1981, 16, 421.
414 P. Pechukas, *J. Chem. Phys.*, 1980, 73, 993.
415 E.T.-Y. Hsieh and A.W. Castleman, jun., *Int. J. Mass Spectrom. Ion Phys.*, 1981, 40, 295.
416 D.R. Bates, *Chem. Phys. Lett.*, 1981, 82, 396.
417 I. Koyano and K. Tanaka, *J. Chem. Phys.*, 1980, 72, 4858.
418 F.A. Houle, S.L. Anderson, D. Gerlich, T. Turner, and Y.T. Lee, *Chem. Phys. Lett.*, 1981, 82, 392.
419 R.M. Bilotta and J.M. Farrar, *J. Chem. Phys.*, 1981, 74, 1699.
420 K. Tanaka, J. Durup, T. Kato, and I. Koyano, *J. Chem. Phys.*, 1981, 74, 5561.

2

Structures and Dynamics of Gas-phase Ions – A Theoretical Approach

BY J. C. LORQUET

1 Introduction

What is the mechanism of a molecular dissociation? Is it related to structural considerations, and how? No simple answer can be given to that question because the problem has two facets. It is an established fact that chemical reactivity is an inherently dynamic process that cannot be expected to reduce to a simple corollary of structural chemistry. Yet, predisposing circumstances clearly exist. It is the purpose of this paper to try to convince the experimentalist that theory can provide guidelines on this problem that are better than those commonly used by researchers. These guidelines will emerge from an analysis of some obvious questions that can be formulated at this juncture.

First of all, what does one mean exactly by 'structure'? This term can have two different meanings: it is used to designate either the equilibrium nuclear geometry (*i.e.* the molecular shape) or the electronic structure. The latter term has a very loose meaning and is usually taken to represent the occupation numbers and characteristics of the relevant molecular orbitals (MOs) and the reactivity indices derived from them, such as bond orders, *etc.* Both concepts have limitations that will be analysed in the first sections of this chapter.

Secondly, why is it so difficult to relate structure and reactivity? The pitfall to be avoided consists basically of an extrapolation process. In essence, a structure is information concerning a particular point on a potential-energy surface, whereas a reaction can be visualized as a trajectory leading from reactants to products, *i.e.* as a succession of structures. A structure can thus at most provide an initial tendency that may be altered and even reversed by the complicated nature of a reaction mechanism. Examples will be given below.

Thirdly, does this mean that one must abandon any hope of relating reactivity to structural considerations? Not necessarily. There exist cases in which, to a first approximation, the outcome of a chemical process is controlled by, or at least depends on, the properties of a particular point of the potential-energy surface or surfaces. This happens in two important circumstances: (i) if the reaction is adiabatic (*i.e.* takes place in a single electronic state) when the transition-state model is valid and (ii) in non-adiabatic processes (*i.e.* in which two electronic states are involved) that are governed by localized surface crossings or interactions.

Unfortunately, the nature and position of these transition states and crossings are not known *a priori*, and their determination requires an in-depth analysis of the reaction. Therefore, many mass spectrometrists look for a much more qualitative correlation. For reasons that have already been alluded to briefly and will be analysed presently, this Reporter cannot agree with them.

2 Molecular Shapes

Let us consider the first use of the term structure, *i.e.* in the study of nuclear conformations and especially of equilibrium geometries. A few remarks are appropriate.

First of all, the study of stereochemistry (interatomic distances and angles), of the stability of molecular aggregates (dissociation energies), in short, the experimental basis of the concepts of valence and the chemical bond, is based on the implicit study of a potential-energy curve (for a diatomic molecule) or a potential-energy surface (for a polyatomic molecule).[1] The existence of such a potential rests on the possibility of treating the electronic and nuclear motions as being independent. However, the argument developed by Born and Oppenheimer cannot be expected to remain valid if the density of electronic states is large.[2-4] This is known to be the case for most ionized molecules.

Secondly, assuming the validity of the Born-Oppenheimer approximation, *i.e.* assuming that a potential-energy surface does exist, one soon runs into problems connected with the fact that many molecular ions are non-rigid molecules. A typical system for which reliable calculations exist is the vinyl cation, which oscillates between a classical and a bridged structure[5] (Figure 1).

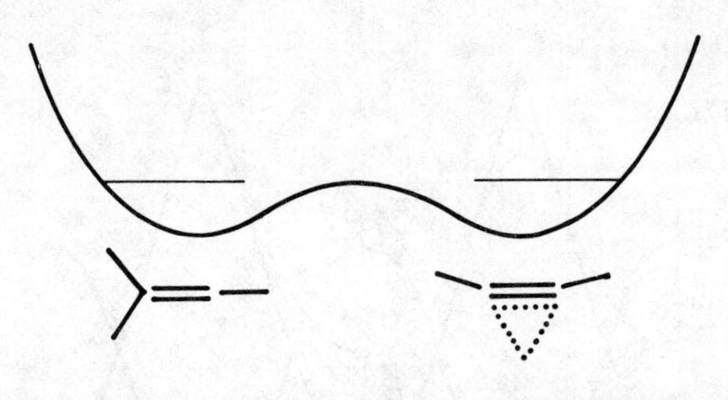

Figure 1 Section of the potential-energy surface corresponding to the optimum transformation path of the classical to the bridged vinyl cation. The height of the barrier is estimated to be of the order of the zero-point energy of the corresponding vibrational motion (*i.e.* a few hundredths of an eV)

The energy barrier is calculated to be of the order of a few hundred wavenumbers, which is of the same order of magnitude as the zero-point energy of the CH bending vibration, which converts one isomer into the other. It is obvious that the concept of nuclear structure loses its conventional meaning if the internal energy of the ion is higher than the energy barrier that separates two possible isomers. Furthermore, a rapid conversion from one nuclear geometry to the other will bring about a corresponding oscillation between two electronic structures, as will be discussed in a later section of this article.

Thirdly, it must be pointed out that there exist cases where an ionic state has no structure at all. An example can be found in the case of the $C_2H_4^+$ ion; it is brought about by a possibility of internal rotation of the two methylene groups. It turns out that for the $\tilde{C}^2B_2$ state of $C_2H_4^+$ a value of the angle of twist, θ, equal to zero corresponds to an energy maximum (more properly to a saddle point) of the potential-energy surface[6] (Figure 2). A relaxation of the nuclear geometry in a search for the energy minimum is found to lead to a straightforward correlation with the ground state $\tilde{X}^2B_3$.

Finally, Woolley has recently pursued this analysis at a much greater depth and has shown that the concept of molecular structure has limits of validity imposed on us by the quantum-mechanical

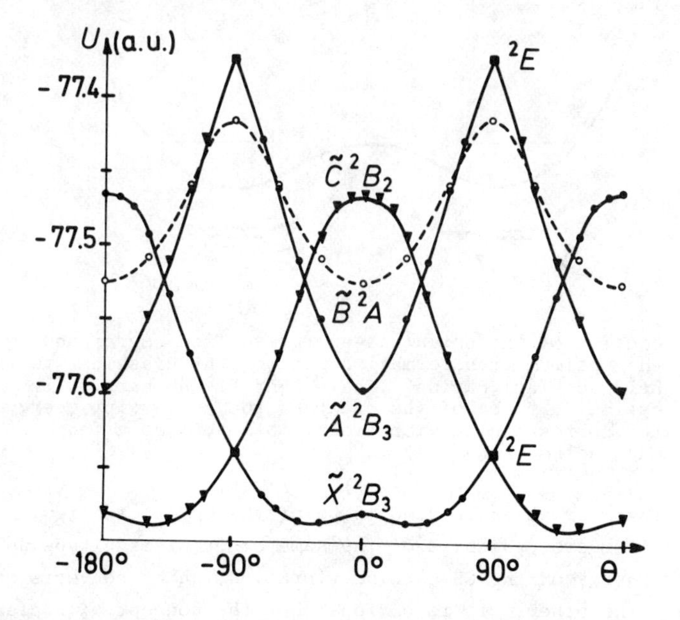

Figure 2 Potential-energy curves of the lowest electronic states of $C_2H_4^+$ as a function of the angle of twist θ

nature of the universe.[7] The concept breaks down in the case of an isolated molecule (as in the context of a beam experiment) studied at high-energy resolution. The point raised by Woolley is that, under such conditions, an experimental measurement leads to information on the *state* under study and not on the molecule. Now, a quantum-stationary state of an isolated molecule has no structure at all; it has no size or shape because no extension in space or time can be defined for it. However, Wilson has challenged this argument.[8]

3 Wavefunctions

Before we delve into the matter any further, it is necessary to emphasize an important distinction of a technical nature. When assessing the usefulness and reliability of a particular quantum-chemical calculation, it is essential to keep in mind the distinction that has to be made between self-consistent field (SCF) (sometimes also termed Hartree-Fock) and configuration interaction

(CI) wavefunctions. In the former the wavefunction is expressed as a single Slater determinant (represented by a single electronic configuration) whereas in the latter it is expressed as a linear combination of several Slater determinants in order to take into account the correlation among electrons, which is neglected in the first method. For that reason, SCF methods are said to be uncorrelated and are much simpler and much more easily calculated.[9-13] While they may be adequate for some purposes (essentially for a calculation of equilibrium geometries), they suffer from basic deficiencies[14,15] that make them unreliable in a study of the dynamic behaviour of electronically and vibrationally excited states.

The validity of SCF wavefunctions is often restricted to the neighbourhood of the equilibrium position. They cannot describe bond-stretching processes because they give rise to incorrect dissociation asymptotes.[16] This is due to the fact that the electronic configuration frequently changes along the reaction path. The configuration appropriate at the equilibrium geometry may be completely different from that close to the dissociation asymptote. It must be added that these problems are less acute when one studies dissociation processes of ionized molecules if the study is made in the open-shell formalism (the so-called unrestricted Hartree-Fock (UHF) method[17]).

The dissociation energies of ionized molecules are frequently so low that it might be reasonable to compare them to a dimer or cluster weakly bound by interaction that is intermediate between van der Waals interaction and genuine chemical bonding. A correct appraisal of the electron correlation is then essential. This can be done only if one goes beyond the SCF approximation.

Thirdly, even in the study of isomerization (processes in which the bond lengths do not change too much) it has been found[5,14,15] that a correct estimate of the correlation energy (that part of the molecular energy which, by definition, is neglected in the SCF approximation) is essential in order to calculate with quantitative accuracy energy differences between isomers.

Since as a rule several electronic states are involved, a study of a reaction mechanism requires simultaneous consideration of several potential-energy surfaces,[4] which should be of comparable quality. SCF wavefunctions do not necessarily meet this requirement. In addition, they are unable to describe the interactions between two crossing surfaces.[4,18-22]

For all these reasons, single-configuration approaches, whether *ab initio* (Hartree-Fock) or semi-empirical (CNDO, MINDO, HAM, *etc.*), can only provide results whose usefulness is restricted to *static* properties (equilibrium geometries). For a study of unimolecular reaction *dynamics*, *ab initio* configuration interaction (and multi-configuration) methods, which are free from these defects, are of much greater value, but they are much more tedious and expensive and, unfortunately, not yet very numerous.

4 Electronic Structures

Let us now go back to the study of electronic structures. An electronic configuration such as $(1a_1)^2(1b_2)^2(2a_1)^2(1b_1)^1$ for the ground state of H_2O^+ consists of a set of occupation numbers of MOs. Contrary to what is often asserted in the literature, *there is as a rule no correlation between the bonding or antibonding nature of a particular MO and the fragmentation of the ionic state that results from electron removal from this MO.* The correlation does not exist simply because there is little reason why it should exist. As already stated in the introduction, the argument is specious because the bonding or non-bonding nature is calculated at the equilibrium position of each state, whereas the nature of the wavefunction at the dissociation asymptote may be entirely different. In such a case, no correlation is to be expected between electronic structure and reactivity. Let us consider a few examples.

The uppermost occupied orbital of formaldehyde (the so-called lone-pair orbital), whose group-theoretical symbol is $2b_2$, is actually CH bonding,[23] and the ground-state ion, $\tilde{X}^2B_2$ $(2b_2)^{-1}$, dissociates into HCO^+ + H, in apparent conformity with the conventional rule. But the next MO ($1b_1$ π_{CO}) is purely CO bonding[20] and yet the first excited state of H_2CO^+, $\tilde{A}^2B_1$ $(1b_1)^{-1}$, dissociates by two competing channels[24-26] to $CO^+ + H_2$ or HCO^+ + H.

The first electronic state of CO_2^+ which dissociates when excited by electron or photon impact is the $\tilde{C}^2\Sigma_g^+$ $(4\sigma_g)^{-1}$ state.[27] Yet it is created by removal of an electron from the $4\sigma_g$ MO, which is slightly CO antibonding.[23] This results from the fact that its internal energy is not used to bring about dissociation. The ion finds a way[28] to cascade down to the ground $\tilde{X}^2\Pi_g$ $(1\pi_g)^{-1}$ state, which has enough energy to dissociate to CO^+ + O. At the equilibrium geometry, the $1\pi_g$ MO is strictly CO non-bonding for symmetry reasons.[23]

The CF_4^+ ion literally explodes into CF_3^+ + F since no trace of a molecular ion can be detected in the mass spectrum and yet the uppermost occupied MO, $1t_1$, is CF non-bonding.[23] The next occupied MO is the slightly CF bonding $4t_2$ orbital. The CF bonding effect is due to the $2t_2$ and $3t_2$ MOs, which are much higher in energy.

Further examples could be given, and a more thorough discussion of the case of $MeOH^+$ will be given in the last section of this paper. The above examples are sufficient to allow us to detect a basic flaw in the often repeated statement according to which the removal of a lone-pair electron does not induce any significant effect. Similarly, it is not reasonable to expect that the removal or excitation of an electron supposed to be localized on a particular bond will produce effects localized on that bond. Arguments of this kind, although often made, are unreliable. They are likely to be correct if one restricts oneself to static properties involving phenomena for which the positions of the nuclei remain practically constant (*e.g.* spectroscopic properties).[29] But, as soon as one considers phenomena for which nuclear motions are very important, such as dissociation processes, the static discussion is no longer reliable as a criterion, and it becomes essential to study the readjustment of the electronic structure during the nuclear motion.

As a matter of fact, a theory has been developed in which chemical reactivity is related to the reorganization undergone by the electronic structure along the reaction path.[30] The more extensive the relaxation, the lower the energy barrier and the easier the reaction.

In general, the problem is far less crucial in the case of the ground state of a neutral molecule because a deep energy well entails a stiff nuclear structure and thus is associated with a well defined electronic configuration. But the situation is entirely different for electronically excited states and is especially acute for ionized molecules. A labile nuclear structure implies a geometry-sensitive electronic structure. One has to deal with a sloppy structure, in both senses of the word structure.

In summary, in the Reporter's opinion, the so-called molecular-orbital theory of mass spectra can be expressed by a single sentence: "The more one forgets about molecular orbitals when discussing fragmentation processes in a mass spectrometer, the better". More correct arguments will now be introduced.

5 Adiabatic Reactions

We may consider the theories that establish a certain link between chemical reactivity and structure. For adiabatic reactions (processes taking place on a single potential-energy surface), such an approach leads to the transition-state theory. In RRKM-QET theory, the rate constant can be written as in equation (1).

$$k(E) = \alpha G^{*}(E - E^{o})/hN(E) \qquad (1)$$

In this expression, E^{o} is the threshold energy for reaction, $N(E)$ is the density of states at energy E (the number of states per unit energy), and $G^{*}(E - E^{o})$ is the integrated density of states, *i.e.* the total number of states between 0 and $(E - E^{o})$. The asterisk indicates that the property refers to the transition state, a system of $(3N - 7)$ degrees of freedom, whereas absence of an asterisk indicates a property of the reactant molecules, a system of $(3N - 6)$ degrees of freedom. The term α is the reaction-path degeneracy. A number of interesting (but rather obvious) consequences result from the fact that both N and G are increasing functions of the energy:

(1) The lower E^{o} is, the larger k is. Reaction takes place preferentially *via* low-energy transition states.
(2) The higher in energy the equilibrium conformation of a particular isomer is, the smaller $N(E)$ is. Thus, high-energy structures react more easily than low-energy ones.
(3) If one of the fragments can exist in several isomeric forms of similar energy, then there exists a different transition state associated with each structure. Each one contributes to the fragmentation and the total rate constant is given by the summation in equation (2).

$$k^{tot}(E) = \sum_{i} k_{i}(E) = \left(\sum_{i} \alpha_{i} G_{i}^{*}(E - E_{i}^{o})\right)/hN(E) \qquad (2)$$

The larger the number of possible exit channels, the larger the volume of phase space associated with the transition states and the larger the rate constant. In addition, the fact that, for example, a hydrogen atom can migrate easily implies the presence in the transition state of a low CH bending frequency. This circumstance increases G^* and hence increases the rate constant above its normal value. In more chemical terms, it may be said that the possibility of a low-energy isomerization process increases the rate constant and hence the probability of the overall reaction. This may account for the 'anchimeric effect' of organic chemists and supports the view, defended by Schwarz,[32] that reactions involving a hydrogen rearrangement are favoured over other straightforward possibilities, even though the existence of the rearrangement may remain unsuspected. The interpretation suggested here provides an alternative mechanism to the usual explanation considered by liquid-phase organic chemists, according to which the neighbouring-group effect consists in a solvent-sensitive stabilization of a transition state in which a charged centre interacts with a functional group possessing weakly bound electrons.

6 Crossings and Funnels

For non-adiabatic reactions (processes involving a transition between two potential-energy surfaces) the connection between structure and reactivity is usually straightforward. Chemical reactivity turns out to be controlled by specific points or regions of space on the potential-energy surfaces. Photochemists, who have to deal with many of the same problems as mass spectrometrists, have coined a new word and designate these particular points as 'funnels'.[2] In brief, a funnel is a curve crossing or a localized region of strong non-adiabatic interaction between two potential-energy surfaces. The word takes its origin in a hydraulic analogy to account for the resulting transfer of population (Figure 3). Dewar talks about a Born-Oppenheimer hole in the surface.[3] As a matter of fact, the process consists of an internal conversion of electronic energy into vibrational energy, and requires a coupling between electronic and vibrational energies. Since the energy gap is small, the Born-Oppenheimer approximation breaks down and a zone of strong non-adiabatic interaction appears.[4] The hydraulic analogy with an ordinary funnel is somewhat misleading because a study of the probabilities of non-adiabatic transition shows that, the larger the width of the zone of non-adiabatic interaction (the larger the diameter of the Born-Oppenheimer hole or the calibre of the funnel), the less likely the transition probability.

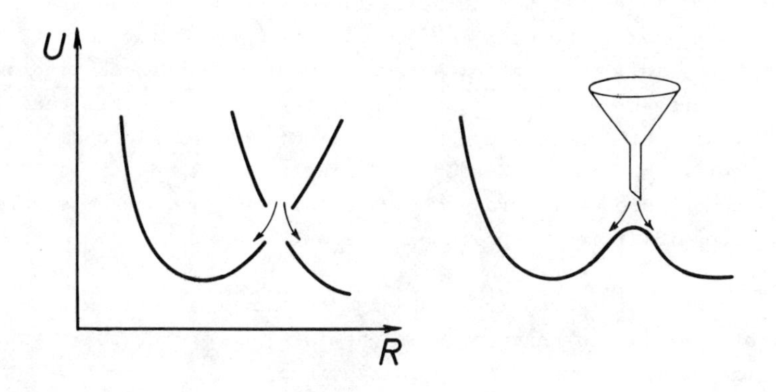

Figure 3 Pair of potential-energy surfaces separated by a small energy gap and giving rise to a strong non-adiabatic interaction. The pictures show the reactive flux through the 'Born-Oppenheimer hole' and the analogy with a funnel

As structural features, crossings and funnels have a much more direct connection with chemical reactivity than equilibrium geometries and electronic configurations. A classification of them can be established (see following four sections).

Spin-forbidden Transitions.- These processes arise when a crossing takes place between two surfaces of different multiplicity, *e.g.* doublet and quartet. Their existence has been recognized in a number of cases: HBr^+,[34] H_2S^+,[35] CO_2^+,[28] CS_2^+,[36,37] N_2O^+.[36,38,39] Such a transition is called intersystem crossing (if both states are stable) or spin-forbidden predissociation (if the final state is repulsive). A transition is made possible by spin-orbit coupling, but this provides a slower process than a spin-allowed process. Predissociation lifetimes in the range 10^{-7} - 10^{-8}s have been suggested for CO_2^+ [28] and N_2O^+.[39] The exact time for this slackening can vary from case to case and shorter lifetimes could be found in other instances. This depends mainly on two factors. First, the closer the electronic configurations of the intersecting states are, the larger the rate constant is.[40] Second, the presence of a heavy (high *Z*) atom in the ion will considerably speed up the process. This is already expected to be noticeable if the molecular ion contains S, P, or Cl atoms. If very heavy atoms such as I, Hg, or Pb are present, the spin-conservation rule becomes much less stringent.[41]

Jahn-Teller and Renner-Teller Interactions.- Let us assume that the molecular ion has a possibility of adopting a nuclear confrontation with a symmetry axis of order equal to or higher than three (this includes linear conformations characterized by an infinite-fold symmetry axis). The role played by such structures, which need not correspond to an energy minimum, can be determinant in the reaction mechanism. The reason is that a high symmetry brings about electronic degeneracies that allow a mechanism for internal conversion.

Consider the CH_2^+ ion. In its linear geometry, the ground $\tilde{X}^2B_1$ state coalesces with the first excited $\tilde{A}^2A_1$ state to form a doubly degenerate $^2\Pi$ state[42] (Figure 4). This process gives rise to a non-adiabatic coupling between the two surfaces and hence to the possibility of a radiationless transition. It is called the Renner-Teller interaction.

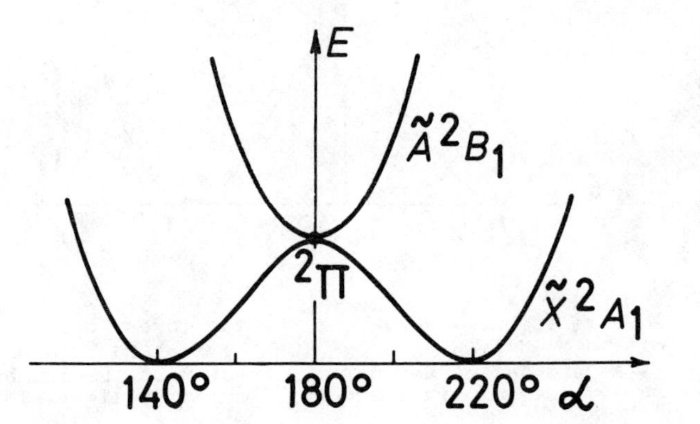

Figure 4 Potential-energy curves of the two lowest energy states of CH_2^+ as a function of the valence angle α. A Renner-Teller degeneracy occurs at the linear confrontation (α = 180°)

A similar process also exists for non-linear polyatomic molecules. Consider the $MeOH^{+\cdot}$ ion and call its COH bond angle ϕ. When ϕ differs from 180°, the molecular ion has a plane as the only element of symmetry. It then belongs to the C_s point group and no degeneracies are possible. The ionic states are labelled A' or A'' according to whether the electronic wave function is symmetrical or antisymmetrical with respect to the symmetry plane. However, when $\phi = 180^o$, the CO and OH bonds are collinear and the ion has a three-fold axis of symmetry (it belongs to the C_{3v} point group). The electronic states are labelled A_1, A_2, E. The degenerate E states result from the coalescence between former A' and A'' states.

Such a process takes place for the ground $\tilde{X}^2A''$ state of $MeOH^{+\cdot}$, which forms a degenerate ^2E state with the first excited $\tilde{A}^2A'$ state when $\phi = 180^o$ (Figure 5).[43] Since the energy of state $\tilde{A}$ is then a minimum, it is readily understood that the population of state $\tilde{A}$ can easily undergo internal conversion to state $\tilde{X}$. The two states, $\tilde{X}$ and $\tilde{A}$, are said to be coupled by a Jahn-Teller interaction.

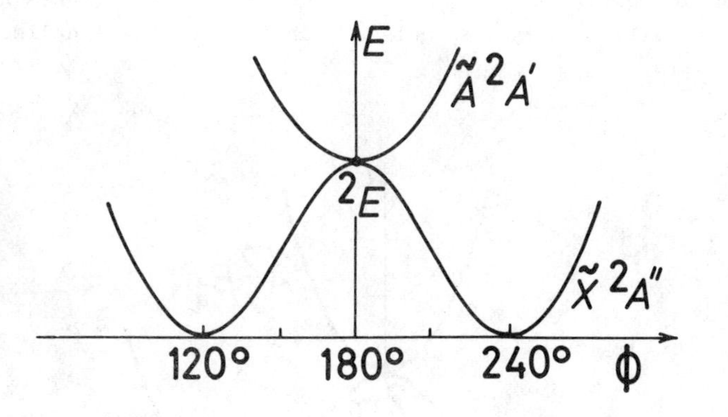

Figure 5 Potential-energy curves of the two lowest energy states of the $MeOH^+$ ion as a function of the COH valence angle ϕ. A Jahn-Teller degeneracy occurs when the ion has a threefold symmetry axis (C_{3v} symmetry) at $\phi = 180^o$

Degeneracy also exists in the case of the $C_2H_4^{+\cdot}$ ion in its D_{2d} point group when the two methylene groups are perpendicular to each other. A four-fold rotation-reflection axis S_4 exists, and the ion can exist in a degenerate ^2E state (Figure 2), which connects[6] states $\tilde{X}^2B_3$ and $\tilde{C}^2B_2$, as already pointed out on page 65.

Conical Intersections.- The Reporter has already discussed this concept in several previous review papers[6,20-22,25,44,45] and has emphasized its bearing on dynamical processes. Suffice it to repeat here that a complication in the reaction path may bring about a reduction in the activation energy.[25] This suggests that, for complex molecular ions, a rearrangement process may be an essential part of a dissociation mechanism.[44-45] Furthermore, conical intersections may lead to a possibility of internal conversion[6,46-48] or, conversely, to a suppression of the non-adiabatic character of a chemical reaction.[25,28] They can also play the role of a bottleneck along a reaction path and hence of a possible transition state.[6,47]

Conical intersections are encountered extremely frequently. They have been found to play a leading part in the dissociation mechanism of nearly all polyatomic molecular ions for which reaction paths have been calculated (H_2O^+,[46] CO_2^+,[28] CH_4^+,[49] H_2CO^+,[25] $C_2H_4^+$,[6,47,48] $MeOH^+$ [43,50]).

Avoided Crossings.- The two surfaces come close but do not cross. In such a situation there will be a certain transition probability for the population of the upper electronic state to drain down to the lower. The existence of such a process has been recognized in some instances,[19,43,51,52] but no quantitative study is available. In the case of the $MeOH^{+\cdot}$ ion, such an avoided crossing has been shown[43] to be the origin of a dynamical process commonly called 'isolated-state decay'. This will be examined in the next section.

7 Isolated-state Decay: A Case Study

The case of $MeOH^{+\cdot}$ provides a dramatic illustration of the fact that chemical reactivity may be under the control of a structural feature of the dynamic type, a funnel, and not of a static structural feature such as an equilibrium geometry or an electronic structure.

It is sometimes observed[53] that the ion-yield curve of a particular fragment has a shape that roughly resembles the photoelectron band of a particular state (or group of states) and is totally different from that predicted by a QET calculation on the electronic ground state of the ion. Recent analysis of a particular case has shown[43] that this cannot be taken as evidence that there exists an excited state characterized by a repulsive potential-energy surface that escapes internal conversion.[54]

The $MeOH^{+\cdot}$ ion is found to dissociate to a variety of fragments whose ion-yield curve obeys the laws of RRKM (QET) theory, with the exception of the Me^+ fragment, which is obviously produced by a non-statistical mechanism. Instead, its ion-yield curve resembles the third and fourth bands of the photoelectron spectrum. Many authors[55-58] have tried to account for the mechanism in structural terms. However, it is easy to see that, for reasons that have been explained in detail in the previous sections, no satisfactory explanation can be arrived at by looking at the bonding or anti-bonding nature of particular molecular orbitals. The upper $2a''$ MO of MeOH is CH bonding and OH non-bonding.[23] Yet the corresponding $(2a'')^{-1}$ $\tilde{X}^2A''$ state (the ground state) correlates with the MeO^+ + H asymptote and not with CH_2OH^+ + H.[50] Similarly, production of Me^+ + OH fragments takes place from the $(7a')^{-1}$ $\tilde{A}^2A'$ state and not from the $(6a')^{-1}$ $\tilde{B}^2A'$ state, although $(7a')$ is CO antibonding and $(6a')$ is CO bonding. Indeed, state $(6a')^{-1}$ $\tilde{B}^2A'$ is stable[43] with respect to CO bond stretching in spite of the CO bonding nature of the $(6a')$ singly occupied MO. Thus, no correlation can be established with the bonding nature of the MOs. Nor is it *a priori* very illuminating to know that the equilibrium geometry of the $MeOH^{+\cdot}$ ion in its $\tilde{B}^2A'$ state, the state whose photoelectron band resembles that of the ion-yield curve, is characterized by a COH angle equal to 80° only.[43] It turns out that the production of the Me^+ ions is controlled by a very specific funnel, which will now be analysed. Photoionization experiments[57,58] have shown that these ions are produced from the third and fourth bands detected in the photoelectron spectrum. An *ab initio* calculation of the potential-energy surfaces[43] reveals that these bands correspond to a group of several states (states $\tilde{B}$, $\tilde{C}$, and $\tilde{D}$, plus possibly some others), which undergo easy internal conversion to the lowest of them, *i.e.* to the second excited state of the ion, $\tilde{B}^2A'$. Although state $\tilde{B}$ corresponds to the removal of an electron from a CO bonding MO, it is stable and does not dissociate to Me^+ + OH. What takes place instead[43] is a further radiationless transition corresponding to an avoided crossing between states $\tilde{B}$ and $\tilde{A}$. This funnel has very peculiar properties. The minimum energy gap is very small (~0.02 eV) and the zone of non-adiabatic interaction is restricted to a very narrow range of internuclear distances. Thus, there is a very localized and highly efficient funnel, which is responsible for the further evolution of the $MeOH^{+\cdot}$ ion and especially for its peculiar bimodal dissociation.

At this funnel, the potential-energy surfaces are sloping in such a way that a branching in the trajectories takes place. Those nuclear trajectories that have enough energy in the CO bond-stretching co-ordinate rapidly reach the Me^+ + OH asymptote and give rise to methyl ions by a fast, non-statistical, diatomic-like process. The other fraction consists of nuclear trajectories that have a small amount of energy in the CO bond-stretching co-ordinate and most of their energy elsewhere. They escape direct dissociation, and rapidly reach the bottom of state $\tilde{A}$, and have to redistribute their internal energy in order to dissociate statistically. This process gives rise to the RRKM component of the mass spectrum. It leads to preferential production of CH_2OH^+ + H fragments since these correspond to the lowest dissociation asymptote.

In summary, the two components of the mass spectrum are produced from the same electronic state ($\tilde{A}^2A'$), but by two different processes. This corresponds to a branching in the reactive flux, as represented schematically in Figure 6.

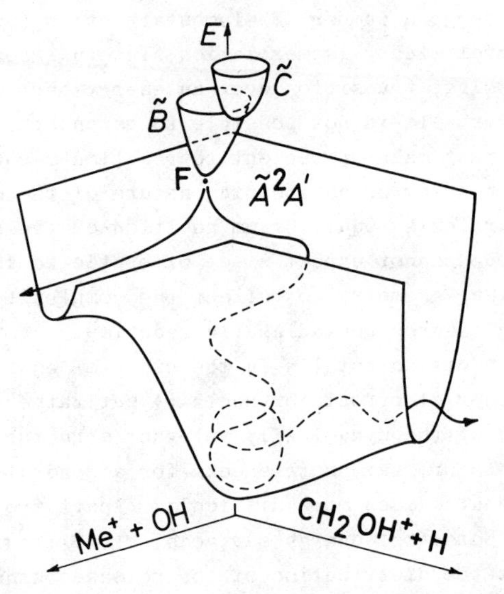

Figure 6 Schematic view of the mechanism of isolated-state decay. The reactive flux branches at the funnel *F* (the avoided crossing) into a RRKM component and a direct fragmentation path to Me^+ ions

The phenomenon, which is commonly referred to as 'isolated-state decay', thus requires a dynamical interpretation. It is associated not with an isolated, decoupled, repulsive electronic state, as has been commonly assumed so far, but with the non-randomized population fraction that dissociates upon single-bond stretching by a fast, non-statistical, diatomic-like process. The peculiar bimodal dissociation of the $MeOH^{+\bullet}$ ion is seen to be controlled by a structural feature, a funnel or avoided crossing, which acts as a kind of switch in the branching of the reactive flux.

8 Conclusions

In summary, it has been repeatedly observed that a distinction has to be established between, on the one hand, equilibrium structures that provide a basis for the understanding of static properties and, on the other hand, transition states and funnels that are the key features controlling the dynamic properties and thus the chemical reactivity. The complexity of a chemical reaction (at least in the context of mass-spectrometric experimentation) results from the fact that it often consists of a complicated and specific sequence of elementary reactions.[59] In order to dissociate, the system has to undergo a number of elementary steps (*viz* specific deformation or preliminary isomerization, internal conversion or intersystem crossing, transition over an energy barrier, *etc.*) in a prescribed order. It is not possible to reconstruct the sequence of requirements that make up the entire reaction mechanism from an analysis of the bonding or antibonding nature of the molecular orbitals at a particular equilibrium position of the nuclear co-ordinates. One cannot expect a set of static indices or an equilibrium nuclear geometry to reflect the complexity inherent in a complicated multi-step non-adiabatic reaction.

Future prospects do not lie in the study of equilibrium structures (whether electronic or nuclear) but rather in the characterization of the dynamically relevant structures: transition states and funnels or, say, bottlenecks for a general term. This approach is of course much more difficult. Apart from *ab initio* calculations of potential-energy surfaces, the Reporter sees hope in the fact that the distribution of the released kinetic energy reflects the structure of the last bottleneck along the reaction path. The relationship has been analysed in two cases: when the transition state consists of a barrier of potential[60] or rotational[61,62] origin. Such studies should be extended to non-adiabatic reactions.

Acknowledgement.- The author is indebted to Professor P. Laszlo for a helpful discussion on the anchimeric effect. This work has been supported by the Belgian Government (Action de Recherche Concertée) and by the Fonds de la Recherche Fondamentale Collective.

References

1 C.A. Coulson, 'Valence', Oxford University Press, 1969.
2 J. Michl, *Top. Curr. Chem.*, 1974, **46**, 1.
3 M.J.S. Dewar and R.C. Dougherty, 'The PMO Theory of Organic Chemistry', Plenum, New York, 1975.
4 J.C. Lorquet, *Gazz. Chim. Ital.*, 1978, **108**, 145.
5 J. Weber, M. Yoshimine, and A.D. McLean, *J. Chem. Phys.*, 1976, **64**, 4159.
6 C. Sannen, G. Raseev, C. Galloy, G. Fauville, and J.C. Lorquet, *J. Chem. Phys.*, 1981, **74**, 2402.
7 R.G. Woolley, *J. Am. Chem. Soc.*, 1978, **100**, 1073.
8 E.B. Wilson, *Int. J. Quantum Chem., Symp.*, 1979, **13**, 5.
9 L. Radom and J.A. Pople in 'Theoretical Chemistry', MTP International Review of Science, Physical Chemistry Series One, ed. W. Byers Brown, Butterworths, London, 1972, Vol. 1, p. 71.
10 W.A. Lathan, L.A. Curtiss, W.J. Hehre, J.B. Lisle, and J.A. Pople, *Prog. Phys. Org. Chem.*, 1974, **11**, 175.
11 J.A. Pople, *Int. J. Mass Spectrom. Ion Phys.*, 1976, **19**, 89.
12 W.J. Hehre in 'Modern Theoretical Chemistry, Applications of Electronic Structure Theory', ed. H.F. Schaefer, tert., Plenum, New York, 1977, Vol. 4, p. 277.
13 J.A. Pople in 'Modern Theoretical Chemistry, Applications of Electronic Structure Theory', ed. H.F. Schaefer, tert., Plenum, New York, 1977, Vol. 4, p. 1.
14 R.H. Nobes, W.R. Rodwell, W.J. Bouma, and L. Radom, *J. Am. Chem. Soc.*, 1981, **103**, 1913.
15 K. Raghavachari, R.A. Whiteside, J.A. Pople, and P. von R. Schleyer, *J. Am. Chem. Soc.*, 1981, **103**, 5649.
16 R. Daudel, R. Lefebvre, and C. Moser, 'Quantum Chemistry', Interscience, New York, 1959, p. 406.
17 M. Krauss, *Annu. Rev. Phys. Chem.*, 1970, **21**, 30.
18 J.C. Tully in 'Dynamics of Molecular Collisions'. ed. W.H. Miller, Plenum, New York, 1976, Part B, p. 217.
19 C. Galloy and J.C. Lorquet, *J. Chem. Phys.*, 1977, **67**, 4672.
20 M. Desouter-Lecomte, C. Galloy, J.C. Lorquet, and M. Vaz Pires, *J. Chem. Phys.*, 1979, **71**, 3661.
21 J.C. Lorquet, D. Dehareng, C. Sannen, and G. Raseev, *J. Chim. Phys. Phys.-Chim. Biol.*, 1980, **77**, 719.
22 J.C. Lorquet, A.J. Lorquet, and M. Desouter-Lecomte in 'Quantum Theory of Chemical Reactions', ed. R. Daudel, Reidel, 1980, p. 241.
23 L.C. Snyder and H. Basch, 'Molecular Wave Functions and Properties', Wiley, New York, 1972.
24 M.P. Guyon, W.A. Chupka, and J. Berkowitz, *J. Chem. Phys.*, 1976, **64**, 1419.
25 M. Vaz Pires, C. Galloy, and J.C. Lorquet, *J. Chem. Phys.*, 1978, **69**, 3242.
26 R. Bombach, J. Dannacher, J.P. Stadelmann, and J. Vogt, *Chem. Phys. Lett.*, 1980, **76**, 429; *Int. J. Mass Spectrom. Ion Phys.*, 1981, **40**, 275; *Chem. Phys. Lett.*, 1981, **77**, 399.
27 J.H.D. Eland, *Int. J. Mass Spectrom. Ion Phys.*, 1972, **9**, 397; J.H.D. Eland and J. Berkowitz, *J. Chem. Phys.*, 1977, **67**, 2782.
28 M.T. Praet, J.C. Lorquet, and G. Raseev, submitted to *J. Chem. Phys.*
29 G. Herzberg, 'Molecular Spectra and Molecular Structure. I. Spectra of Diatomic Molecules', Van Nostrand, Princeton, U.S.A., 1966, p. 367.

30 R.F.W. Baker, *Can. J. Chem.*, 1962, 40, 1164; J.C. Lorquet, *Mol. Phys.*, 1966, 10, 489; L. Salem, *Chem. Phys. Lett.*, 1969, 3, 99; L. Salem and J.S. Wright, *J. Am. Chem. Soc.*, 1969, 91, 5947; R.G. Pearson, *Theor. Chim. Acta*, 1970, 16, 107; *Acc. Chem. Res.*, 1971, 4, 152; *J. Am. Chem. Soc.*, 1972, 94, 8287.
31 W. Forst, 'Theory of Unimolecular Reactions', Academic Press, New York, 1973.
32 H. Schwarz, *Top. Curr. Chem.*, 1981, 97, 1.
33 M. Desouter-Lecomte and J.C. Lorquet, *J. Chem. Phys.*, 1979, 71, 4391.
34 M.J. Haugh and K.D. Bayes, *J. Phys. Chem.*, 1971, 75, 1472; G. Mathieu, H. Wankenne, and J. Momigny, *Chem. Phys. Lett.*, 1972, 17, 260; J. Delwiche, P. Natalis, J. Momigny, and J.E. Collin, *J. Electron Spectrosc. Relat. Phenom.*, 1972, 1, 219.
35 J.H.D. Eland, *Int. J. Mass Spectrom. Ion Phys.*, 1979, 31, 161; G. Hirsch and P.J. Bruna, *Int. J. Mass Spectrom. Ion Phys.*, 1980, 36, 37.
36 J. Momigny, *J. Chim. Phys. Phys.-Chim. Biol.*, 1980, 77, 725.
37 J. Momigny, G. Mathieu, H. Wankenne, and M.A.A. Ferreira, *Chem. Phys. Lett.*, 1973, 21, 606; *Adv. Mass Spectrom.*, 1974, 6, 923.
38 R.G. Orth and R.C. Dunbar, *J. Chem. Phys.*, 1977, 66, 1616; D.G. Hopper, *J. Am. Chem. Soc.*, 1978, 100, 1019.
39 I. Nenner, P.M. Guyon, T. Baer, and T.R. Govers, *J. Chem. Phys.*, 1980, 72, 6587; D. Klapstein and J.P. Maier, *Chem. Phys. Lett.*, 1981, 83, 590.
40 A.J. Lorquet, J.C. Lorquet, H. Wankenne, J. Momigny, and H. Lefebvre-Brion, *J. Chem. Phys.*, 1971, 55, 4053.
41 S.P. McGlynn, T. Azumi, and M. Kinoshita, 'Molecular Spectroscopy of the Triplet State', Prentice-Hall, 1969, p. 261.
42 C.F. Bender and H.F. Schaefer, *J. Mol. Spectrosc.*, 1971, 37, 423.
43 C. Galloy, C. Lecomte, and J.C. Lorquet, submitted to *J. Chem. Phys.*
44 J.C. Lorquet, *Adv. Mass Spectrom.*, 1980, 8A, 3.
45 J.C. Lorquet, *Org. Mass Spectrom.*, 1981, 16, 469.
46 D. Dehareng, X. Chapuisat, J.C. Lorquet, C. Galloy, G. Raseev, and A. Nauts, to be published.
47 M. Desouter-Lecomte, C. Sannen, and J.C. Lorquet, to be published.
48 H. Köppel, L.S. Cederbaum, and W. Domcke, *J. Chem. Phys.*, in press.
49 E.F. Van Dishoeck, W.J. Van der Hart, and M. Van Hemert, *Chem. Phys.*, 1980, 50, 45.
50 J. Momigny, H. Wakenne, and C. Krier, *Int. J. Mass Spectrom. Ion Phys.*, 1980, 35, 151.
51 S. Sakai, S. Kato, K. Morokuma, and I. Kusunoki, *J. Chem. Phys.*, 1981, 75, 5398.
52 P. Rosmus, P. Botschwina, and J.P. Maier, *Chem. Phys. Lett.*, 1981, 84, 71.
53 J.P. Stadelmann and J. Vogt, *Adv. Mass Spectrom.*, 1980, 8A, 47.
54 C. Lifshitz, *Adv. Mass Spectrom.*, 1978, 7A, 3.
55 P. Wilmenius and E. Lindholm, *Ark. Fys.*, 1962, 21, 97.
56 S. Ikuta, K. Yoshihara, and T. Shiokawa, *Mass Spectrom.*, 1974, 22, 233; *ibid.*, 1975, 23, 61.
57 J. Berkowitz, *J. Chem. Phys.*, 1978, 69, 3044.
58 T. Nishimura, Y. Niwa, T. Tsuchiya, and H. Nozoye, *J. Chem. Phys.*, 1980, 72, 2222.
59 J.C. Lorquet, C. Sannen, and G. Raseev, *J. Am. Chem. Soc.*, 1980, 102, 7976.

60 S. Kato and K. Morokuma, *J. Chem. Phys.*, 1980, 72, 206; *ibid.*, 1980, 73, 3900.
61 C. Klots, *Z. Naturforsch., Teil A*, 1972, 27, 553.
62 W.J. Chesnavich and M.T. Bowers, *J. Am. Chem. Soc.*, 1977, 99, 1705.

3

Ion/Molecular Beams Chemistry

BY S. A. SAFRON

1 Introduction

To study reaction dynamics is to examine the most intimate and subtle details of molecular interactions and structure. It requires the close collaboration of theoreticians and experimentalists and demands of them patience, ingenuity, and finesse. Ion/molecular beams experiments are one avenue of approach to this general problem. The reactions are of widespread occurrence in nature and are interesting in themselves. They also allow for the development and testing of models for systems exhibiting a broad spectrum of electronic properties and for reactions of these systems over a wide range of experimental conditions.

There has been a number of reviews recently on ion/molecule reactions,[1-3] on theoretical reaction dynamics,[4-7] and on potential-energy surface calculations.[8] In this article a very brief description of the basics for crossed-beams-scattering experiments is presented together with a short examination of recent experimental results on chemical reactions.

2 Theoretical Considerations

It is probably best to begin with a brief description of what one actually does measure in a crossed-beams-scattering experiment and then follow with an equally brief description of what one would observe with certain idealized kinds of molecular interactions. The construction of models is quite instructive in that one is able to associate the prominent features of the scattering with the simplified reaction dynamics without going through laborious calculations.

Cross-sections.- One finds in any undergraduate physical-chemistry text that the number of collisions per second per unit volume between molecules A and B (or ion A^+ and neutral B) is given by equation (1):

$$Z_{AB} = \sigma_{AB} v N_A N_B \qquad (1)$$

where σ_{AB} is the cross-section for collisions between A and B, v is the relative velocity, N_A is the number density of A, and N_B is the number density of B. The whole physics of the scattering is buried in the quantity σ_{AB}.

Some fraction of these collisions, f, results in the elementary bimolecular reaction $A + B \rightarrow C + D$. Hence, the number of C's (or D's) formed per second per unit volume is given by equation (2):

$$\frac{dN_c}{dt} = \sigma_n v N_A N_B \qquad (2)$$

where $\sigma_n = f\sigma_{AB}$ is the reactive cross-section. In usual chemical kinetics the bimolecular rate constant is given by $k = \langle \sigma v \rangle$, the average being taken over collision velocities and the internal-state distributions of the reactants.

In the same way, one can calculate the number of C's per second per unit volume that are produced in a particular internal state by multiplying (2) by $f(j)$, the fraction of C's formed in the state j, that is $\sigma(j) = f(j)\sigma_n$.

We are interested here in the fraction of products scattered in a particular direction with a particular speed. Because these are continuous properties, the pertinent quantity is the differential cross-section:

$$\frac{d^3\sigma_n}{dv'_L d^2\Omega} = I_L(v'_L, \Theta, \Phi)$$

Normally Θ is the polar angle of v' measured from the ion-beam direction (the z-axis) and Φ is the azimuthal angle. If the apparatus has solid-angle resolution $\Delta\Omega$, velocity or speed resolution Δv, and a reaction volume V (determined by the intersection of the beams), then the number of C's per second reaching the detector should be given by equation (3):

$$\frac{dN_c}{dt} = I_L(v_2', \Theta, \Phi) v N_A N_B V(\Delta\Omega)(\Delta v) T \qquad (3)$$

where T is a transmission factor that may depend on the apparatus geometry and the nature of the velocity analyser. A determination of I_L requires a knowledge of the ion-beam flux ($\approx vN_A$) and neutral-beam density in the intersection volume V, as well as the apparatus factors $\Delta\Omega$, ΔV, and T. The total cross-section may then be found by integrating, as shown in equation (4):

$$\sigma_n = \int I_L(v_L', \Theta, \Phi) dv_L' d^2\Omega \qquad (4)$$

The quantities v_L', Θ, and Φ for the molecule C are measured with respect to the laboratory (LAB) framework. Since the experiment is carried out under field-free conditions (ignoring gravity), the molecular interactions are more naturally described in a co-ordinate system that depends only on the co-ordinates of the molecules relative to each other. This is a moving co-ordinate system with the origin at the centre-of-mass (c.m.) of the reactants. In either classical mechanics or quantum mechanics the motion of the c.m. remains constant under field-free conditions and is therefore also the c.m. of the products. The relation between the LAB system and the barycentric or c.m. system is a simple vectorial one: the LAB velocity of C is the vector sum of the velocity of the c.m. in the LAB system plus the c.m. velocity of C. These relations are illustrated in Figure 1. (It is necessary to keep in mind that each point in this diagram, called a Newton diagram by Herschbach,[9] refers to a velocity not a position co-ordinate!) The c.m. differential cross-section, $I_B(u', \theta, \phi) = \frac{d^3\sigma_n}{du' d^2\omega}$, and is therefore obtained from $I_L(v_L', \Theta, \Phi)$ by equation (5):

$$I_B(u', \theta, \phi) du' d^2\omega = I_L(v_L', \Theta, \Phi) dv_L' d^2\Omega \qquad (5)$$

Because the volume elements in 'velocity space' must be equal, $u'^2 du' d^2\omega = v_L'^2 dv_L' d^2\Omega$. The relation (5) thus becomes

$$I_B(u', \theta, \phi) = \left(\frac{u'^2}{v_L'^2}\right) I_L(v_L', \Theta, \Phi) \qquad (6)$$

Equation (6) suffers from the fact that I_B must be calculated

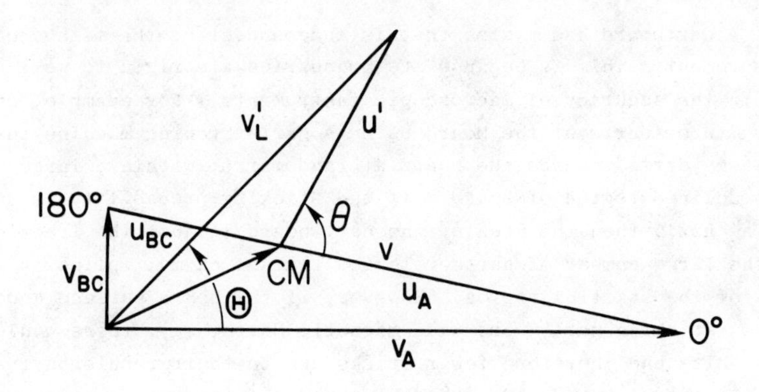

Figure 1 Vector diagram showing the relation between the LAB and the c.m. velocities. 0° and 180° refer to forward and backward directions in the c.m. system with respect to the initial *A* beam c.m. direction. In this diagram the ratio of LAB velocities V_A to V_{BC} is taken as 5 to 1, and the masses m_A and m_{BC} are taken to be 30 and 70, respectively

using a quantity, u', that is not determined experimentally.[10] However, one can convert the function $I_L(v'_L,\Theta,\Phi)$ from polar to Cartesian velocity co-ordinates, obtaining $I_C(v'_x,v'_y,v'_z)$, by using the relation $v'^2_L dv'_L d^2\Omega = dv_x dv_y dv_z$. Then, using the analogous relation for the c.m. functions, one obtains the c.m. Cartesian differential cross-section (equation 7):[10]

$$I_C(u'_x,u'_y,u'_z) = \left(\frac{1}{v'^2_L}\right) I_L(v'_L,\Theta,\Phi) \tag{7}$$

Actually, because of the difficulty in determining experimentally all the other quantities in equation (3), the investigator usually reports I_C normalized to the maximum product flux. This is usually done in the form of a contour map: equal values of I_C connected together as contours in a plot of scattering angle and speed.

Cross-sections and Potential-energy Surfaces.- The general theoretical calculation[11] of reactive cross-sections is well beyond the scope of this review.[6,8] However, it is important to examine qualitatively the connection between cross-sections and the forces of interaction between molecules, the understanding of which is one aspect of these experiments.

A dartboard has a size that is independent of the method of measurement. This is because its boundaries are rigid to well within the accuracy of macroscopic instruments. For example, one could throw darts at the board using a dart-throwing machine that releases darts towards the board at random from within a large well defined region of space. If the sticking probability is unity for a 'hit', then the area of the dart board is just the fraction of the large number of darts released that stick multiplied by the area of the starting region. However, if the board surface and the darts' sticking parts were made magnetic North, say, there would be fewer hits and therefore fewer sticks due to their repulsion than for the normal board. On the other hand, if the one were magnetic North and the other were magnetic South, there would be more sticks. That is, the cross-section of the dartboard to hits by a dart is smaller than geometric in the former situation and larger in the latter. Since the magnetic force is known, it is possible to calculate the sticking cross-section by calculating the trajectories from classical mechanics. Then the cross-section is just the fraction of the random trajectories that lead to the board multiplied by the area of the starting region.

If one measured the area of the board above by this method, not knowing that the materials were magnetized, could one determine the force? The answer is that just determining the sticking cross-section does not provide enough information. Clearly, many different kinds of forces could lead to the same total cross-section. One needs to examine the outcomes of a great number of throws. Even then, unless the force is particularly simple, two darts starting from different spots might well end up in the same place on the board although each trajectory is uniquely determined by the initial conditions, the force and Newton's Laws. The inversion of scattering data to uncover the force or potential energy is not a trivial matter.[12] The more detailed the information one has on the effects of the force, the better one can infer the nature of the force.

In the case of molecules the size depends strictly on the interaction; there is no 'independent' measurement that can be used as a standard. Thus, the photoabsorption cross-sections, the crystal-lattice dimensions, the van der Waals radii, and the transport coefficients all provide only the scantest guide to the size of the reactive cross-section between molecules A and B. The bulk rate constant gives the cross-section averaged over the whole range of collision energies and reactant states and provides only the crudest approximation to the intermolecular forces that lead to reaction. Total reactive cross-sections as a function of collision energy provide more information particularly about barriers, but still yield only a rough guide to the reaction dynamics. Very much better are the specific product-state cross-sections and particularly the differential cross-sections as functions of angle and velocity.

To calculate cross-sections from the intermolecular forces one would just take the fraction of starting trajectories of A's relative to B's that lead to the formation of C's and D's and multiply by the area spanned by the range of impact parameters. (The impact parameter for a given trajectory is the distance of closest approach between the c.m. of A and the c.m. of B if there were no molecular interaction.) In general for a given impact parameter there is some probability for reaction that depends on orientation and other initial conditions. In classical mechanics the differential cross-section can be thought of as the infinitesimal area corresponding to the range of impact parameters from b to $b + \mathrm{d}b$, which leads to reactive scattering with product C (or D) appearing in the solid angle defined by θ to $\theta + \mathrm{d}\theta$ and ϕ to $\phi + \mathrm{d}\phi$ and with speed u' to $u' + \mathrm{d}u'$, divided by $\mathrm{d}u'\mathrm{d}^2\omega$. In relative terms, the more trajectories that lead to the same interval in u' and ω, the larger the differential cross-section.

The real world, however, is described by quantum mechanics. Trajectories and impact parameters for molecules are not intrinsically well defined concepts; the relative motion of A and B must be described by a wavefunction. Yet, the picture is in many ways very similar. In the usual treatment,[11] the scattered wave is broken down into partial waves or 'wavelets' characterized by the orbital angular momentum (*i.e.* the wave is expressed as a superposition of eigenfunctions of the orbital angular momentum). The cross-section is determined by the probability of reaction for each wavelet multiplied by its cross-section, $\pi(2l + 1)\lambda\!\!\!^{-2}$, where l is the angular momentum of the wavelet and $\lambda\!\!\!^{-}$ is the de Broglie wavelength for the relative motion of A and B, divided by 2π. The difference from classical mechanics is that each wavelet contributes to scattering at all angles whereas a particular trajectory will always lead to the same angle. Even so, it is found that when many wavelets are required to describe the scattering, which is the usual case in chemical reactions, each contributes principally to a small range of product-scattering angles. One can then identify the angular momentum with the classical impact parameter. (Classically $L = \mu v b$). The upshot is that, for chemical reactions, a semiclassical treatment works very well.[13-17]

If the Born-Oppenheimer approximation is valid, then the electronic energy at each nuclear position becomes the potential energy of the molecular motion. It is a function of the $3n$ co-ordinates of the n atoms in the system minus 3 degrees of freedom for the c.m. that remains at the origin in the c.m. co-ordinate system and minus 3 degrees of freedom for overall rotation that does not change the positions of the atoms relative to each other. The potential energy can thus be represented as a hypersurface in $3n - 5$ dimensional space. In most texts the potential-energy surface for triatomic systems is depicted as a contour diagram of equipotentials as a function of two distance co-ordinates (r_{AB} and r_{BC}) at fixed A-BC orientations. We will refer to the 'entrance channel' of the potential-energy surface as the region corresponding to the approach of A and B from infinite separation and to the 'exit channel' as the region corresponding to the departure of products to infinite separation.

The fact that the scattering data cannot be simply inverted to give the potential-energy surface means that simple and straightforward models having guidelines for applicability become absolutely essential. Calculations on model potential-energy surfaces are also very instructive. Polanyi and co-workers have carried out a series of model-surface trajectory calculations for triatomic systems.[18] These show what general features in the scattering one should observe for different kinds of surface features, such as barriers, depressions, and sharply varying potentials, and for different combinations of masses of the reactant molecules.

That a chemical reaction is carried out on a single potential-energy surface (the usual textbook picture of reaction dynamics) may prove to be more the exception rather than the rule. We will discuss this point later.

Idealized Models.- *Spectator Stripping.* The spectator stripping model[19,20] (SSM) for the reaction $A + BC \rightarrow AB + C$ is based on the idealized and simple picture that C is totally unaffected by what goes on between A and B. Thus C retains its original c.m. velocity, and A and B, now stuck together, move with the velocity of their c.m. This means that the products are scattered exactly forward in the c.m. system, really undeflected at all, as is shown in Figure 2. In a real system approximated by the SSM one would expect to see a distribution of scattered products with the peak in I_B near the SSM predicted value.

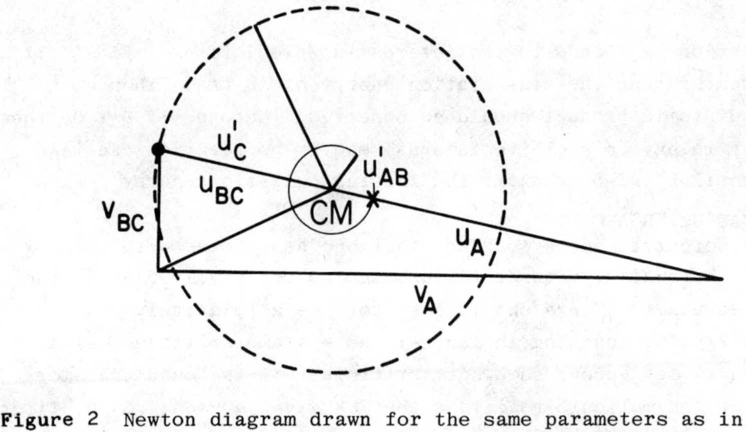

Figure 2 Newton diagram drawn for the same parameters as in Figure 1, showing the loci of product velocities according to the spectator models. The positions of *X* and *O* mark the expected SSM velocities for the products *AB* and *C*, respectively. For elastic scattering of the products, *AB* would be confined to the solid circle and *C* to the dashed circle

It is easy to calculate the relative translational energy of the products from the above conditions.[20] First, the c.m. translational energy or collision energy of the reactants is just $E = \frac{1}{2}\mu v^2$, with $\mu = m_A m_{BC}/m_{ABC}$. Similarly, the relative collision energy of A and B is $E^* = \frac{1}{2}\mu^* v^2$, with $\mu^* = m_A m_B/m_{AB}$. Since A and B stick together, E^* must remain with them as internal energy. Energy conservation requires that the relative translational energy of the products be given by $E' = \frac{1}{2}\mu' v'^2 = E - E^*$ or, as shown in equation (8):

$$E' = \frac{m_A m_B}{m_{AB} m_{BC}} E \qquad (8)$$

where $\mu' = m_C m_{AB}/m_{ABC}$. If the reaction is exoergic by amount D (or endoergic for $D < 0$), then this energy is also retained by the product AB, because C is just a spectator, and its internal energy becomes W' (equation (9)):

$$W' = E\left(\frac{m_B m_{ABC}}{m_{AB} m_{BC}} + \frac{D}{E}\right) \qquad (9)$$

Finally, if the reactant BC had internal energy W initially, energy conservation requires that it also go into product translation so that equation (8) must be modified to (10):

$$E' = W + \frac{m_A m_B}{m_{AB} m_{BC}} E \qquad (10)$$

Equation (9) leads to two interesting predictions. First, if W' is greater than the dissociation energy of AB (D_0), then no stable diatomic product should be observed. Second, if $D < 0$, then in order to have a positive internal energy no reaction can take place until $E^* = -D$ or until the initial collision energy $E = -Dm_{AB}m_{BC}/m_B m_{ABC}$.

The potential-energy surface that one associates with this model is attractive between A and B and is relatively flat in the B-C direction.[18] There should therefore be a relatively good probability for reaction in large-impact-parameter trajectories. Although it may appear somewhat artificial, it is found that for very many ion/molecule reactions the SSM gives a good, rough, first approximation to the experimental results in the sense that the most probable product velocity is near the SSM prediction. Herschbach has also suggested a modification to the SSM: elastic scattering of the spectator C by the AB molecule.[21] The effect of this is to preserve the energetics given above in equations (8) - (10) but to allow the products to appear over a range of c.m. angles depending on the force responsible for the elastic scattering. Again, in a real system one expects to see a distribution of products peaking with the SSM value shown in Figure 2. The simplest elastic scattering is 'hard-sphere' scattering, which is like three-dimensional billiards. The products in this case would appear at all angles with equal probability, *i.e.* isotropically.[22]

Ideal Knockout. The ideal knockout model[23] (IKM) is the other side of the coin from the SSM. In this case A collides with the C end of the molecule, knocking it away, and then picks up the remaining B. As before, one can work out fairly easily the angular distribution and energetics of the products from these conditions.[24,25] For a head-on collision C is knocked directly backwards in the c.m. system. In the reduced system consisting of just A and C the relative translational energy $E^* = \frac{1}{2}\mu^* v^2 = Em_C m_{ABC}/m_{AC} m_{BC}$, and the velocity of C relative to the c.m. of this system $u^* = vm_A/m_{AC}$. After the collision, this vector is rotated by 180° about the AC c.m. With respect to the actual c.m. of the entire system the new velocity of C $u' = a + u^*$, where $a = vm_A m_C/m_{AB} m_{ABC}$. For collisions other than straight head-on, u' is given by expression (11):

$$u' = a\cos\theta + (u^{*2} - a^2\sin^2\theta)^{\frac{1}{2}} \qquad (11)$$

The relation between these quantities is illustrated in Figure 3. In a real system approximated by the IKM, one expects to see a distribution of products peaking at the IKM value, predominantly in the backward direction.

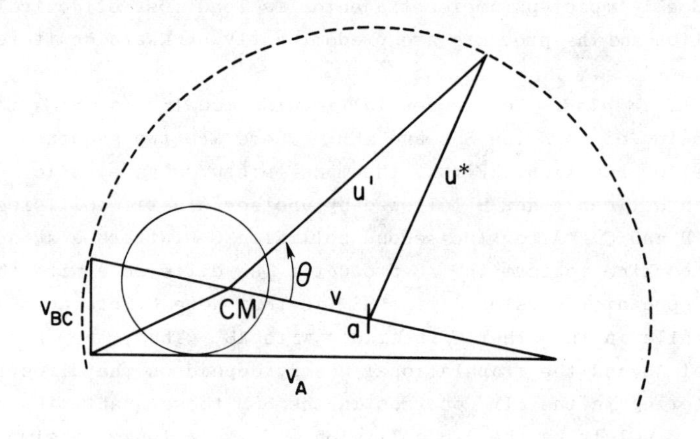

Figure 3 Newton diagram drawn for the same parameters as in Figure 1, with $m_B = 50$ and $m_C = 20$, showing the loci of velocities according to the IKM. The *AB* product is confined to the solid circle while *C* is restricted to the dashed circle (partly shown). The other quantities are as defined in the text

Finally, if the reaction is exo- or endo-ergic by amount D, then u^* must be modified by multiplication of $(1 + D/E^*)^{\frac{1}{2}}$. The value of u' is still given by equation (11). Momentum conservation in the c.m. system then gives the relative velocity of the products from the expression $v' = u'm_{ABC}/m_{AB}$ and the relative translational energy from $E' = \frac{1}{2}\mu' v'^2$. Energy conservation gives the product internal energy as $W' = E + W - E' + D$, where W is the reactant internal energy.

Because of the requirements that u^* should be positive, in an endoergic reaction the minimum $E^* = -D$, so that the minimum collision energy for an endoergic reaction is predicted to be $E = -Dm_{AC}m_{BC}/m_C m_{ABC}$. As before, to form stable AB products, W'

must be less than D_0.

The potential-energy surface for a knockout process must be steeply repulsive between A and C, particularly in the region where A gets close to the BC molecule. A collision with the B end of the molecule should not lead to reaction. That is, a surface for this orientation will not have an accessible exit channel. A collision with the C end of the molecule leads to an exit valley that is again fairly flat after the steep decline to the energy corresponding to the reaction exo- or endo-ergicity. The model in effect says that there must be substantial momentum transfer to C. Small-impact-parameter trajectories lead most effectively to reaction and the products are predominantly backward scattered.

Sequential Impulse. The sequential impulse model[26,27] (SIM) is an extension of both the SSM and IKM. There are two sequences possible for reaction. First, the model pictures an elastic collision between A and B followed by another elastic collision between B and C. After the second collision C scatters away and A and B combine to form the AB product. The difference with the elastic spectator version of the SSM is that here C interacts with B while in the other C interacts with AB. Thus, the internal energy of AB and the translational energy depend on the direction of scattering in the SIM, whereas in the SSM these quantities are determined solely by the A-B collision. C is no longer a simple spectator.

The second sequence has A colliding first with C, which then collides with B as it scatters off. C is knocked out, but there is some momentum transfer to B in the process.

This model attempts to add a bit of 'realism' to the previously described processes. The price for this is increased mathematical complexity. Mahan and co-workers[27] have worked out the relations for purely elastic collisions when A strikes B first. The result is a distribution of product velocities that peaks at the SSM value. This is modified somewhat at high collision energies when the predicted internal energy of AB exceeds its D_0 value.

Other Direct Models. Other models in which the reaction takes place *directly*, as opposed to ones that form *collision complexes*, have been proposed.[28] These are basically the moderation of the extremely simplified dynamics represented by the models presented so far by the introduction of realistic potentials. Of these we mention only direct interaction with product repulsion (DIPR), developed by Kuntz *et al*.[29,28] The central feature is a short-range force between the atoms of the reactant diatomic BC as they separate. This force gives rise to an impulse that changes the velocity of the departing atom C and therefore plays the controlling role in the angular distribution of the products. Because such forces are not very well known, the impulse can be treated as an empirical parameter to be determined by fitting the data.

Collision Complexes. There are several different versions of collision complex models[30-34] (CCM), depending on the assumptions made about the structure of the complex and the approximations that go into evaluating the energy partitioning among its degrees of freedom.[35]

The basic idea is that the reactants A and BC come together to form the collision complex (ABC), which then survives for long enough to 'forget' how it was formed, usually taken to be several rotational periods.[36] The dissociation of the complex to the products (or perhaps back to reactants) then just depends on its internal energy and angular momentum, which must be conserved, and on its structure. It is important to note that the complex must dissociate because its total energy, the sum of its kinetic and potential energies, is always positive. This is so since it is formed from two reactants initially so far away that the total energy is just the collision energy plus the reactant internal energy, a positive quantity.

Without going into the details of any specific collision complex model, the principal predictions of the CCM are that the angular distribution of products is symmetric with respect to 90° in the c.m. system and that the energy distribution at each angle is determined by the statistical partitioning of the available energy among the degrees of freedom of the complex, consistent with energy and angular-momentum conservation. (There may be additional dynamical, *i.e.* non-statistical, effects due to barriers in the exit valley.[37,38])

In ion/molecular beams experiments, the normal criterion for having observed that a reaction proceeds *via* long-lived complex formation is the symmetry of the measured angular distribution. This is probably reliable with complementary experimental or theoretical information about the system. One should remember that hard-sphere scattering is isotropic and hence also symmetric with respect to 90°.

Calculations on model surfaces have shown that complex formation is possible when the potential-energy surface has a 'well' in the region corresponding to an ABC 'molecule'.[39] Further, the collision energy must not be so great that the dissociation time becomes less than the rotational period.

Multiple Potential-energy Surfaces.- Until this point we have been discussing reactions as if they occurred on a single potential-energy surface. In many, perhaps even in most reactions, careful consideration of the electronic state of the reactants shows that it does not correlate with the ground state of the products.[40] Moreover, in ion/molecule reactions for a given set of products, there are at least two surfaces involved: A^{+} + B and A + B^{+}. In the most obvious case, the total product spin is different from the reactant spin. Such a reaction may proceed as a direct reaction if there is very strong spin-orbit coupling. With weaker spin-orbit coupling the reaction may still proceed *via* long-lived complex formation,[41] depending on the shapes of the surfaces involved.

The more common case is that the reactants and products have the same spin states and electronic-state symmetries, but the molecular orbitals of the products correlate with excited reactants, or *vice versa,* that is they occur on different *diabatic* surfaces. The probability for reaction depends on the accessibility of the reactants to the region where the diabatic surfaces cross each other and on the coupling between them.[42,11] If the coupling is strong enough, then a transition between surfaces has a reasonable probability of occurring, and so does the reaction. From another view, if the coupling is strong enough, the *adiabatic* surfaces will be far apart in the region of the 'avoided' crossing, and the reaction will occur on the single adiabatic surface since a transition between the adiabatic surfaces is unlikely. The obverse case where the reactants must 'hop' to another adiabatic surface is also possible with the proper coupling. This kind of process is important in high-energy collisions where the products appear in excited states.

Another possibility is that the surfaces of the reactants and products may be correlated by the symmetry of a particular configuration of the atoms in the region corresponding to an ABC 'molecule'. In other words, when (or if) the reactants attain a configuration of a certain symmetry, then the symmetry of the molecular orbitals allows for transitions as above, which can lead to product formation.[43]

If ABC is a molecule for which the electronic spectroscopy is known, one can predict or estimate the barriers to reaction due to the levels of the 'matching' surfaces. Even if this precise information is lacking, an analysis of this kind can reveal whether one ought to expect symmetry-imposed barriers to reaction.[40] Hence, it may be possible in high-resolution ion/molecular beams experiments on simple systems to obtain rough spectroscopic information about a molecule ABC that is not stable enough to be prepared.

3 Experimental Considerations

The apparatus required for ion/molecular beams experiments consists of four basic components: the ion source, the neutral-beam source, the detector (including energy or velocity analyser), and the vacuum chamber to house them. This is not the place for a detailed examination of the design of scattering experiments.[44] Each of the components is briefly described, along with some of the major considerations.

The Ion Source.- Positive ions are normally produced by electron bombardment.[45] In principle all manner of ions can be made from a suitable parent compound by this technique, and this is its great advantage. One needs just to select the ion desired in the experiment.

There are some drawbacks. The principal one is that to produce a particular ion one must employ an electron beam with substantially more energy than the threshold energy where the formation cross-section is very small. This leads to an uncertainty in the distribution of electronic or in the case of molecular ions also the vibrational energies of the reactant ions.[46]

Plasma- and discharge-ion sources have also been used for some systems with the advantage over electron bombardment that sometimes a much more intense beam of ions can be produced and with less electronic excitation.[47,48] It has the disadvantage that it is effective only for a limited number of compounds. A mass spectrometer is required as with electron bombardment.

Surface ionization[49-51] has the advantage over the above sources in that the ions are produced thermally and hence are in their ground electronic states. However, the kinds of ions that can be produced this way are severely limited to the most electropositive elements: the alkali metals and possibly a few others such as barium. For these, mass-selection is unnecessary because of the selectivity.

Other techniques, such as photoionization,[52,53] do not yield sufficient numbers of ions at present to carry out crossed-beam measurements. Because of the specificity, photoionization could be a powerful addition to the experimentalist's arsenal.

Negative ions have been produced in discharge sources or plasmas.[54,55] The intensities are generally much lower than in positive-ion beams, although one expects the ions all to be in the same state (except for degenerate spin states). A problem is that, for some molecules, the electron autodetaches, so causing the lifetime of the negative ion to be limited.

After the positive or negative ions are produced and mass-selected, they are accelerated or decelerated to the proper energy and carefully focused into a beam. The energy and angular spreads in the ion beam largely control the resolution of the experiment. It is not difficult to produce narrow ion beams with a spread in velocities of a few per cent at beam energies above a few eV's. One must be extremely careful to achieve this with ion beams below 1 eV, for which the ion optics must be skillfully constructed and properly handled.[43,60]

The Neutral-beam Source.- The neutral beam is normally made by allowing a vapour to enter the scattering region by effusive flow from a small orifice through a set of collimating slits. Under these conditions the velocity distribution of the molecules is Maxwellian, characterized by the temperature of the source. This technique has been used for decades.[56] Although such a beam has a proportionally broad velocity distribution, thermal velocities are usually so much smaller than the ion-beam velocity that the resolution of the experiment is not affected.

For experiments at low energy it is better to use a nozzle source for the neutral.[57] In this case the vapour, usually mixed with an inert carrier gas at high pressure (such as He at several hundred torr), is allowed to expand into the vacuum chamber through a very small orifice. The neutral beam formed this way has a peak velocity somewhat greater than Maxwellian at the same temperature, but more importantly it has a very much narrower velocity spread.[58,43]

The resolution of the experiment depends on the angular width of the neutral beam. This is because the reaction volume or scattering zone depends on the intersection of the two beams. A wide neutral beam means that collisions take place over an extended distance along the ion-beam path. Two products scattered to the same angle relative to the directions of the reactants would appear at different angles in the detector if the positions of their formative collisions were far apart.

In higher-energy experiments it may be possible to use a gas cell for the neutral without broadening the relative collision energy very much.[59] The angular resolution may not be compromised either, provided the cell is not too large. The advantage is that in a cell the partial pressure of the neutral can be made larger than in a beam. According to equation (1), the number of collisions is proportional to the gas density or pressure, and hence the scattered intensity should increase in direct proportion; however, the pressure must not become so great that a second collision becomes probable! In total reactive cross-section measurements this beam/cell arrangement is the typical one.

The Detector.- The detector unit for the product ions consists of an energy analyser, a mass analyser, and a particle detector connected to electronic equipment for pulse counting and data processing.

The energy analysis of the products can be carried out in several different ways. The stopping-potential technique[60,51] works by measuring the transmission of product ions at each scattering angle as a function of retarding voltage. As the voltage is increased, the number of ions measured will decrease as fewer will have sufficient translational energy to overcome the potential energy to reach the detector. The energy distribution at each angle is the slope of the transmission curve at each stopping energy, *i.e.* $P(E) = \mathrm{d}I(E)/\mathrm{d}E$.

Other techniques employ focusing lenses for ions of a particular energy to obtain $P(E)$. The most common of these are the spherical electrostatic focusing lenses[43,48,61] and the 127° cylindrical-sector field.[49]

The advantage of the first technique is that it is a simple, straight-through device that can be made practically as small as one has room for. Its disadvantages are that the energy distributions are not obtained directly but only after differentiation of the transmission curves. The advantages and disadvantages of the focusing methods just complement these.

Other techniques are possible, including the venerable time-of-flight[62] method. This does require a long flight path, which necessitates a larger apparatus. It should perhaps be pointed out that what is measured here is $P(t)$, the arrival-time distribution.

Mass analysis is normally accomplished by the usual magnetic or quadrupole mass spectrometer.

The actual ion detector is either a low dark-current electron multiplier, such as a Channeltron, or a scintillation type of detector,[63] which acts as a β^- detector. The pulses are finally counted by a scaler, stored in the computer or multi-channel analyser, and processed to give differential cross-sections.

The Vacuum System.- The major consideration for the vacuum system is that it should be just large and good enough to carry out the experiments. In principle, the larger the vacuum system the more flexibility one has. This view, however, must be tempered by the costs.

Consideration of equation (1) shows that scattering leads to a decrease in ion-beam intensity, which follows a Beer's law relationship. For ion flux, $F = N_A v$, and neutral density, N_B, equation (12) holds:

$$F = F_0 \exp(-x/\lambda) \tag{12}$$

where $\lambda = (\sigma N_B)^{-1}$ is the mean free path. Thus, the path lengths of the beams should be as short as possible. For background gas at 10^{-6} Torr and a total cross-section of 30 Å^2, the beam drops by 0.2% in 20 cm. As the pressure increases, this changes rapidly so that the pressure needs to be kept at most to this range in the scattering region. It may be necessary to pump separately the ion source and the neutral-beam nozzle source.

The main source of the pressure in the scattering chamber is the neutral beam. For effusive flow the mean free path of the gas should be less than the exit-slit dimensions and the flux of molecules per unit area per second is $\frac{1}{4} N_B \langle v_B \rangle$. If the orifice diameter is about 0.5 mm, the gas-oven pressure must be of the order of 0.1 Torr or less. From this one can estimate that the pumps must have a pumping speed of several thousand litres per second to keep the scattering chamber at about 10^{-6} Torr.

Product Intensities.- Since the gas density is the flux divided by the speed, the pressure of the neutral reactant just outside its oven exit slit is 1/4 the pressure in the neutral 'oven'. Now, the neutral gas molecules expand out from the slit so that the pressure decreases by a factor of approximately 10^2 after about 2 cm. Therefore the neutral pressure in the scattering zone should be of the order of 10^{-4} Torr. The collimating slits limit the extent of the scattering zone to a few millimetres on a side. Thus, one can estimate that only a few tenths of a per cent of the ions will make collisions with the neutral molecules and that, with a reactive cross-section of 10 $Å^2$, an ion flux of 10^{13} ions s^{-1} cm^{-2} would give about 10^9 reactive collisions s^{-1}. If the angular resolution of the detector is 10^{-5} steradian and the products are scattered isotropically, one could expect a signal of 10^3 ions s^{-1} to be counted. This naturally depends on the transmission of the detector svstem as indicated in equation (3). Energy analysis reduces this intensity still further. As one can see from these typical numbers, one cannot afford to lose many product ions along the way.

From the measured energy distributions, $P(E)$, at several angles one can calculate the relative differential cross-section, $\bar{I}_L(v'_L,\Theta,\Phi)$. The absolute differential cross-sections could be obtained from equation (3). However, it is easy to see that several of these quantities may be difficult to determine with any precision. One usually makes do with the relative cross-sections, which contain nearly all the useful information.

To get the c.m. differential cross-sections from the LAB data, one could use a computer extraction routine such as the one composed by Siska,[64] which is very valuable for rather precise data. Otherwise, one converts the relative $\bar{I}_L$ values by means of equation (7) to Cartesian, $\bar{I}_C$, values, which are the same in both LAB and c.m. systems.

4 Reactions Involving Hydrogen

It is natural to begin a review of ion/molecular beams experiments by considering reactions involving hydrogen. The 'simplicity' of these reactions has fostered the very close collaboration between theoreticians and experimentalists that approaches the scientific ideal. To chemical physicists, H is the archetypal atom. Hydrogenic orbitals are the models for all atomic orbitals, the molecular orbitals of H_2^+ are the prototypes for other molecular orbitals, and the potential-energy surface for the reaction of H with H_2 was for a long time the standard surface for chemical reactions.[65]

(Three-atom)$^+$ Systems.- *Noble-gas/Hydrogen Reactions.* Next to the reaction of H^+ with H^2, the 'simplest' ion/molecule reactions are the noble-gas/hydrogen reactions. For the noble gas X, there are two ways [reactions (13) and (14)] that XH^+ can be formed:

$$X^+ + H_2 \rightarrow XH^+ + H \quad (13)$$

$$X + H_2^+ \rightarrow XH^+ + H \quad (14)$$

The first is an H atom transfer and is isoelectronic with the reaction of a halogen atom with H_2. The second is a proton transfer without a direct analogy among neutral/neutral gas-phase reactions, but which may resemble some aspects of solution acid/base reactions. Examination of the correlation diagrams for this system[40] shows that it is the second set of reactants that correlates with the products. The first set correlates with the dissociative ($X + H + H^+$) state, and, in order to react, there must first be a surface hopping event (a charge transfer).

The most elementary of these reactions occurs when X = He. This reaction has been examined experimentally[66] and theoretically[67] in some detail. For the two reactions above, the first is exoergic and the second is endoergic. Despite this, the first reaction has a very small rate constant while the second goes readily if there is sufficient reactant energy.[40] The reason for this is that the two adiabatic surfaces are sufficiently far apart to make transitions unlikely. The crossed-beam experiments for the proton-transfer reaction showed that, although the reaction is endoergic, the SSM prediction is approximately achieved.[66] This has been attributed to the relatively large vibrational excitation of the H_2^+ produced in an electron-bombardment source.

Qualitatively, the theoretical considerations for proton transfer to neon and argon have certain similarities to those for helium. To test these, Farrar and co-workers have carried out a detailed investigation of reaction (14) with X = Ne[68] or Ar.[69,46] They employed a nozzle source for the noble-gas beam and produced the ions by electron bombardment with varying source pressures so as to control the vibrational excitation. Some of their results for the system H_2^+ + Ne, with a mean vibrational excitation of 0.89 eV, are shown in Figure 4. The scattering in all cases is primarily forward with respect to the initial direction of the Ne. This is not exactly spectator stripping since the velocity of the peak in the distribution is considerably greater than the SSM prediction and, one should note, this reaction is endoergic by 0.56 eV. Moreover, as the energy of collision increases, there is more backward scattering, *i.e.* HNe^+ is scattered in the initial H_2^+ direction. Normally, a spectator type of reaction becomes even more sharply forward peaked as the energy is increased.

Quenching the vibrational excitation of the reactant ion increases the backward scattering and decreases the cross-section. Their interpretation of this result is that the effectiveness of the translational motion in overcoming the reaction endoergicity is greatest for small-impact-parameter collisions, while vibrationally 'hot' H_2^+ ions can react at larger impact parameters. This tends to weight the cross-sections for vibrationally excited ions. One expects from this analysis that, if a purely ground-state reactant H_2^+ beam could be made, one would observe a substantial flux of backward-scattered products.

The fact that they observed much more translational energy in the products than predicted by the SSM suggested to the investigators that the potential energy at close distances becomes energy of compression in the H-H co-ordinate, which is released as repulsion between the hydrogens as they separate. Theoretical calculations for the Ne reactions[70] show the critical region of the surface controlling the reaction to be the 'hard wall' at small separations between atoms. However, it was found that the DIPR model did not give a particularly good fit to the angular distribution.

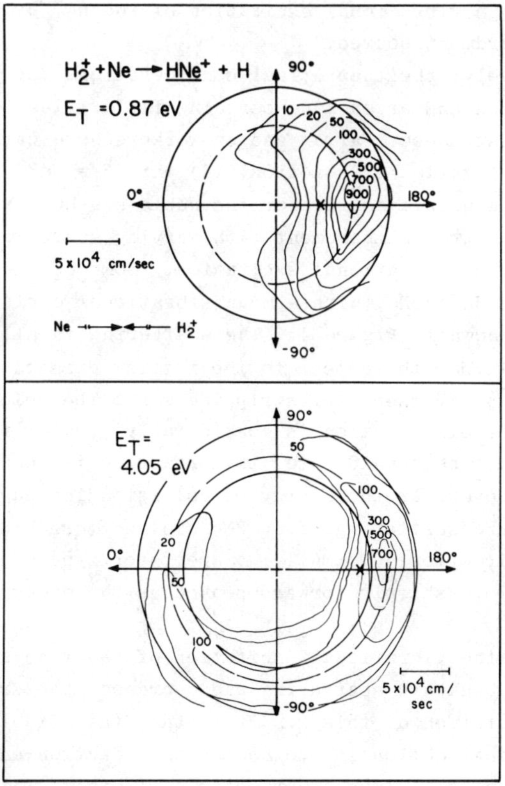

Figure 4 Contour maps of $\bar{I}_B$ for HNe^+ from experiments with collision energies of 0.87 and 4.05 eV. The $+$ in the centre of each diagram represents the centre-of-mass of the system. The *X* in each marks the SSM values. The dashed circle locates the experimental peak velocity; the outer solid circle shows the maximum velocity allowed by energy conservation. In the lower diagram the inner solid circle shows the minimum translational energy that permits stable NeH^+ to form. Note that the scattering must be symmetric about the relative velocity vector, or about the line from 0° to 180°

(Reproduced with permission from *J. Chem. Phys.*, 1981, 75, 1776)

The reaction with Ar differs from the Ne reaction in that it is exoergic, by 1.3 eV. The investigators observed that, over the range of collision energies from 0.45 to 7.75 eV, product HAr^+ ions are scattered primarily in the direction in which the Ar atoms are initially moving, that is forward scattered, as is the case in the Ne and He reactions. However, unlike Ne, at each energy the product intensity reaches a maximum near the SSM predicted velocity.

Particularly interesting in this reaction was that calculation of the internal energy of the product HAr^+ from energy conservation showed that it exceeded the dissociation energy, even for reaction at collision energies as low as 2 eV. Since excessive vibrational energy in a diatomic ion leads to very rapid dissociation, the conclusion was that this energy must be product rotation. For the highest-energy experiments this result implies astonishingly large rotational quantum numbers of about 60 $\hbar$.

In experiments to test the effect of vibrational excitation of the reactant, it was observed that, for the same total energy, increasing the fraction of translational energy of the reactants increased the translational energy of the products. In addition, at a fixed total energy they found very little change in the proton-transfer cross-section as the ratio of translational energy to vibrational energy was varied.

A comparison of the Ne and Ar reactions appears to show substantial differences. In the former, vibrational excitation of the H_2^+ seems to have a great effect on reaction probability and on the angular distribution of products, while in the latter there is almost no effect. In the former the product energy distributions peak considerably higher than the predictions of the SSM while in the latter the peaks are very near these predictions. In the latter, the angular distributions become more sharply peaked as the collision energy is increased but less sharply peaked in the former. Energy conservation requires that the products in the Ar reaction should be rotationally highly excited, but there is no evidence for similar excitation in the Ne reaction.

These disparities are probably more apparent than real. In both reactions, the surfaces favour products with large fractions of the total available energy partitioned into *kinetic* energy. In the endoergic Ne case, this translational energy is due largely to repulsion in a nearly collinear configuration as the products separate. This configuration is also more favourable for the utilization of vibrational energy to promote reaction. In the exoergic case, the H_2^+ bond is similarly compressed but in a bent configuration, which leads to rotational excitation. In this case, the shape of the configuration probably is not so critical for reaction, but rather it is more likely to occur in random collisions.

The results of the He experiments and calculations are in general agreement with this picture. The surfaces are thus probably similar in general form but differ merely in the energy-level spacings.

$N^+ + H_2$. The thermoneutral reaction (15) is one of the most thoroughly studied ion/molecule reactions.[71] The system is

$$N^+ + H_2 \rightarrow NH^+ + H \tag{15}$$

isoelectronic with CH_2, and there is known to be a very deep well associated with the NH_2^+ species. Thus, the early beam experiments which showed that the reaction followed the SSM predictions down to collision energies of 2 eV were somewhat surprising. One would have suspected with such a deep well that one would have observed collision complex behaviour.

Mahan and co-workers[48] reinvestigated the reaction with a carefully designed apparatus, particularly in the use of discharge sources for the N^+ beam. This was done to ensure the production of N^+ in the ground ^3P-term state. Electron bombardment produces significant amounts of metastable excited terms such as ^1D.

Figure 5 shows some of their measured intensity maps for this reaction. At low energies the distributions were found to be symmetric about the c.m. (within experimental error). At higher collision energies the scattering shifts to the forward direction, with the peak near the SSM value at the highest energies. These workers were also able to observe that, at low energies, the non-reactively scattered N^+-ion distributions showed a backward peak. Taken together, this is very suggestive that the reaction does take place *via* complex formation. The backward N^+ peak is evidently the result of a non-reactive dissociation of the complex. While such non-reactive scattering has been observed in neutral collision complexes, this is the first report of it from ion/molecule complexes (probably because of the experimental difficulty in detecting it).

The investigators' explanation of the observed behaviour is contained in a rather elegant correlation diagram for the many relevant electronic states of this system. The ground-state reactants $N^+(^3P) + H_2$ correlate with both the ground state of NH_2^+, the $^3B_1$ state in C_{2v} symmetry, and a higher state, the $^3A_2$ state.

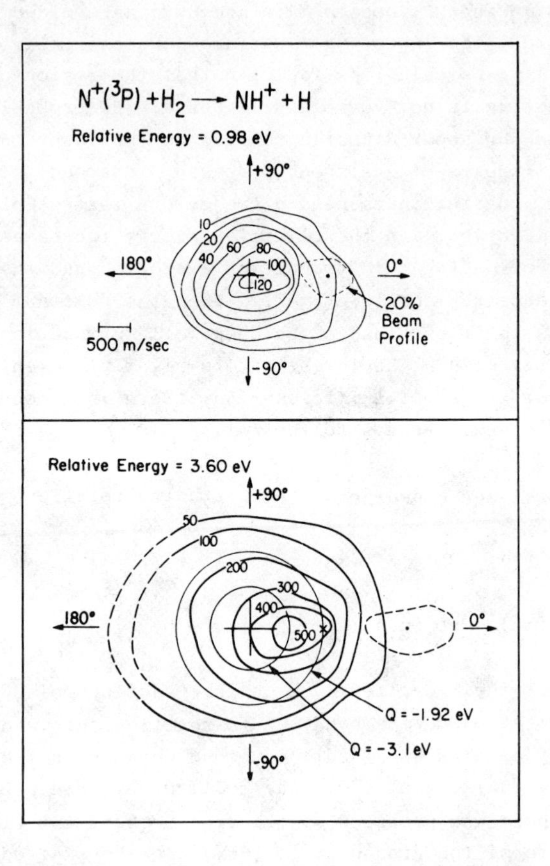

Figure 5 Contour maps of $\bar{I}_C$ for the NH^+ product at relative collision energies of 0.98 and 3.60 eV. The + in the centre represents the position of the centre-of-mass of the system. In the lower diagram the *X* and the circle going through it locate the spectator predicted velocity; the inner circle is drawn for the actual peak velocity

(Reproduced with permission from *J. Chem. Phys.*, 1980, 73, 3750)

To reach the latter, the surface has a shallow depression but, to reach the former, there is a barrier of nearly 3 eV. Thus, at low collision energy with the H-H bond perpendicular to the approaching N^+, the reactants cannot get into the deep well but can only gain access to the shallow excited NH_2^+ well. When the svmmetry of the collision is not quite C_{2v} but rather C_s, the ground-state reactants have an adiabatic path to the deep well since both $^3A_2$ and $^3B_1$ become $^3A''$ and have an avoided crossing. In other words, the lowest $^3A''$ surface connects the reactants to the deep well. Calculations indicate that the region of this avoided crossing is not very large. Hence, the probability of hopping from the upper diabatic surface to the lower one is increased if the reactants spend a reasonable amount of time in the region, *i.e.* the reactants need a trajectory that samples a good part of this region. As the collision energy increases, the probability of making this transition decreases and the reactants just experience the shallower well. When the reactant N^+ is in the ^1D state, a transition to the deep-well surface is unlikely because of the spin. The investigators rightly comment that to have CCM behaviour it is not sufficient that there is a deep well; the reactants must have access to the well.

$B^+ + D_2$. Koski and co-workers[61] have studied reactions (16) and (17)

$$B^+(^1S) + D(^1\Sigma^+) \rightarrow BD^+(^2\Sigma^+) + D(^2S) \qquad (16)$$

$$B^+(^1S) + D(^1\Sigma^+) \rightarrow BD^+(^2\Pi) + D(^2S) \qquad (17)$$

to complement their earlier work on the reactions of $B^+(^3P)$.[72] BH^+ is isoelectronic with BeH_2. Both reactions are endoergic, the first by about 2.5 eV and the second by about 5.8 eV. The experiment actually employed a gas cell rather than a D_2 beam. The B^+ ion beam was produced by electron bombardment of BF_3 and had a mixture of the ground state (65%) and the excited ^3P state (35%). The scattering from each of these states could be resolved by their energetics.

Their results show that the excited BD^+ scattering intensity [reaction (17)] is symmetric about 90° in the c.m. system at energies just above threshold and becomes more forward scattered as the collision energy is increased. This behaviour is suggestive of collision complex formation. For the ground-state product [reaction (18)] the results are similar, but it was not possible to measure angular distributions as close to the threshold of reaction as was done for reaction (17). The results suggest that there is no barrier to either reaction beyond the endoergicity.

As in the case of the $N^+ + H_2$ reaction discussed above, spin and symmetry analysis of the connecting potential-energy surfaces of reactions (17) and (18) is made complicated by the number of possibilities. The ground-state reactants do not lie on the same diabatic surface with the lowest state of the ($B^+ + D_2$) intermediate. The lowest state of the intermediate species appears to correlate with $B^+(^1P)$. Hence, the investigators believe that the reaction takes place *via* a transition at the intersection of the diabatic surfaces, as with $N^+ + H_2$. In C_s symmetry, both diabatic surfaces are $^1A'$ and in linear configuration both are $^1\Sigma$. To form the excited product, the investigators concluded that the reactants must intersystem-cross to a triplet surface. Until trajectory calculations are carried out for this system, it is not clear what the probabilities for this supposition are.

(Four-atom)$^+$ System.- The simplest four-atom reaction is between H_2^+ and H_2. Because H_4^+ is not a stable species like H_3^+, the result is either a proton transfer or atom transfer to form H_3^+. Hierl and Herman[73] have examined the isotopic variation reaction (18).

$$H_2^+ + D_2 \rightarrow D_2H^+ + H \tag{18}$$

Since the other possible product, H_2D^+, has the same integer mass as the charge-exchange product D_2^+, the two could not be distinguished.

The ions for this experiment were produced by electron bombardment and have a mean vibrational excitation of about 0.8 eV. The product D_2H^+ ions were found to peak in the forward direction with respect to the initial D_2 beam and with a velocity just slightly greater than the SSM value. These distributions are rather broad and energy conservation requires that there should be a fair amount of internal excitation in the products.

For the lowest-energy experiment there appeared to be substantial backward scattering as well. As calculations do not show a deep well for H_4^+ and as the apparatus resolution was not very good in this region of velocity space, the investigators discounted the possibility of complex formation.

The total cross-section with state-selected reactants experiments of Lee and co-workers[52,74] showed that the D_2H^+ formed from either $D_2^+ + H_2$ or $H_2^+ + D_2$ were so similar that they appeared to originate from the same process. So, if complex formation is not responsible for this behaviour, the alternative is that the reaction takes place after multiple charge transfers, which 'scramble' the initial charge 'location'. Then D_2H^+ is formed by proton transfer from H_2^+. This sequence of events is a likely one because the entrance channels $H_2^+ + D_2$ and $D_2^+ + H_2$ are virtually degenerate.

(Five-atom)$^+$ Systems.- $H_2^+ + H_2O$. The proton-transfer reaction (19)

$$H_2^+ + H_2O \rightarrow H_3O^+ + H \quad (19)$$

might be taken as a model for proton transfer in solution. It is very exoergic by about 4.6 eV. Bilotta and Farrar[75] have investigated both it and isotopic variations with D_2^+ and with D_2O. They also examined the charge-transfer process which is competitive with reaction (19).

In these experiments, the hydrogen ions were formed by electron bombardment and have a mean internal energy of nearly 0.9 eV. Over the range of collision energies from 0.7 to 6.9 eV the results showed that proton transfer is a direct process, the product H_3O^+ being scattered forward with respect to the H_2O direction, but with a peak velocity somewhat greater than the SSM value.

Experiments with D_2^+ replacing H_2^+, or with D_2O replacing H_2O, yielded some DOH^+ product in addition to the proton- or deuteron-transfer product from reaction (19). This means that some of the H_2O^+, formed apparently by charge transfer, is really the result of a proton (deuteron) transfer followed by a unimolecular decomposition of H_3O^+. But this is not all. The investigators found that the angular distributions of the DOH^+ appear to be a superposition of a symmetric distribution, which is expected for a unimolecular decay, plus another component which, in the higher-energy experiments, does not have forward-backward symmetry. This result seems to imply that there is yet another channel for scrambling hydrogens, which the investigators postulate to be the four-centre exchange (20), where one H attaches to the O as

$$H_2^+ + D_2O \rightarrow HD + DOH^+ \quad (20)$$

the other H forms a bond with one of the D atoms. All four of these atoms would lie in the same plane. Thus, reaction (19) really represents only one channel of several that are possible in this 'simple' system.

H^+ + D. Gentry and co-workers[76] have examined reaction (21) as

$$H_3^+ + D_2 \rightarrow HD_2^+ + H_2 \tag{21}$$

part of a series of studies of hydrogen reactions using the merging-beams technique.[77] This technique is similar to crossed-beams experiments, but it is not suited to measuring the angular distribution of products. It is extremely useful in determining the product-energy distribution and the total cross-sections, even down to the thermal-energy range, which is essentially impossible for crossed ion/neutral beams.

Analysis of the data shows that the reaction dynamics are consistent with the SSM. Further corroboration of this deduction is that, if a collision complex were formed at low energy, one would expect to observe a statistical mixture of H_3^+, H_2D^+, and HD_2^+ due to scrambling, but such is not the case. Calculations on this system show that a weakly bound complex may be stable. Vibrational excitation in the reactant D_2 (produced by electron bombardment to form D_2^+ and then neutralized) again appears to prohibit the reactants from sampling the lowest part of the potential-energy surface.

Comments.- Because these reactions serve as models for more complicated systems, it is worth noting a few interesting observations. Firstly, of course, is the importance of the accessibility of the diabatic surface crossings in a large number of reactions. This hardly needs stressing.

Secondly, charge-transfer, proton-transfer, and hydrogen-atom-transfer reactions are very widespread and important in gas- and condensed-phase chemistry. They are by no means 'well understood' yet, although much progress has been made.

Thirdly, it is clear in many of these experiments that the vibrational excitation of the reactants has been the dominant factor in determining the character of the reaction. It may be that the SSM works well for so many ion/molecule reactions because vibrational excitation of a reactant is a very common experimental condition. This aspect needs to be investigated theoretically and experimentally.

Finally, a rough guide to the reaction dynamics of a particular reaction can be obtained by looking at the results for isoelectronic systems, keeping in mind their obvious differences. This is hardly a new idea in chemistry. For example, the reaction of $H_2^+ + D_2$ is fairly similar to the reaction of $H_2^+ + He$, considering the different energetics. Also, the $H_2^+ + H_2O$ reaction is similar to the $H_2^+ + Ne$ reaction as far as proton transfer goes. It is just that the complexity of H_2O compared to Ne allows for other reaction channels. One might use this analogy to guess intelligently the most important reaction channel.

5 Other Positive-ion Reactions

With increasing numbers of atoms and electrons in the reactants, it becomes much more difficult to analyse the correlation of reactant and product surfaces. The interpretation of the scattering then rests much more on comparisons with models and with hydrogen systems that behave similarly.

Smaller Systems.- $Ar^+ + H_2O$. Herman and co-workers[78] have examined reaction (22) and the charge-transfer reaction that

$$Ar^+ + H_2O \rightarrow ArH^+ + OH \qquad (22)$$

competes with it. Reaction (22) is estimated to be exoergic by about 0.8 eV, and the electron-recombination energies of the two spin-orbit states of Ar^+ are more than 3 eV greater than that of the ground state of H_2O^+.

Their results very simply show that the ArH^+ ion is scattered primarily in the forward direction with somewhat broader distributions at lower energies than at higher energies. The peaks in the distribution are found to be near but slightly greater than the SSM values. On the whole it is typical 'stripping' behaviour. The charge-transfer experiments showed very much the same result. The H_2O^+ distribution peaked in the forward direction for H_2O motion very near to the resonant charge-transfer velocity (*i.e.* with no conversion of translational to internal energy). There is also a much smaller backward peak, which again comes near the resonant charge-transfer velocity.

These workers' interpretation of the reactive scattering is very simple. The reaction is considered to be direct and a hydrogen atom is transferred from water to Ar^+ as the reactants approach.

However, in line with earlier comments, one notes that this

reaction is 'like' that for Ar^+ + Ne, except that here the ionization energy for water is much less than that for Ar while that of Ne is greater. One suspects that charge transfer might not be just a competitive process but, rather, it might dominate the reaction dynamics. That is, one wonders whether the reaction could really be a two-step process in which a charge exchange to the $^2A_1$ electronic state of water (with vibrational excitation) is followed by proton transfer. One might note also that the reaction is analogous to that for the system Cl + H_2O, so that H atom transfer is certainly possible. It is worth calling attention to other possible reaction modes.

CO^+ + NO. Hirst and co-workers have looked at reactions (23)[79] and 24.[80] The first of these reactions is exoergic but the second is endoergic by about 2.7 eV.

$$CO^+ + NO \rightarrow CO_2^+ + N \qquad (23)$$

$$CO^+ + NO \rightarrow NCO^+ + O \qquad (24)$$

Their experimental results yielded angular distributions for CO_2^+ that were forward peaked over the whole range of collision energies from 2.5 to 13 eV. At low energies some significant backward-scattered product was observed as well. At high collision energies there appears to be some conversion of CO^+ to $O^+(^4S)$ and $CO(^1\Sigma^+)$. For the second reaction, the products were mostly forward scattered, but at higher energies the amount of backward scattering increased.

The interpretation made by the investigators is that these reactions are direct and in a sense complementary. One normally associates forward scattering with large-impact-parameter collisions, such as is envisioned in stripping processes, and backward scattering with small-impact-parameter collisions, as in knock-out processes. Hence, in this system at low energy only the exoergic reaction can occur for both large- and small-impact-parameter collisions. As the energy is increased, the endoergic process can also occur and, perhaps, dominate for small-impact-parameter collisions. A similar result for a different process was mentioned earlier in the H_2^+ + Ne reaction. The authors ascribe forward scattering in the endoergic reaction to vibrational excitation in the CO^+ beam (produced by electron impact). This kind of effect has been previously commented on. The investigators also suggest that their data are consistent with the SIM for the exoergic reaction.

$Xe^+ + CH_4$. Similar dynamics have been reported by Hierl and co-workers for the competing reactions (25) and (26).[81,82]

$$Xe^+ + CH_4 \rightarrow XeH^+ + CH_3 \quad (25)$$

$$Xe^+ + CH_4 \rightarrow XeCH_3^+ + H \quad (26)$$

Earlier, these workers reported similar reactions with Kr^+.[83] Their apparatus uses electron bombardment to produce the Xe^+ ions but employs a gas cell for the methane rather than a beam. There are two spin-orbit states of Xe^+, which differ by 1.61 eV.

The results for the angular and energy distributions of the XeH^+ ions over the range 0.43 - 7.31 eV show a gradual shift from a contour map that has substantial forward and backward scattering to one that is nearly spectator stripping. In contrast, the $XeCH_3^+$ ion was observed to be predominantly backward scattered. The investigators interpreted this to imply, as in the case above with CO^+ + NO, that the first reaction takes place at large impact parameters while the second reaction takes place at small impact parameters. Thus, the branching ratio for these reactants is not determined by the statistical weights of the products, a common assumption when the dynamics are unknown. It is determined, rather, only by the trajectory impact parameter.

Larger Systems.- In the Reporter's laboratory, endoergic reactions (27) of Cs^+ with various aromatic molecules (PhX; X = F, Cl, Br, I, or NO_2) have been investigated.[25] The scattered phenyl-

$$Cs^+ + C_6H_5X \rightarrow CsX + C_6H_5^+ \quad (27)$$

cation distribution for reaction (27) with X = Br is shown in Figure 6. The other reactive systems have virtually the same phenyl-cation distributions. The peak in the distribution is distinctly backward with respect to the initial bromobenzene direction. In fact, very little forward scattering was observed in any of these reactions. Maximum recoil and appearance energies were found to be consistent with the IKM.

The interpretation of these results was in terms of two potential-energy surfaces, one a 'covalent' surface that describes

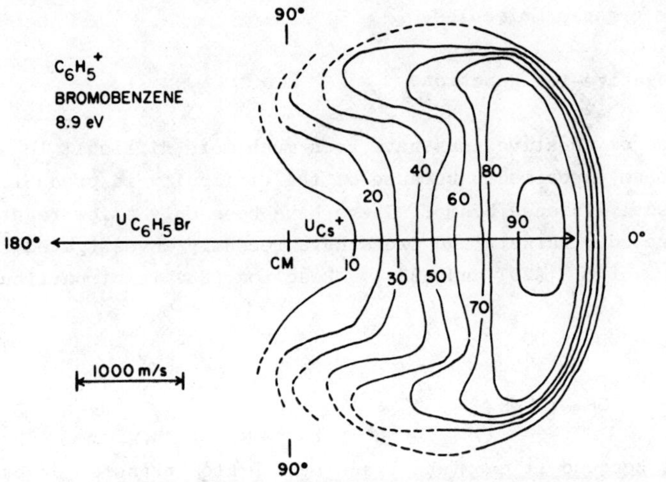

Figure 6 Contour map of $\bar{I}_C$ for the phenyl cation from the reaction at 8.9 eV collision energy. For the phenyl cation the forward direction is toward 180° because that is the initial c.m. direction of the bromobenzene

(Reproduced with permission from *J. Am. Chem. Soc.*, 1981, 103, 6333)

the reaction $Cs^+ + C_6H_5X \rightarrow CsX^+ + C_6H_5\cdot$ and the other an 'ionic' surface for the reaction $Cs^+ + (C_6H_5)^+X^- \rightarrow CsX + C_6H_5{}^+$. The first surface does not lead to stable products for most of these systems (CsI^+ is barely stable with respect to dissociation). The second would correspond to highly excited reactants. One can estimate roughly the 'crossing point' from thermodynamic data, dipole moments, and other properties of the aromatic molecules. For these phenyl halides this point comes at very small separations. Hence, the 'prediction' is that this region is very repulsive and the products should be strongly backward scattered. A similar treatment of the benzyl halides or of t-butyl chloride yields a crossing point much farther out, and the respective products of these systems, though also endoergic, were found to be primarily forward scattered.[51,84] Earlier experiments suggest that access to the crossing may be influenced by ring substituents.[85] This treatment seems to correlate also with the behaviour of such compounds in solution in that the phenyl compounds rarely form ionic intermediates whereas the benzyl and t-butyl ones readily do so. Cs^+ ions play the same electrostatic role in the gas phase that solvent molecules do in solution to stabilizing ionic forms of these organic molecules.

6 Negative-ion Reactions

Reactions of negative ions have been much more difficult to study in ion-beam experiments because of the difficulty in producing sufficiently intense beams. There have been only a few reactions investigated. Karnett and Cross have recently investigated two reactions of O^- [(28) and (29)]. Reaction (28) is of particular

$$O^- + D_2O \rightarrow OD^- + DO \tag{28}$$

$$O^- + C_3H_4 \rightarrow OH^- + C_3H_3 \tag{29}$$

interest because it can take place by a D atom transfer or by a D^+ ion transfer.[54] It is an endoergic reaction by about 0.36 eV and the experiments covered the range of collision energies from 3.2 to 10.5 eV.

It was found that the scattering distribution was predominantly forward, with the peaks of the distributions near the SSM limit for an atom transfer rather than a D^+ ion transfer. These processes are separable because the D_2O was moving with only thermal energies, much slower than the O^- ions. Since the product OD^- ion is found moving rapidly, the implication is that the atom rather than the ion has been picked up. The investigators point out that this is not completely certain as it is possible that the process could be a D^+ ion transfer followed by charge transfer. Since this system is isoelectronic with the F + H_2O system, a hydrogen-atom transfer seems more reasonable. (See also the comment for Ar^+ + H_2O for which charge transfer is not possible.) One cannot exclude the two-step mechanism without some theoretical calculations, although isotopic labelling of the reactant O^- ions, an expensive proposition, would help unravel this puzzle.

A somewhat similar situation exists for reaction (29).[86] Again, the products were observed to be scattered near the SSM values over the entire range of collision energies used, 4.6 to 10.8 eV. No carbanions were observed so that proton stripping, which yields OH + CH^-, did not appear to take place. Since this latter reaction has been found at thermal energies (*i.e.* $C_3H_3^-$ ions were observed), it is possible that charge transfer is again involved.

7 Conclusions

Ion/molecular beams experiments are being carried out to examine the dynamics of reactions at different levels of complexity. The experiments with hydrogen or hydrogen ions are the most detailed and best suited for comparison with theoretical calculations. The knowledge so obtained can be applied to the problems of the dynamics of reactions of larger systems, which in turn act as a bridge to the understanding of more conventional chemical kinetics.

The general conclusions that one can draw from these experiments are that we are learning how to predict which potential-energy surfaces are likely to play a role in the interaction between reactants and what factors control the relative importance of each role. Excitation energy in the reactants is one such factor and the effect of substituents is another for large molecules. The models for idealized reaction dynamics that have been developed give remarkably good descriptions of the scattered product distributions, probably better than is deserved.

More detailed work needs to be carried out in scattering experiments, particularly with state-selected reactants. This probably means that laser-photoionization techniques will have to be developed for beam sources. Experiments with state-analysed products should also be carried out, but as this is very difficult the complementary experiments measuring total cross-sections for the formation of specific product states will probably have to suffice. At the other end of the spectrum of molecular complexity, experiments with a greater variety of large molecules need to be carried out, including studies of substituent effects.

More sophisticated models for predicting the influence of various factors on the dynamics are necessary if one is going to be able to direct the course of a reaction. To be useful and effective these models must be readily applicable from information readily available.

References

1 'Kinetics of Ion-Molecule Reactions', ed. P. Ausloos, Plenum Press, New York, 1979.
2 'Gas Phase Ion Chemistry', ed. M.T. Bowers, Academic Press, New York, 1979, Vols. 1 and 2.
3 T.O. Tiernan and C. Lifshitz in 'The Excited State in Chemical Physics', ed. J.W. McGowen, John Wiley and Sons Inc., New York, 1981.
4 'Dynamics of Molecular Collisions', ed. W.H. Miller, Plenum Press, New York, 1976.
5 B.H. Mahan in 'International Review of Science, Phys. Chem. Ser. II, Vol. 9', ed. D.R. Herschbach, Butterworths, London, 1976.
6 'Atom-Molecule Collision Theory', ed. R.B. Bernstein, Plenum Press, New York, 1979.
7 R.B. Bernstein, 'Chemical Dynamics via Molecular Beam and Laser Techniques', Oxford University Press, New York, 1982.
8 'Potential Energy Surfaces and Dynamics Calculations', ed. D.G. Truhlar, Plenum Press, New York, 1981.
9 D.R. Herschbach, *Vortex*, 1961, 22.
10 R. Wolfgang and R.J. Cross, *J. Phys. Chem.*, 1969, 73, 743.
11 M.S. Child, 'Molecular Collision Theory', Academic Press, New York, 1974.
12 M. Shapiro, R.B. Gerber, U. Buck, and J. Schleusener, *J. Chem. Phys.*, 1977, 67, 3570.
13 D.G. Truhlar and J.T. Muckerman in ref. 8, Ch. 16.
14 W.H. Miller and T.F. George, *J. Chem. Phys.*, 1972, 56, 5637.
15 W.H. Miller, *Adv. Chem. Phys.*, 1974, 25, 69.
16 E.J. Heller, *J. Chem. Phys.*, 1975, 62, 1544.
17 E.J. Heller, *Chem. Phys. Lett.*, 1975, 34, 321.
18 J.C. Polanyi and J.L. Schreiber in 'Physical Chemistry', ed. Wilhelm Jost, Academic Press, New York, 1974, Vol. 6A.
19 K. Lacmann and A. Henglein, *Ber. Bunsenges. Phys. Chem.*, 1965, 69, 286.
20 D.R. Herschbach, *Adv. Chem. Phys.*, 1966, 10, 319.
21 D.R. Herschbach, *Appl. Opt. Suppl.*, 1965, 2, 128.
22 R.D. Levine and R.B. Bernstein, 'Molecular Reaction Dynamics', Oxford University Press, New York, 1974, p. 53.
23 R.J. Cross and R. Wolfgang, *J. Chem. Phys.*, 1961, 35, 2002.
24 G.D. Miller and S.A. Safron, *J. Chem. Phys.*, 1976, 64, 5065.
25 S.A. Safron, G.A. King, and R.C. Horvat, *J. Am. Chem. Soc.*, 1981, 103, 6333.
26 T.F. George and R.J. Suplinskas, *J. Chem. Phys.*, 1971, 54, 1037.
27 B.H. Mahan, W.E. Ruska, and J.S. Winn, *J. Chem. Phys.*, 1976, 65, 3888.
28 D.G. Truhlar and D.A. Dixon in ref. 6, Ch. 18.
29 P.J. Kuntz, M.H. Mok, and J.C. Polanyi, *J. Chem. Phys.*, 1969, 50, 4623.
30 P. Pechukas, J.C. Light, and C. Rankin, *J. Chem. Phys.*, 1966, 44, 794.
31 W.B. Miller, S.A. Safron, and D.R. Herschbach, *Faraday Discuss. Chem. Soc.*, 1967, 44, 108; S.A. Safron, N.D. Weinstein, D.R. Herschbach, and J.C. Tully, *Chem. Phys. Lett.*, 1972, 12, 564.
32 C.E. Klots, *J. Chem. Phys.*, 1976, 64, 4269.
33 L. Holmlid and K. Rynefors, *Chem. Phys.*, 1977, 19, 261.
34 W.J. Chesnavich and M.T. Bowers, *J. Chem. Phys.*, 1977, 66, 2306.
35 J.C. Light in ref. 6, Ch. 19.

36 G.A. Fisk, J.D. McDonald, and D.R. Herschbach, *Faraday Discuss. Chem. Soc.*, 1967, 44, 228.
37 J.D. McDonald and R.A. Marcus, *J. Chem. Phys.*, 1976, 65, 2180.
38 G. Worry and R.A. Marcus, *J. Chem. Phys.*, 1977, 67, 1636.
39 P. Brumer and M. Karplus, *Faraday Discuss. Chem. Soc.*, 1973, 55, 80.
40 B.H. Mahan, *Acc. Chem. Res.*, 1975, 8, 55.
41 J.C. Tully, *J. Chem. Phys.*, 1974, 61, 61.
42 J.C. Tully and R.K. Preston, *J. Chem. Phys.*, 1971, 55, 562.
43 For example, see R.M. Bilotta, F.N. Preuninger, and J.M. Farrar, *J. Chem. Phys.*, 1980, 72, 1583.
44 W. Gentry in ref. 2, Ch. 15.
45 For example, see H. Udseth, C.F. Giese, and W.R. Gentry, *Phys. Rev. A*, 1973, 8, 2483.
46 For example, see R.M. Bilotta and J.M. Farrar, *J. Chem. Phys.*, 1981, 74, 1699.
47 M. Menzinger and I. Wahlin, *Rev. Sci. Instrum.*, 1969, 40, 102.
48 S.G. Hansen, J.M. Farrar, and B.H. Mahan, *J. Chem. Phys.*, 1980, 73, 3750.
49 J. Schoettler and J.P. Toennies, *Z. Phys.*, 1968, 214, 472.
50 W.R. Gentry, Y.T. Lee, and B.H. Mahan, *J. Chem. Phys.*, 1968, 49, 1758.
51 S.A. Safron, G.D. Miller, F.A. Rideout, and R.C. Horvat, *J. Chem. Phys.*, 1976, 64, 5051.
52 S.L. Anderson, F.A. Houle, D. Gerlich, and Y.T. Lee, *J. Chem. Phys.*, 1981, 75, 2153.
53 For metal ions, see R.C. Burnier, G.D. Byrd, and B.S. Freiser, *J. Am. Chem. Soc.*, 1981, 103, 4360.
54 M.P. Karnett and R.J. Cross, *Chem. Phys. Lett.*, 1981, 82, 277.
55 M.S. Huq, D.S. Fraedrich, L.D. Doverspike, and R.L. Champion, *J. Chem. Phys.*, 1982, 76, 4952.
56 For example, see R.C. Miller and P. Kusch, *J. Chem. Phys.*, 1956, 25, 860.
57 For example, see R.E. Smalley, D.H. Levy, and L. Wharton, *J. Chem. Phys.*, 1976, 64, 3266.
58 For example, see K.L. Saenger, *J. Chem. Phys.*, 1981, 75, 2467.
59 For example, see J.R. Wyatt, L.W. Strattan, S.C. Snyder, and P.M. Hierl, *J. Chem. Phys.*, 1975, 62, 2555.
60 Z. Herman, J.D. Kerstetter, T.L. Rose, and R. Wolfgang, *Rev. Sci. Instrum.*, 1969, 40, 538.
61 N.A. Sondergaard, I. Sauers, J. Kaufman, and W.S. Koski, *J. Phys. Chem.*, 1982, 86, 533.
62 For example, see H.E. van den Bergh, M. Faubel, and J.P. Toennies, *Faraday Discuss. Chem. Soc.*, 1973, 55, 203; for a cross-correlation time-of-flight method, see V.L. Hirschy, M.S. Thesis, The Florida State University, 1969.
63 Y.T. Lee, J.D. McDonald, P.R. Le Breton, and D.R. Herschbach, *Rev. Sci. Instrum.*, 1969, 40, 1402.
64 P.E. Siska, *J. Chem. Phys.*, 1973, 59, 6052.
65 Henry Eyring, John Walter, and George E. Kimball, 'Quantum Chemistry', John Wiley and Sons Inc., New York, 1944, p. 301.
66 V. Pácak, U. Havemann, Z. Herman, F. Schneider, and L. Zuelicke, *Chem. Phys. Lett.*, 1977, 49, 273.
67 N. Sathyamurthy, J.W. Duff, C. Stroud, and L.M. Raff, *J. Chem. Phys.*, 1977, 67, 3563.
68 R.M. Bilotta and J.M. Farrar, *J. Chem. Phys.*, 1981, 75, 1776.
69 R.M. Bilotta, F.N. Preuninger, and J.M. Farrar, *J. Chem. Phys.*, 1980, 73, 1637.
70 C. Stroud and L.M. Raff, *Chem. Phys.*, 1980, 46, 313.
71 See J.A. Fair and B.H. Mahan, *J. Chem. Phys.*, 1975, 62, 515, and the references cited in ref. 48.

72 N.A. Sondergaard, I. Sauers, A.C. Jones, J.J. Kaufman, and W.S. Koski, *J. Chem. Phys.*, 1979, 71, 2229.
73 P.M. Hierl and Z. Herman, *Chem. Phys.*, 1980, 50, 249.
74 F.A. Houle, S.L. Anderson, D. Gerlich, T. Turner, and Y.T. Lee, *J. Chem. Phys.*, 1982, 77, 748.
75 R.M. Bilotta and J.M. Farrar, *J. Phys. Chem.*, 1981, 85, 1515.
76 C.H. Douglass, G. Ringer, and W.R. Gentry, *J. Chem. Phys.*, 1982, 76, 2423.
77 W.R. Gentry and G. Ringer, *J. Chem. Phys.*, 1977, 67, 5077; C.H. Douglass, D.J. McClure, and W.R. Gentry, *J. Chem. Phys.*, 1977, 67, 4931.
78 J. Glosík, B. Friedrich, and Z. Herman, *Chem. Phys.*, 1981, 60, 369.
79 M.F. Jarrold and D.M. Hirst, *Mol. Phys.*, 1980, 40, 1197.
80 M.F. Jarrold and D.M. Hirst, *Mol. Phys.*, 1981, 42, 97.
81 G.D. Miller, L.W. Strattan, C.L. Cole, and P.M. Hierl, *J. Chem. Phys.*, 1981, 74, 5082.
82 G.D. Miller, L.W. Strattan, and P.M. Hierl, *J. Chem. Phys.*, 1981, 74, 5093.
83 J.R. Wyatt, L.W. Strattan, and P.M. Hierl, *J. Chem. Phys.*, 1975, 63, 5044.
84 R.C. Horvat and S.A. Safron, *Chem. Phys. Lett.*, 1980, 69, 177.
85 S.A. Safron and R.C. Horvat in 'State-to-State Chemistry', ACS Symp. Ser. 56, ed. P.R. Brooks and E.F. Hays, American Chemical Society, Washington, 1977.
86 M.P. Karnett and R.J. Cross, *Chem. Phys. Lett.*, 1981, 84, 501.

4

Stuctures and Reactions of Gas-phase Organic Ions

BY I. HOWE

1 Introduction

Although the most significant recent advances in mass spectrometry have been instrumental (with associated biological applications), there continues to be a firm undercurrent of fundamental research that probes the behaviour of ions in the mass spectometer. In this basic research, principles of physical organic chemistry are applied to elucidate the structures and energy states of ions (i) at their instant of formation and (ii) at different periods of passage along their flight path in the mass spectrometer. The Reporter is fortunate in having a combination of various instrumental techniques potentially at his disposal. For example, several different methods exist for each of the following: (i) ionization, (ii) ion separation, (iii) the addition of energy to ions after formation, and (iv) ion detection. In addition, there are many variations in computerized data systems that can be used for mass-spectrometer control and data analysis. Appropriate selection from this battery of techniques, augmented for example by isotopic labelling of the neutral precursor molecules, enables structures and reactions of gas-phase organic ions to be investigated in detail. This large variety of approaches is assembled diagramatically in the Figure, and relevant general introductory texts are listed in refs. 1 - 6. For alternative review articles, see the appropriate sections of refs. 7 and 8.

The aim of this chapter is to review publications in the recent literature that have employed the methods represented in the Figure to investigate the structures, rearrangements, and decompositions of organic ions produced in a mass spectrometer. Because of the wide scope of the review, considerable selection has

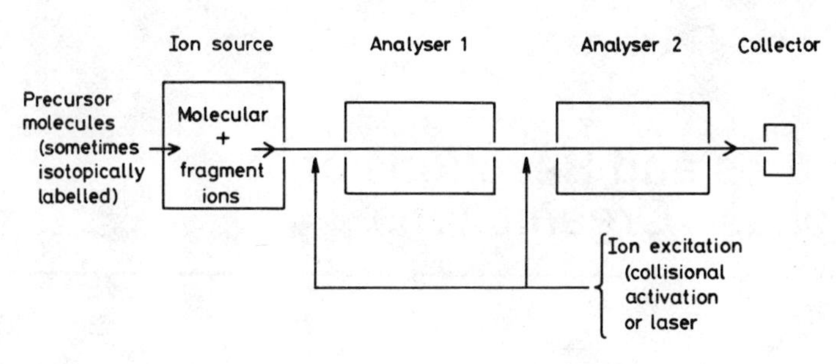

Ionization techniques

Electron impact
 ($\rightarrow M^+$, sometimes M^-)
Photoionization
 ($\rightarrow M^+$)
Chemical ionization
 ($\rightarrow MH^+$, $(M-H)^-$,
 condensed ions, ion/
 molecule reactions,
 equilibrium studies)
Field ionization/desorption
 (field-ionization kinetics)
Charge exchange
 (fixed-energy transfer)
Fast atom bombardment

Analyser combinations

Electrostatic/magnetic
 (metastables, c.a., linked scans)
Magnetic/electrostatic
 (m.i.k.e.s., metastables, c.a.,
 linked scans)
Triple analysers
 (without quadrupoles - high-
 resolution studies: c.a.,
 metastables, m.i.k.e.s.)
Analysers including quadrupoles
 (ion-abundance studies, c.a.)
Ion-cyclotron-resonance analysers
 (ion/molecule reactions)

Related variables

Source temperature
Electron energy } ΔH_f determinations
Photon energy }
C.i. reagent gas

Figure Experimental variations used for structural studies on gas-phase organic ions

been necessary. Particular emphasis has been given to those publications that incorporate innovative principles and for which the subject matter does not overlap excessively with that of other chapters.

The topics in this review are divided broadly into three sections. The thermochemistry of ion formation is covered first. In the two subsequent sections the two types of chemical changes undergone by the ions (namely isomerization and decomposition reactions) are reviewed. The division between these two sections is somewhat arbitrary for certain publications, since techniques that employ decomposition reactions (*e.g.* collisional activation) inevitably convey information about isomerizations preceding these decompositions. The chapter is concluded by a summary.

2 Thermochemistry

Ideally the structure of an organic ion may be inferred from its heat of formation, and this parameter is the one most commonly employed for the identification of gas-phase ion structures. The progressive research in this field continues to be concerned with the measurement of more reliable critical energies and hence heats of formation, using mass-spectrometric methods. Molecular-orbital calculations of ion enthalpies can supplement such data.

Many authors have noted and it has been emphasized in a recent review[9] that heats of formation obtained from mass-spectrometric measurements are frequently not the optimum values. However, discrepancies are minimized by use of photoionization or of ionizing electron beams of narrow energy.

An example of the use of photoionization methods to obtain reliable heats of formation has been demonstrated for the 2-butyl cation.[10] The photoion-yield curves for Bu^+ were measured for a series of precursors MeCHXEt and, where necessary, were corrected for contributions from $^{13}CC_3H_8$. In the calculations of the heat of formation of Bu^+ the stationary-electron convention was assumed.[11] Because the fragmentation process $AB + h\nu \rightarrow A^+ + B + e^-$ is a unimolecular decomposition, it is necessary to use equation (1) (ΔH°_{298} represents $H^\circ_{298} - H^\circ_0$) to calculate the heat of formation at 298 K for the ion.

$$\Delta H_{f,298}(A^+) = AE - \Delta H_{f,298}(B) + \Delta H_{f,298}(AB) + \Delta H^\circ_{298}(A^+) + \Delta H^\circ_{298}(B) - 5RT/2 \quad (1)$$

A careful study, employing photoionization efficiency curves and requiring high accuracy, has revealed that hydrogen migrations are prevalent in the acetonitrile molecular ion.[12] Measured appearance energies for the fragment ions Me^+ and CH_2^+ (and deuteriated analogues), together with their relative intensities, strongly indicate that rapid hydrogen shifts occur prior to Me^+ formation.

Electron monochromators have long been employed to obtain reliable heats of formation, and the recent literature contains various examples of their use.[13-16] Sometimes large discrepancies between these and other data are revealed. For example, the appearance energy (12.54 ± 0.03 eV) of the $C_6H_4^{+\cdot}$ ion from ionized benzonitrile obtained by this method is at least 1 eV below most other previous measurements.[15] Clearly there is a low-energy fragmentation of very slow rate of onset, which is not revealed by some traditional methods for onset potential measurement. In another study using this technique, appearance-energy measurements suggest that loss of HCl from ionized alkyl chlorides (see for example reaction (2) below) is a 1,3-elimination reaction, forming an ionized cyclopropane.[16]

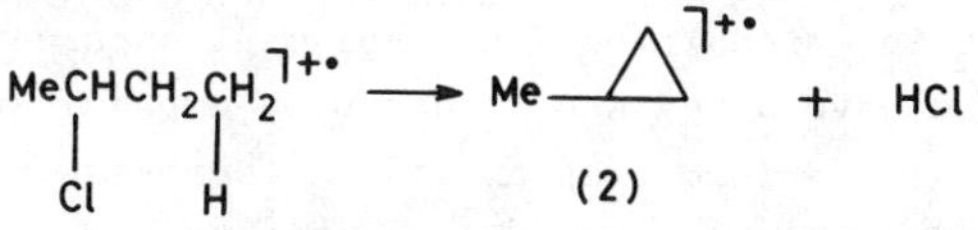

$$\mathrm{MeCH(Cl)CH_2CH_2(H)}^{+\bullet} \longrightarrow \mathrm{Me{-}c\text{-}C_3H_5}^{+\bullet} + \mathrm{HCl} \qquad (2)$$

(1) (2)

Thermodynamic quantities for uncharged radicals can also be determined by mass-spectrometric methods if relevant ionic heats of formation are known. For example, by the use of an electron monochromator the heats of formation and hence the stabilization energies for radicals (3) - (5) were found to be remarkably large (42, 71, and 84 kJ mol^{-1}, respectively).[13] These high values underlie many of the chemical and physical properties of α-amino-alkyls.

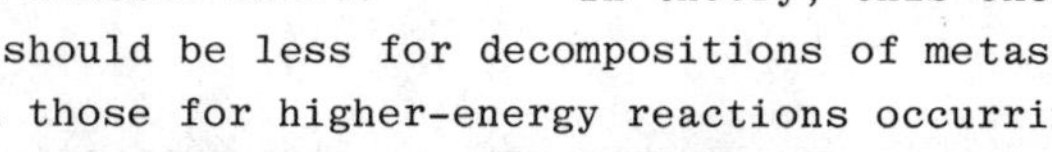

$H_2N\dot{C}H_2$ (3) $MeNH\dot{C}H_2$ (4) $Me_2N{-}\dot{C}H_2$ (5)

Appearance-energy measurements may be compromised by the presence of a kinetic shift.[1-4,12] In theory, this excess of non-fixed energy should be less for decompositions of metastable ions compared with those for higher-energy reactions occurring in the source. A method for measuring the appearance energies of 'metastable peaks' has been described;[17] it is claimed to be simple, accurate, and reproducible. Good energy resolution of the ion beam is not required and the technique involves comparison of the intensity of the 'metastable peak' over a small range of ionizing energies above their onsets. It is shown that in most cases the derived appearance-energy values are not much in excess of those obtained from monochromatic electron-impact and photoionization methods.

Enol-keto tautomerism in gas-phase ions formed from appropriate keto or alcohol precursors is an extensively studied phenomenon, and heats of formation have proved to be valuable parameters in these investigations.[18-21] It is common that the enolic ions are thermodynamically more stable than their keto counterparts, and this has been demonstrated for some $C_4H_6O^{+\cdot}$ [19] and $C_8H_8O^{+\cdot}$ [20] isomers.

In most studies of enol-keto tautomerism, the enol ion is generated *via* dissociative ionization rather than as a molecular ion by direct ionization. In a recent study, however, comparison has been effected by direct ionization of stable isomeric ketones and enols (see for example compounds (6) and (7)) as well as by fragmentation (*cf.* loss of CH_2CO from the molecular ion of compound (8)).[21] In this investigation it was established that

Mes(Ph)CH—C(=O)—Mes (6) Mes = 2,4,6-trimethylphenyl

Mes(Ph)C=C(OH)Mes (7)

Mes(Ph)C=C(OCOMe)Mes (8)

(i) the ionized enol (7) is 61 kJ mol^{-1} more stable than the ionized keto form (6), (ii) the $[M-CH_2CO]^+$ ion from the acetate (8) is identical with the ionized enol (7), and (iii) the ionized enol (7) undergoes a rate-determining isomerization in which the intermediate ionized ketone (6) is formed with an internal energy at least 2 eV above the dissociation limit. The critical-energy data were supplemented by measurements of the translational energy released in metastable transitions.

The identity of particular hydrocarbon-ion isomers can often be inferred from the determination of heat of formation.[22-26] For example, a detailed study of some gaseous alkyne ions has revealed some surprising decomposition reactions.[22] Cyclization, with accompanying hydrogen rearrangements, occurs readily to form the most stable structure. Loss of an ethyl radical from the molecular ions of 2- and 3-heptyne (reactions (3) and (4)) probably yields the cyclopentenyl ion (9). The respective linear ions (10) and (11), which would be formed if simple cleavage occurred, have heats of formation that are too high to be identified with the above-mentioned $[M-Et]^+$ ions. Note that the heats of formation of $[M-Et]^+$ ions are maximum values due to their formation with excess of energy.

In contrast, heats of formation can be employed to gain mechanistic information about ring-cleavage reactions in hydrocarbon ions. The mass-spectrometric fragmentation pattern of the

ΔH_f/kJ mol^{-1}

$$MeC{\equiv}CCH_2CH_2CH_2Me]^{+\cdot} \longrightarrow [M-Et]^+ \quad 912 \qquad (3)$$

$$MeCH_2C{\equiv}CCH_2CH_2Me]^{+\cdot} \longrightarrow [M-Et]^+ \quad 912 \qquad (4)$$

(9) 832

$MeC{\equiv}CCH_2CH_2^+$ 1087

(10)

$MeCH_2C{\equiv}CCH_2^+$ 1104

(11)

five *cis*/*trans* isomeric 1,2,4,5-tetramethylcyclohexanes (12) is influenced by their configuration, and correlations have been noted between certain fragment-ion abundances and enthalpy

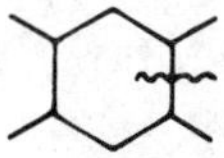

(12)

differences between the neutral isomers.[23] As the steric strain increases in the cylohexanes (12), there is an increased tendency for the molecular ion to undergo fragmentations (*e.g.* loss of an ethyl radical) that involve ring cleavage, presumably between the two positions of substitution as indicated. In an independent study, employing photoionization and charge exchange, it has been established that the degree of ring cleavage in cyclopentane and alkylcyclopentane molecular ions depends critically on their internal energy.[26]

More thermochemical data have been forthcoming on the structures of ionized toluene isomers. Formation of the methylene cyclohexadiene ion (15) has long been depicted as shown in Scheme 1, and appearance-energy measurements now lend support to this. Using the procedure outlined above[17] the appearance energies of $C_7H_8^{+\cdot}$ ions from metastable ions (13) and (14) have been measured.[24] The derived heat of formation (960 kJ mol^{-1}) excludes acyclic ions (which have higher heats of formation) and is somewhat higher than

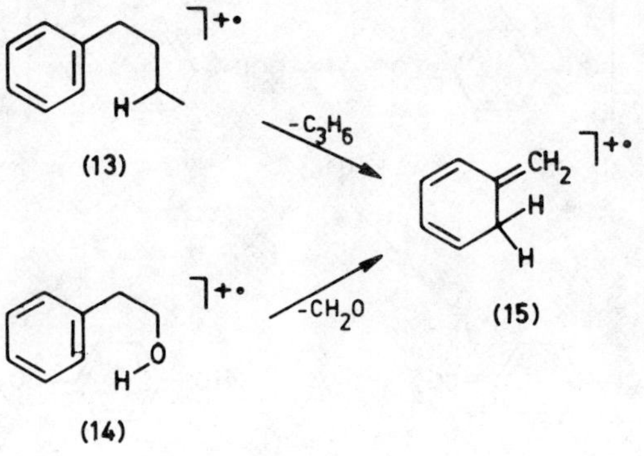

(13) $\xrightarrow{-C_3H_6}$ (15) $\xleftarrow{-CH_2O}$ (14)

Scheme 1

values obtained for the toluene ion and for $C_7H_8^{+\cdot}$ ions formed from (13) and (14) in the ion source. Collisional-activation data substantiate these conclusions. A separate study using an ion-cyclotron-resonance spectrometer has determined the heat of formation of ion (15) to be 75 kJ mol^{-1} above that of the toluene molecular ion.[25]

Thermochemical parameters for gas-phase ions can be obtained from measurements on chemical equilibria established in ion-cyclotron-resonance or high-pressure mass spectrometers,[4,27,28] and a 25-year compilation of such parameters has been published recently.[28]

An alternative method for determination of proton affinities has been described;[29] it utilizes the decompositions of proton-bound dimers. The method relies on the generation and decomposition, in a reverse-geometry mass spectrometer, of a dimeric positively or negatively charged ion $A_1HA_2^{+(-)}$, where A_1 and A_2 are species whose proton affinity is of interest. Examples of such a dimer and its dissociations are shown in Scheme 2. In the m.i.k.e. spectrum of ion (16), ions (17) and (18) are formed in a ratio of about 4 to 1, with the latter ion having the larger proton affinity. For a series of alkylamines, substituted anilines, and substituted pyridines, a linear relationship exists between $\log([A_1H^+]/[A_2H^+])$ *vs.* proton affinity, in the dissociation of $A_1HA_2^+$. In general, proton affinities obtained by such methods agree within 1.3 kJ mol^{-1} with values from equilibrium measurements,

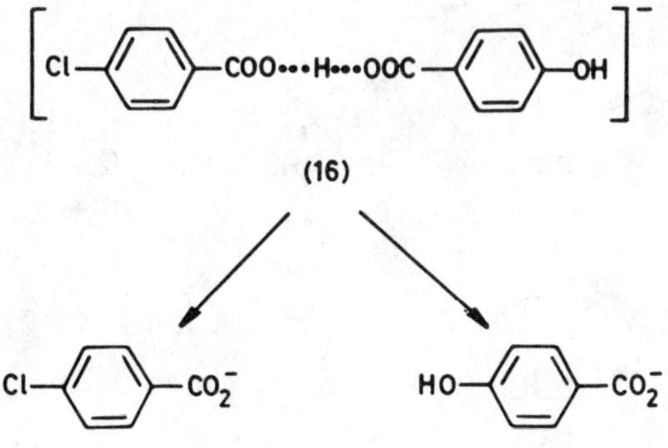

$$[Cl-C_6H_4-COO\cdots H\cdots OOC-C_6H_4-OH]^-$$

(16)

$$Cl-C_6H_4-CO_2^-$$ (17)

$$HO-C_6H_4-CO_2^-$$ (18)

Scheme 2

although careful choice of reference compounds is required. The technique can be extended to the measurement of gas-phase acidities, hydride affinities, and metal-cation affinities.

The decomposition reactions of protonated molecular ions formed *via* chemical ionization can now be correlated adequately with the thermochemistry of the ionization process.[30-32] For example, the chemical-ionization spectra of substituted halobenzenes RC_6H_4X (X = halogen) have been shown to vary according to the proton affinities (shown in brackets) of the reagent gases H_2 (423 kJ mol^{-1}), N_2 (477), CH_4 (536), and CO_2 (540).[30] The decomposition reactions of the benzenes, RC_6H_4X, induced when using the reagent gas, CO_2H^+, are similar to those following protonation by CH_5^+, while the reactions induced by N_2H^+ are intermediate between those observed for reaction with CH_5^+ and with H_3^+. Various factors operate to control the breakdown pathways, and intermediate ion/molecule complexes can be invoked, particularly in the case of protonated alkyl benzenes.[31] In addition, it is clear from several recent studies[33-37] that the breakdown pattern of protonated molecular ions can be directly related to the site of protonation.

Heats of formation for hypothetical product ions and neutrals from mass-spectral reactions can be employed to explain and predict preferred decompositions.[38,39] The principles involved in this type of investigation are well illustrated by the fragmentation of alkyl benzoate molecular ions. In this publication, heats of formation, low-voltage spectra, and deuterium-labelling data were carefully utilized to construct decomposition pathways for 13 alkyl benzoates. One of the simpler deductions is illustrated in Scheme 3. A six-centre

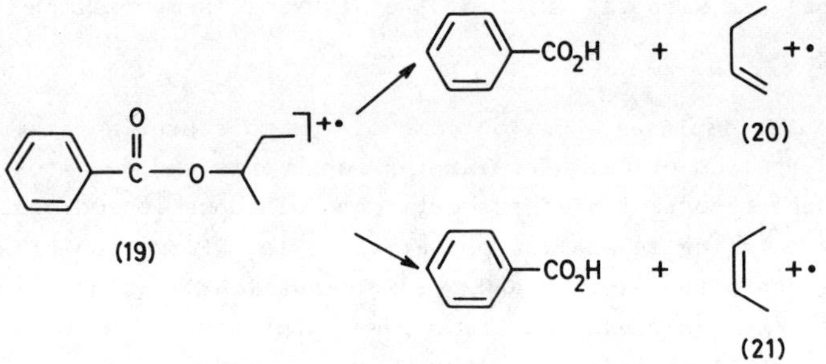

Scheme 3

rearrangement from ionized 2-butyl benzoate (19) yields $C_4H_8^{+\cdot}$ ions, which could correspond to either the 1-butene (20) or 2-butene (21) structure. The latter is energetically preferred (ΔH_f 54 kJ mol^{-1} lower), and transfer of a hydrogen from the 3-position of the butyl chain is confirmed by deuterium labelling.

The research reported in this section forms only a fraction of the total where thermochemical measurements have been made, but it does provide many leading references. Many of the publications reported below have also used such data in part.

3 Isomerization

The tendency of gas-phase organic ions to undergo isomerization reactions along the flight path of the mass spectrometer is well established as a result of extensive research (see refs. 1 - 6 for relevant bibliography). Several reaction channels may be open to an isolated organic ion, but the accessibility and relative importance of these various channels are a function of internal energy. Most organic ions possessing low excess of internal energy in the mass spectrometer are inert, existing in a potential well, whereas they can usually undergo isomerization and/or decomposition reactions at higher energy. The competition between isomerization and decomposition reactions is an important topic in gas-phase ion chemistry, and relevant discussions are found in refs. 1 - 6.

Energy measurements (either translational or internal) can be used to identify isomerization reactions in the mass spectrometer. It is commonplace simply to measure the decomposition reactions undergone by the isomerizing ion. Techniques that involve mass-selection of ions prior to decomposition (*e.g.* by collisional activation) are also valuable for the study of isomerization reactions.

Ion/Molecule Complexes.- One outcome of the more precise evaluation of the energetics of unimolecular rearrangements and decompositions has been the recognition of the existence of loose ion/molecule complexes existing in shallow potential wells. Formation of such complexes can often explain and predict, qualitatively or quantitatively, rearrangement reactions occurring in isolated organic ions prior to decomposition.[40]

Some intricate rearrangements, *via* ion/molecule complexes, have been identified in order to explain the complex reactions of ionized alkanes and their branched-chain isomers.[41,42] The explanatory mechanisms propose stretching of a carbon-carbon bond to form a radical co-ordinated to an inchoate carbonium ion, as shown in structure (23). Rearrangement of this ion/molecule complex can then occur to form a more stable species (24). Scheme 4 illustrates such a rearrangement for ionized n-hexane and accounts for the involvement of the 3-carbon atom in the elimination of a Me radical.[42]

Major fragmentation reactions of protonated alkyl benzenes have also been rationalized by invoking ion/molecule complexes as intermediates.[31] The two main decomposition reactions are olefin elimination to form a protonated arene (30) and arene elimination to form an alkyl ion (31). From the effects of parent-molecule structure and protonation exothermicity it was concluded that rearrangement of primary alkyl groups to the more stable secondary structure occurs prior to alkyl-ion formation (Scheme 5).

The translational energy (T) released during the decomposition of an ion/molecule complex is a useful parameter for identifying rearrangements. In particular, comparison of T values for isomeric ions derived from different sources yields information about interconversions between these ions. For example, the T value for loss of water from the solvated s-butyl ion (32) is much greater than that from its t-butyl isomer (33).[43] It was

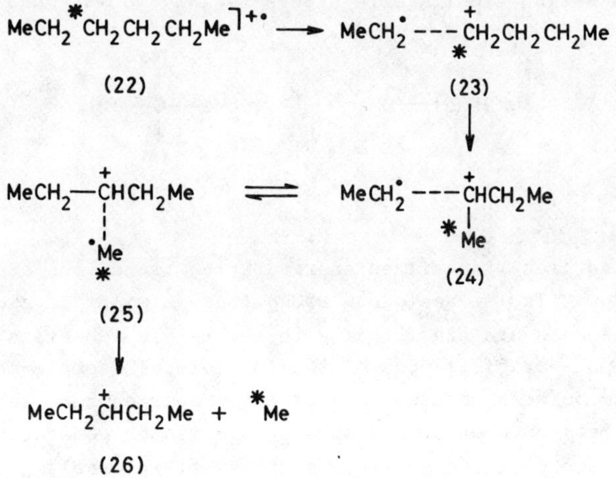

$MeCH_2\overset{*}{C}H_2CH_2CH_2Me]^{+\bullet}$ (22) → $MeCH_2^{\bullet}$ --- $\overset{+}{\underset{*}{C}}H_2CH_2CH_2Me$ (23)

(23) → $MeCH_2^{\bullet}$ --- $\overset{+}{C}HCH_2Me$ with $^{*}Me$ (24) ⇌ $MeCH_2$—$\overset{+}{C}HCH_2Me$ ··· $^{\bullet}\overset{*}{Me}$ (25)

(25) → $MeCH_2\overset{+}{C}HCH_2Me$ + $^{*}Me$ (26)

Scheme 4

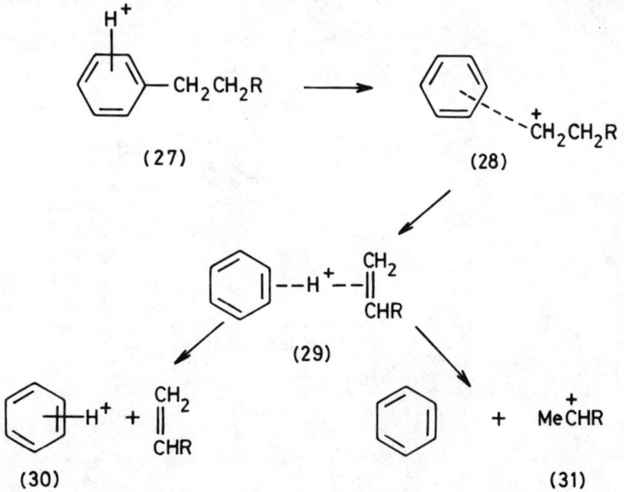

(27) (protonated $C_6H_5CH_2CH_2R$, H^+) → (28) (C_6H_6 ··· $\overset{+}{C}H_2CH_2R$) → (29) (C_6H_6 --H^+-- $CH_2{=}CHR$)

(29) → (30) $C_6H_6H^+$ + $CH_2{=}CHR$

(29) → C_6H_6 + $Me\overset{+}{C}HR$ (31)

Scheme 5

concluded that an exothermic, irreversible isomerization of the incipient s-$C_4H_9^+$ to its more stable t-$C_4H_9^+$ isomer occurs, with

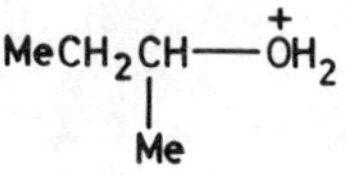

$MeCH_2CH(Me)-\overset{+}{O}H_2$ **(32)** $Me_3C-\overset{+}{O}H_2$ **(33)**

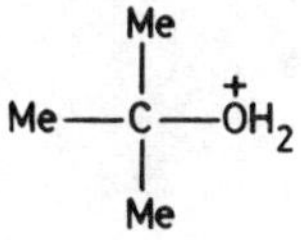

a resulting higher non-fixed energy in the transition state for loss of H_2O. In the analogous $PrOH_2$ ions, similar *T* values for the two isomers are consistent with reversible isomerization due to ion/dipole stabilization of the ion/molecule complexes. The difference between the two sets of ions reported here seems to depend merely on whether the more stable isomer generated by isomerization contains sufficient excess of internal energy to overcome the attractive ion/dipole interaction. The case of the isobutanol molecular ion (34) also provides an illustration of the utility of this concept in explaining complex rearrangements.[44] The preferential formation of $C_3H_5^{\cdot}$ and $MeOH_2^+$ is rationalized in terms of the stability of ions (35) and (36) conferred by ion/dipole interaction. The overall reaction (Scheme 6) requires the

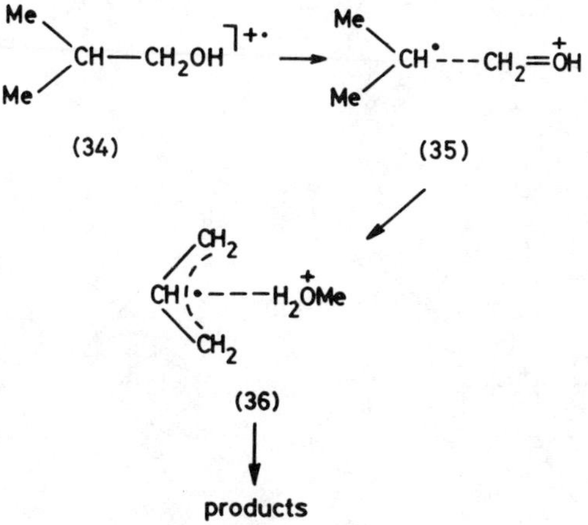

$[Me_2CH-CH_2OH]^{+\cdot}$ **(34)** → $Me_2\dot{C}H\text{---}CH_2{=}\overset{+}{O}H$ **(35)** → $CH_2{=}CH{-}\dot{C}H_2\text{---}H_2\overset{+}{O}Me$ **(36)** → products

Scheme 6

transfer of two hydrogen atoms, and such a double-hydrogen transfer in molecular ions of 4-pyridyl ethers has been suggested as occurring *via* an ion/molecule complex.[45]

Publications discussed above include examples of exothermic, irreversible conversions to more stable isomers, followed by fast decomposition with an accompanying large release of translational energy (T). A similar type of result that requires a different explanation has been demonstrated for the decompositions of the $(M-Me)^+$ ions from t-butyl methyl ketone (37).[46] Deuterium labelling confirms that two mechanisms for methyl loss from the ketone (37) occur (Scheme 7). Further loss of CO is accompanied

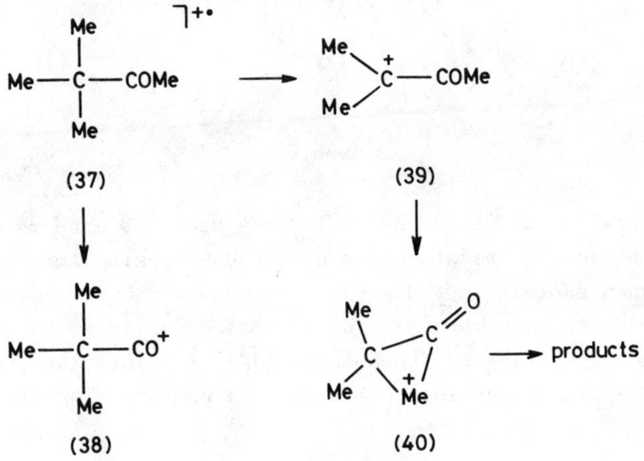

Me
Me—C—COMe
Me
]+•
(37)

Me
Me—C—CO+
Me
(38)

Me
+
C—COMe
Me
(39)

Me
C
C=O
+
Me
Me
(40)
products

Scheme 7

by a T value of 16 meV from ion (38) and 365 meV from (39). Although no exact description of the processes involved can be formulated without knowledge of the potential-energy hypersurface of the system, it appears that the acyl carbenium ion (39) isomerizes to its isomer (40), or a closely related structure. Collapse to products then occurs with a large release of translational energy over a potential surface well above the critical energy for decomposition. It is necessary to venture an explanation of this kind since ions (38) and (39) have similar enthalpies.

Ring-closure reactions have been inferred to proceed *via* ion/

molecule complexes.[47] Evidence has been forthcoming in part from the capability of detecting neutral products from mass-spectrometric reactions in an electron-bombardment (e.b.) flow reactor. For example, the congruence of the mass-spectrometric deuterium-labelling and e.b. flow results for the elimination of $PhOH^{+\cdot}$ from compound (41) was taken as support for an ion/molecule complex (42) as the source of cyclopentene (Scheme 8). The bulk of the neutral products were linear dienes.

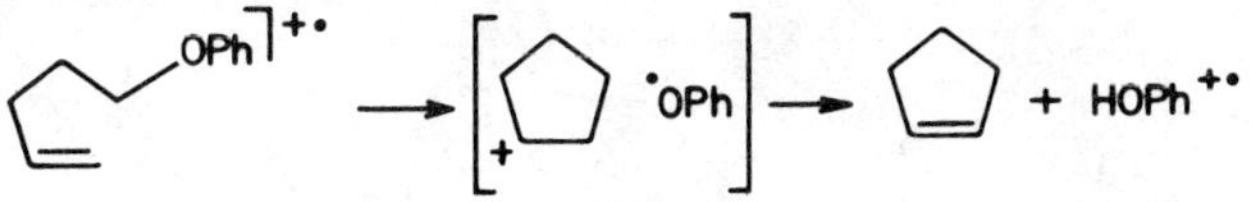

(41) (42) (43)

Scheme 8

In general, analysis of neutral decay products appears to be a useful pointer to the structure of the decomposing ion. For example, when radiolysis and g.c. techniques were used, no neutral products were observed that contained a linear C_3H_5 part from the decomposition of the complex (44).[48] Hence the complex (44) must retain the cyclopropyl ring as a component of its structure.

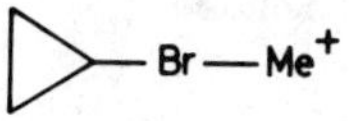

(44)

Many ion/molecule complexes described in mass-spectrometric literature are depicted as two neutral fragments bound by a solvating proton,[29,31,49,50] and various thermochemical evaluations of the stability of such complexes have been attempted. An equation for predicting energy barriers has been applied to the estimation of stabilities of proton-bound dimers of neutral molecules and anions in the gas-phase.[50] The average deviation between calculated and experimental well depths of 51 such dimers is about

2.5 kJ mol^{-1}. The energy of a proton-bound dimer $(A-H-B)^+$ can be equated in most cases to the average of the energies of the two symmetrical dimers $(A-H-A)^+$ and $(B-H-B)^+$.

'Intramolecular' proton-bound dimers have been identified thermochemically as being the most stable species in protonated α,ω-hydroxyalkylamines (45).[35] The two most prevalent reaction pathways from cyclic ions (45) involve loss of H_2O and NH_3 and occur *via* ring-opening isomerizations on either side of the solvating proton. A striking contrast is observed between dissociations occurring in the ion source and those occurring following collisional activation, but the differences can be rationalized thermochemically.

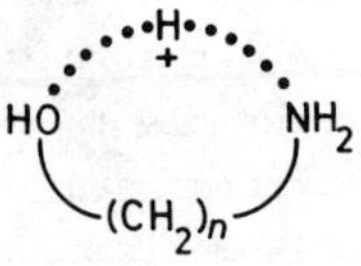

(45)

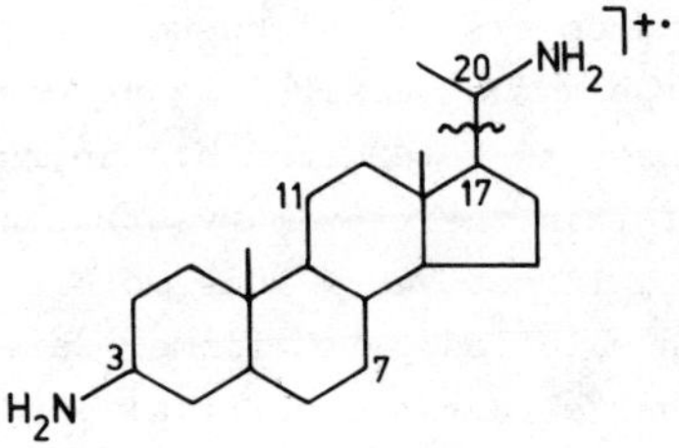

(46)

There has been some speculation concerning the existence of long-lived ion/molecule complexes in diamino-steroids. In the 3,20-diaminopregnane molecular ion (46) such a complex has been postulated, resulting from cleavage of the 17-20 bond, as shown.[51] If the two fragments move apart with low relative velocity then a rotational interaction can occur and would explain hydrogen exchange between the 3- and 20-positions (normally 1 nm apart) prior to formation of completely separate fragments. It is contended that an intra-ionic hydrogen-transfer mechanism (involving B-ring cleavage) can be invoked to explain fragment ions in the mass spectra of 3,7- and 3,11-diaminopregnanes.[52]

It is evident from the research reviewed in this section that many mass-spectrometric isomerization reactions are explicable by invoking intermediate ion/molecule complexes. Suggested structures for these complexes are often speculative but are more convincing when supported by isotope-labelling studies, energy measurements, and even lifetime determinations.[53]

Hydrocarbon Rearrangements.- In this section, those isomerization reactions of hydrocarbon ions are considered that have not specifically been identified to occur *via* ion/molecule complexes as discussed above. Future research may show that this division between the various publications is somewhat arbitrary.

A new technique has been employed to investigate the isomerization reactions of low-energy, stable hydrocarbon ions.[54] Radicals are formed from hydrocarbon ions by ion-electron recombination in a chemical-ionization plasma and are trapped in a fast gas-phase reaction with tetraquinodimethane, prior to analysis by collisional activation. Isomerization reactions occurring in selected hydrocarbon ions are delineated and are found to be consistent with most gas-phase data obtained by critical-energy measurements, as discussed in the section on thermochemistry above. For example, a significant percentage of n-butyl and n-propyl ions, as determined by the radical-trapping technique, appear to retain their primary structure. The ion structures observed for larger alkyl ions more closely resemble those often produced in solution, with isomerization occurring readily. This suggests that either the critical energy for isomerization is lower for larger ions or the internal energy imparted to the ions during formation is higher. The radical-trapping method is also able to distinguish between phenonium (47) and α-phenylethyl (48) ions.

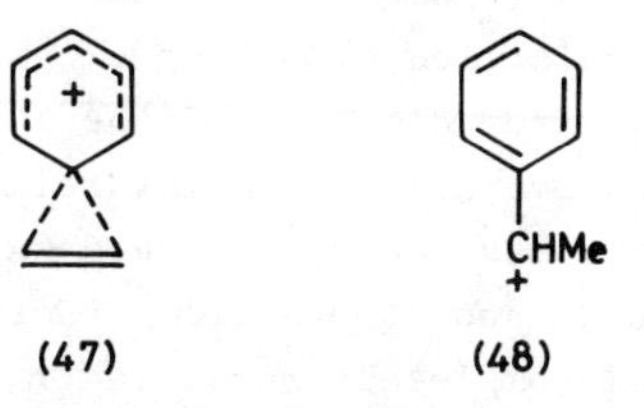

The above examples illustrate that alternative procedures are available for distinction between isomeric gas-phase hydrocarbon ions where conventional methods indicate extensive isomerization to common structures. The crucial variable in determining the extent of isomerization is the internal energy of the ions sampled, and the utilization of photodissociation methods neatly confirms this fact.[55] Stable $C_5H_8^{+\cdot}$ ions produced from isomers (49) - (52) were excited by the use of photons of different energies, and the ensuing fragmentations were monitored by scanning an electric sector. While all the fragmentation spectra were

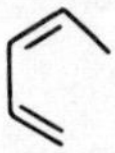

$CH_2{=}C{=}CH{-}CH_2Me$ $MeC{\equiv}C{-}CH_2Me$

(49) (50) (51) (52)

similar using 515 nm (2.4 eV) photons, wavelengths of 357 nm (3.5 eV) achieved substantial differences. For these particular ions, stable structurally distinct $C_5H_8^{+\cdot}$ ions can be excited from their potential wells to decompose before isomerizing. Further reference to this promising technique can be found in the section on decomposition reactions below, and more fundamental work on the physics of photon and multi-photon excitation appears in Chapter 1.

A detailed assessment of the extent of isomeric interconversions between gas-phase ions as a function of internal energy can be achieved from charge-exchange spectra.[56] Such an approach has been applied to fourteen C_6H_{12} isomers. From the charge-exchange data, breakdown graphs that present an impressive overall picture of interconversions between isomeric ions as a function of energy were constructed. The isomeric cycloalkanes (53) and (54) gave qualitatively similar breakdown graphs, suggesting that conversion to linear isomers occurs prior to fragmentation. However, dissimilarity between the cycloalkane [(53), (54)] and hexene [(55) - (57)] breakdown graphs seems to preclude such a mechanism.

(53) 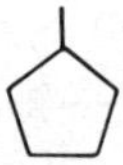 (54)  (55)  (56)  (57)

Isomerization mechanisms for hydrocarbon ions can only be probed in depth if deuterium and/or ^{13}C labelling is employed, as has been illustrated for the molecular ion of 2-methyl-1-hexane.[57] Variations in excess of internal energy were achieved by sampling ions of different lifetimes.

Molecular-orbital calculations continue to be used to augment mass-spectrometric studies of gas-phase hydrocarbon ions,[58-61] but the detailed evaluation of the methods employed is beyond the scope of this review and, in any case, is reviewed in Chapter 4. Most significantly, molecular-orbital calculations have suggested the intermediacy of pyramidal carbo-cations in hydrocarbon-ion chemistry.[58] One example is found in the elimination of C_2H_4 from $C_6H_{11}^+$ ions. M.i.k.e. spectra of doubly ^{13}C-labelled $C_6H_{11}^+$ ions generated from appropriately labelled compounds (58) - (61) establish that randomization of the carbon atoms occurs prior to

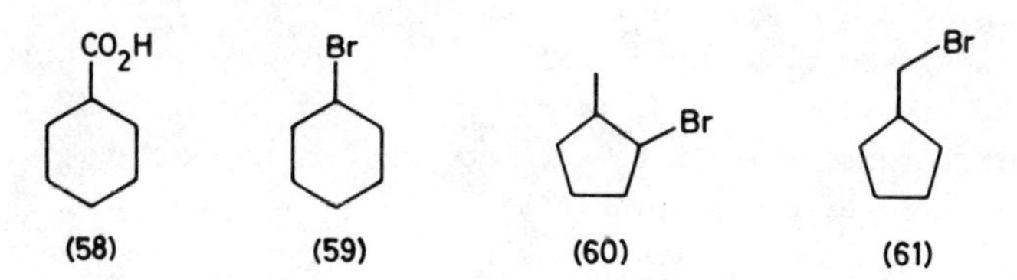

elimination of C_2H_4, regardless of the source of the $C_6H_{11}^+$ ion. The randomization of the carbon atoms can be represented as an equilibrium between the cyclohexyl (62) and 1-methylcyclopentyl (66) ions, involving both protonated cyclopropane derivatives (63) and the pyramidal cations (64) and (65) (see Scheme 9).

Scheme 9

Heteroatomic Ions.- Often, isomerization reactions in organic ions containing heteroatoms can be well characterized because skeletal or hydrogen randomization reactions in these ions are usually less than those observed for hydrocarbons. Several publications covering gas-phase isomerization reactions of heteroatomic ions are discussed elsewhere in this chapter in the sections on ion/molecule complexes and release of translational energy. This section describes further examples of such reactions, with emphasis on the variety of experimental techniques that can be brought to bear on their investigation.

The traditional method of comparing metastable peak heights has been employed to investigate interconversions between a large number of isomeric $C_nH_{2n+1}O^+$ ions.[62] It has been shown that lower homologues behave exceptionally and that, in the higher homologues (n = 4 - 13), the ions generally retain their original identity to a high degree. Interconversions do occur, but generally within a narrow range of closely related structures. Enol-keto interconversions have been extensively studied by critical-energy measurements,[18-21] and metastable decompositions of isotopically labelled compounds also continue to provide valuable insights.[63,64] From such measurements on $C_5H_{10}O^{+\cdot}$ ions it has been suggested[63] that the 3-pentanone ion (67) is formed by isomerization of enol ions with incomplete randomization of internal energy, flouting the tenets of the quasi-equilibrium theory (QET) of mass spectra. However, this type of conclusion has been contested for the $C_3H_6O^{+\cdot}$ ion.[64] Discussion of QET and non-QET behaviour of ions appears in Chapters 1 and 4.

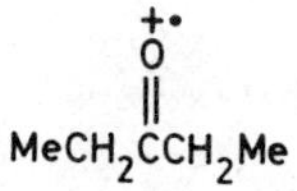

$MeCH_2C(=O^{+\cdot})CH_2Me$

(67) 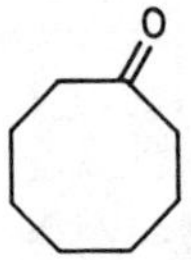 (68) 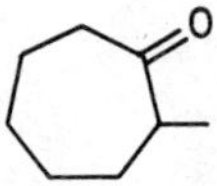 (69) 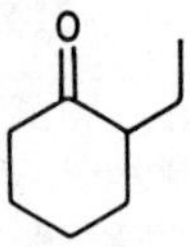 (70)

Ring-opening isomerization reactions in the molecular ions of isomeric cyclanones, $C_8H_{14}O^{+\cdot}$, have been identified by isotopic labelling and m.i.k.e. spectometry.[65] Isomerization of the cyclooctanone (68) and 2-methylcycloheptanone (69) ions occurs to form 2-ethylcyclohexanone (70) prior to elimination of C_2H_4.

Methods that sample decompositions as a function of ion lifetime can be employed to evaluate isomerization reactions as a function of internal energy. Field-ionization kinetics (FIK) have been used to select ions of short lifetime from isomeric alcohols,[66,67] and it has been shown that collision-induced dissociation, with resolution of scattering angle, gives data comparable to those produced by FIK.[67] Appropriate sampling of consecutive reactions occurring in the field-free regions of a reversed-geometry mass spectrometer offers another means of investigating isomerization reactions (in this case of $C_5H_5O^+$ ions) at different internal energies.[68]

The occurrence of isomerization reactions can frequently be inferred simply by observation of unexpected decomposition reactions. This has been illustrated in some immonium ions, which violate the even-electron rule by eliminating methyl radicals.[69] Proposed isomerizations were supported by energy and isotope-labelling experiments. The structure of $C_2H_4N^+$ ions formed from various precursors has been investigated by collisional activation as a function of internal energy.[70] The association reaction between Me^+ and HCN yields an ion that rearranges to protonated MeCN and not protonated MeNC.

4 Decompositions

Proximity Effects.- Direct interaction between two substituents in adjacent (*ortho*) positions of aromatic rings can lead to the so-called *ortho* effect. Energetically and entropically favoured reactions occur that are not found for the corresponding *meta* and *para* isomers and result in electron-impact mass spectra that are diagnostically valuable. These *ortho* decompositions are often unpredictable from first principles, as with the loss of OH from ionized *ortho*-nitrotoluene, and this has led to extensive investigations of their mechanisms since the first examples were observed. Traditionally, *ortho* effects have been studied by isotopic labelling and measurements of heats of formation but, in recent years, various other techniques have been applied to the problem. This section reports on some of the progress made in understanding *ortho* effects in organic mass spectrometry. For a full review see reference 71.

Nitro-aromatics frequently exhibit complex *ortho* effects.[72-77] Loss of OH from the molecular ion is a common reaction[72-75] whose occurrence remains somewhat puzzling. The hydrogen atom always originates from a sterically convenient position on the *ortho* substituent and is invariably accompanied by a metastable peak (which illustrates its low-energy characteristics). The most extensively studied (collisional activation, deuterium labelling, field ionization, and critical-energy measurements) example in the recent literature establishes that the loss of OH from *ortho*-nitroethylbenzene (71) involves specifically the α-hydrogens of the ethyl group, has a very low critical energy of activation (0.05 eV), and occurs over a wide range of ion lifetimes.[72] The presence of a sulphur atom in addition to a nitro group renders *ortho* effects even more bizarre and, in the spectra of 2-nitrothiobenzamides (72),

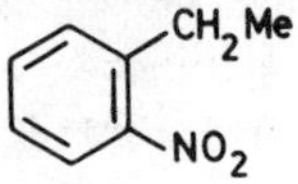

(71) 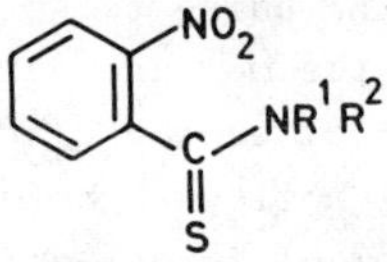 **(72)**

all major fragment ions (including $(M-OH)^+$) arise from rearrangement reactions. *ortho*-Substituted alkyl benzoates also show a propensity to fragment in the mass spectrometer *via* rearrangements,[78,79] and the dipole moments of interacting groups have been considered as factors in these reactions.[78] It appears that certain substituents favour unusual *ortho* effects for energetic reasons that are not yet fully understood.

Another type of proximity effect that has been frequently recognized in gas-phase organic ions is the intramolecular anchimeric assistance of one substituent group in the elimination of another.[80-86] The identification of such a reaction necessitates gaining some information about its transition state, and measurements of the translational energy released in the decomposition have been particularly useful in this respect (for discussions see below). Various groups have been implicated in these reactions, and the anchimerically assisting bonds may be formed from aromatic π-electrons (73)[82] or from the non-bonding electrons of nitrogen (74),[83] oxygen (75),[81] or sulphur (76).[86]

Release of Translational Energy.- Mass-spectrometric ionic decompositions are accompanied by a release of translational energy T, a useful parameter for the extraction of information on ion structures and decomposition mechanisms. If standard means are employed to extract the experimental T value, this parameter is

(73) 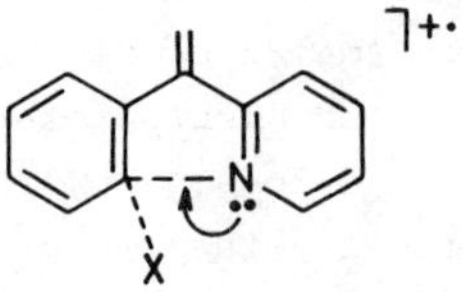 (74) 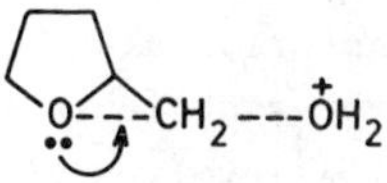 (75) 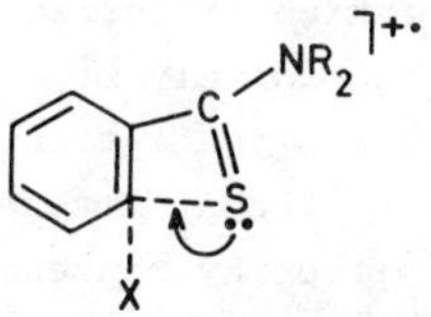 (76)

reproducible in different laboratory systems. T can be measured readily from metastable peak widths on magnetic sector instruments,[2-5] including those of Mattauch-Herzog geometry.[87]

Most simply, the measurement of this parameter enables a direct comparison of isomeric structures as a criterion for identity or non-identity to be made. Ions having the same structure but generated from different parent compounds will have similar lifetimes and internal energies when they decompose unimolecularly in a field-free region (with the exception of ions formed *via* a rate-determining isomerization). Therefore, equal T values are in general consistent with identical structures, and this criterion has been extensively applied (see for example refs. 5, 19 - 21, 39, 88, 89). Conversely, dissimilar T values for isomeric species indicate different structures (again excepting the involvement of rate-determining isomerizations). This property has been employed to distinguish between epimeric steroids.[90]

The fraction of excess of internal energy that appears as translational energy in the products from an ionic decomposition can vary in individual cases from values close to zero to greater than 50%. Several authors have contended that the 'energy-partitioning quotient' (the fraction of reverse critical energy released as translation in the products) is related to the position of the transition state on the reaction co-ordinate. Therefore, measurement of this parameter can yield detailed information about ionic reaction mechanisms. A large energy-partitioning quotient is observed for reactions with 'symmetrical' or 'late' transition states, whereas most of the excess of energy remains as internal energy of the products for reactions having 'early' transition states. This is illustrated in the decompositions of ionized *ortho*-substituted *NN*-dimethylthiobenzamides (76).[84] When X = H or Me in structure (76), a large fraction (greater than 0.5) of the transition-state energy is released as translational energy for the elimination of X. (Broad dish-topped metastable peaks are observed.) From this and other evidence it appears that the transition state is late, consistent with an intermediate structure such as (77). In contrast, when X = Cl, Br, or I, an early transition state such as (76) is indicated by the low energy-partitioning quotient.

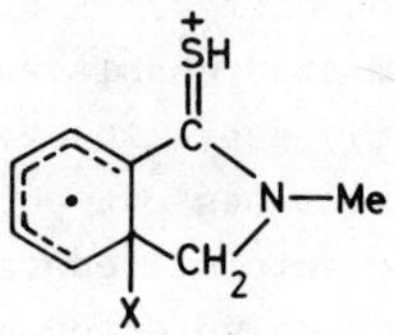

(77)

Energy-partitioning accompanying the metastable ion dissociations of Bu^+ ions has been investigated by the use of MINDO/3 calculations.[91] Transition-state geometries and reaction profiles were determined. Quantitative conversion of reverse critical energy into translational energy is associated with a tight pyramidal complex structure analogous to CH_5^+. Intermediate solvated ion structures may be involved.

The research covered in this section, and in the above section on rate-determining isomerizations, has shown that there is a variety of factors influencing the T value for a reaction. The overall situation is not well understood, but a better understanding of the factors affecting T should be forthcoming from careful determinations of the transition-state geometrics for individual decomposing ions.

Excitation of Ions.- Ions leaving an electron-impact or chemical-ionization source normally have lifetimes of at least 1μs. They comprise mainly stable ions, having insufficient energy to fragment, together with some metastable ions, which can undergo unimolecular decomposition between source and collector. The internal energy of the ions along the flight path can be increased by collisional activation[1-5] or *via* interaction with laser radiation (see Chapter 1, refs. 305 and 306, and this chapter, refs. 92 and 93). This additional excitation energy promotes decomposition reactions, and information on the precursor-ion structure and its fragmentation mechanisms can be obtained from the ion products and their relative abundances. Individual types of ion can be selected for investigation and their reactions studied by appropriate scanning of magnetic, electric, or quadrupole fields.[94-97] This section is concerned with the structural information obtainable for gas-phase ions from flight-path interactions with collision gases or photons.

Collision between neutral gas molecules and the ion beam converts a fraction of the translational energy of the incident ion into internal excitation energy. A broad range of internal energies is added, from a most probable value of 1 - 5 eV up to values (with low probability) above 10 eV. Indeed, collision-induced charge-stripping reactions can occur.

The extent to which the internal energy of the collisionally activated ions can be varied or selected has been the subject of much research. The possible effects of ion-source pressure,[70] multiple collisions,[98] consecutive reactions,[99] and scattering angle[100] have been investigated. The efficiency of the collisional-activation process as a function of the target gas has been assessed.[100] Helium is the most efficient target gas.

There is little doubt from the large number of studies to date that the collisional-activation (c.a.) spectrum in most cases provides a characteristic fingerprint of the stable-gas phase structures(s) of the ion under study (see refs. 19, 20, 22, 37, 64, 82, 86, 101, and 102 for recent representative publications). This fact has been used on many occasions to identify unknown ions by employing characteristic c.a. spectra of ions from model compounds. For example, the structures of the abundant $C_8H_8O^{+\cdot}$ ions (80) formed by loss of C_2H_4 from the molecular ions (78) and (79) (Scheme 10) have been investigated by comparing the c.a. spectra of

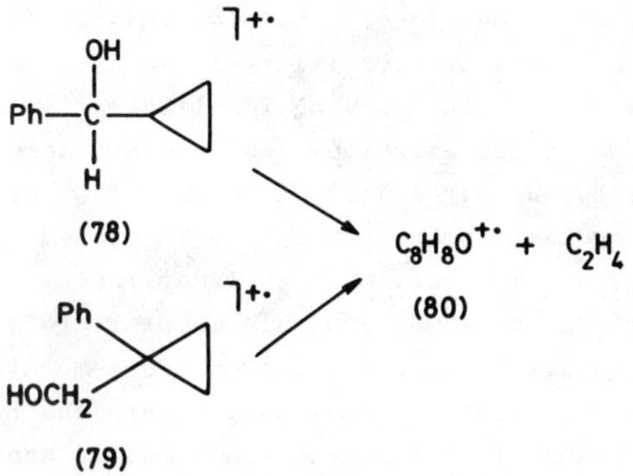

Scheme 10

these and the model ions (81) - (85).[20] The dissimilarity between the c.a. spectra of ions (81) - (85) and the close identity between the c.a. spectra of the test $C_8H_8O^{+\cdot}$ ions (80) and (84) strongly suggest that the stable (non-decomposing) ions (80) have the structure of the enol of phenylacetaldehyde (84). Data from measurements of heat of formation are broadly in agreement with the c.a. results (both techniques sample *stable* ions). B/E linked-scan spectra reveal that the corresponding metastable ions (higher energy) are not identical but show some similarities.

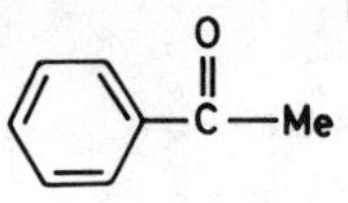

$PhC(=O)Me^{+\cdot}$ (81) $PhC(OH)=CH_2^{+\cdot}$ (82) $PhCH_2CHO^{+\cdot}$ (83)

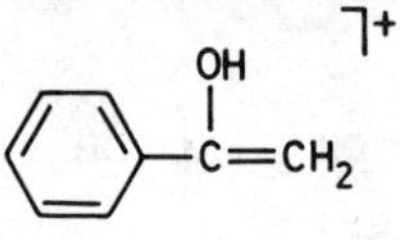

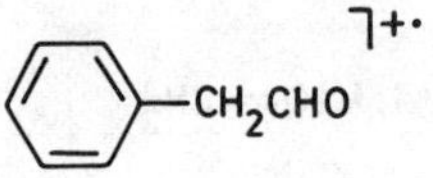

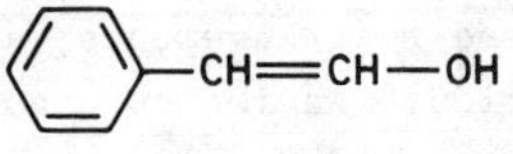

$PhCH=CH-OH^{+\cdot}$ (84) 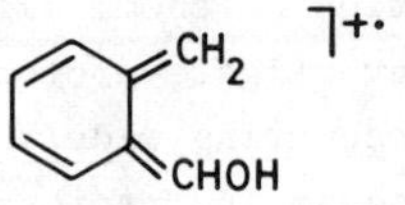 (85)

$MeCH_2OCH_2CH_2CH_2Me$ (86) $MeCH_2OCH_2CHMe_2$ (87)

There are examples of observations which are the converse of the above, *i.e.* metastable ions from different sources having the same structure but the corresponding stable ions being different. The m.i.k.e. spectra of the ionized ethers (86) and (87) indicate that their metastable ions have isomerized to a common structure whereas the c.a. spectra reveal that stable ions exist in potential wells that retain the initial structure of the molecular ions.[101] Participation of the ether oxygen is invoked to explain the rearrangements in metastable ions.

When the c.a. process adds a sufficiently large excess of internal energy, a doubly charged ion is formed whose fragmentation (charge-stripping) mass spectrum is characteristic of the structure of the original singly charged ion. This technique can be utilized to determine ion structures on occasions where the interpretation of c.a. results is ambiguous. It has been indicated[3] that the inverse relationship between collision cross-section and energy transferred results in the preferential sampling of ions close to the lowest decomposition threshold by the c.a. method. These are the ions most likely to isomerize. In contrast, when charge-stripping occurs, the larger amount of energy transferred means that ions of lower average internal energy are sampled. Such ions are less likely to isomerize. This principle is illustrated in the charge-stripping mass spectra of various EtO^+ ions.[103] Four distinct species (88) - (91) were identified from eight separate precursors whereas the corresponding c.a. spectra did not permit such fine distinction.

$MeO^+{=}CH_2$ (88) $MeCH{=}OH^+$ (89) (90) $CH_2{=}CH{-}OH_2^+$ (91)

A combination of chemical ionization (c.i.) and c.a. can be utilized to investigate proton-catalysed rearrangements, a class of reactions occurring widely in solution. An ion such as CH_5^+ or Bu^+ is employed as the protonating agent, and prospective product ions of the rearrangement are compared with model ions by c.a. mass spectrometry. For example, evidence for the gas-phase acid-catalysed aldol condensation of acetaldehyde (92) to form protonated crotonaldehyde (95) (Scheme 11) has been presented using c.i. and

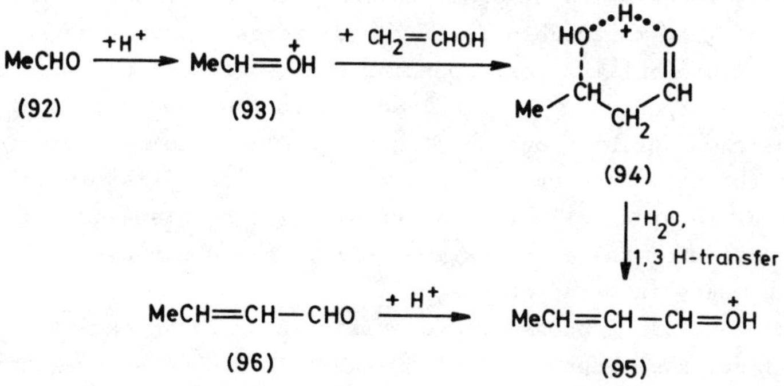

MeCHO (92) $\xrightarrow{+H^+}$ $MeCH{=}OH^+$ (93) $\xrightarrow{+ CH_2{=}CHOH}$ (94) $\xrightarrow{-H_2O,\ 1,3\ H\text{-transfer}}$ $MeCH{=}CH{-}CH{=}OH^+$ (95)

$MeCH{=}CH{-}CHO$ (96) $\xrightarrow{+H^+}$ $MeCH{=}CH{-}CH{=}OH^+$ (95)

Scheme 11

c.a. techniques.[104] That the ion (95) is indeed protonated crotonaldehyde was confirmed by establishing the identity of the c.a. spectra of ions at m/z 71 formed *via* the proposed aldol reaction (92) - (95) and *via* protonation of crotonaldehyde (96) itself. Incidentally, the occurrence of the overall reaction (92) - (95) poses the question to what extent $(2M + H)^+$ ions formed from aldehydes and ketones in c.i. sources have the protonated dimer structure MHM^+.

The c.a. process adds a broad range of internal energies to an ion that may already possess an internal energy spread of up to several electron volts. This can be an advantage when it is necessary to discount any differences of internal energy between identical structures, but conversely it is a problem where the effects of differences in internal energy need to be precisely evaluated. The use of photons to excite ions along the flight path adds an exact but variable amount of internal energy and has provided some useful complementary information on ion structures.

The translational energy released in such a photo-induced reaction, measured as a function of the photon energy, can be used[105,106] as a structural probe (see Chapter 1). For example, *para*-xylene is readily distinguishable from the other two xylene isomers by this technique.[105] The rearrangements and decompositions occurring in the xylene ions and their $C_8H_{10}^{+\cdot}$ isomers have recently been extensively investigated[107] by determinations of heat of formation and isotope labelling. The situation is complex, but it appears that p-xylene (97), methylcycloheptatriene (98), and ethylbenzene (99) ions can interconvert at energies below the

(97) 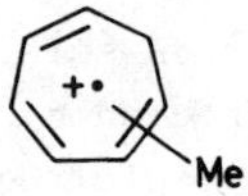 (98) 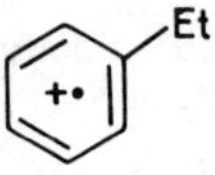 (99) 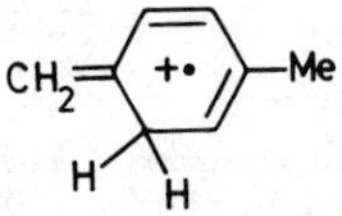 (100)

threshold for loss of a methyl radical. Hydrogen migrations and skeletal rearrangements occur independently of each other, and the p-xylene isomer (97) appears to be the most stable. MNDO calculations employed in this publication[107] identify a variety of stable cyclic intermediates. For example, the isomer (100) has a calculated heat of formation equal to that of the ethylbenzene molecular ion (99).

From the early results on photodissociation studies using ion beams it is evident that several different structurally important parameters are obtainable. Measurements both of T values[105,106] and of relative product-ion abundances[55,92] can be used to characterize ion structures. In addition, the cross-section for dissociation of an ion is a characteristic parameter and is obtainable by these methods.

5 Summary

The chemistry of ions in the gas phase is ideally investigated by mass spectrometry. Over the past twenty years various techniques have been accumulated that probe the intricacies of the chemical reactions occurring in the mass spectrometer from the instant of ionization, through the passage of ions along the flight path, to their arrival at the collector. If this whole process is considere as a chemical experiment, there is a variety of changes that can be examined, in terms of energy input and method of observation, to attain information about the reactions occurring. First, the ions under study can be generated by different ionization techniques. Second, the ions may be generated by fragmentation of molecular or parent ions. Third, the input molecules can be structurally labelled with isotopes (usually ^{2}H or ^{13}C). Fourth, the energy involved in the ionization process (*e.g.* energy of ionizing electrons or proton affinity of a chemical-ionization reagent) can be varied. A valuable measurement that ensues from controlled changing of the electron energy is the heat of formation of an ion. Fifth, excitation of ions along the flight path can be achieved by collisional activation or interaction with laser radiation. Sixth, the ions (and even neutral products) can be selected and analysed in a number of ways and can be interconnected by suitable combinations of electric, magnetic, and radiofrequency fields. Peak heights or areas and shapes are reproducibly measurable as well as are the masses of ions that they represent. A valuable parameter obtainable from peak shapes is the translational energy released during fragmentation.

Appropriate selection from the methods listed above can yield fundamental information about any particular ion or ionic reaction mechanism. In most cases more than one type of measurement should be acquired in order to obtain a satisfactory answer to a specific problem in gas-phase ion chemistry. This is amply illustrated in many of the publications reported in this chapter, where authors have substantiated their arguments by use of a combination of methods.

Although *fundamental* advances in the methods applicable to gas-phase chemistry of large organic ions have remained elusive during the period covered by this Report, there has been significant progress in certain areas, particularly for small ions (Chapter 1). It is being recognized increasingly that gas-phase ions can undergo intricate rearrangements *via* complex intermediates and that the mass spectrometrist has methods available to probe intermediate species. It is noticeable in this respect that discussions in many publications incorporate energy profiles for the ions involved. One reason for this is the increased reliability of determinations of heat of formation (either experimental or calculated). It appears that the time may be arriving when detailed decompositions of gas-phase ions can be mapped with a fair degree of certainty.

References

1 I. Howe in 'Mass Spectrometry', ed. R.A.W. Johnstone (Specialist Periodical Reports), The Royal Society of Chemistry, London, 1981, Vol. 6, p. 59.
2 I. Howe, D.H. Williams, and R.D. Bowen, 'Mass Spectrometry: Principles and Applications', McGraw-Hill, 1981.
3 K. Levsen, 'Fundamental Aspects of Organic Mass Spectrometry', Verlag Chemie, 1978.
4 'Gas-Phase Ion Chemistry', ed. M.T. Bowers, Academic Press, 1979, Vols. 1 and 2.
5 J.L. Holmes and J.K. Terlouw, *Org. Mass Spectrom.*, 1980, 15, 383.
6 H. Schwarz, *Top. Curr. Chem.*, 1981, 97, 1.
7 A.L. Burlingame, A. Dell, and D.H. Russell, *Anal. Chem.*, 1982, 54, 363R.
8 'Mass Spectrometry Reviews', ed. G.R. Waller and O.C. Dermer, Wiley, New York, 1982 *et seq.*
9 A. Maccoll, *Org. Mass Spectrom.*, 1982, 17, 1.
10 J.C. Traeger, *Org. Mass Spectrom.*, 1981, 16, 193.
11 J.C. Traeger and R.C. McLoughlin, *J. Am. Chem. Soc.*, 1981, 103, 3647.
12 D.M. Rider, G.W. Ray, E.J. Darland, and G.E. Leroi, *J. Chem. Phys.*, 1981, 74, 1652.
13 D. Griller and F.P. Lossing, *J. Am. Chem. Soc.*, 1981, 103, 1586.
14 F.P. Lossing, Y-T. Lam, and A. Maccoll, *Can. J. Chem.*, 1981, 59, 2228.
15 A. Maccoll and D. Mathur, *Org. Mass Spectrom.*, 1981, 16, 261.
16 M. Maccoll and D. Mathur, *Org. Mass Spectrom.*, 1980, 15, 483.
17 P.C. Burgers and J.L. Holmes, *Org. Mass Spectrom.*, 1982, 17, 123.
18 G. Depke, C. Lifshitz, H. Schwarz, and E. Tzidony, *Angew. Chem., Int. Ed. Engl.*, 1981, 20, 792.
19 J.K. Terlouw, W. Heerma, J.L. Holmes, and P.C. Burgers, *Org. Mass Spectrom.*, 1980, 15, 582.

20 J.W. Dallinga, N.M.M. Nibbering, and G.J. Louter, *Org. Mass Spectrom.*, 1981, 16, 183.
21 S.E. Biali, C. Lifshitz, Z. Rappoport, M. Karni, and A. Mandelbaum. *J. Am. Chem. Soc.*, 1981, 103, 2896.
22 W. Wagner-Redeker, K. Levsen, H. Schwarz, and W. Zummack, *Org. Mass Spectrom.*, 1981, 16, 361.
23 R. Herzschuh, G. Mann, Y. Werner, and E. Mende, *Org. Mass Spectrom.*, 1981, 16, 358.
24 P.C. Burgers, J.K. Terlouw, and K. Levsen, *Org. Mass Spectrom.*, 1982, 17, 295.
25 J.E. Bartness, *J. Am. Chem. Soc.*, 1982, 104, 335.
26 L.W. Sieck, M. Meot-Ner, and P. Ausloos, *J. Am. Chem. Soc.*, 1980, 102, 6866.
27 M. Meot-Ner, *J. Am. Chem. Soc.*, 1982, 104, 5.
28 R. Walder and J.L. Franklin, *Int. J. Mass Spectrom. Ion Phys.*, 1980, 36, 85.
29 S.A. McLuckey, D. Cameron, and R.G. Cooks, *J. Am. Chem. Soc.*, 1981, 103, 1313.
30 W.G. Liauw, M.S. Lin, and A.G. Harrison, *Org. Mass Spectrom.*, 1982, 16, 381.
31 J.A. Herman and A.G. Harrison, *Org. Mass Spectrom.*, 1981, 16, 423.
32 M. Colosimo and E. Brancaleoni, *Org. Mass Spectrom.*, 1982, 17, 286.
33 W.G. Liauw and A.G. Harrison, *Org. Mass Spectrom.*, 1981, 16, 388.
34 T. Keough and A.J. DeStefano, *Org. Mass Spectrom.*, 1981, 16, 527.
35 D.V. Davis and R.G. Cooks, *Org. Mass Spectrom.*, 1981, 16, 176.
36 Y.K. Lau, K. Nishizawa, A. Tse, R.S. Brown, and P. Kebarle, *J. Am. Chem. Soc.*, 1981, 103, 6291.
37 E.E. Kingston, J.S. Shannon, V. Diakiw, and M.J. Lacey, *Org. Mass Spectrom.*, 1982, 16, 428.
38 H.E. Audier and A. Milliet, *Org. Mass Spectrom.*, 1980, 15, 477.
39 G. Bouchoux and J. Dagaut, *Org. Mass Spectrom.*, 1981, 16, 246.
40 R.D. Bowen and D.H. Williams, *J. Am. Chem. Soc.*, 1980, 102, 2752.
41 J.F. Wendelboe, R.D. Bowen, and D.H. Williams, *J. Am. Chem Soc.*, 1981, 103, 2333.
42 J.F. Wendelboe, R.D. Bowen, and D.H. Williams, *J. Chem. Soc., Perkin Trans. 2*, 1981, 958.
43 H. Schwarz and D. Stahl, *Int. J. Mass Spectrom. Ion Phys.*, 1980, 36, 285.
44 R.D. Bowen and D.H. Williams, *J. Chem. Soc., Chem. Commun.*, 1981, 836.
45 H.W. Biermann, W.P. Freeman, and T.H. Morton, *J. Am. Chem. Soc.*, 1982, 104, 2307.
46 H.F. Grützmacher, A.-M. Dommröse, and U. Neuert, *Org. Mass Spectrom.*, 1981, 16, 279.
47 D.G. Hall and T.H. Morton, *J. Am. Chem. Soc.*, 1982, 104, 2307.
48 M. Colosimo and R. Bucci, *J. Chem. Soc., Chem. Commun.*, 1981, 659.
49 J.G. Liehr, A.G. Brenton, J.H. Beynon, J.A. McCloskey, W. Blum, and W.J. Richter, *Helv. Chim. Acta*, 1981, 64, 835.
50 D.E. Magnoli and J.R. Murdoch, *J. Am. Chem. Soc.*, 1981, 103, 7465.
51 P. Longevialle and R. Botter, *J. Chem. Soc., Chem. Commun.*, 1980, 823.
52 H. Budziciewicz, W. Ockels, and A.C. Campbell, *Org. Mass Spectrom.*, 1982, 17, 107.
53 R.R. Squires, C.H. DePuy, and V.M. Bierbaum, *J. Am. Chem. Soc.*, 1981, 103, 4256.

54 C.N. McEwen and M.A. Rudat, *J. Am. Chem. Soc.*, 1981, 103, 4355.
55 W. Wagner-Redeker and K. Levsen, *Org. Mass Spectrom.*, 1981, 16, 538.
56 J.A. Herman, Y.-H. Li, and A.G. Harrison, *Org. Mass Spectrom.*, 1982, 17, 143.
57 A.M. Falick, T. Gäumann, A. Heusler, H. Hirota, D. Stahl, and P. Tecon, *Org. Mass Spectrom.*, 1980, 15, 440.
58 H. Schwarz, *Angew. Chem., Int. Ed. Engl.*, 1981, 20, 991.
59 W. Francke and H. Schwarz, *J. Org. Chem.*, 1981, 46, 2806.
60 M. Tasaka, M. Ogata, and H. Ichikawa, *J. Am. Chem. Soc.*, 1981, 103, 1885.
61 S.K. Pollack, B.C. Raine, and W.J. Hehre, *J. Am. Chem. Soc.*, 1981, 103, 6308.
62 U.I. Zahorsky, *Org. Mass Spectrom*, 1982, 17, 253.
63 D.J. McAdoo, W. Farr, and C.E. Hudson, *J. Am. Chem. Soc.*, 1980, 102, 5165.
64 R.C. Heyer and M.E. Russell, *Org. Mass Spectrom.*, 1981, 16, 236.
65 T. Gäumann, H. Schwarz, D. Stahl, and J.-C. Tabet, *Helv. Chim. Acta*, 1981, 64, 2782.
66 J.E. Gonzalez and H.-F. Grützmacher, *Int. J. Mass Spectrom Ion Phys.*, 1981, 38, 181.
67 D.J. Burinsky, G.L. Glish, R.G. Cooks, J.H. Zwinselman, and N.M.M. Nibbering, *J. Am. Chem. Soc.*, 1981, 103, 465.
68 C.J. Porter, C.J. Proctor, E.A. Larka, and J.H. Beynon, *Org. Mass Spectrom.*, 1982, 17, 331.
69 R.D. Bowen and A.G. Harrison, *Org. Mass Spectrom.*, 1981, 16, 180.
70 A.J. Illies, S. Liu, and M.T. Bowers, *J. Am. Chem. Soc.*, 1981, 103, 5674.
71 H. Schwarz, *Top. Curr. Chem.*, 1978, 73, 231.
72 M.A. Baldwin, D.M. Carter, and J. Gilmore, *Org. Mass Spectrom.*, 1982, 17, 45.
73 K. Clausen, S. Scheibye, S.-O. Lawesson, J.H. Bowie, and T. Blumenthal, *Org. Mass Spectrom.*, 1980, 15, 640.
74 W.C.M.M. Luijten and J. Van Thuijl, *Org. Mass Spectrom.*, 1982, 17, 299.
75 W.C.M.M. Luijten and J. Van Thuijl, *Org. Mass Spectrom.*, 1982, 17, 304.
76 C. Paradisi, M. Prato, G. Scorrano, B. Ciommer, and H. Schwarz, *Org. Mass Spectrom.*, 1982, 17, 199.
77 R.J. Goldsack and J.S. Shannon, *Org. Mass Spectrom.*, 1980, 15, 545.
78 S. Tajima, T. Yanagisawa, T. Azami, and T. Tsuchiya, *Org. Mass Spectrom.*, 1980, 15, 609.
79 D.V. Ramana and N. Sundaram, *Org. Mass Spectrom.*, 1981, 16, 499.
80 H. Schwarz, *Top. Curr. Chem.*, 1981, 97, 1.
81 H. Thies, R. Wolfshütz, G. Frenking, J. Schmidt, and H. Schwarz, *Tetrahedron*, 1982, 38, 1647.
82 Y. Apeloig, W. Franke, Z. Rappoport, H. Schwarz, and D. Stahl, *J. Am. Chem. Soc.*, 1981, 103, 2770.
83 R. Schubert and H.-F. Grützmacher, *J. Am. Chem. Soc.*, 1980, 102, 5323.
84 D.V. Ramana and H.-F. Grützmacher, *Org. Mass Spectrom.*, 1981, 16, 227.
85 B. Schaldach, B. Grotmeyer, J. Grotemeyer, and H.-F. Grützmacher, *Org. Mass Spectrom.*, 1981, 16, 410.
86 H.-F. Grützmacher, *Org. Mass Spectrom.*, 1981, 16, 448.
87 Z.V.I. Zaretskii and P. Dan, *Org. Mass Spectrom.*, 1981, 16, 372.
88 R.D. Bowen, *J. Chem. Soc., Perkin Trans. 2*, 1982, 403.
89 R.D. Bowen, *J. Chem. Soc., Perkin Trans. 2*, 1982, 409.
90 E.A. Larka, I. Howe, J.H. Beynon, and Z.V.I. Zaretskii, *Org. Mass Spectrom.*, 1981, 16, 465.

91 R.J. Day and R.G. Cooks, *Int. J. Mass Spectrom. Ion Phys.*, 1980, 35, 293.
92 E.S. Mukhtar, I.W. Griffiths, F.M. Harris, and J.H. Beynon, *Int. J. Mass Spectrom. Ion Phys.*, 1981, 37, 159.
93 D.C. McGilvery and J.D. Morrison, *Int. J. Mass Spectrom. Ion Phys.*, 1978, 28, 81.
94 C.J. Porter, J.H. Beynon, and T. Ast, *Org. Mass Spectrom.*, 1981, 16, 101.
95 A. Maquestiau, P. Meyrant, and R. Flammang, *Org. Mass Spectrom.*, 1982, 17, 96.
96 D.F. Hunt, J. Shabanowitz, and A.B. Giordani, *Anal. Chem.*, 1980, 52, 386.
97 J.H. Beynon, F.M. Harris, B.N. Green, and R.H. Bateman, *Org. Mass Spectrom.*, 1982, 17, 55.
98 P.J. Todd and F.W. McLafferty, *Int. J. Mass Spectrom. Ion Phys.*, 1981, 38, 371.
99 C.J. Proctor, B. Kralj, A.G. Brenton, and J.H. Beynon, *Org. Mass Spectrom.*, 1980, 15, 619.
100 J.A. Laramée, D. Cameron, and R.G. Cooks, *J. Am. Chem. Soc.*, 1981, 103, 12.
101 H.E. Audier, G. Bouchoux, Y. Hoppilliard, and A. Milliet, *Org. Mass Spectrom.*, 1982, 17, 382.
102 C.E. Hudson and D.J. McAdoo, *Org. Mass Spectrom.*, 1982, 17, 366.
103 P.C. Burgers, J.K. Terlouw, and J.L. Holmes, *Org. Mass Spectrom.*, 1982, 17, 369.
104 C. Wesdemiotis and F.W. McLafferty, *Org. Mass Spectrom.*, 1981, 16, 381.
105 E.S. Mukhtar, I.W. Griffiths, F.M. Harris, and J.H. Beynon, *Org. Mass Spectrom.*, 1981, 16, 51.
106 I.W. Griffiths, E.S. Mukhtar, F.M. Harris, and J.H. Beynon, *Int. J. Mass Spectrom. Ion Phys.*, 1981, 38, 333.
107 J. Grotemeyer and H.-F. Grützmacher, *Org. Mass Spectrom.*, 1982, 17, 353.

5

Reactions of Negative Ions in the Gas Phase

BY J. H. BOWIE

1 Introduction

This review is primarily concerned with the formation and reactivity of organic negative ions in the gas phase. It does not cover the ion chemistry of atomic negative ions, nor of negative ions derived from inorganic molecules. The policy adopted in the last review[1] has been retained, *i.e. selective* references are listed in each section, but only those of *particular* interest in the context of this chapter are considered in any detail.

During the review period the major advances in negative-ion mass spectrometry have been in studies of negative-ion/molecule reactions, the development of negative-ion chemical-ionization mass spectrometry, and the introduction of fast-atom-bombardment mass spectrometry. Other areas of particular interest are laser applications and negative-ion field desorption.

Reviews[2-16] concerning various aspects of negative-ion chemistry have been published in the last two years. The review by Christophorou,[4] which deals with negative-ion states of polyatomic molecules, and those by DePuy[12] and Nibbering[13] on ion/molecule reactions are particularly recommended.

2 Negative Ions Formed by Electron Impact (or Dissociative Electron Impact): Fragmentation Mechanisms

Several reviews of this area have been published.[5,15] Negative-ion spectra have been measured for the following systems: carbon monoxide,[17] sulphonic acids,[18] nitrogen-containing explosives,[19] 1-amino-2-benzothiazoles,[20] barbiturates,[21] some citraconic anhydride derivatives of *Aspergillus wentii* Wehmer,[22] derivatives of usnic acid,[23,24] bis-(1,1,1,5,5,5-hexafluoropentane-2,4-dionato)-metal complexes,[25] and various bis-chelates of Ni^{II}.[26]

The positive-ion spectra of 15-oxohexadecylcitraconic anhydride (1) and related compounds show weak molecular ions and complex fragmentation pathways, whereas the negative-ion spectra show abundant molecular anions and little fragmentation.[22] The negative-ion spectra of ring-C-cleaved usnic acid derivatives show pronounced molecular anions and characteristic fragmentation patterns. For example, compound (2) undergoes the processes $M^{-\cdot} \rightarrow [M - (MeCO^{\cdot} + CH_2CO)]^-$ and characteristic rearrangement of the side chain, illustrated in structures (2) and (3).

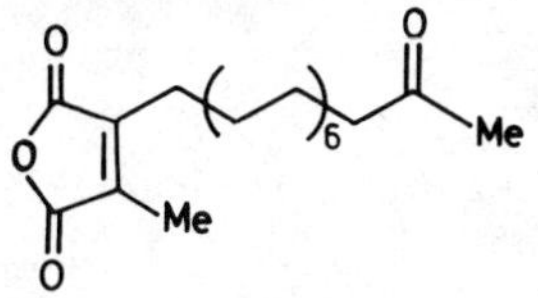

(1)

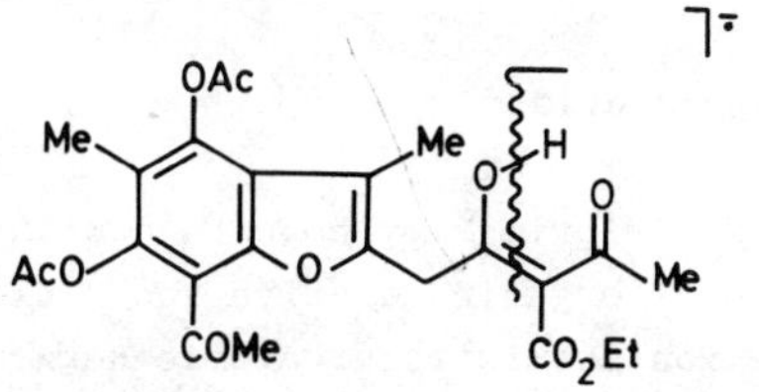

(2)

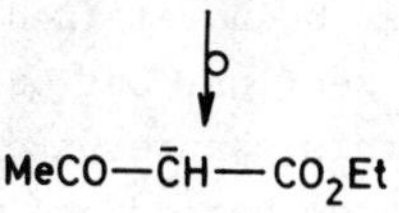

MeCO—$\bar{C}$H—CO_2Et

(3)

3 Field Desorption, Laser Applications, Secondary-ion Mass Spectrometry, and Californium Plasma Spectrometry

Further development is reported of field-desorption mass spectrometry of negative ions.[27-30] The low field strength required for field electron emission from metals and semiconductors and secondary effects such as cathode sputtering and gas discharges has hindered the development of the technique. It has now been shown[27-29] that, if samples are admixed with a small amount of some salt and viscous polymers such as polyethylene oxides, the field strength necessary to desolvate negative ions from the condensed phase is reduced. In this way, anion-cluster spectra may be obtained for a variety of organic compounds including sterols,[27] sugars,[27,28] fatty acids,[28] and adenosines.[30]

Laser-induced electron photodetachment is an important

technique that provides precise information on negative ions,[31-36] in particular on electron affinities.[31-35] Certain negative ions may be photodissociated by means of low-intensity infrared-laser radiation; for example the negative ion [F^-(HOMe)] has been photodissociated under collisionless conditions.[37] Multi-photon electron detachment is a convenient method for obtaining infrared spectra of negative ions, thus allowing differentiation between isomeric structures. For example, the benzyl, cycloheptatrienyl, and norbornadienyl negative ions (produced by deprotonation of the appropriate neutral by strong bases in an ion-cyclotron-resonance cell) show quite different infrared spectra.[38] Laser-induced mass spectrometry has been used for the determination of airborne particles,[39] and anion attachment reactions in laser-desorption mass spectrometry have been investigated.[40]

Further work has been reported on negative-secondary-ion mass spectrometry (s.i.m.s.).[41,43,44] In s.i.m.s., a solid is bombarded by a kilovolt-energy ion beam, often Ar^+, and the mass spectrum of the sputtered secondary ions is recorded.[42] The method has been used for the structure determination of amines,[43] amides,[43] and organic acids.[44] Acids show $M^{-\bullet}$ and $(M - H)^-$ ions as the dominant high-mass species.[44]

Evidence has been provided which suggests that ^{252}Cf fission fragment mass spectrometry involves an ionization process that is primarily a high-energy one, but that there is also a low-energy component involved in the formation of quasi-molecular ions.[45] For example, the ^{252}Cf negative-ion spectrum of guanosine shows an $(M - H)^-$ ion of small abundance whereas the base peak of the spectrum is produced by CN^-, an ion that can only be formed *in this case* by high-energy processes.[45]

4 Fast-atom-bombardment Spectrometry

Recently, a new mass-spectrometric technique called fast atom bombardment (f.a.b.) has been developed for dealing with involatile, thermally labile molecules.[46,47] The f.a.b. technique involves bombarding the sample in the ion source with argon atoms having large translational energies (2 - 10 keV). F.a.b. spectrometry can be routinely used in both the positive- and negative-ion modes, and molecular weights in the 1000 - 2000 dalton range may be routinely determined by the use of less than 1 nmol of material. Published work has appeared on the negative-ion f.a.b. spectra of

[52,55] polar antibiotics,[50] nucleoside phosphates,[50] [53] monosaccharides,[54] and ecdysone phosphates.[56] Much current work is still in press, and bulletins from manufacturers of f.a.b. sources (which refer to the unpublished work) promise even more spectacular advances.

Negative-ion f.a.b. spectra are particularly useful for molecular-weight determination since they usually contain pronounced $(M - H)^-$ ions (with minimal fragmentation), thus complementing the positive-ion spectra that are often characterized by intense fragment cations.[52] It is possible that $(M - H)^-$ ions are formed by proton-transfer reactions that may occur as the molecules are bombarded. $(M - H)^-$ ions have to date been observed for peptides containing up to 21 amino-acids,[50,55] for tetranucleoside triphosphates,[50] and for polysaccharides.[57] The negative-ion f.a.b. spectrum of γ-cyclodextrin $[(C_6H_{10}O_5)_8$, molecular weight 1296 daltons$]$ shows a pronounced $(M - H)^-$ ion together with a smaller fragment ion $[(M - H) - C_6H_{10}O_4]^-$.[57] The most spectacular application of f.a.b. to date has been its use by Rinehart and colleagues[52] for the determination of the structures of the Zervamicin and Emerimicin peptide antibiotics.

5 Negative-ion Chemical-ionization Mass Spectrometry

It is arguable whether the term 'chemical-ionization mass spectrometry' should be used to describe electron-capture or dissociative-electron-capture spectra obtained when the sample under investigation captures a low-energy electron produced by ionization of a suitable neutral that is also present in the ion source. Such spectra are, however, usually determined in chemical-ionization ion sources, and they have been classified 'chemical ionization' in an earlier review.[58] Electron attachment to halocarbons from nitrogen and argon/nitrogen mixtures has been described.[59] Methane 'chemical-ionization negative-ion mass spectra' have been reported for polyhalogenated hydrocarbons,[60] chlorinated 2-phenoxyphenols,[61] hydroxypolychlorodibenzofurans,[62] polychlorinated dibenzofurans,[63] organic nitriles,[64-67] penicillins,[68] 1,3-dihydro-5-phenyl-1,4-benzodiazepin-2-ones,[69,70] aflatoxins,[71] and mycotoxins.[71] Isobutane 'chemical-ionization negative-ion mass spectra' have been described for amines,[72] xenobiotic chemicals including polychlorodibenzo-p-dioxins,[73] peptides,[74] and for direct mixture analysis.[75]

Atmospheric-pressure chemical-ionization negative-ion mass spectrometry (using ambient air) has been used to form $M^{-\bullet}$ and $(M - H)^-$ ions from dinitrotoluene[76] and the trichlorophenoxide negative ion from trichlorophenoxyacetic acid metabolites.[77]

Several aspects of this work need to be highlighted. Negative-ion mass spectrometry is an extremely sensitive method for the screening of environmental contamination, particularly by polyhalogenated aromatic compounds. Extracts of fish from certain rivers and lakes in the United States show the presence of polychlorodiphenyls, naphthalenes, and the extremely hazardous dibenzofurans (4) and dibenzodioxins (5).[73] Polychlorodibenzofurans

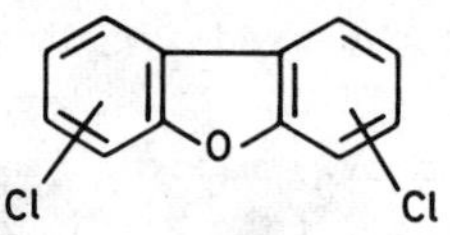

(4) 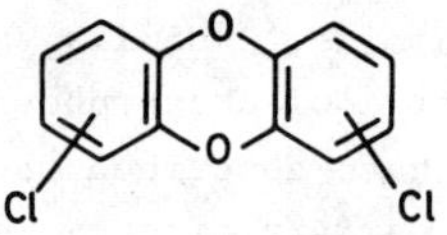 (5)

have also been detected in turtles (from the Hudson River, U.S.A.) and in grey seals (from the Gulf of Bosnia, Sweden).[63] In these cases methane or isobutane/oxygen (10:1) was used as the 'reagent' gas. Negative-ion spectrometry (methane) may be used to monitor the presence of aflatoxin B_1 in peanut oil and corn and aflatoxin M_1 in milk, at levels as low as 10 p.p.b.[71]

In the last review[1] we recorded that 7,7,8,8-tetracyanoquinodimethane (TCNQ) under CH_4/n.i.c.i. conditions yields $(M + H)^-$, $(M + Me)^-$, and $(M + Et)^-$ ions formed by radical reactions, since H^-, Me^-, and Et^- are not present in the methane plasma.[65] It has now been shown that TCNQ reacts sufficiently fast with carbon radicals so that the reaction products dominate the n.i.c.i. mass spectrum at low TCNQ pressures.[66] Collisional activation (c.a.) of certain negative ions resulting from the reaction of TCNQ with radicals provides c.a. spectra that are dominated by fragmentation through that part of the molecule formed from the original radical. Thus the spectrum is interpretable in terms of the radical structure. For example, the negative-ion collision-induced decomposition of $[(TCNQ + Bu^s) - CN]^{-\bullet}$ (6) is shown by structures (6) to (7), thus identifying the radical structure.[67]

$(M + Cl)^-$ peaks are observed in the CF_2Cl_2 n.i.c.i. spectra of a number of compounds of biological interest including prostaglandins

(6) $\xrightarrow{CA}$ (7)

and some alkaloids;[78] similar peaks are observed for dichloromethane n.i.c.i. spectra of organophosphorus pesticides.[79] Tiernan and colleagues[80] have described the F^- (from NF_3) n.i.c.i. mass spectra of a variety of organic molecules; $(M - H)^-$ ions are observed for carboxylic acids, phenols, ketones, and aldehydes, whereas $(M + F^-)$ negative ions are observed for carboxylic acids, nitriles, ketones, and aldehydes.

HO^- n.i.c.i. mass spectra have been reported for aromatic components in liquid fuels,[81] acyl chlorides, diacyl chlorides, and perfluorodiacyl chlorides,[82] alcohols,[83-86] carbonyl compounds,[87,88] peptides,[89] and nitro compounds.[90,91] Polyperfluoropropylene oxide has been recommended as a calibration substance for n.i.c.i. mass spectrometry.[92]

Alkoxide (RO^-) and 'solvated' alkoxide ($RO^- \cdots HOR$) negative ions derived from monoalcohols (formed by HO^- n.i.c.i. mass spectrometry) do not undergo exchange with D_2; $(M - H)^-$ and $(2M - H)^-$ ions derived from dialcohols do exchange with D_2, and the mechanism involving a change from structure (8) to (9) has

(8) $\xrightarrow{D_2}$ $\longrightarrow$ (9) + HD

been proposed to explain the exchange in $(M - H)^-$ ions.[86] In an important paper Hunt[88] has described the collision-induced spectra derived from the $(M - H)^-$ ions of carbonyl compounds. Some rearrangements of acyclic ketones are shown in the fragmentation of ion (10) to (11) and in (12) to (13). The $(M - H)^-$ negative

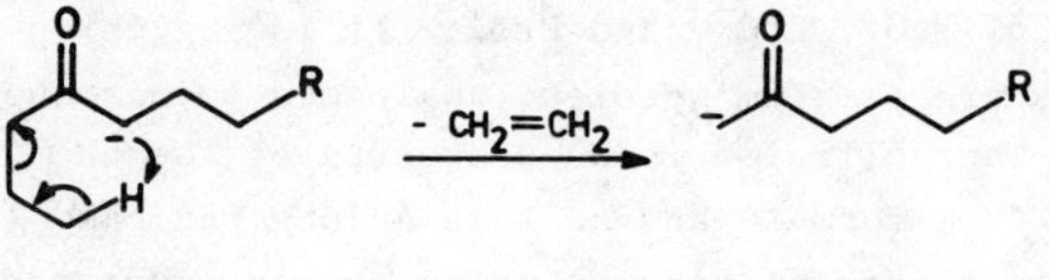

(10) $\xrightarrow{-CH_2=CH_2}$ (11)

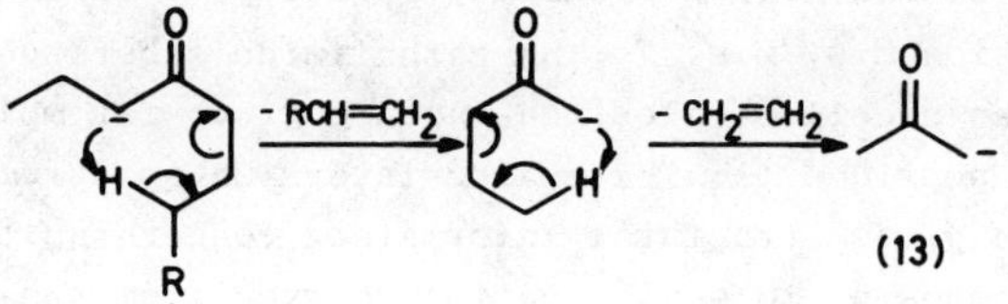

(12) $\xrightarrow{-RCH=CH_2}$ $\xrightarrow{-CH_2=CH_2}$ (13)

ion derived from cyclohexanone undergoes the retro Diels-Alder reaction of structure (14) to give (15).[86] The collision-induced spectra derived from $(M - H)^-$ ions from tripeptides show the fragmentations illustrated in structure (16).[89]

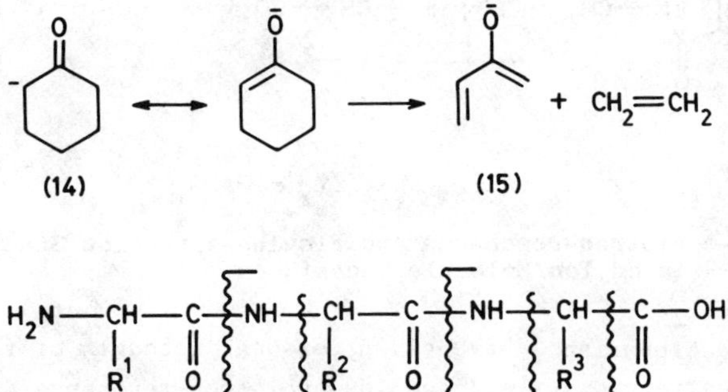

(14) ⟷ ⟶ (15) + $CH_2=CH_2$

$H_2N-CH(R^1)-C(=O)-NH-CH(R^2)-C(=O)-NH-CH(R^3)-C(=O)-OH$

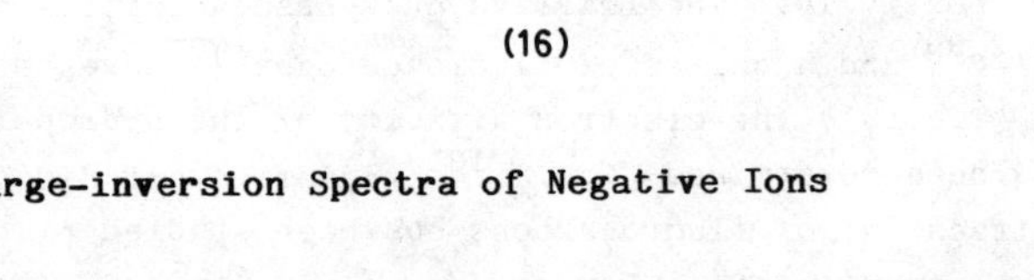

(16)

6 Charge-inversion Spectra of Negative Ions

This topic was discussed in Volume 6 of this series (ref. 1, p. 240). Thiocarboxylate cations $(RCOS^+)$, formed by charge inversion from $(RCOS)^-$, fragment to form R^+, $(COS)^{+\cdot}$, and RCO^+ as major product ions.[93] The positive ions formed on charge reversal of methoxide, ethoxide, n-propoxide, and isopropoxide negative ions do not have the same structures as $CH_2{=}\overset{+}{O}H$, $MeCH{=}\overset{+}{O}H$, $MeOCH_2^+$, $Me_2C{=}\overset{+}{O}H$, $EtOCH_2^+$, and $MeCH{=}OMe^+$, which have been identified as the major stable forms of MeO^+, EtO^+, and PrO^+. It is suggested that the new positive ions are oxenium species, analogous to carbenes and nitrenes.[94] The collision-induced spectra of the charge-reversed ions derived from formate and acetate anions indicate that no structural rearrangement occurs during charge inversion.[95] Polycyclic aromatic hydrocarbons may be analysed during n.i.c.i. charge-inversion m.s./m.s.[96] The phthalimido substituent has been found to be an excellent electron-capture group for molecular-anion formation of peptides.[97] The charge-inversion spectrum of that molecular ion gives structural information concerning the peptide, *e.g.* the decomposing molecular cation derived from the molecular anion of the Gly-Gly-Gly derivative (17) fragments as shown.[97] The charge-inversion spectra of the $(M - H)^-$ ions derived from lactic and hippuric acid have been recorded, using SF_6 as the collision gas.[98]

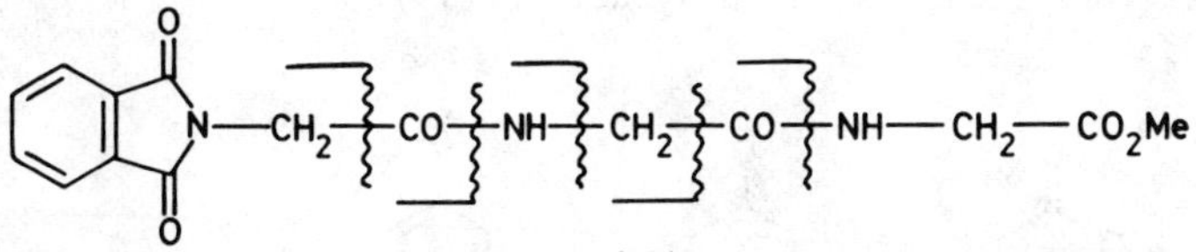

N—CH2—CO—NH—CH2—CO—NH—CH2—CO2Me

(17)

7 Ion-cyclotron-resonance and Flowing-afterglow Studies, and Related Ion/Molecule Techniques

Depuy[12] and Nibbering[13] have given personal accounts of their work on flowing-afterglow (f.a.) and ion-cyclotron-resonance (i.c.r.) studies, respectively. The relative gas-phase acidities of 81 organic acids[99] and a series of fluoroacetones[100] have been determined (i.c.r.), the electron affinity of the hydroperoxy radical has been determined (f.a.),[101] radiative and dissociative electron attachment of fluorocarbons has been studied (i.c.r.),[102] a theoretical study of the H^-/MeCl reaction has been reported,[103] and reactions of $O^{-\cdot}$ with allene (crossed beam)[104] and of HO^- with allene (f.a.)[105] have been described.

Ion/molecule reactions of the cyclopentadienyl anion ($C_5H_5^-$) have been studied (f.a.); for example, it reacts with alcohols to form 'solvated' species $[C_5H_5^-(HOR)]$, and it undergoes conjugate addition to acrylonitrile to yield $C_5H_5{-}CH_2{-}\bar{C}HCN$.[106] Further H-D exchange reactions of carbanions have been reported (f.a.);[107] rates of proton abstraction from conjugate dienes (i.c.r.)[108] and by HO^- from alcohols and carbonyl compounds (f.a.)[109] have been determined. The rate constant for reaction (1) decreases markedly

$$[HO^-(H_2O)_n] + MeBr \longrightarrow [Br^-(H_2O)_n] + MeOH \qquad (1)$$

as n increases from 0 to 3 (f.a.); semi-quantitative reaction-co-ordinate profiles have been constructed for these reactions.[110] The HO^- negative ion reacts with diethyl ether as in equation (2) (i.c.r.).[111]

$$HO^- + H{-}CH_2{-}CH_2{-}OEt \longrightarrow [EtO^-(H_2O)] + C_2H_4 \qquad (2)$$

The reactions of the phenylnitrene radical anion have been studied by f.a.[112,113] and i.c.r. techniques.[114] The proton affinity of $PhN^{-\bullet}$ is estimated to be 372 ± 2 kcal mol^{-1};[113] $PhN^{-\bullet}$ reacts with phenyl azide to produce ions (18) and (19),[112] with alcohols by the sequence of bimolecular reactions shown in reaction (3), and with propionaldehyde by the overall sequence

$$[Ph{-}N{=}N{-}N{=}N{-}Ph]^{-\bullet} \quad (18) \qquad [Ph{-}N{=}N{-}Ph]^{-\bullet} \quad (19)$$

shown in reaction (4).[114] The reactions of a number of nucleophiles

$$PhN^{-\bullet} + ROH \longrightarrow [PhN^{-\bullet}(HOR)] \longrightarrow PhNH^\bullet + [RO^-(HOR)] \longrightarrow [RO^-(HOR)_x] \qquad (3)$$

$$PhN^{-\bullet} + EtCHO \longrightarrow \left[PhN{-}\overset{\overset{\displaystyle \bar{O}}{|}}{\underset{\underset{\displaystyle Et}{|}}{C}}{-}H\right] \longrightarrow Ph{-}\bar{N}{-}CHO + Et^\bullet \qquad (4)$$

with alkyl nitrites have been described under f.a. conditions, and these results have been compared with those of earlier i.c.r. studies (ref. 1, p. 239).[115] For example, the fluoride negative ion reacts with methyl nitrite to produce NO_2^-, NO^-, and $[NO^-(HF)]$, while carbanions undergo the usual nitrosation reaction (*e.g.* equation (5); see also ref. 1, p. 239). The reactions of many carbanions with alkyl nitrites suggest the initial formation of a complex that may undergo one or more rearrangements; an example is shown in reaction (6).[115]

$$CH_2{=}CH{-}CH_2^- + RONO \longrightarrow CH_2{=}CH{-}CH{=}NO^- + ROH \qquad (5)$$

$$Me{-}C(=O){-}CH_2^- + RONO \longrightarrow [Me{-}C(=O){-}CH_2^-(RONO)] \longrightarrow Me{-}C(=O)OR + CH_2NO^- \qquad (6)$$

$$RO^- + MeCO_2Pr \longrightarrow {}^-CH_2CO_2R + PrOH \qquad (7)$$

The study of the reactions of gas-phase negative ions with carbonyl-containing molecules[116-125] continues to be a fruitful area of research. Reaction (7) is characteristic of alkyl acetates (i.c.r.); the reaction is thought to proceed through a tetrahedral intermediate (20), with the hydrogen-transfer step being rate determining.[118] *Ab initio* studies confirm that the acetate negative ion may react with acetic anhydride to produce the tetrahedral intermediate (21) but that the subsequent γ-hydrogen rearrangement produces the 'solvated' acetate ion (22), not a tetrahedral product ion.[119] Alkoxide negative ions undergo

RO OPr Ō H (20)   Me O Ō O O Me H (21) $\longrightarrow [MeCO_2^-(HO_2CMe)]$ (22)

nucleophilic attack at both a methyl or the carbonyl group of dimethylcarbonate (i.c.r.).[120] It has been suggested that strongly basic nucleophiles (*e.g.* RO^-, F^-) react with CH—CO— systems to form an initial intermediate (23) that rearranges to a *stable* product (24), whereas less basic anions (*e.g.* $MeCO_2^-$) react to give (25), which rearranges to a tetrahedral intermediate (26).[120,121] In an important i.c.r. study, Bartmess[123] has shown that enolate negative ions react with esters in the gas phase by an addition-elimination deprotonation mechanism similar to that of the Claisen condensation in solution. Possible reaction-co-ordinate profiles

NuH---C⁻—C(=O)— ⟶ (24)

(23) (24)

Nu⁻---H—C—C(=O)— ⟶ HC—C(O⁻)(Nu)—

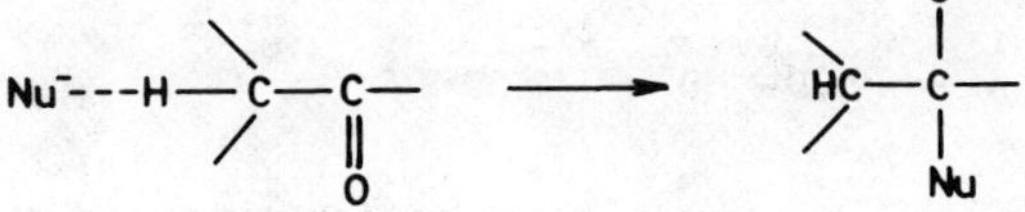

(25) (26)

have been proposed, but some intermediates are questionable; a simplified version of the mechanism is given in the sequence shown in reaction (8).

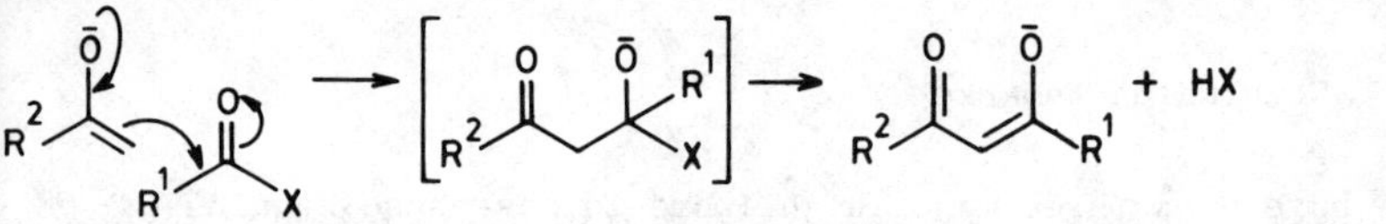

$$R^2C(O^-)=CH_2 + R^1C(O)X \longrightarrow [R^2C(O)CH_2C(O^-)(R^1)X] \longrightarrow R^2C(O)CH=C(O^-)R^1 + HX \quad (8)$$

Many gas-phase anions react with carbonyl sulphide to form RS^- ions (f.a.).[126] Reactions of various nucleophiles with 1,3-dithianes and 1,3-dithiolanes have been reported (i.c.r.);[127] for example the methoxide negative ion reacts with 1,3-dithiane to form $(M - H)^-$ and $CH_2=CH-CH_2S^-$ ions. Both f.a.[128-130] and i.c.r.[131] studies of silanes have been described. A number of stable pentavalent silicon negative ions are observed for the reactions (f.a.) of F^- with silanes (*e.g.* the stable adduct (27) is formed between F^- and 1,1-dimethylsilacyclobutane) and for the reactions (i.c.r.) of RO^- with silanes.[131] The decomposing forms of the trigonal-bipyramidal adducts decompose to stable

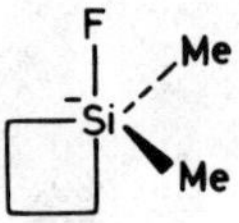

(27) 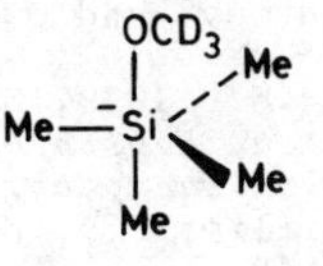 (28) 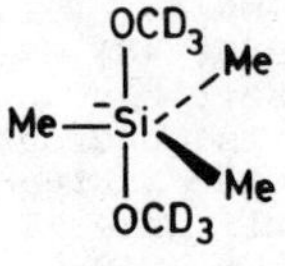 (29)

tetrahedral negative ions by elimination of neutral molecules; for example ion (28) eliminates CH_3CD_3, whereas (29) eliminates both CH_4 and CH_3CD_3.[131] The nucleophilic displacement reaction between Me_3SiR and F^- [128,129] or MeO^- [131] yields R^- ions that are often difficult to produce by conventional ionization procedures. Reactions of the R^- ions can lead to information concerning the structure of that ion. For example, reaction (9) demonstrates that the non-decomposing benzyl negative ion does not undergo

$$PhCD_2^- + MeONO \longrightarrow PhCD{=}NO^- + MeOD \qquad (9)$$

benzyl hydrogen/phenyl hydrogen equilibration prior to the fragmentation reaction.[131] Similarly, the allyl anion reacts as a symmetrical species, and the 'propargyl' anion $C_3H_3^-$ may react as either $HC{\equiv}C{-}CH_2^-$ or $^-HC{=}C{=}CH_2$.[131] Silyl enol ethers react with F^-, NH_2^-, or HO^- to yield enolate negative ions (f.a.),[132] and various negative ions react with trimethylphosphate to produce $(MeO)_2PO_2^-$ i.c.r.[133]

8 Concluding Remarks

There have been three major highlights in gas-phase studies of negative ions over the past two years: firstly, the studies of the mechanisms of ion/molecule reactions occurring between negative ions and organic molecules and the development of a more quantitative approach to this problem, secondly, the further development of the n.i.c.i. technique, in particular its application to environmental problems, and, thirdly, the introduction and promise of f.a.b. mass spectrometry.

References

1 J.H. Bowie in 'Mass Spectrometry', ed. R.A.W. Johnstone (Specialist Periodical Reports), The Royal Society of Chemistry, London, 1981, Vol. 6, p. 233.
2 A.A. Solov'ev, V.I. Kadentsev, and O.S. Chizhov, *Russ. Chem. Rev.*, 1979, 48, 631.
3 R.W. Kiser, *Top. Curr. Chem.*, 1979, 85, 89.
4 L.G. Christophorou, *Environ. Health Perspect.*, 1980, 36, 3.
5 J.H. Bowie, *Environ. Health Perspect.*, 1980, 36, 89.
6 A.L. Burlingame, T.A. Baillie, P.J. Derrick, and O.S. Chizhov, *Anal. Chem.*, 1980, 52, 214R.

7 H. Brandenberger in 'Recent Developments in Mass Spectrometry in Biochemistry and Medicine', Elsevier Sci. Publ. Co., Amsterdam, 1980, Vol. 6, p. 391.
8 E.C. Horning, D.I. Carroll, R.N. Stillwell, and I. Dzidic in 'Recent Developments in Mass Spectrometry in Biochemistry and Medicine', Elsevier Sci. Publ. Co., Amsterdam, 1980, Vol. 6, p. 453.
9 H.-J. Stan, D. Quantz, and B. Abraham in 'Recent Developments in Mass Spectrometry in Biochemistry, Medical and Environmental Research', Elsevier Sci. Publ. Co., Amsterdam, 1981, Vol. 7, p. 335.
10 M. Cooke and A.J. Dennis, 'Polynuclear Aromatic Hydrocarbons, Chemical Analysis and Biological Fate', Battelle Press, Columbus, Ohio, 1981.
11 L.G. Christophorou in 'Photon, Electron and Ion Probes of Polymer Structure and Properties', ed. D.W. Dwight, T.J. Fadish, and H.R. Thomas, American Chemical Society, Washington D.C., 1981.
12 C.H. DePuy and V.M. Bierbaum, *Acc. Chem. Res.*, 1981, 14, 146.
13 N.M.M. Nibbering, *Recl.: J. R. Neth. Chem. Soc.*, 1981, 100, 297.
14 W.C. Brumley and J.A. Sphon, *Biomed. Mass Spectrom.*, 1981, 8, 390.
15 H. Budzikiewicz, *Angew. Chem., Int. Ed. Engl.*, 1981, 20, 624.
16 N.H. Mahle and L.A. Shadoff, *Biomed. Mass Spectrom.*, 1982, 9, 45.
17 R. Abouaf, D. Teillet-Billy, and S. Goursaud, *J. Phys. B*, 1981, 14, 3517.
18 Z. Przybylski and A. Borkowska, *Sb. Vys. Sk. Chem.-Technol. Praze, Technol. Paliv*, 1980, D42, 39.
19 F. Daubin, *Combust. Flame*, 1981, 40, 105.
20 S. Claude, R. Tabacchi, L. Duc, and J.-F. Marrel, *Helv. Chim. Acta*, 1981, 64, 1545.
21 L.V. Jones and M.J. Whitehouse, *Biomed. Mass Spectrom.*, 1981, 8, 231.
22 A. Selva, P. Traldi, L. Camarda, and G. Nasini, *Biomed. Mass Spectrom.*, 1980, 7, 148.
23 S. Huneck and J. Schmidt, *Biomed. Mass Spectrom.*, 1980, 7, 301.
24 J. Schmidt, S. Huneck, and P. Franke, *Biomed. Mass Spectrom.*, 1981, 8, 293.
25 D.R. Dakternieks, I.W. Fraser, J.L. Garnett, and I.K. Gregor, *Aust. J. Chem.*, 1979, 32, 2405.
26 J.L. Garnett, I.K. Gregor, M. Guilhaus, and D.R. Dakternieks, *Inorg. Chim. Acta*, 1980, 44, LI2I.
27 K.H. Ott, F.W. Röllgen, J.H. Zwinselman, R.H. Fokkens, and N.M.M. Nibbering, *Org. Mass Spectrom.*, 1980, 15, 419.
28 J.J. Zwinselman, R.H. Fokkens, N.M.M. Nibbering, K.H. Ott, and F.W. Röllgen, *Biomed. Mass Spectrom.*, 1981, 8, 312.
29 K.H. Ott, F.W. Röllgen, J.J. Zwinselman, R.H. Fokkens, and N.M.M. Nibbering, *Angew. Chem., Int. Ed. Engl.*, 1981, 20, 111.
30 K.H. Ott, F.W. Röllgen, P. Dähling, J.J. Zwinselman, R.H. Fokkens, and N.M.M. Nibbering, *Org. Mass Spectrom.*, 1981, 16, 336.
31 R.L. Jackson, A.H. Zimmerman, and J.I. Brauman, *J. Chem. Phys.*, 1979, 71, 2088.
32 B.K. Janousek, A.H. Zimmerman, K.J. Reed, and J.I. Brauman, *J. Am. Chem. Soc.*, 1978, 100, 6142.
33 A.H. Zimmerman, R.L. Jackson, B.K. Janousek, and J.I. Brauman, *J. Am. Chem. Soc.*, 1978, 100, 4674.

34 A.H. Zimmerman, R. Gygax, and J.I. Brauman, *J. Am. Chem. Soc.*, 1978, 100, 5595.
35 B.K. Janousek, K.J. Reed, and J.I. Brauman, *J. Am. Chem. Soc.*, 1980, 102, 3125.
36 F. Breyer, P. Frey, and H. Hotop, *Z. Phys. A - At. Nucl.* 1981, 300, 7.
37 R.N. Rosenfeld, J.M. Jasinski, and J.I. Brauman, *J. Am. Chem. Soc.*, 1979, 101, 3999.
38 C.A. Wight and J.L. Beauchamp, *J. Am. Chem. Soc.*, 1981, 103, 6499.
39 P. Wieser, R. Wurster, and H. Seiler, *Atmos. Environ.*, 1980, 14, 485.
40 K. Balasanmugam, T.A. Dang, R.J. Day, and D.M. Hercules, *Anal. Chem.*, 1981, 53, 2296.
41 W.L. Baum, *Pure Appl. Chem.*, 1982, 54, 323.
42 A. Benninghoven and W. Sichtermann, *Anal. Chem.*, 1978, 50, 1180.
43 S.E. Unger, R.J. Day, and R.G. Cooks, *Int. J. Mass Spectrom. Ion Phys.*, 1981, 39, 231.
44 L.K. Liu, S.E. Unger, and R.G. Cooks, *Tetrahedron*, 1981, 37, 1067.
45 B.T. Chait, W.G. Agosta, and F.H. Field, *Int. J. Mass Spectrom. Ion Phys.*, 1981, 39, 339.
46 D.J. Surman and J.C. Vickerman, *J. Chem. Soc., Chem. Commun.*, 1981, 324.
47 M. Barber, R.S. Bordoli, R.D. Sedgwick, and A.N. Tyler, *J. Chem. Soc., Chem. Commun.*, 1981, 325; M. Barber, R.S. Bordoli, G.J. Elliott, R.D. Sedgwick, and A.N. Tyler, *Anal. Chem.*, 1982, 54, 645A; see also F.M. Devienne and J.-C. Roustan, *Org. Mass Spectrom.*, 1982, 17, 173.
48 M. Barber, R.S. Bordoli, D.R. Sedgwick, and L.W. Tetler, *Org. Mass Spectrom.*, 1981, 16, 256.
49 M. Barber, R.S. Bordoli, G.V. Garner, D.B. Gordon, R.D. Sedgwick, L.W. Tetler, and A.N. Tyler, *Biochem. J.*, 1981, 197, 401.
50 D.H. Williams, C. Bradley, G. Bojesen, S. Santikarn, and L.C.E. Taylor, *J. Am. Chem. Soc.*, 1981, 103, 5700.
51 D.H. Williams, G. Bojesen, A.D. Auffret, and L.C.E. Taylor, *Fed. Eur. Biochem. Soc. Lett.*, 1981, 128, 37.
52 K.L. Rinehart, L.A. Gaudioso, M.L. Moore, R.C. Pandey, J.C. Cook, M. Barber, R.D. Sedgwick, R.S. Bordoli, A.N. Tyler, and B.N. Green, *J. Am. Chem. Soc.*, 1981, 103, 6517.
53 M. Barber, R.S. Bordoli, R.D. Sedgwick, A.N. Tyler, B.N. Green, V.C. Parr, and J.L. Gower, *Biomed. Mass Spectrom.*, 1982, 9, 11.
54 M. Barber, R.S. Bordoli, R.D. Sedgwick, and J.V. Vickerman, *J. Chem. Soc., Faraday Trans. 1*, 1982, 78, 1291.
55 D.H. Williams, C.V. Bradley, S. Santikarn, and G. Bojesen, *Biochem. J.*, 1982, 201, 105.
56 R.E. Isaac, M.E. Rose, H.H. Rees, and T.W. Goodwin, *J. Chem. Soc., Chem. Commun.*, 1982, 249.
57 'Fast Atom Bombardment Spectra', Kratos Bulletin, A300-0981, 1982, p. 2.
58 K.R. Jennings in 'Mass Spectrometry', ed. R.A.W. Johnstone (Specialist Periodical Reports), The Chemical Society, London, 1977, Vol. 4, p. 203.
59 P. Popp and J. Leonhardt, *Beitr. Plasmaphys.*, 1981, 21, 293.
60 F.W. Crow, Å. Bjorseth, K.T. Knapp, and R. Bennett, *Anal. Chem.*, 1981, 53, 619.
61 K.L. Busch, A. Norström, C.-A. Nilsson, M.M. Bursey, and J.R. Hass, *Environ. Health Perspect.*, 1980, 36, 125.
62 M. Deinzer, D. Griffin, T. Miller, J. Lamberton, P. Freeman, and V. Jonas, *Biomed. Mass Spectrom.*, 1982, 9, 85.

63 C. Rappe, H.R. Buser, D.L. Stalling, L.M. Smith, and R.C. Dougherty, *Nature (London)*, 1981, 292, 524.
64 K.L. Bush, C.E. Parker, D.J. Harvan, M.M. Bursey, and J.R. Hass, *Appl. Spectrosc.*, 1981, 35, 85.
65 C.N. McEwan and M.A. Rudat, *J. Am. Chem. Soc.*, 1979, 101, 6470.
66 C.N. McEwan and M.A. Rudat, *J. Am. Chem. Soc.*, 1981, 103, 4343.
67 M.A. Rudat and C.N. McEwen, *J. Am. Chem. Soc.*, 1981, 103, 4349.
68 J.I. Gower, C. Beaugrand, and C. Sallot, *Biomed. Mass Spectrom.*, 1981, 8, 36.
69 W.A. Garland and B.J. Miwa, *Environ. Health Perspect.*, 1980, 36, 69.
70 B.J. Miwa, W.A. Garland, and P. Blumenthal, *Anal. Chem.*, 1981, 53, 793.
71 W.C. Brumley, S. Neshelm, M.W. Trucksess, E.W. Trucksess, P.A. Drelfuss, J.A.G. Roach, D. Andrzejewski, R.M. Eppley, A.E. Pohland, C.W. Thorpe, and J.A. Sphon, *Anal. Chem.*, 1981, 53, 2003.
72 E.A. Bergner and R.C. Dougherty, *Biomed. Mass Spectrom.*, 1981, 8, 204.
73 R.C. Dougherty, M.J. Whitaker, L.M. Smith, D.L. Stalling, and D.W. Kuehl, *Environ. Health Perspect.*, 1980, 36, 103.
74 D.F. Hunt, A.M. Buko, J.M. Ballard, J. Shabanowitz, and A.B. Giordani, *Biomed. Mass Spectrom.*, 1981, 8, 397.
75 D. Zakett, A.E. Schoen, R.W. Kondrat, and R.G. Cooks, *J. Am. Chem. Soc.*, 1979, 101, 6781.
76 M.J.F. Asselin and J.J.R. Paré, *Org. Mass Spectrom.*, 1981, 16, 275.
77 R.K. Mitchum, J.R. Althaus, W.A. Korfmacher, K.L. Rowland, K. Nam, and J.F. Young, *Biomed. Mass Spectrom.*, 1981, 8, 539.
78 A.K. Bose, H. Fujiwara, and B.N. Pramanik, *Tetrahedron Lett.*, 1979, 4017.
79 R.C. Dougherty and J.D. Wander, *Biomed. Mass Spectrom.*, 1980, 7, 401.
80 T.O. Tiernan, C. Chang, and C.C. Cheng, *Environ. Health Perspect.*, 1980, 36, 47.
81 L.W. Sieck, K.R. Jennings, and P.D. Burke, *Anal. Chem.*, 1979, 51, 2232.
82 J.R. Lloyd, W.C. Agosta, and F.H. Field, *J. Org. Chem.*, 1980, 45, 1614.
83 G. Caldwell and J.E. Bartness, *Int. J. Mass Spectrom. Ion Phys.*, 1981, 40, 269.
84 F.J. Winkler and D. Stahl, *J. Am. Chem. Soc.*, 1978, 100, 6779.
85 R. Houriet, D. Stahl, and F.J. Winkler, *Environ. Health Perspect.*, 1980, 36, 63.
86 J.R. Lloyd, W.C. Agosta, and F.H. Field, *J. Org. Chem.*, 1980, 45, 3483.
87 D.F. Hunt, S.K. Sethi, and J. Shabanowitz, *Environ. Health Perspect.*, 1980, 36, 33.
88 D.F. Hunt, J. Shabanowitz, and A.B. Giordani, *Anal. Chem.*, 1980, 52, 386.
89 C.V. Bradley, I. Howe, and J.H. Beynon, *J. Chem. Soc., Chem. Commun.*, 1980, 562.
90 W.J. Bouma and K.R. Jennings, *Org. Mass Spectrom.*, 1981, 16, 331.
91 A.M. Reddy, M.S.B. Nayar, G.S. Reddy, and K.G. Das, *Org. Mass Spectrom.*, 1982, 17, 42.
92 A.P. Bruins, *Biomed. Mass Spectrom.*, 1980, 7, 454.
93 V.C. Trenerry, D.J.M. Stone, J.H. Bowie, K. Clausen, S. Scheibye, and S.-O. Lawesson, *Org. Mass Spectrom.*, 1981, 16, 451.
94 M.M. Bursey, J.R. Hass, D.J. Harvan, and C.E. Parker, *J. Am. Chem. Soc.*, 1979, 101, 5485.

95 M.M. Bursey, D.J. Harvan, C.E. Parker, L.G. Pedersen, and J.R. Hass, *J. Am. Chem. Soc.*, 1979, 101, 5489.
96 D. Zakett, J.D. Cuipek, and R.G. Cooks, *Anal. Chem.*, 1981, 53, 723.
97 R.N. Hayes and J.H. Bowie, *Spectrosc. Int. J.*, 1982, in press.
98 D.J. Douglas and B. Shushan, *Org. Mass Spectrom.*, 1982, 17, 199.
99 J.E. Bartmess, J.A. Scott, and R.T. McIver, *J. Am. Chem. Soc.*, 1979, 101, 6046.
100 R. Faird and T.B. McMahon, *Can. J. Chem.*, 1980, 58, 2307.
101 V.M. Bierbaum, R.J. Schmitt, C.H. DePuy, R.D. Mead, P.A. Schulz, and W.C. Lineberger, *J. Am. Chem. Soc.*, 1981, 103, 6262.
102 R.L. Woodin, M.S. Foster, and J.L. Beauchamp, *J. Chem. Phys.*, 1980, 72, 4223.
103 E.R. Talaty, J.J. Woods, and G. Simons, *Aust. J. Chem.*, 1979, 32, 2289.
104 M.P. Karnett and R.J. Cross, *Chem. Phys. Lett.*, 1981, 84, 501.
105 C.I. Mackay and D.K. Bohme, *Org. Mass Spectrom.*, 1980, 15, 593.
106 R.N. McDonald, A.K. Chowdhury, and D.W. Setser, *J. Am. Chem. Soc.*, 1980, 102, 6491.
107 R.R. Squires, C.H. DePuy, and V.M. Bierbaum, *J. Am. Chem. Soc.*, 1981, 103, 4256.
108 F.K. Meyer, M.J. Pellerite, and J.I. Brauman, *Helv. Chim. Acta*, 1981, 64, 1058.
109 S.D. Tanner, G.I. Mackay, and D.K. Bohme, *Can. J. Chem.*, 1981, 59, 1615.
110 D.K. Bohme and G.I. Mackay, *J. Am. Chem. Soc.*, 1981, 103, 978.
111 R. van Doorn and K.R. Jennings, *Org. Mass Spectrom.*, 1981, 16, 397.
112 R.N. McDonald and A.K. Chowdhury, *J. Am. Chem. Soc.*, 1980, 102, 5118.
113 R.N. McDonald, A.K. Chowdhury, and D.W. Setser, *J. Am. Chem. Soc.*, 1981, 103, 6599.
114 M.J. Pellerite and J.I. Brauman, *J. Am. Chem. Soc.*, 1981, 103, 676.
115 G.K. King, M.M. Maricq, V.M. Bierbaum, and C.H. DePuy, *J. Am. Chem. Soc.*, 1981, 103, 7133.
116 J.C. Sheldon, *Aust. J. Chem.*, 1981, 34, 1189.
117 A.J. Noest and N.M.M. Nibbering, *Int. J. Mass Spectrom. Ion Phys.*, 1980, 34, 383.
118 G. Klass, D.J. Underwood, and J.H. Bowie, *Aust. J. Chem.*, 1981, 34, 507.
119 J.C. Sheldon and J.H. Bowie, *Nouv. J. Chim.*, 1982, in press.
120 M.F. Dottore and J.H. Bowie, *J. Chem. Soc., Perkin Trans. 2*, 1981, 283.
121 G. Klass, J.C. Sheldon, and J.H. Bowie, *Aust. J. Chem.*, 1982, in press.
122 R.N. McDonald and A.K. Chowdhury, *J. Am. Chem. Soc.*, 1982, 104, 901.
123 J.E. Bartmess, R.L. Hays, and G. Caldwell, *J. Am. Chem. Soc.*, 1981, 103, 1338.
124 J.E. Bartmess, *J. Am. Chem. Soc.*, 1980, 102, 2483.
125 R.N. McDonald and A.K. Chowdhury, *J. Am. Chem. Soc.*, 1980, 102, 6146.
126 C.H. DePuy and V.M. Bierbaum, *Tetrahedron Lett.*, 1981, 5129.
127 J.E. Bartmess, R.L. Hays, H.N. Khatri, R.N. Misra, and S.R. Wilson, *J. Am. Chem. Soc.*, 1981, 103, 4746.
128 C.H. DePuy, V.M. Bierbaum, L.A. Flippin, J.J. Grabowski, G.K. King, and R.J. Schmitt, *J. Am. Chem. Soc.*, 1979, 101, 6443.

129 C.H. DePuy, V.M. Bierbaum, L.A. Flippin, J.J. Grabowski, G.K. Kim, R.J. Schmitt, and S.A. Sullivan, *J. Am. Chem. Soc.*, 1980, 102, 5012.
130 S.A. Sullivan, C.H. DePuy, and P. Damrauer, *J. Am. Chem. Soc.*, 1981, 103, 480.
131 G. Klass, V.C. Trenerry, J.C. Sheldon, and J.H. Bowie, *Aust. J. Chem.*, 1981, 34, 519.
132 R.R. Squires and C.H. DePuy, *Org. Mass Spectrom.*, 1982, 17, 187.
133 R.V. Hodges, S.A. Sullivan, and J.L. Beauchamp, *J. Am. Chem. Soc.*, 1980, 102, 935.

6

Developments and Trends in Instrumentation

BY T. R. KEMP

1 Introduction

Every so often, there comes to an analytical technique a development that changes either its compass of application or the direction of its further development. Mass spectrometry has witnessed several such milestones: the relationship between elemental composition and accurate mass, which led to the development of high-resolution double-focusing instruments in the early 1960s; on-line gas chromatography, which opened up the path to mixture analysis by mass spectrometry and paved the way for the widespread application of the quadrupole mass analyser in organic analysis; field ionization (f.i.) and chemical ionization (c.i.), the first of the 'soft' ionization techniques, and the forerunners of the desorption techniques, field desorption (f.d.) and desorption chemical ionization (d.c.i.); on-line acquisition and processing of mass-spectral data by minicomputer, without which modern mass spectrometry and its associated techniques would be far less effective. Such another milestone has been passed during the period under review: fast atom bombardment, blessed with the acronym f.a.b., has had such an impact, not only on the application of mass spectrometry to molecules of extreme polarity and extraordinary size, but also on the design of commercial instrumentation, that it merits a chapter of its own in this volume. As a consequence, consideration of f.a.b. will be limited here to the instrumental aspects of the technique and its effects on instrument design.

2 Ionization Techniques

Fast atom bombardment was first described in early 1981,[1] and reviews of the technique,[2] and its applications,[3] have issued from the same laboratory more recently. The technique employs a beam of fast neutral atoms, typically of argon of between 2 and 8 keV of translational energy, to sputter ions from a sample. The sample is dispersed in a viscous liquid medium of low volatility, glycerol being used extensively for this purpose.[4] The original method of production of fast atoms, which has continued to be the most widely used, involves focusing a beam of argon ions, produced by a saddle-field ion gun, into a collision chamber containing a high pressure (10^3 - 10^4 mbar) of argon gas. Resonant charge exchange takes place with little loss of forward momentum, resulting in a fast-moving beam of argon neutrals. An alternative method of neutral production utilizes a Townsend discharge source,[5] which has a higher electrical efficiency than the saddle-field gun, so that there is no noticeable heat production. Additionally, this latter atom gun is sufficiently small to permit installation within the source housing. An instrument designed for secondary-ion mass spectrometry (s.i.m.s.) has had its ion gun modified to produce neutral species by mounting a perforated metal plate at the end of the gun.[6] A charge-exchange reaction takes place at the metal surface, and recombination of argon ions to argon neutrals of about 25% has been observed.

The bombardment of a surface by energetic atoms has obvious ancestry in s.i.m.s., a technique that is well established in surface science but, until recently, has found less favour in organic mass spectrometry, partly because of the operational difficulties associated with focusing charged ion beams into the high-voltage source of a high-performance mass spectrometer, and partly because of the very low secondary-ion yields. The relationship between f.a.b. and s.i.m.s. was described very early in the development of f.a.b.,[7] and these techniques, together with several other ionization methods of similar character, were the subject of a vigorous seminar.[8] A more recent report suggests that a higher depth resolution can be achieved for depth profiling using neutral oxygen or argon atoms than in the conventional s.i.m.s. experiment,[9] although it has been reported that caesium-ion bombardment is at least as effective as xenon neutral-atom bombardment in sputtering biological molecules of intermediate molecular weight from a glycerol matrix.[10] A technique surprisingly similar to f.a.b., called molecular-beam solid analysis (m.b.s.a.), was described in 1966, although the use of a liquid matrix was not reported in those experiments.[11]

One of the products of the development of f.a.b. is a resurgence of interest in other ionization techniques using particle bombardment for the study of thermally labile or non-volatile materials. The application in organic mass spectrometry of static, or low-damage, s.i.m.s. (despite the difficulties), in which a reduced primary-ion current density is used to moderate the sputter rate, was pioneered by Benninghoven in the late 1960s, and two recent reviews describe the current status of the technique.[12] In particular, the coupling of s.i.m.s. with other analytical techniques has given promising results. Direct identification of compounds present as separated components on a thin-layer chromatogram has been achieved,[13] and continuous ribbon interfaces of silver,[14] and high-purity nickel,[15] have been used as support for the analysis of liquid-chromatography eluents by s.i.m.s. The capabilities of the technique for the analysis of organic samples are enhanced by mixing the sample with ammonium chloride.[16] High-performance mass spectrometers have been designed or adapted specifically to overcome some of the difficulties associated with the s.i.m.s. technique and, more specifically, to cope with the singularly high molecular weights of the ion species produced. An MAT 311A mass spectrometer has achieved a resolution of 8000 in a s.i.m.s. experiment,[17] and a much modified CEC 21-110, originally a spark-source instrument,[18] has produced ions with masses in excess of 18000.[19] A time-of-flight mass spectrometer with a pulsed primary-ion source emitting alkali-metal ions[20] has produced significant quasi-molecular ions from biomolecules up to m/z 1355.

The ionization of large, polar molecules is also demonstrated by the ^{252}Cf plasma-desorption ion source, from which a molecular ion at m/z 12637 ± 10 has recently been reported.[21] The low ion yields and extreme mass range produced by this technique necessitate the special characteristics of the time-of-flight analyser, which has an almost unlimited mass range but an inherently low resolution.

Lasers have been used increasingly in the ionization process, both as the primary particle beam and to assist other methods. A recent review article indicates the scope of the application of the laser-ion source to the analysis of solids.[22] In the combination of laser desorption with chemical ionization,[23] neutral molecules are first desorbed by the laser, then ionized by the reagent gas in the normal way; later experiments show that cationized species are also produced directly by laser desorption.[24] Laser assistance in field desorption not only improves the sensitivity, due to more even heating of the emitter, but also enables higher temperatures to be achieved because of the lower mechanical strain placed on the emitter. This is demonstrated by the laser-assisted field desorption of inorganic and organometallic compounds[25] and metal powders.[26] The use of a laser as the sole ionizing medium has the advantages that virtually no sample preparation is necessary, that the amount of energy deposited on the sample can be controlled by the laser intensity chosen, and therefore that the degree of fragmentation in laser-induced spectra can be considerably influenced.[27] A commercial instrument is based on a pulsed laser-ion source and a time-of-flight analyser[28] and has been used for the analysis of both organic and inorganic materials.[29] The analyser embodies an ion-reflector system, so that differences in drift times in the field-free portions of the analyser due to different initial ion energies[30] can be compensated for by different path lengths in the ion reflector (this technique has been called energy/time focusing). One of the benefits of this arrangement is that the length of this instrument is half that of a conventional linear drift time-of-flight mass spectrometer. Despite the problems of synchronizing a scanning mass spectrometer with a pulsed laser, a technique has been developed using selected-ion monitoring with a magnetic-sector instrument.[31] A system has also been developed for continuous sample introduction,[32] employing a continuous moving stainless-steel belt on to which the sample is electrosprayed. Ionization is achieved by focusing a tunable dye laser on to the belt, and the method has been successful for the analysis of several classes of 'non-volatile' bimolecules. The mechanism of laser desorption has been studied by mass-analysed ion kinetic-energy spectroscopy (m.i.k.e.s.), using a reverse-geometry double-focusing instrument,[33] whereby remarkable similarities in fragmentation behaviour between laser desorption and s.i.m.s. were found. Experiments have been designed to determine cation production and quasi-molecular-ion production as a function of substrate temperature.[34]

Other thermal techniques have had some success for the ionization of involatile molecules. A standard quadrupole mass analyser has been modified for thermal ionization using a specially constructed probe;[35] the system has been used not only for the determination of the isotope ratios of urinary calcium but also for the study of quaternary ammonium salts[36] and other non-volatile samples.[37] The desorption of quasi-molecular ions from an electrically heated stainless-steel ribbon has been observed,[38] and the mechanisms of thermal desorption have been studied by evaporating samples from an indirectly heated metal plate on to a directly heated tungsten wire placed at a few millimetres distance and on the optical axis of a quadrupole mass analyser.[39] Some interesting developments concern the production of ions from the liquid phase; the electrospray technique involves charging an aerosol from which the solvent is then partially evaporated,[40] leaving charged, solvated ions. Electrohydrodynamic ionization of polyethylene glycols in glycerol solution gave singly charged ions up to m/z 1582 and triply charged species up to about mass 645.[41] During the development of an on-line l.c.-m.s. system,[42] a method of ionizing samples directly from solution was discovered whereby the ions are produced as a result of passing the solution (the l.c. eluent) through a heated stainless-steel capillary tube and directing the resulting aerosol through a skimmer jet on to a heated, nickel-plated copper probe.[43] This method has been successful for the ionization of simple peptides, nucleotides, and dinucleotides;[44] the spectra obtained are very similar to those obtained by f.d.

Atmospheric-pressure ionization (a.p.i.), using a corona discharge formed typically at the rounded tip of a wire of small diameter,[45] is generally used for the analysis of volatile samples in the gas phase.[46] However, the system can be modified to accept samples in solution;[47] the solution enters the corona discharge either through a stainless-steel capillary tube, providing a sufficiently low flow rate is used, or by deposition of the solution on the end of a metal rod, which is inserted into the discharge chamber. The major species produced by this method are protonated and cluster ions,[48] and fragmentation can be influenced by control of the voltage at the extraction pinhole. An alternative method for the analysis of solutions uses a 2 MHz ultrasonic vibrator to nebulize the solution in admixture with nitrogen gas.[49] Amino-acids and nucleosides analysed by this method give spectra containing protonated molecular ions with very little fragmentation.

Elemental analysis is possible from solution using an

inductively coupled argon plasma (ICP).[50] The line-of-sight plasma is sampled using a two-stage extractor comprising a water-cooled stainless-steel skimmer and an orifice of diameter 50 µm.[51] Modifications to align the plasma at 20° to the ion axis of a high-performance quadrupole instrument and sample through a 70 µm orifice have given significant improvement.[52] An alternative method of plasma production is microwave induction,[53] which gives a comparable detection limit.

Notwithstanding the emergence of new ionization methods with several attractive features, the established techniques of e.i. (electron ionization), c.i., and f.d. have continued to attract development interests. In particular, the 'in-beam' methods are used increasingly, and a recent review describes their application to non-volatile samples.[54] Commercial instrumentation is relatively easy to modify to produce results using such techniques.[55] E.i. is the oldest method in use for organic samples, and yet important developments continue. The use of low work-function lanthanum hexaboride filaments not only reduces the operating temperature of the ion source but also increases the efficiency of ionization.[56] The adaptation of a hot-cathode Penning-discharge ion source to a quadrupole mass analyser has produced high-intensity ion beams with acceptable energy spread and good ionization efficiency.[57] Instead of the normal magnetic collimation of the electron beam, a fine wire electrode has been placed along the path of the electron beam;[58] maintained at a positive potential relative to the surrounding ion chamber, it plays a dual role as both electron-beam collimator and ion repeller. A dual e.i./c.i. source has been developed with interchangeable ion volumes,[59] each built specifically for e.i. or c.i., and designed for removal and replacement through the direct-probe vacuum lock. This arrangement ensures the optimum efficiency in each mode, and also reduces contamination, thus extending the operating time between routine cleaning.

Chemical ionization has also been the subject of some development effort; in a c.i. source in which a Townsend discharge, mounted in a direct probe,[60] is used to produce the ionizing electrons, the advantages over a conventional hot-cathode ion source are seen in the ability to use oxidizing reagent gases,[61] which drastically reduce the lifetime of a hot metal filament. Problems of electrical discharges occurring in the high-pressure regions of the high-voltage c.i. source of a magnetic-sector mass spectrometer[62] have been overcome by insertion of a plug of glass wool between the sample and source. A similar method provides an electrically safe interface for a Baratron capacitance manometer for the measurement of source pressures.

Mixing of gases for c.i. is possible using a proportional flow valve system[63] developed primarily for application in gas chromatography. The simple design allows the proportions of reagent gases to be varied whilst keeping constant the total pressure of gas in the ion source.

Desorption c.i. is a relatively recently established technique which, because of its value in the analysis of thermally labile molecules, is still undergoing development in several areas, among the more important being the form and material of the emitter and the method of sample loading. A method of coating the d.c.i. emitter wire with polyimide film has been developed,[64] which increases the range of compounds accessible to mass spectrometry, and at greater sensitivity than is possible with conventional wire emitters. Silanized emitters provide a protonating surface for either d.c.i. or f.i., resulting in increased probability of production of protonated molecular ions.[65] A probe, vacuum lock, and current-control unit have been constructed to enable activated f.i. emitters to be used for d.c.i. in a commercial quadrupole instrument.[66] A simple but effective extension to a standard direct probe, using gold gasket wire which is heated by means of the probe heater,[67] provides a d.c.i. probe; a similar construction has produced 'direct e.i.' spectra.[68] The advantages of a porous ceramic sample holder and those of a metal coil have been combined by fitting a chromel wire coil inside a closed ceramic tube. This probe will absorb samples in solution and can be cleaned by flaming, and its heating rate is also sufficient for d.c.i.[69]

Field desorption, despite its reputation as a 'black art' (or perhaps because of it), still has its devotees, and there is no doubt that the proponents of the art are capable of achieving astonishing results. One of the problems of the technique is the conservation of the small amount of sample available on the emitter, which entails, *inter alia*, the achievement as quickly and efficiently as possible of the best anode temperature (BAT). Naturally, the method of loading the emitter is also an important factor, best results being obtained with an even coating over the whole activated surface. This is not easy to achieve using the customary techniques of either dipping the emitter in a few microlitres of sample solution, then air-drying the solvent, or depositing several microlitres by microsyringe. However, loading with an electrospray gives the desired even deposition.[70] This method of sample preparation is also used in ^{252}Cf-p.d.m.s.[71] The attainment of BAT has been examined in some detail, and the result has been the development of several methods of automatic emitter-current (heating) control. One device operates on the principle that the resistance of a tungsten wire varies as a known function of temperature;[72] a feedback system employing the Wheatstone bridge principle maintains a constant resistance in the emitter, and thus the temperature also remains constant. A more conventional approach, using total ion-current feedback to regulate the emitter current,[73] has the additional facility of permitting the ion current at any mass to be substituted for the TIC, and the unit also has the facility for linear programmed operation.

In the activation of f.d. emitters, some compromise must be made between the complexity of the activation process and the strength and durability of the microneedles. At one end of the scale, what appears to be a definitive programme for the production of durable, high-temperature emitters has been described,[74] although this process takes between 12 and 24 hours. In an attempt to reduce activation time, various organic compounds as activators have been examined;[75] acceptable emitters, which are not inferior in emission characteristics, can be achieved in a shorter time using indene or naphthalene. A 15-minute process based on the cathodic growth of metallic needles produces emitters of lower efficiency;[76] pulsed electrodeposition of silver dendrites from silver nitrate solution has also been described.[77] Emitters devoid of microneedles of any type, but with a relatively smooth surface covered with a thin layer of a 'tungsten bronze',[78] can be produced by electrochemical etching. These emitters are mechanically fragile, having a diameter of only about 1 μm.

Improvements and additions are also being carried out to f.d. and f.i. sources. A combined f.d./f.i./e.i. source fitted on a magnetic-sector mass spectrometer continues to be improved;[79] latest developments, including some mechanical changes, give increased sensitivity and provide easier maintenance. The volcano ion source has also been subject to improvement.[80] The emitter of this source is a circular blade, 2 μm in diameter, which may be treated as a point source of ions, giving a resolution of 3500 under optimum working conditions. Sensitivity is as high as 10^{-8} C μg^{-1}, although there is a practical limit to sample flow of about 10 μg $(30\ min)^{-1}$; larger samples have a deleterious effect on resolution and lifetime of the source. A Bonn-type ion source,[81] which enables full three-dimensional positioning of emitters of many different types, has been designed for installation on a mass spectrometer of vertical layout. An overall increase in sensitivity by a factor of 16 has been achieved for a conventional f.i./f.d. source,[82] both by turning the emitter through 90° so that it sits at right angles to the mass-spectrometer entrance slit and by widening the aperture in the counter-electrode.

With interest in the production and detection of negative ions increasing, particularly in the c.i. mode,[83] the instrumental aspects of negative-ion mass spectrometry have been well considered,[84] particularly for the case of the quadrupole mass analyser. Some ionization techniques are reported to yield positive ions and negative ions with equal facility; this characteristic has been reported for both f.a.b. and laser-induced ionization.[85]

3 High-mass Analysis

The second major product of the development of f.a.b. is the increase in molecular weights of samples capable of routine analysis by mass spectrometry and the resultant demand for instrumentation capable of routine high-mass determination. A high-field magnet for the Kratos MS50, using high-saturation alloy pole tips and a reduced magnet gap,[86] has been available for some time, and ions up to m/z 3500 have been observed from polybutadiene by the use of a 'grand-scale' mass spectrometer.[87] A secondary-ion mass spectrometer of Mattauch-Hertzog geometry, using xenon-ion bombardment of alkali halides, has yielded cluster ions up to m/z 18000,[88] though with limited resolution. Another research instrument, the Wien filter, currently in use with the volcano f.i. source,[89] is reported to be operable over any mass range and at any accelerating voltage. A resolution of 15000 is theoretically possible using 30 keV ion energy. However, such instruments are either not available commercially or, if they are, are generally beyond the budgetary means of most establishments. This has led to the development of a new breed of medium-performance mass spectrometers capable of mass ranges in excess of 2000 at full ion energy. Two such instruments are currently available, one in which third-order terms and non-ideal fringe fields have been considered in the development of a magnet of complex design,[90] using segmented inhomogeneous fields, for the Kratos MS80RF. The VG Analytical 7070E has been developed from a medium-performance g.c.-m.s. instrument by the simple expedient of increasing the radius of the magnetic sector and using a reduced sector angle; this ensures that the size of the magnet is not significantly increased.[91] A relatively simple chemical modification to the sample prior to f.d.m.s., cationization with, for instance, Ca^{2+} or Ba^{2+}, induces the production of doubly charged species,[92] thereby effectively doubling the mass scale.

The medium-performance mass spectrometer must also be capable of routine analytical functions, such as g.c.-m.s., where the increased use of capillary columns has caused a demand for magnetic-sector instruments with scan speeds comparable with those of the quadrupole mass analyser. Using fully laminated core magnets with low-impedence, high-current coils,[93] scan rates down to 100 msec $decade^{-1}$ can be achieved. The low hysteresis effects exhibited by these magnets permit control by micro-processor.[94] An alternative approach to fast scanning with a magnetic-sector instrument is to scan the accelerating voltage. Under normal conditions, this approach is undesirable because of the reduction in sensitivity at low accelerating voltage. However, an instrument has been developed that uses a retarding ion source to increase the extraction efficiency at low accelerating voltages; scan rates better than 50 μsec $m.u.^{-1}$ have been achieved.[95]

4 Reaction Studies

The study of the gas-phase reactions of ions in the mass spectrometer began with the observation of metastable ions in early single-focusing magnetic-sector instruments, where the appearance of a 'metastable peak' permitted the determination (albeit with some ambiguity) of possible fragmentation pathways. A recent review[96] describing the current state of metastable-ion studies is ample demonstration of the advances in this area. In a single-focusing instrument, metastable peaks arise as a result of spontaneous decompositions occurring in the field-free region between the ion exit slit of the source and the entrance of the magnetic analyser. In a double-focusing mass spectrometer of conventional geometry, similar metastable peaks arise from decompositions occurring in the field-free region between the two analysers (the second field-free region). Those decompositions occurring in the first field-free region, before the electrostatic analyser (ESA), result in ions having insufficient energy to traverse the ESA. (It is interesting to note that the reverse-geometry instrument, like the quadrupole, displays no metastable peaks in its normal mass spectrum.) Recently, methods have been devised for studying the reactions occurring in the first field-free region of a double-focusing instrument by performing linked scans of the electrostatic (E) and magnetic (B) sectors.[97] The type of information provided by the linked scan depends upon the relationship between B and E: a scan of constant B/E results in all the fragment (daughter) ions of a chosen parent ion; a scan of B^2/E gives all the precursors of a chosen fragment. The application of such linked scans in analytical mass spectrometry is exemplified by the study of a peptide using f.d. with collisional activation.[98]

For all types of linked scan, however, including a recently developed one[99] that selectively records all ions representing loss of the same neutral fragment, the presence of artefact peaks may lead to ambiguities in interpretation.[100] Even for mass spectrometers of reverse geometry, where artefact peaks in linked scans present less of a problem,[101] their identification can be difficult. Although computer simulation can explain some occurrences of interferences,[102] the most successful method for the resolution of ambiguities is by production of a three-dimensional ion-current surface,[103] which can also provide a record of sequential fragmentation processes.[104] Even under computer control,[105] this is a laborious process; an adequate compromise is the partial three-dimensional map produced by using two scanning modes in alternation,[106] by, for example, adding a.c. modulation to either the accelerating voltage or the ESA voltage, with detection based on a phase-locked amplifier. Alternatively, some ambiguities are suitable for resolution by variation of the ionizing energy.[107] The precursor scan, B^2/E, retains the kinetic-energy information of the dissociation process, resulting in a broadening of the peak and a reduction of mass resolution. A new scan, B^2E,[108] which gives similar information but without the energy component, and therefore with increased mass resolution, is applicable only to instruments of reverse geometry.

The selectivity of quantitative mass spectrometry has been increased by monitoring metastable decompositions.[109] Modification of a standard linked-scan system to utilize a digital peak-switching accessory allows two reactions to be monitored, including, for instance, a stable-isotope-labelled standard.[110]

More detailed studies of the reactions of ions can be made by placing two mass spectrometers in sequence, a technique called, variously, two-dimensional mass spectrometry,[111] sequential mass spectrometry,[112] tandem mass spectrometry,[113] and mass spectrometry/ mass spectrometry.[114] Despite Beynon's humorous comments,[115] for reasons of brevity the abbreviation m.s.-m.s. will be used here to denote the addition of a second analyser after the mass analyser of a normal mass spectrometer. The reverse-geometry double-focusing mass spectrometer has been providing m.s.-m.s. for several years,[116] under the pseudonyms m.i.k.e.s. or d.a.d.i. (direct analysis of daughter ions), and, indeed, has recently been shown to be capable of providing extremely comprehensive analytical facilities.[117] Similar analytical versatility is offered by the multiple-quadrupole instrument, a tandem mass spectrometer which, in its simplest form, comprises two mass analysers separated by a collision region. The first quadrupole acts as a separation device, the second quadrupole as the analyser; in the daughter-ion experiment, the first quadrupole is set to transmit the chosen parent ion, which undergoes collisionally induced dissociation (c.i.d.) in the collision region (usually an area of high pressure), and the product ions are monitored by scanning the second quadrupole. In the precursor-ion experiment, the first quadrupole is scanned over the whole mass range, so that each ion sequentially undergoes c.i.d. in the collision cell. The second quadrupole is set to transmit only a selected ion, so that only those ions that dissociate to the selected one will produce a signal at the final collector. The constant neutral loss experiment is readily achieved by scanning the two analysers with a constant mass difference.

Such a double-quadrupole instrument,[118] with the first analyser operating in the 'notch filter' mode (which, by rejecting narrow bands of ions, enables major components of a mixture to be discarded) and the second analyser scanning normally, has been used for the analysis of trace components of mixtures. A similarly simple instrument, making use of standard, commercially available components, has the two quadrupoles separated by a simple gas cell comprising a grounded plate with a central aperture, and a radial channel for the introduction of the collision gas.[119] The m.s.-m.s. experiment is carried out by setting the second quadrupole to transmit all masses (this is achieved by removing the d.c. component of the quadrupole field, leaving the r.f. component only) and scanning the first analyser to locate the ion of interest. This ion is then transmitted while the second quadrupole is scanned in the mass-analysis mode.

Collision cells for m.s.-m.s. instruments have been the subject of some scrutiny; in a comparative study of quadrupolar and electrostatic cells,[120] the former exhibited several advantages, including a degree of focusing and a reduction of scattering, which contributed to significantly greater daughter-ion intensity. The use of an intermediate quadrupolar collision cell has led to the development of several 'triple-quadrupole' instruments, in which the second (collision) quadrupole is operated in the r.f.-only mode. The instrument, originally described in 1979,[121] has formed the basis of current commercially available instruments, which differ from it only in detail.[122] Although the development of such instruments has been spurred mainly by the many potential analytical applications,[123] an instrument fitted with an a.p.i. source has been used to study c.i.d. mechanisms,[124] demonstrating the suitability of this type of instrumentation for physical-chemical studies. In a variation of the triple-quadrupole theme, a double-quadrupole instrument has been constructed with a tube of ferrite ceramic inserted at the junction of the two analysers.[125] This ceramic has a ratio of conductivity to dielectric constant such that its interior is fully shielded from the d.c. component of the quadrupole field, whilst permitting the r.f. component, suitably attenuated, to penetrate. The advantage of this system over the standard triple-quadrupole arrangement (apart from the reduction in both size and cost) is that an accelerating or decelerating field is also available along the axis of the collision cell.

The choice of a quadrupole mass analyser as the first stage necessarily limits the resolution (and, to some extent, the mass range) available for separation. A hybrid instrument has been developed[126] in which the first stage comprises a single-focusing magnetic-sector instrument, with a quadrupolar collision cell and a quadrupole mass analyser as the second stage. (The nomenclature BQQ is used to describe this instrument, source and collector being at left and right, respectively.) This instrument is truly versatile, capable of operation in several other modes, including interchange of source and collector (QQB). Kinetic studies using this hybrid instrument in the BQQ mode[127] demonstrate the energy dependence of collision-induced mass spectra. The substitution of a double-focusing mass spectrometer as the first stage of such a hybrid instrument (EBQQ or BEQQ) offers the advantage of parent-ion selection at significantly higher resolution.[128]

The attachment to a double-focusing instrument of an electrostatic second stage,[129] giving either EBE or BEE, with a collision cell placed in the third field-free region,[130] can give extremely high resolution in the first stage for parent-ion selection.[131] However, the choice of an electrostatic sector as the analysis stage not only limits the mass resolution for daughter-ion detection but also increases the susceptibility of the analysis to artefact peaks, as in m.i.k.e.s.[132] A third electrostatic sector added after the second stage serves as a beam deflector and assists in the elimination of such artefact peaks.[133] Such an instrument has demonstrated its capability in experiments requiring better than 10000 resolution for parent-ion selection.[134] In addition, the instrument has been used to study consecutive collision-induced reactions by utilizing the first two sectors to select a product of a specific reaction occurring in the first field-free region, performing c.i.d. on this ion in the third field-free region, and studying the resulting fragments using the third sector.[135]

The substitution of a magnet as the second stage, giving EBB' or BEB', gives increased daughter-ion resolution whilst maintaining the high-resolution selectivity for parent ions. The preferred configuration is BEB', where both BE and EB' are double focusing. This gives either high parent-ion resolution with moderate daughter-ion resolution, using BE for parent-ion selection and B' for daughter-ion analysis, or moderate parent-ion resolution with high daughter-ion resolution, using B for parent-ion selection and EB' for daughter-ion analysis.[136]

The 'perfect' m.s.-m.s. system is one that combines high-resolution parent-ion selection with high-resolution daughter-ion analysis. Such an instrument has been built,[137] using double-focusing mass spectrometers for both stages, separated by a collision region that permits acceleration of the parent ions prior to collision from 10 keV energy to 30 keV. The molecular-beam collision cell is of a design which, despite reducing the length of the collision region, increases substantially the efficiency for collisional activation. A collision region having similar properties and of simple design[138] comprises a hypodermic needle fixed in place with epoxy resin; the resin is covered with silver by vapour deposition to ensure that no insulating surfaces are visible to the ion beam. In an even simpler experiment, collisional activation has been achieved in the first field-free region of an MAT 731 mass spectrometer by loosening one of the bolts holding the slit-tilting mechanism.[139] Needless to say, reproducible results were difficult to achieve by this method. The photodissociation of ions has been studied[140] by the use of a reverse-geometry instrument fitted with a modified flight tube that allows a laser to be focused into the second field-free region. Ion kinetic-energy spectra were recorded with the aid of a phase-sensitive detector.

A tandem mass spectrometer comprising two magnetic analysers, BB', has also been constructed.[141] Predictably, this instrument exhibits moderate resolution for both parent and daughter ions, but it has the unique property in tandem instruments of a flat focal plane within the second analyser, thus permitting simultaneous detection of a 4:1 ratio in mass, with a concomitant increase in sensitivity. Additionally, the two stages are mutually electrically isolated, allowing a 60 kV differential between them in order to study charge inversion.

It is possible that the use of mass spectrometers in sequence represents another milestone in mass spectrometry; certainly several enthusiastic reviews of the technique have appeared,[142] and one in particular cites 46 examples of the application of m.s.-m.s.[143] However, a note of caution is sounded in the American Chemical Society's biennial review,[144] which comments on the limited number of applications that have been reported 'despite impressive instrumental developments' and points out that, as a tool for the characterization of a total mixture, the technique is possibly as time-consuming as chromatography-m.s. For the rapid screening of target compounds, or for those samples that are not amenable to chromatography, m.s.-m.s. is an invaluable technique.

5 Detection Systems

Detection systems for commercial instrumentation have undergone little change for some time, the most important development being the modifications made to permit the detection of negative ions.[145] The incorporation of a conversion dynode maintained at a high positive potential converts the negative ions to secondary positive ions by sputtering, fragmentation, and charge stripping; stray electrons do not effect secondary emission under these conditions, so baseline noise is significantly reduced. The proper positioning of two such conversion dynodes of opposite polarity at the exit of a quadrupole mass analyser enables the selection and detection of either positive or negative ions by the same collector system. The instrumentation necessary for the detection of negative ions has been described in some detail,[146] including pulse-counting techniques. Such techniques have been used to improve detection levels,[147] particularly in quantitative experiments using single-ion monitoring.

The disadvantage of sequential detection, using scanning mass spectrometers with electron multiplier or Faraday plate detectors, is that the majority of ions produced are discarded. The natural alternative is a simultaneous detector, such as an ion-sensitive photoplate, which can detect ions over a relatively wide mass range. The photoplate is still widely used in certain areas, such as spark-source mass spectrometry, and a method for mass measurement from the photoplate has been developed,[148] which needs only the presence of two ions and a knowledge of their mass difference. However, the photoplate has several disadvantages, not the least of which is its lack of immediacy. Consequently, several attempts have been made to develop an 'electronic photoplate',[149] using a string of two-dimensional channel electron-multiplier arrays (CEMA). One approach uses a fibre-optical window to convey the signal from the CEMA to a photodiode array. Loss of secondary electrons in the stray magnetic field of a Mattauch-Hertzog instrument is prevented by tilting the CEMA and phosphor with respect to the magnetic field; the tilted line images are then re-aligned by special fibre-optic rotator devices. In an electro-optical ion detector (EOID) of similar construction, used for g.c.-m.s.,[150] it was concluded that the resolution of the device, even when used in conjunction with a high-resolution mass spectrometer, was limited to about 1000, and that dynamic range was also limited. As a consequence, this type of detector cannot be used with any advantage for g.c.-m.s. The CEMA in use with the BB' m.s.-m.s. instrument is preceded by quadrupolar lenses to maintain a flat, rectangular focal plane.[151]

In an alternative construction, the phosphor-coated anode is substituted with a thin, shaped resistive plate.[152] The electronic charge of each pulse produced by the ions impinging on the CEMA is divided among four leads attached to the plate, and from the charges collected by the four leads two voltages are produced whose amplitudes are proportional to the x and y co-ordinates of the electron pulse. The x-axis signal is also used to accumulate the spectrum in a multi-channel analyser.

In a quantitative assessment of the use of a CEMA in a spark-source mass spectrometer,[153] sensitivity was higher than the photoplate, but resolution was lower; the non-linear response and restricted dynamic range made it less suitable than the photoplate for this application. A system using two channelplates of diameter 70 mm with holes of diameter 25 μm was shown to be suitable for low count-rate signals,[154] and a system incorporating a CEMA has been optimized for the detection of high-mass ions produced by ^{252}Cf-p.d.m.s. and using a time-of-flight mass spectrometer.[155]

Detection of ions using an electron multiplier with a multi-channel analyser has formed the basis of a general method for the acquisition of low-intensity spectra at very high mass.[156] By using a narrow scan of the accelerating voltage instead of the magnetic field, a higher reproducibility is achieved; a real-time oscilloscope is interfaced with the multi-channel analyser, and this enables the operator to determine visually the point at which a satisfactory signal-to-noise ratio has been achieved. This oscilloscope also permits changes in instrumental parameters, as, for instance, source tuning, with an immediate visual feedback. The system has been applied to field-desorption analyses above m/z 1500; subsequent acquisition of a reference compound under the same scan conditions permits simple chemical mass marking. The acquisition of data at high mass (above m/z 3000) and high resolution (more than 30000) demonstrates the success of this technique. A variant of the technique,[157] under software control using a minicomputer, has increased versatility.

6 Fourier-transform Mass Spectrometry

A recent development which is reported to display many of the attributes of the perfect mass spectrometer (high resolution, high mass range, rapid scanning, sensitivity, simultaneous detection) is the Fourier-transform mass spectrometer (FT m.s.).[158] The basic principle of this method is the cyclotron (circular) motion of ions in a uniform magnetic field; an ion's cyclotron orbit can be expanded by the application of an alternating electric field normal to both the magnetic field and the cyclotron orbit, and of the same frequency as the cyclotron frequency. Thus, in the FT m.s. experiment, ions are generated by a short pulse from an electron beam, then subjected to a pulse of r.f. field whose frequency is varied linearly across its duration. The frequency range covers the entire range of cyclotron frequencies of interest. At the end of the pulse, the alternating image currents induced by the orbiting ions are digitized and stored in a computer. The Fourier transform of this 'time-domain' transient signal is computed to yield the frequency-domain spectrum, which can be related to the mass spectrum. The signal-to-noise ratio of the spectrum can be increased by summing several time-domain transients. One of the major advantages of the FT m.s. method is that all ions are detected within so short a time period that it can be said to be instantaneous. As with all such detection methods, this implies high sensitivity as well as speed.

An inherent feature of FT m.s. is the availability of extremely high resolution; in one report[159] a resolution of 760000 is demonstrated at m/z 78, and in another[160] a resolution of 1.5×10^6 has been achieved by the use of a superconducting magnet of 4.7 T. (Both reports use the peak width at half-height definition.) Unlike in other forms of mass spectrometry, there is no trade-off in FT m.s. between resolution and signal-to-noise. Indeed, making use of signal averaging, signal-to-noise and resolution can be seen to increase simultaneously. However, resolution does decrease with increasing mass, although with the superconducting magnet a resolution of 150000 is still possible at m/z 1660, and 1.5×10^7 is theoretically possible at m/z 16.

Resolution in FT m.s. is also critically dependent upon pressure.[161] This makes the coupling of a gas chromatograph one of the most challenging of design problems.[162] Using an interface comprising either a direct line or a glass jet separator made up with 20 ml min^{-1} helium, it has been possible to perform g.c.-FT m.s. with a SCOT column.[163] The mass spectrometer used in this study was specially designed with a high-conductance vacuum chamber that allowed the mass analyser to operate at a sufficiently low pressure to permit adequate mass resolution.

Because of the disadvantages of operation at elevated pressures, c.i. with a high pressure of reagent gas, such as is used in other mass spectrometers, is not practical with the FT m.s. instrument. Instead, a technique of ion trapping is used to increase the residence time of the sample ions in the source, effecting an increase in the number of collisions between sample and reagent ions prior to mass analysis. A much lower pressure of reagent gas thus suffices to permit the observation of c.i. reactions.[164] In experiments designed to determine the appropriate conditions for routine and reproducible c.i., reagent-gas pressures of about 10^{-6} mbar were used with a partial vapour pressure of sample in the order of 5×10^{-8} mbar.

Another experiment that would appear, in principle, to be incompatible with the low pressures necessary in FT m.s. is collision-induced dissociation. However, an ion may be selected by irradiation of the sample to the chosen ion's cyclotron frequency.[165] It is then accelerated into the target gas, present at a pressure of, typically, 10^{-5} mbar, when c.i.d. takes place.[166] This, like the triple-quadrupole experiment, is a low-energy technique with a relatively high efficiency. Variation of the collision energy is possible over a limited range by adjustment of the level of irradiating r.f.

7 Other Techniques

Despite the proliferation of new, exciting techniques for the analysis of complex or intransigent mixtures, many people are committed to maintaining an existing, perhaps ageing instrument. The challenge of restoring such instruments to current specifications is being met both at the 'professional' level, by manufacturers and specialist groups, and at the laboratory level, where an expedient that meets the several needs of cost, speed, and convenience must be found. Two methods have been reported for reducing the cost of replacement of high-vacuum gold seals; one method uses an aluminium gasket, prepared from wire of 0.5 mm or 1 mm diameter, in a similar way to the preparation of gold gaskets.[167] The aluminium seals, which offer a significant cost saving, can be used several times and do not deteriorate under baking. The other method uses a PTFE 'rope' made from 0.25 inch wide PTFE tape of the type used by plumbers.[168] This method is adequate to a vacuum of 10^{-7} mbar or better, but the gaskets are not re-usable.

The replacement of valve amplifiers by solid-state devices, which offer greater reliability and lower power consumption as well as smaller size, has always had high priority in any update plan. A low-cost solid-state preamplifier replacement not only provides extra facilities, such as automatic gain change under computer control, but also prolongs the life of the electron multiplier.[169] Replacement of the power supplies for both magnet and electrostatic analyser with programmable solid-state devices[170] enables rapid and reproducible scanning to be performed under either manual or computer control, and a solution to the problem of magnet instability during selected-ion recording is provided by superimposing a sinusoidal sweep on the accelerating voltage.[171] The energy resolution of the electrostatic sector of a CEC 21-110B instrument has been improved by the addition of a three-element Einzel lens followed by a variable β-slit.[172]

Alterations to currently available commercial equipment can also yield improvement, and examples of developments of ion sources and inlet systems have already been reported. Other examples include the design and construction of a mass marker for many types of scan, linked scans, and m.i.k.e.s.[173] This device measures the ESA voltage and converts it into a mass reading, which is output on a digital voltmeter; drivers are included for event markers on a chart output. A VG Analytical ZAB reverse-geometry instrument has also been modified by the addition of a slit system to allow angular collimation of the beam by eliminating those ions not moving along the optical axis of the instrument.[174] This modification permits the selection of the products of fragmentation of metastable ions on the basis of their internal energies. A multiple-ion detection unit, which incorporates a 'lock-mass' procedure to compensate for drift, has been modified to enable automatic, accurate mass measurement of seven ions simultaneously under computer control.[175] The modified unit also permits signal averaging so that the signal-to-noise ratio is improved, increasing mass-measurement accuracy, particularly on small peaks.

The system of control of sample evaporation using feedback

from the TIC, used extensively in f.d., has also been applied in the microdistillation of crude oils from a probe under e.i. conditions.[176] The probe used in this study was also capable of operation with a linear temperature program. A temperature-programmed fractionator for the study of polymers has been constructed.[177] This low-cost device permits fractionation under a stream of helium, and it is interfaced with the mass spectrometer *via* a single-stage jet separator. The pyrolysis of polymers, with particular emphasis on forensic applications, has also been described.[178] Pyrolysis was effected using specially constructed pyrolysis probes, which enabled pyrolysers of either the Curie-point or filament type to be connected directly with the ion source. A Curie-point pyrolyser has been mounted in a specially modified f.i. source so that the sample, contained in a quartz capillary and surrounded with a cylinder of ferromagnetic material, can be transferred into the centre of the high-frequency coil through the direct introduction system.[179] An automatic sample-introduction system has been designed using a rotary cassette system from which small ampoules containing the samples are injected sequentially into the ion source.[180] The ampoules can be loaded automatically as, for example, from an autosampler attached to a liquid chromatograph.[181] In an on-line method for liquid chromatography, a microbore l.c. column has been incorporated into the direct probe of a high-voltage magnetic-sector instrument operating in c.i. mode.[182] Since the probe is at 8 kV, the authors recommend that the high voltage is switched off before an injection is made!

Rapid and convenient analysis of compounds separated by thin-layer chromatography has been generally elusive, necessitating elution of the sample from the plate prior to mass spectrometry. The use of a polyamide-based absorbent, however, makes the elution process unnecessary;[183] mass spectrometry of the absorbent shows only low background, which has little effect on analysis. A more complex approach gives an on-line analysis, using c.i. reagent gas to sweep the compound off the heated plate.[184] A loss of spatial resolution is caused by the thermal memory of the plate, but r.f. values show good reproducibility.

Developments in the interfacing of thermal analytical methods (d.t.a., t.g.a.) with mass spectrometers have been reported; these include the use of a constant-volume sampling interface,[185] giving reproducible sample introduction for qualitative analysis, and a totally automated system,[186] using an evenly heated Monel tube with a specially built thermo-analyser furnace. Several other interfacing methods have been tried,[187] including stainless-steel capillary tubes of several different dimensions. One of the more successful interfaces uses a jet separator with a variable by-pass valve, which allows the correct amount to leak through to the mass spectrometer, excess of gas being vented at the by-pass valve.

A membrane inlet system similar to the Llewellyn-Littlejohn separator has been used with a small quadrupole mass analyser for the detection of sub-p.p.m. levels of organic vapours in air.[188] One of the important uses of this type of system is in breath analysis,[189] where a fast response time must be coupled with a detection level down to 0.05 mg m^{-3} for selected substances. Additionally, there must be no loss in sensitivity or interference caused by high water-vapour levels. Selective permeation through thin membranes has also been used to study electrolysis products[190] and to study reaction kinetics in biological systems.[191]

In a system for the analysis of particulate pollutants,[192] an air sample passes through a two-stage inlet system, with skimmer jet and collimator. Particulate matter passes through the skimmer system and impinges on a hot filament, where it is ionized. This system is one of several that are transportable.[193] Of the many experiments available in such a system, one is selected-ion monitoring with a mobile instrument, which gives the concentration of the targeted compound as a function of distance; this provides a useful method for the assessment of the extent of penetration of noxious emissions.

Finally, an instrument has been described that produces mass spectra without the need for a mass spectrometer, at a cost of less than £100.[194] Unfortunately, this is a description not of a radically new instrument but of a teaching device based on a simple microprocessor, which simulates a mass spectrometer or, with some simple modification, any other type of spectrometer. 'Unknowns', comprising spectra of 80 or more peaks, are stored in the memory of the instrument.

References

1 M. Barber, R.S. Bordoli, R.D. Sedgwick, and A.N. Tyler, *J. Chem. Soc., Chem. Commun.*, 1981, 325.
2 M. Barber, R.S. Bordoli, G.J. Elliott, R.D. Sedgwick, and A.N. Tyler, *Anal. Chem.*, 1982, 54, 645A.
3 M. Barber, R.S. Bordoli, R.D. Sedgwick, and A.N. Tyler, *Nature (London)*, 1981, 293, 270.
4 L.C.E. Taylor, *Ind. Res. Dev.*, 1981, 23, 124.
5 R.A. McDowell, A. Dell, and H.R. Morris, presented at the 29th Annual Conference on Mass Spectrometry and Allied Topics, Minneapolis, 24 - 29 May, 1981.
6 G. Borchardt, H. Scherrer, S. Weber, and S. Scherrer, *Int. J. Mass Spectrom. Ion Phys.*, 1980, 34, 361.
7 D.J. Surman and J.C. Vickerman, *J. Chem. Soc., Chem. Commun.*, 1981, 324.
8 C.J. McNeal, *Anal. Chem.*, 1982, 54, 43A.
9 A. Iino and A. Mizuike, *Bull. Chem. Soc. Jpn.*, 1981, 54, 1975.
10 A.L. Burlingame, A. Dell, and D.H. Russell, *Anal. Chem.*, 1982, 54, 363R.
11 F.M. Devienne and J.-C. Roustan, *Org. Mass Spectrom.*, 1982, 17, 173.
12 R.J. Colton, *J. Vac. Sci. Technol.*, 1981, 18, 737; N.H. Turner and R.J. Colton, *Anal. Chem.*, 1982, 54, 293R.
13 S.E. Unger, A. Vincze, R.G. Cooks, R. Chrisman, and L.D. Rothman, *Anal. Chem.*, 1981, 53, 976.
14 A. Benninghoven, A. Eicke, M. Jimak, W. Sishterman, J. Krizek, and H. Peters, *Org. Mass Spectrom.*, 1980, 15, 459.
15 R.D. Smith, J.E. Burger, and A.L. Johnson, *Anal. Chem.*, 1981, 53, 1603.
16 L.K. Liu, K.L. Busch, and R.G. Cooks, *Anal. Chem.*, 1981, 53, 109.
17 K.D. Klöppel, and G. Von Bünau, *Int. J. Mass Spectrom. Ion Phys.*, 1981, 39, 85.
18 R.J. Colton, J.E. Campana, T.M. Barlak, J.J. DeCorpo, and J.R. Wyatt, *Rev. Sci. Instrum.*, 1980, 51, 1685.
19 J.E. Campana, T.B. Barlak, R.J. Colton, J.J. DeCorpo, J.R. Wyatt, and B.I. Dunlop, *Phys. Rev. Lett.*, 1981, 47, 1046.
20 B.T. Chait and K.G. Standing, *Int. J. Mass Spectrom. Ion Phys.*, 1981, 40, 185.
21 C.J. McNeal and R.D. Macfarlane, *J. Am. Chem. Soc.*, 1981, 103, 1609.
22 R.J. Conzemius and J.M. Capellen, *Int. J. Mass Spectrom. Ion Phys.*, 1980, 34, 197.
23 R.J. Cotter, *Anal. Chem.*, 1980, 52, 1770.
24 R.J. Cotter, *Anal. Chem.*, 1981, 53, 719.
25 H.-R. Schulten, B.B. Monkhouse, and R. Müller, *Anal. Chem.*, 1982, 54, 654.
26 H.-R. Schulten, R. Müller, and D. Haaks, *Fresenius' Z. Anal. Chem.*, 1980, 304, 15.
27 H.J. Heinen, *Int. J. Mass Spectrom. Ion Phys.*, 1981, 38, 309.
28 E. Denoyer, R. van Grieken, F. Adams, and D.F.S. Natusch, *Anal. Chem.*, 1982, 54, 26A.
29 S.W. Graham, P. Dowd, and D.M. Hercules, *Anal. Chem.*, 1982, 54, 649.
30 H. Vogt, H.J. Heinen, S. Meier, and R. Wechsung, *Fresenius' Z. Anal. Chem.*, 1981, 308, 195.
31 F. Heresch, E.R. Schmid, and J.F.K. Huber, *Anal. Chem.*, 1980, 52, 1803.
32 E.D. Hardin and M.L. Vestal, *Anal. Chem.*, 1981, 53, 1492.

33 D. Zackett, A.E. Schoen, R.G. Cooks, and P.H. Hemberger, *J. Am. Chem. Soc.*, 1981, 103, 1295.
34 G.J.Q. van der Peyl, K. Isa, J. Haverkamp, and P.G. Kistemacher, *Org. Mass Spectrom.*, 1981, 16, 416.
35 A.L. Yergey, N.E. Vieira, and J.W. Hansen, *Anal. Chem.*, 1980, 52, 1811.
36 R.J. Cotter and A.L. Yergey, *J. Am. Chem. Soc.*, 1981, 103, 1596.
37 R.J. Cotter and A.L. Yergey, *Anal. Chem.*, 1981, 53, 1306.
38 R. Stoll and F.W. Roellgen, *Org. Mass Spectrom.*, 1981, 16, 72.
39 J. Schade, R. Stoll, and F.W. Roellgen, *Org. Mass Spectrom.*, 1981, 16, 441.
40 J.V. Iribarne and B.A. Thomson, *U.S. Patent*, No. 4,300,044.
41 S.-T.F. Lai, K.W. Chan, and K.D. Cook, *Macromolecules*, 1980, 13, 953.
42 C.R. Blakey, J.J. Carmody, and M.L. Vestal, *Anal. Chem.*, 1980, 52, 1636.
43 C.R. Blakey, J.J. Carmody, and M.L. Vestal, *J. Am. Chem. Soc.*, 1980, 102, 5931.
44 C.R. Blakey, J.J. Carmody, and M.L. Vestal, presented at the 28th Annual Conference on Mass Spectrometry and Allied Topics, New York, 25 - 30 May, 1980.
45 V. Caldecourt, D. Zackett, and J. Tou, presented at the 29th Annual Conference on Mass Spectrometry and Allied Topics, Minneapolis, 24 - 29 May, 1981.
46 D.A. Lane, *Environ. Sci. Technol.*, 1982, 16, 38A.
47 M. Tsuchiya and T. Taira, *Int. J. Mass Spectrom. Ion Phys.*, 1980, 34, 351.
48 M. Tsuchiya, K. Seita, and T. Taira, *Shitsuryo Bunseki*, 1980, 28, 235.
49 H. Kambara, *Anal. Chem.*, 1982, 54, 143.
50 A.L. Gray and A.R. Date, *Dyn. Mass Spectrom.*, 1981, 6, 252.
51 R.S. Houk, V.A. Fassel, G.D. Flesch, H.J. Svec, A.L. Gray, and C.E. Taylor, *Anal. Chem.*, 1981, 52, 2283.
52 A.R. Date and A.L. Gray, *Analyst (London)*, 1981, 106, 1255.
53 D.J. Douglas and J.B. French, *Anal. Chem.*, 1981, 53, 37.
54 R.J. Cotter, *Anal. Chem.*, 1980, 52, 1589A.
55 M. Ohashi, K. Tsujimoto, S. Tamura, N. Nakayama, Y. Okumura, and A. Sakurai, *Biomed. Mass Spectrom.*, 1980, 7, 153.
56 L. Kelner, H.M. Fales, S.P. Markey, and C.K. Crawford, presented at the 29th Annual Conference on Mass Spectrometry and Allied Topics, Minneapolis, 24 - 29 May, 1981.
57 S.L. Koontz and M.B. Denton, *Int. J. Mass Spectrom. Ion Phys.*, 1981, 37, 227.
58 H. Ezoe and K. Watanabe, *Shitsuryo Bunseki*, 1981, 29, 389.
59 G.C. Stafford, D.C. Bradford, and D.R. Stephens, presented at the 29th Annual Conference on Mass Spectrometry and Allied Topics, Minneapolis, 24 - 29 May, 1981.
60 R.C. Schiebel, O.P. Tanner, and K.V. Wood, *Anal. Chem.*, 1981, 53, 550.
61 R.C. Schiebel, O.P. Tanner, and K.V. Wood, presented at the 28th Annual Conference on Mass Spectrometry and Allied Topics, New York, 25 - 30 May, 1980.
62 A.J. Illies, M.T. Bowers, and G.G. Meisels, *Anal. Chem.*, 1981, 53, 1557.
63 G.D. Price and D.A. Carlson, *Anal. Chem.*, 1981, 53, 564.
64 V.N. Reinhold and S.A. Carr, *Anal. Chem.*, 1982, 54, 499.
65 D.F. Fraley and M.M. Bursey, *Anal. Chem.*, 1981, 53, 1546.
66 J.A.G. Roach, A.J. Malatesta, J.A. Sphon, W.C. Brumley, D. Andrzejewski, and P.A. Dreifuss, *Int. J. Mass Spectrom. Ion Phys.*, 1981, 39, 151.

67 D.I. Carroll, J.G. Nowlin, R.N. Stillwell, and E.C. Horning, *Anal. Chem.*, 1981, 53, 2007.
68 P. Traldi, *Org. Mass Spectrom.*, 1982, 17, 245.
69 F.A. Bencsath and F.H. Field, presented at the 29th Annual Conference on Mass Spectrometry and Allied Topics, Minneapolis, 24 - 29 May, 1981.
70 R.C. Murphy, K.L. Clay, and W.R. Mathews, *Anal. Chem.*, 1982, 54, 336.
71 C.J. McNeal, S.A. Narang, R.D. Macfarlane, H.M. Hsuing, and R. Brousseau, *Proc. Natl. Acad. Sci. U.S.A.*, 1980, 77, 735.
72 D.F. Fraley, W.S. Woodward, and M.M. Bursey, *Anal. Chem.*, 1980, 52, 2290.
73 D.L. Smith, J.A. McCloskey, and J.K. Mitchell, *Anal. Chem.*, 1981, 53, 1130.
74 W.D. Lehmann and R. Fischer, *Anal. Chem.*, 1981, 53, 743.
75 M. Rabrenovic, T. Ast, and V. Kramer, *Int. J. Mass Spectrom. Ion Phys.*, 1981, 37, 297.
76 F. Okuyama, *Int. J. Mass Spectrom. Ion Phys.*, 1981, 38, 255.
77 I.V. Goldenfeld and H.J. Veith, *Int. J. Mass Spectrom. Ion Phys.*, 1981, 40, 361.
78 W.A. Schmidt, I.V. Goldenfeld, and A.I. Helal, *Int. J. Mass Spectrom. Ion Phys.*, 1981, 38, 241.
79 A.M. Hogg and J.D. Payzant, presented at the 29th Annual Conference on Mass Spectrometry and Allied Topics, Minneapolis, 24 - 29 May, 1981.
80 W. Aberth, *Biomed. Mass Spectrom.*, 1980, 7, 367.
81 F. Okuyama, K. Ishikawa, M. Chida, and Y. Yamato, *Anal. Chem.*, 1980, 52, 1987.
82 H.U. Winkler and H.D. Beckey, *Int. J. Mass Spectrom. Ion Phys.*, 1981, 39, 111.
83 R.C. Dougherty, *Anal. Chem.*, 1981, 53, 625A.
84 M. McKeown, *Environ. Health Perspect.*, 1980, 36, 97; G.C. Stafford, *ibid.*, 1980, 36, 85.
85 See, for example, refs. 2 and 27.
86 H.R. Morris, A. Dell, and R.A. McDowell, *Biomed. Mass Spectrom.*, 1981, 8, 463.
87 G.A. Craig, P.G. Cullis, and P.J. Derrick, *Int. J. Mass Spectrom. Ion Phys.*, 1981, 38, 297.
88 J.E. Campana, T.M. Barlak, R.J. Colton, J.J. DeCorpo, J.R. Wyatt, and B.I. Dunlop, *Phys. Rev. Lett.*, 1981, 47, 1046.
89 See ref. 80.
90 S. Evans, L.C.E. Taylor, D.R. Denne, H.J.M. Fitches, C.J. Wakefield, and K.R, Compson, presented at the 29th Annual Conference on Mass Spectrometry and Allied Topics, Minneapolis, 24 - 29 May, 1981; D. Hazelby, S. Evans, D.R. Denne, and L.C.E. Taylor, Kratos Analytical Instruments, Technical Bulletin.
91 R.H. Bateman, J.C. Bill, B.N. Green, and T.R. Kemp, 1982 Pittsburgh Conference on Analytical Chemistry and Applied Spectroscopy, Paper 097; R.H. Bateman, B.N. Green, and V.C. Parr, VG Analytical, Technical Notes.
92 G.W. Wood and W.F. Sun, *Biomed. Mass Spectrom.*, 1980, 7, 399.
93 P. Burns, B.N. Green, and D.S. Millington, presented at the 28th Annual Conference on Mass Spectrometry and Allied Topics, New York, 25 - 30 May, 1980; see also ref. 90.
94 T.R. Kemp, *Int. Lab.*, 1981, July/Aug., 46.
95 T. Takeda, S. Shibata, and H. Matsuda, *Shitsuryo Bunseki*, 1980, 28, 217.
96 J.L. Holmes and J.K. Terlouw, *Org. Mass Spectrom.*, 1980, 15, 383.

97 R.K. Boyd and J.H. Beynon, *Org. Mass Spectrom.*, 1977, 11, 163.
98 T. Matsuo, H. Matsuda, I. Katakuse, Y. Shimonishi, Y. Maruyama, T. Higuchi, and E. Kubota, *Anal. Chem.*, 1981, 53, 416.
99 W.F. Haddon, *Org. Mass Spectrom.*, 1980, 15, 539.
100 W. Heerma, M.M. Sarneel, and J.K. Terlouw, *Org. Mass Spectrom.*, 1981, 16, 325.
101 B. Shushan and R.K. Boyd, *Anal. Chem.*, 1981, 53, 421.
102 C.G. Macdonald and M.J. Lacey, *Int. J. Mass Spectrom. Ion Phys.*, 1981, 37, 87.
103 B. Shushan and R.K. Boyd, *Int. J. Mass Spectrom. Ion Phys.*, 1980, 34, 37; M.J. Lacey and C.G. Macdonald, *Anal. Chem.*, 1982, 54, 135.
104 R.K. Boyd and B. Shushan, *Int. J. Mass Spectrom. Ion Phys.*, 1981, 37, 355.
105 G.A. Warburton, R.S. Stradling, R.S. Mason, and M. Farncombe, *Org. Mass Spectrom.*, 1981, 16, 507.
106 M.J. Lacey and C.G. Macdonald, *Org. Mass Spectrom.*, 1980, 15, 484.
107 B. Shushan and R.K. Boyd, *Int. J. Mass Spectrom. Ion Phys.*, 1981, 37, 369.
108 R.K. Boyd, C.J. Porter, and J.H. Beynon, *Org. Mass Spectrom.*, 1981, 16, 490.
109 E.K. Chess and M.L. Gross, *Anal. Chem.*, 1980, 52, 2057.
110 D.A. Durden, *Anal. Chem.*, 1982, 54, 666.
111 P.F. Bente, tert. and F.W. McLafferty, *Pract. Spectrosc.*, 1980, 3, 253.
112 P.H. Dawson, J.B. French, J.A. Buckley, D.J. Douglas, and D. Simmons, *Org. Mass Spectrom.*, 1982, 17, 205.
113 F.W. MCLafferty, *Science*, 1981, 214, 280.
114 R.G. Cooks and G.L. Glish, *Chem. Eng. News*, 1981, Nov., 40.
115 J.H. Beynon, *Org. Mass Spectrom.*, 1981, 16, 280.
116 J.H. Beynon, R.G. Cooks, J.W. Amy, W.E. Baitinger, and T.Y. Ridley, *Anal. Chem.*, 1973, 45, 1023A.
117 C.J. Porter, T. Ast, and J.H. Beynon, *Org. Mass Spectrom.*, 1981, 16, 101.
118 R.E. Reinsfelder and M.B. Denton, *Int. J. Mass Spectrom. Ion Phys.*, 1981, 37, 241.
119 D. Zackett, R.G. Cooks, and W.J. Fies, *Anal. Chim. Acta*, 1980, 119, 129.
120 C.A. Boitnott, U. Steiner, M.S. Story, and R.D. Smith, *Dyn. Mass Spectrom.*, 1981, 6, 71.
121 R.A. Yost and C.G. Enke, *Anal. Chem.*, 1979, 51, 1251A.
122 D.F. Hunt, J. Shabanowitz, and A.B. Giordani, *Anal. Chem.*, 1980, 52, 386.
123 J.R.B. Slayback and M.S. Story, *Ind. Res. Dev.*, 1981, Feb., 129.
124 D.J. Douglas, *J. Phys. Chem.*, 1982, 86, 185.
125 M.W. Siegel, *Anal. Chem.*, 1980, 52, 1790.
126 G.L. Glish, S.A. McLuckey, T.Y. Ridley, and R.G. Cooks, *Int. J. Mass Spectrom. Ion Phys.*, 1982, 41, 157.
127 S.A. McLuckey, G.L. Glish, and R.G. Cooks, *Int. J. Mass Spectrom. Ion Phys.*, 1981, 39, 219.
128 J.H. Beynon, F.M. Harris, B.N. Green, and R.H. Bateman, *Org. Mass Spectrom.*, 1982, 17, 55.
129 D.H. Russell, E.H. McBay, and T.R. Mueller, *Int. Lab.*, 1980, April, 49.
130 A. Maquestiau, P. Meyrant, and R. Flammang, *Bull. Soc. Chim. Belg.*, 1981, 90, 173.
131 D.H. Russell, D.H. Smith, R.J. Warmack, and L.K. Bertram, *Int. J. Mass Spectrom. Ion Phys.*, 1980, 35, 381.

132 M.J. Lacey, C.G. Macdonald, K.F. Donchi, and P.J. Derrick, *Org. Mass Spectrom.*, 1981, 16, 357.
133 M.L. Cross, E.K. Chess, P.A. Lyon, F.W. Crow, S. Evans, and H. Tudge, *Int. J. Mass Spectrom. Ion Phys.*, 1982, 42, 243.
134 E.K. Chess, M.L. Gross, P.A. Lyon, and F.W. Crow, presented at the 29th Annual Conference on Mass Spectrometry and Allied Topics, Minneapolis, 24 - 29 May, 1981.
135 D.R. Burinsky, R.G. Cooks, E.K. Chess, and M.L. Gross, *Anal. Chem.*, 1982, 54, 295.
136 See ref. 128.
137 F.W. McLafferty, P.J. Todd, D.C. McGilvery, and M.A. Baldwin, *J. Am. Chem. Soc.*, 1980, 102, 3360.
138 G.L. Glish and P.J. Todd, *Anal. Chem.*, 1982, 54, 842.
139 D. Desiderio and J.Z. Sabbatini, *Biomed. Mass Spectrom.*, 1981, 8, 565.
140 E.S. Mukhtar, I.W. Griffiths, F.M. Harris, and J.H. Beynon, *Int. J. Mass Spectrom. Ion Phys.*, 1981, 37, 159.
141 G.J. Louter, A.J.H. Boerboom, P.F.M. Stalmeier, H.H. Tuithof, and J. Kistemacher, *Int. J. Mass Spectrom. Ion Phys.*, 1980, 33, 335.
142 See refs. 111, 114, 121, and 123.
143 In ref. 113.
144 See ref. 10.
145 G.C. Stafford in ref. 84.
146 M. McKeown in ref. 84.
147 D. Picart, F. Jacolot, F. Berthou, and H.H. Floch, *Biomed. Mass Spectrom.*, 1980, 7, 464.
148 W.V. Ligon, jun., *Int. J. Mass Spectrom. Ion Phys.*, 1980, 36, 387.
149 R.A. Britten, C.E. Giffen, and R.A. Johansen, presented at the 28th Annual Conference on Mass Spectrometry and Allied Topics, New York, 25 - 30 May, 1980.
150 B. Hedfjäll and R. Ryhage, *Anal. Chem.*, 1981, 53, 1641.
151 See ref. 141.
152 W. Aberth, *Int. J. Mass Spectrom. Ion Phys.*, 1981, 37, 379.
153 D.L. Donohue, J.A. Carter, and G. Mamantor, *Int. J. Mass Spectrom. Ion Phys.*, 1980, 35, 243.
154 P.J.C.M. Nowak, H.H. Holsboer, W. Heubers, R.W. Wijnaendts van Resandt, and J. Los, *Int. J. Mass Spectrom. Ion Phys.*, 1980, 34, 375.
155 J.E. Hunt, R.D. Macfarlane, J.J. Katz, and R.C. Doughtery, *Proc. Natl. Acad. Sci. U.S.A.*, 1980, 77, 1745.
156 W.V. Ligon, jun., *Int. J. Mass Spectrom. Ion Phys.*, 1982, 41, 213.
157 C.R. Snelling, jun., J.C. Cook, jun., R.M. Milberg, and K.L. Rinehart, jun., presented at the 29th Annual Conference on Mass Spectrometry and Allied Topics, Minneapolis, 24 - 29 May, 1981.
158 R.T. McIver, jun., *Int. Lab.*, 1981, Jan./Feb., 17.
159 R.L. White, E.B. Ledford, jun., S. Ghaderi, C.L. Wilkins, and M.L. Gross, *Anal. Chem.*, 1980, 52, 1525.
160 M. Allermann, H.P. Kellerhals, and K.P. Wanczek, *Chem. Phys. Lett.*, 1980, 328.
161 E.B. Ledford, jun., R.L. White, S. Ghaderi, C.L. Wilkins, and M.L. Gross, presented at the 28th Annual Conference on Mass Spectrometry and Allied Topics, New York, 25 - 30 May, 1980.
162 C.L. Wilkins and M.L. Gross, *Anal. Chem.*, 1981, 53, 1661A.
163 E.B. Ledford, jun., R.L. White, S. Ghaderi, C.L. Wilkins, and M.L. Gross, *Anal. Chem.*, 1980, 52, 2450.
164 S. Ghaderi, P.S. Kulkarni, E.B. Ledford, C.L. Wilkins, and M.L. Gross, *Anal. Chem.*, 1981, 53, 428.

165 R.B. Cody, R.C. Burnier, and B.S. Freiser, *Anal. Chem.*, 1982, 54, 96.
166 R.B. Cody and B.S. Freiser, *Int. J. Mass Spectrom. Ion Phys.*, 1982, 41, 199.
167 S.B. Davies, *Org. Mass Spectrom.*, 1981, 16, 466.
168 A.I. Mallet and P. Tippett, *Org. Mass Spectrom.*, 1980, 15, 274.
169 B.D. Soltmann, *Org. Mass Spectrom.*, 1982, 17, 53.
170 G.J. Greenwood, J.F. Stewart, and D.M. Cantor, presented at the 28th Annual Conference on Mass Spectrometry and Allied Topics, New York, 25 - 30 May, 1980.
171 L. Gruenke, J.C. Craig, and D.M. Bier, *Biomed. Mass Spectrom.*, 1980, 7, 381.
172 R.J. Ryba and G.J. Mains, *Int. J. Mass Spectrom. Ion Phys.*, 1982, 42, 169.
173 C.J. Porter, A.G. Brenton. J.H. Beynon, and A. Popplestone, *Org. Mass Spectrom.*, 1980, 15, 564.
174 C.J. Porter, C.J. Proctor, and J.H. Beynon, *Org. Mass Spectrom.*, 1981, 16, 62,
175 D.J. Harvan, J.R. Hass, and D. Wood, *Anal. Chem.*, 1982, 54, 332.
176 L.R. Schronk, R.D. Grigsby, and S.E. Scheppele, *Anal. Chem.*, 1982, 54, 748.
177 K.C. Chan, R.S. Tse, and S.C. Wong, *Anal. Chem.*, 1982, 54, 1238.
178 I. Jane, presented at the 28th Annual Conference on Mass Spectrometry and Allied Topics, New York, 25 - 30 May, 1980.
179 H.-R. Schulten, U. Bahr, and W. Goertz, *J. Anal. Appl. Pyrol.*, 1981, 3, 137.
180 C. Brunee, L. Delgmann, and G. Martens, *Ger. Offen.*, No. 3,002,575.
181 C. Brunee, *Ger. Offen.*, No. 3,007,538.
182 J.J. Brophy, D. Nelson, and M.K. Withers, *Int. J. Mass Spectrom. Ion Phys.*, 1980, 36, 205.
183 R. Kraft, A. Otto, A. Makower, and G. Etzold, *Anal. Biochem.*, 1981, 113, 193.
184 L. Ramalay, W.D. Jamieson, and R.G. Ackman, presented at the 28th Annual Conference on Mass Spectrometry and Allied Topics, New York, 25 - 30 May, 1980.
185 J. Chiu and A.J. Beattie, *Thermochim. Acta*, 1981, 50, 49.
186 H.K. Yuen, G.W. Mappes, and W.A. Grote, *Thermochim. Acta*, 1982, 52, 143.
187 E. Clarke, *Thermochim. Acta*, 1981, 51, 7.
188 T.W. Ottley, *Dyn. Mass Spectrom.*, 1981, 6, 212.
189 H.K. Wilson and T.W. Ottley, *Biomed. Mass Spectrom.*, 1981, 8, 606.
190 W.J. Pinnick, B.K. Lavine, C.R. Weisenberger, and L.B. Anderson, *Anal. Chem.*, 1980, 52, 1102.
191 K.C. Calvo, C.R. Weisenberger, L.B. Anderson, and M.H. Klapper, *Anal. Chem.*, 1981, 53, 981.
192 J.J. Stoffels, *Int. J. Mass Spectrom. Ion Phys.*, 1981, 40, 217.
193 D.A. Lane, *Environ. Sci. Technol.*, 1982, 16, 38A; see also ref. 189.
194 J.C. Traeger, *J. Chem. Educ.*, 1981, 58, 411.

7

Gas Chromatography–Mass Spectrometry and High-performance Liquid Chromatography–Mass Spectrometry

BY M. E. ROSE

1 Introduction

When taking over a periodical review such as this, it is tempting to stamp one's own personality on it. However, the format of the chapter has been well established over the past twelve years and is largely retained by this Reporter. There is one departure from previous practice: the addition of a short critical appraisal of one specialized area within the sphere of gas chromatography-mass spectrometry (g.c.-m.s.) and liquid chromatography-mass spectrometry (l.c.-m.s.). The topic selected for this Report is metastable-ion techniques in combination with chromatography-mass spectrometry. The aim of this new section is not full literature coverage but rather an assessment of the area and an indication of future trends. In this part of the chapter, the Reporter will feel less bound by the review period of the Report (June 1980 - June 1982) and draw on any literature considered as pertinent.

With well over 2000 publications on chromatography-mass spectrometry in the review period, fully comprehensive coverage is not possible. The policy adopted in previous reports[1] is continued here. The full range of applications of the combined techniques is still represented, but the citations within each category are restricted to those which, in the Reporter's opinion, are of most importance. Advances in methodology and novel applications are emphasized whereas the more mundane applications of g.c.-m.s. are largely omitted. Where the same work is described in full publications, preliminary communications, and/or conference reports, only the full paper is cited here. Similarly, space has been saved by referring only to the latest publications in long series of papers.

Where papers have been published in journals not read by or accessible to the Reporter, the sources for references have been the two excellent abstracting journals *Gas Chromatography-Mass Spectrometry Abstracts*[2] and *CA Selects: Mass Spectrometry*. Users and potential users of l.c.-m.s. should note the launch of a new abstracting journal, *Liquid Chromatography-Mass Spectrometry Abstracts*.[3] The first issue of this timely and useful publication appeared in November 1981, having the same format as its sister journal, *Gas Chromatography-Mass Spectrometry Abstracts*. It is to these two publications[2,3] that the reader is referred for succinct information on all papers concerning combined chromatography-mass spectrometry.

Specialized books and reviews are cited at the beginning of the appropriate section. Some general books and reviews germane to combined chromatography-mass spectrometry are listed here: general textbooks including aspects of g.c.-m.s. and/or l.c.-m.s.,[4-9] a biennial review of mass spectrometry,[10] gas-chromatography texts,[11-16] quantitative mass spectrometry (selected-ion monitoring),[17-23] reviews of g.c.-m.s.,[24-27] ionization techniques for g.c.-m.s. and/or l.c.-m.s.,[28,29] marine natural products,[30] stable isotopes,[31-35] accuracy and sensitivity of analytical data,[36] analysis of glycosides by mass spectrometry and g.c.-m.s.,[37,38] and biochemical, medical, and environmental applications.[39-48] A review of contaminants commonly encountered in mass spectrometry is recommended reading.[49] A new quarterly journal, with the self-explanatory title of *Mass Spectrometry Reviews*,[50] has been introduced.

Being pertinent to combined chromatography-mass spectrometry but not in context in any of the categories below, two reports[51,52] of thin-layer chromatography-mass spectrometry are noted here. Components separated on polyamide t.l.c. plates may be introduced into a mass spectrometer together with the polyamide adsorbent on a direct insertion probe, thus removing the need for the elution step.[51] True combination of t.l.c. and mass spectrometry has been effected for the analysis of quaternary alkaloids in mushrooms by ion bombardment of the developed t.l.c. plate.[52] Detection limits were not encouraging.

2 Methodology (Gas Chromatography-Mass Spectrometry)

Instrumentation.- Instrumentation for g.c.-m.s. has been reviewed[6,24] and is discussed in this volume (Chapter 6). A survey of the current state of quadrupole mass spectrometry has appeared.[53]

A mass spectrometer and a Fourier-transform infrared spectrophotometer can be coupled in parallel *via* a split valve[54,55] or in series[56] to a gas chromatograph. One interesting, computerized g.c.-FT i.r.-m.s. combination has been shown to enable analysis of model mixtures to be made when mass or i.r. spectral data alone were insufficient.[54] A mixture of alkylbenzenes and an unknown siloxane has been characterized using a capillary (support-coated open tubular, SCOT) column for the first time in such an instrument.[55] It is also possible to link g.c.-m.s. equipment to conventional thermoanalytical apparatus for mass-spectral analysis of evolved gases. Details of the interface have been given.[57] Integration of a gas-flow proportional counter into a commercial computerized g.c.-m.s. system[58] has aided biological radio-isotope incorporation experiments by application of radio-gas chromatography-mass spectrometry.[59] The combined instrument was used for simultaneous measurement of quantity, radio-isotopic content, and identity of analytes.[58]

Ionization techniques suitable for g.c.-m.s. have been reviewed,[28] but advances in this area have been few. During gas chromatography-chemical-ionization mass spectrometry (g.c.-c.i.m.s.), monitoring of the signal due to ions of the reactant gas is an alternative to conventional selected-ion monitoring of sample ions for quantification of eluents.[60,61] Advantages and applications of this 'reactant-ion monitoring' are as yet unclear. Poorly separated eluents from a g.c. can be characterized by injecting small amounts of a solvent such as ethanol into the chromatographic column.[62] The solvent acts as an initiator of chemical ionization and quasi-molecular ions are formed, as for example with toluenesulphonamides.[63] As previously reported,[1c] gas chromatography-photoionization mass spectrometry (g.c.-p.i.m.s.) is insensitive and unpopular relative to electron-impact (e.i.) and chemical-ionization (c.i.) methods. This conclusion remains unchanged after publication of the analysis of sixteen nitrite and nitrate esters by g.c.-p.i.m.s.[64] The simple fragmentation behaviour exhibited was offset by low sensitivity and paucity of molecular ions. Developments of the promising atmospheric-pressure ionization instrument[65] and plasma chromatograph[66] have been disappointingly quiet. The negative-ion mass spectra obtained by both methods have been compared for dicarboxylic acids.[67] Differences in the spectra were associated with the different ion lifetimes in the two techniques.

Of considerable importance are reports of the use of Fourier-transform mass spectrometers[68] as g.c. detectors.[68,69] For g.c.-m.s. the principal advantages of the FT mass spectrometer are its rapid-scanning and high-resolution capabilities. Even high-resolution g.c. separations need not be sacrificed in order to obtain accurate mass measurement of eluents, at least over a limited mass range. For selected-ion monitoring, very fast switching over wide mass ranges and at high resolution is possible.[68] The instrument has considerable potential also for compounds of high mass, making attractive a l.c.-FT m.s. combination (possibly with a moving-belt interface). Using a direct transfer line of glass-lined tubing or a jet separator, a g.c. capillary column was interfaced to an FT mass spectrometer fitted with a high-gas-conductance vacuum chamber. Both e.i. and c.i. mass spectra were available.[69] It has also been shown[70] that an ion-cyclotron-resonance (i.c.r.) spectrometer can be used to measure relative momentum transfer collision frequencies at resonance of g.c. eluents. In this way, the three isomers of difluorobenzene may be distinguished. The effects of ion/molecule reactions on the g.c.-i.c.r. method and possible analytical applications of the technique have been discussed.[70]

Mass spectrometers can be used as (very expensive) specific g.c. detectors for nitrogen-[71] or sulphur-containing[72] compounds by monitoring continuously capillary-column effluents for CH_2N^+ or CHS^+ ions, respectively. It should be noted that the mass 'window' scanned (m/z 28.018 ± 0.006) and the medium resolution (3000) selected for N-specific detection[71] allow not only CH_2N^+ but also $H^{13}CN^{+\cdot}$ and $^{13}C_2H_2^{+\cdot}$ ions to be detected. Both selected-ion monitoring methods[71,72] were used for analysing fuels. Detection of ions by ion counting has been reported to result in improved precision and sensitivity, relative to analogue g.c.-m.s. detection, for quantification of low levels of sterol derivatives[73] and the synthetic pyrethroid insecticide permethrin.[74] An electro-optical detector, comprising a multi-channel electron multiplier, a phosphor-coated fibre-optical window, and a self-scanning array of photodiodes, has been assessed for g.c.-m.s.[75] Being of limited resolution and dynamic range, both mechanical and electronic problems remain to be solved before the detector can be used advantageously for qualitative or quantitative g.c.-m.s. analysis. The dynamic range of a conventional (12-bit) analogue-to-digital converter can be improved by incorporating before it three parallel integrators of different sensitivities. Logic circuitry is used to test the output of data by the converter and then to make decisions on gain range. Applications of the system to g.c.-m.s. have been presented.[76]

Pervaporation is the term applied to the process of evaporation of components of a liquid mixture through a polymeric membrane, one side of which is in contact with the liquid, the other under vacuum. A pervaporation cell, containing a circulating solution, can be coupled directly with a mass spectrometer or g.c.-m.s. instrument to measure the gaseous outflow from the membrane.[77] By comparing the known molar composition of a reference solution with the mass-spectral analysis of pervaporated components, membrane selectivity can be assessed in terms of enrichment factors for given organic compounds. The system can be used with membranes of different selectivities to separate, identify, and quantify volatile organic compounds in water (*e.g.* industrial waste water).[77]

Interfaces.- Since the last Report,[1c] there has been considerable improvement in g.c.-m.s. interfaces with regard to decomposition of eluents on active surfaces. Open-split/direct-coupling interfaces are available for rigid glass or flexible fused-silica capillary columns that boast an absence of dead volume, adsorption, and thermal decomposition.[78-81] The approach by Friedli,[79] using a deactivated pressure-drop capillary as an interface in which the eluent contacts no surfaces other than the column wall, is particularly recommended. Compounds as labile as aflatoxin B can be assayed with this system.[79] Supply of pre-heated helium to the purge port of an open-split interface improves performance, efficiency, and base-line stability in g.c.-m.s.[82]

An extremely simple and popular means of coupling capillary columns (particularly flexible ones) to mass spectrometers is to insert the end of the column directly into the ion source and, if required, right up to the electron beam for obtaining 'in-beam' mass spectra. The main advantages of the method are freedom from dead volume and adsorption on the walls of an interface, simplicity, increased speed of analysis, and improved detection limits. The effect of lowering the outlet pressure below atmospheric has been reported variously to improve g.c. resolution, to have no effect, or to decrease it by up to 30%! In a useful theoretical treatment for open tubular columns, the maximum column efficiency was calculated to decrease by 12.5% at most when changing from atmospheric pressure to vacuum outlet conditions.[83] The loss of chromatographic efficiency expected theoretically as a result of vacuum outlet operation is rarely observed in practice as long as it is recognized that the optimum carrier-gas velocity is higher than that for atmospheric outlet pressure. This has led some workers, with considerable justification, to conclude that the best g.c.-m.s. coupling device is no device at all.[83,84]

By coupling directly a dense (supercritical) gas chromatograph to a mass spectrometer (see the previous Report[1c]), it is hoped to provide an alternative to and maybe an improvement on l.c.-m.s. and g.c.-m.s. for analysis of thermally labile and/or involatile compounds. Current instrumentation,[85] using a supersonic molecular-beam interface and capable of chromatographic pressures of at least 300 atm and temperatures of 10 - 60 ^{o}C, falls well short of the stated aim, but further development is recommended.

Data Processing.- Since some form of data system is mandatory for thorough analysis by g.c.-m.s., items of particular interest to the g.c.-m.s. user are covered here.

Extraction from raw, repetitively scanned g.c.-m.s. data of mass spectra free of background and contributions from closely overlapping peaks has been described.[86] The method utilized a calculation of a background ion current at each mass, assumed to be constant over 10 - 12 contiguous scans, and was compared with other published data-enhancement routines.[86] A new algorithm has been reported[87] to improve recognition of small g.c. peaks in g.c.-m.s. data. Detection of peaks is based on a computationally simple function to represent peak sharpness. A method for analysing data acquired from chromatographic systems employing multi-channel detectors is reported to be applicable to resolution of several overlapping g.c.-m.s. peaks.[88] The determination of the number of components in complex, overlapping peaks can be brought about by this technique[88] or by one based on plotting a set of orthogonal vectors from the original vectors representing the mass spectra of test mixtures.[89] The latter method was used to find the number of components in a chromatographic peak of pentadecane isomers.

A number of publications has appeared on the application of factor analysis for processing of repetitively scanned spectra in g.c.-m.s.[90-93] The number of components and the individual, constituent mass spectra of partially resolved or unresolved chromatographic peaks can be determined readily by factor analysis.[90,91] Hidden impurities in single g.c. peaks can also be detected by this methodology.[92,93] Using simulated, partially resolved peaks of Gaussian geometry and in the absence of noise, the detection limit of an impurity by factor analysis was about 1% of the main peak. The detection limit was increased to 3 - 10% when noise was added to the data matrix.[93]

A computer can be used to search sequentially each mass spectrum of a repetitively scanned g.c.-m.s. analysis for a specified and characteristic isotope cluster.[94] In this way, chlorine- and/or bromine-containing compounds in a complex mixture can be located rapidly. The use of a new reverse library search system, restricted to those reference mass spectra falling within a narrow retention index range about the retention index of each sample spectrum, has been described.[95] Two different methods have been compared for detection, determination, and library searching of silylated organic acids in urine.[96] The time required for searching of a large mass-spectral library (over 40000 reference spectra) can be reduced markedly by ordering each entry according to the mass of its most statistically significant peak.[97] When the search is restricted to the most probable matches only, retrieval of all correct identifications in 431 randomly selected cases was achieved when just 7.5% of the data base was searched.[97]

Quantification.- General comments have been made on selected-ion monitoring as regards the selection of ions and monitoring of them, treatment of calibration data, and accuracy, precision, and sensitivity.[6,98] In an interlaboratory precision test using capillary g.c.-m.s.,[99] the average relative standard deviation of response factors was 18.9% The influence of variations in mass-spectrometric operating conditions on long-term precision of isotope-dilution analyses has been evaluated, and data-reduction techniques that account for such changes have been proposed.[100] For selected-ion monitoring, reduction of dwell time on each mass peak monitored is reported to improve precision in the measurement of isotope ratios because it minimizes the mass cycling error.[101] The system was exemplified by the assay of palmitic acid turnover in dogs, using $(1-^{13}C)$palmitic acid. A further study of precision has concerned the quantification of compounds amenable to the inert-gas 'purge-and-trap' sampling method.[102]

Fundamental (non-empirical) criteria for selection of optimum m/z values, amount of internal standard, and data-processing techniques for selected-ion monitoring have been identified.[103] The authors sensibly point out that principles established through theory should be tempered with practical considerations such as a knowledge of background signals due to co-eluting impurities, column bleed, and so on.

Quantification of a substance and its isotopically labelled analogue can be brought about by two separate g.c.-m.s. (selected-ion monitoring) measurements. The ratio of natural to labelled compound is measured first then the sample is supplemented with a known amount of either variant and the isotope ratio determined again. The analysis is exemplified with (^{14}C) analogues of androstenedione and progesterone, about 10 picogrammes of which could be assayed in the presence of 1 nanogramme of the natural substances.[104]

The use of selected-metastable-peak monitoring for quantification is described in Section 5.

Sampling Techniques.- Only publications with emphasis on sampling methods of general interest are discussed here. The remainder are covered in the appropriate sections below. A useful review of preconcentration and solvent-extraction techniques for analysis of traces of organic compounds in environmental samples has been published.[105]

Volatile compounds trapped in adsorption tubes of Tenax-GC can be desorbed sequentially for g.c.-m.s. analysis by application of a heat gradient.[106] The method is said to prevent the masking of minor components by major ones. The polymeric adsorbents Tenax-GC and Amberlite XAD-2 have been evaluated for sampling fossil-fuel combustion products containing nitrogen oxides.[107] Both adsorbents decompose, but the major products from Tenax-GC do not interfere with the subsequent gas-chromatographic analysis of desorbed volatile compounds. Water is thought to fracture some of the resin beads of XAD-2 and allow leakage of alkane impurities.[108] The problem can be minimized by use of a series of solvents to effect displacement of trapped analytes. A further method of preventing foaming in 'purge-and-trap' g.c.-m.s. analyses, using silicone-based anti-foam agents, has been described.[109] The drawback of such a system is a decrease in precision of the g.c.-m.s. results owing to the complex nature of the matrix used.

Cryogenic traps for environmental pollutants have been described.[110,111] Volatile compounds may be trapped cryogenically in the first loop of a fused-silica open tubular column during dynamic headspace and pyrolytic studies of polymers,[111] and simultaneous analysis of all five organic priority pollutant fractions.[112] Sep-Pak^{R} C_{18} cartridges are proving to be simple, rapid, and effective means of extracting organic analytes such as steroids and their conjugates,[113] corticosteroids,[114] prostaglandins,[115] and tricyclic antidepressant drugs,[116] prior to assay by g.c.-m.s. Removal of unbound and protein-bound hydrophobic molecules from aqueous biological solutions can be brought about by use of Lipidex 1000.[117]

An extensive series of liquid-liquid fractionations prior to g.c.-m.s. is reported to aid analysis by separating complex mixtures into the following classes: strong acids, weak acids, bases, polar compounds, non-polar compounds, aldehydes, and ketones.[118] Fractionation into discrete chemical classes can also be brought about by column chromatography, as illustrated for g.c. and g.c.-m.s. study of organic compounds in synthetic fuels.[119] Oleate-replacement ultrafiltration forms a new method for quantitative recovery of organic acids from human plasma.[120,121] In this method, the problem of high protein content of plasma is overcome by addition of oleic acid, which competes for protein binding, thus releasing organic acids present in the plasma for metabolic profiling by g.c-m.s.[121] Finally, gel-permeation chromatography has been assessed for clean-up of xenobiotic chemicals in biological tissue samples.[122,123]

Chromatographic Aspects.- The superiority of capillary columns over packed columns for g.c.-m.s. studies is now widely accepted. The user who opts for a capillary column for his g.c. work now faces a further decision: to use columns constructed of conventional (rigid) glass or (flexible) fused silica. A book comparing the two different types of capillary column is available.[11] For g.c.-m.s., flexible columns have definite advantages. In particular, they are easily and effectively interfaced to mass spectrometers by feeding them directly into the ion source[1c] (see above section on interfaces). The limitations noted in the previous Report[1c] regarding the poor wettability of fused-silica capillary columns have been largely overcome; immobilized stationary phases can now be prepared simply.[124,125] Reasonable coating efficiencies have been achieved even with highly polar liquid phases such as SP-2330 and SP-2340. The range of commercially available fused-silica columns now also includes coatings of SE-30, SE-52, SE-54, OV-17, OV-101, OV-1701, FFAP, Dexsil 300, Dexsil 400, Dexsil 410, and SP-1000.

A very useful check on efficiency of capillary g.c. columns, using a single test mixture, has been described.[126] Yet even when using highly efficient capillary columns, probability calculations[127] show that more components of mixtures being analysed will remain unresolved than many practitioners would like to believe. Fused-silica, wall-coated open tubular (WCOT) columns have been reported to break when held in metal cages.[128] It was proposed[128] that vibration in the g.c. oven from the fan results in rubbing of the metal against the column, causing its polymer coating to weaken. An alternative to metal cages was proposed to prevent mid-column breakages.

Chiral stationary phases can be used to coat capillary columns for gas-chromatographic separation of optical isomers. Application and development of this methodology for assignment of absolute configuration are encouraged. Enantiomeric internal standards form ideal internal standards for assay of optically active analytes such as amino-acids[129,130] and some drugs[131,132] by g.c.-m.s. The enantiomers behave identically during work-up and mass-spectral fragmentation yet are fully resolved on chiral columns, permitting use of the most sensitive method of quantification: single-ion monitoring. Chiral stationary phases include copper(II) complexes[133] and optically active amides (including Chirasil-Val).[1c,129,130,134-136] Enantiomeric amino-acids,[129,130,134] amphetamines,[131,132] 2-hydroxycarboxylic acid esters,[133,136] amines,[134,136] alcohols,[135] and monosaccharides[137] have been separated successfully. The alternative to the use of chiral g.c. columns for resolution of optical isomers is derivatization with a chiral reagent followed by g.c. analysis of the resulting mixture of diastereoisomers on a conventional (achiral) column as exemplified for the racemic drug propranolol[138] and urinary 2-hydroxydicarboxylic acids.[139] The two different approaches have been compared for determination of chiral amphetamine drugs.[131,132]

The topic of injectors for capillary g.c. columns is largely beyond the scope of this review, but the reader should note a comparative study of split, splitless, and on-column injection techniques.[140] The last of these three methods[141] can extend the use of gas chromatography to compounds with retention indices at and above the level of n-$C_{50}H_{102}$. Fused-silica columns in a g.c.-m.s. system have been coupled *via* a 3-way union to an all-glass injector, allowing flush or split injection of solids.[142]

The first report of conventional g.c. analysis of transition-metal porphyrin complexes has appeared.[143] With Kovats retention indices in the range 5200 - 5600, short capillary columns were required. Complexes with different metals were resolved and g.c.-m.s. analysis was used to show that intact chelates were eluted rather than their decomposition products.[143] A small gas-chromatographic pre-column containing co-ordination polymers of the lanthanides has been shown to act as a selective, reversible trap for nucleophilic compounds in mixtures.[144]

Capillary columns are easily overloaded. A packed g.c. column, connected to a capillary column *via* a variable stream splitter and a cold trap, can serve as a means of regulating the quantity of individual components reaching the WCOT column in a g.c.-m.s. instrument.[145] A flame-ionization detector samples one stream (10%) from the first (packed) column and provides the electrical feedback to control the amount of the other stream (up to 90%) transferred to the second (WCOT) column. The quantity passed to the column is inversely proportional to the signal it elicits at the flame-ionization detector. In this way, the need for manual diversion of solvent and careful control of injection sizes is eliminated. This technique (given the absurd title of quantity-optimized two-dimensional gas chromatography[145]) would be particularly useful for capillary g.c.-m.s. analysis of mixtures in which overloading of major components would normally limit the injection volume to levels insufficient for determination of the minor components.

Derivatization.- Chemical derivatization of samples is a powerful and frequently necessary adjunct to analysis by g.c. and g.c.-m.s. Judicious selection of derivative can optimize the value or information content of the ensuing analytical results.[6] Parameters particularly influenced by the choice of derivative are volatility, stability, molecular weight, g.c. resolution, sensitivity (detection limit), and fragmentation behaviour of analytes. Chemical derivatization for g.c.-m.s.,[146] for pesticides,[147] and for g.c. with electron-capture detection[148] has been reviewed. The last of these articles[148] is also relevant to g.c.-m.s. since the electronegative derivatives described impart great mass-spectrometric sensitivity in the negative-ion mode. An excellent first review of cyclic boronate esters as derivatives for g.c.-m.s. analysis of bifunctional compounds has been published.[149] In this section, new and/or general procedures for chemical derivatization are discussed. Other descriptions of derivatives and their formation are referred to in the appropriate sections below.

Silylation of nucleophilic centres remains a widespread derivatization process. The trimethylsilyl (TMS) enol ethers of pentane-2,4-dione or methyl acetoacetate are highly reactive agents for introducing the TMS group into primary, secondary, and tertiary alcohols in good yield at room temperature.[150] The corresponding tert-butyldimethylsilyl (tBDMS) enol ethers also react rapidly with alcohols under very mild conditions.[151] In synthetic studies,[152] 1,2- and 1,3-diols can be protected by reaction with di-tert-butyldichlorosilane to form cyclic derivatives. This method may prove useful for g.c.-m.s. of diols as an alternative to the use of cyclic boronate esters. Silyl derivatives other than TMS include dimethoxymethylsilyl ethers of steroids,[153] alkoxydialkylsilyl[154] and vinyldimethylsilyl[155] ethers of steroids and cannabinoids, isopropyldimethylsilyl[156] and tBDMS[157] ethers of prostaglandins, tBDMS esters of organic acids,[158] triethylsilyl and tBDMS ethers of ketosteroids,[159] n-propyldimethylsilyl ethers of catecholamines,[160] and various sterically crowded trialkylsilyl derivatives of 2'-deoxy-nucleosides.[161] Alternatively, in the field of nucleosides and pyrimidine and purine bases, *NO*-peralkyl,[162] acyl,[161] and perfluoracyl[161,162] derivatives may be used. Phenolalkylamines can be trimethylsilylated and acylated on-column using a double-injection technique.[163]

Diacetates, TMS ethers, and cyclic alkaneboronate esters have been compared for g.c.-m.s. characterization of dihydroarenediols.[164] Whilst trimethylsilylation was the most satisfactory general procedure for 'metabolic' diols,[164] cyclic boronate esters offer several advantages for bifunctional compounds such as greater volatility, low mass increments, some conformational selectivity, and abundant molecular ions.[149,164,165] As examples, ethylene glycol in blood[166] and 2- and 3-hydroxy fatty acids[167] have been characterized as cyclic boronate ester derivatives. Using the same principle, but in reverse, simple diols (propane-1,3-diol and 1,2-dihydroxybenzene) have been used as derivatizing agents for g.c-m.s. analysis of substituted benzeneboronic acids.[168]

An apparatus has been described for preparation of diazomethane on the analytical scale. It was applied to the methylation and g.c.-m.s. analysis of barbiturates.[169] One-step methylation of the hydroxyl and carboxyl groups of bile acids with iodomethane and dimsyl sodium has been detailed.[170] Methylation of gibberellins, carbohydrates, and cytokinins can be brought about by reaction with iodomethane and sodium hydride in dimethylformamide,[171] and prostaglandins have been extracted from aqueous solution and simultaneously alkylated prior to analysis by g.c.-m.s.[172] A useful micro-acetylating technique for alcohols, using only gaseous reagents with neither base nor catalyst, has been illustrated.[173]

Location of double bonds in linear alkenes normally requires the unsaturated group to be 'fixed' by derivatization. The iodine-catalysed addition of dimethyldisulphide to alkenes affords α,β-bis(methylthio)alkanes that are amenable to g.c.-m.s. Cleavage of the carbon-carbon bond between adjacent MeS groups affords ions characteristic of double-bond position.[174] Epoxidation of double bonds in long-chain aldehydes and acetates followed by g.c.-m.s. analysis similarly locates the original sites of unsaturation through preferential cleavage α to the epoxy function.[175] A further publication dealing with location of double bonds in fatty acids by use of pyrrolidide derivatives has appeared.[176] Nicotinates of long-chain alcohols are suitable for g.c.-m.s. and display mass-spectrometric fragmentation diagnostic of chain structure.[177]

The method of ionization intended for a sample is a crucial factor in deciding which derivative to employ for its mass-spectral analysis.[6] For negative-ion chemical ionization (n.i.c.i.) of biogenic amines, some derivatives of high electron affinity have been compared with respect to ease of preparation, selectivity to primary amines, and limits of detection.[178] To detect traces of aliphatic alcohols by g.c.-c.i.m.s. in the positive- and/or negative-ion mode, TMS and various perfluoracyl derivatives have been investigated.[179]

2,4-Dinitrophenylhydrazones of ketones and aldehydes have been investigated by high-performance liquid chromatography (h.p.l.c.) and g.c.-m.s. Experimental conditions such as deactivation and temperature of the g.c. column influenced the extent of *syn-anti* isomerization so that careful control of the conditions facilitated identification of the derivatives.[180] Alternatively, aldehydes can be determined by g.c.-m.s. as their *O*-methyloxime or *O*-benzyloxime derivatives.[181] Several *O*-alkoxime pertrifluoracetyl derivatives of aldoses have been determined by g.c.-c.i.m.s. with a WCOT column.[182] Each compound afforded two peaks, the *syn*- and *anti*-alkoximes. Examples of derivatization techniques for determination of organic acids in urine have been given.[183] Lastly, β-propiolactone can be identified in complex biological matrices by g.c.-m.s. following reaction with octadecylamine and subsequent silylation.[184]

3 Applications (Gas Chromatography-Mass Spectrometry)

Long-chain Compounds.- Addition of formic acid to the g.c. carrier gas permits biological samples to be injected directly for analysis of volatile (C_2 - C_{20}) fatty acids.[185] The formic acid serves two functions: to displace the fatty acids from bound protein or from their sodium salts and to act as c.i. reactant gas for g.c.-c.i.m.s. The method was used to identify phytanic acid in plasma.[185b] G.c.-c.i.m.s. in combination with a 50 m SE-30 SCOT column has been used for the analysis of fatty-acid methyl esters. It was found that use of ammonia or methane as reactant gas allowed differentiation between unsaturated and cyclopropane functions in the hydrocarbon chains, whereas helium c.i. mass spectra were similar to e.i. spectra and enabled location of any branch points.[186] Cyclopentenyl fatty acids have been reported for the first time in leaves of various *Flacourtiaceae*.[187] Discrimination between cyclopropanoid and monoenoic fatty acids, as methyl esters, has also been accomplished by use of vinyl methyl ether as reactant gas for g.c.-c.i.m.s.[188] Polyenoic fatty acids have been identified by g.c.-c.i.m.s. following oxidation by osmium tetroxide and Hakomori permethylation of the resulting diol functions.[189] Diabolic acids are long-chain dicarboxylic acids with vicinal dimethyl branching. They have been shown to occur in both simple-stomached and ruminant animals.[190] Fatty acids esterified with di- and tri-terpene alcohols have been characterized in the latex of *Euphorbia lathyris*.[191] For analysis of hydroxy fatty acids as TMS ether or heptafluorobutyrate methyl esters, capillary g.c.-c.i.m.s. is superior to g.c.-e.i.m.s.[192] The number and position of the hydroxyl functions can be deduced from methane or isobutane c.i. spectra. Using the heptafluorobutyryl derivatives, highest sensitivity was achieved by use of negative-ion c.i. mass spectrometry,[192] the detection limit being 10^{-15} g. Mycolic acids in the range C_{22} - C_{39} have been isolated from *Bacterionema matruchotii* and analysed by g.c-m.s. These β-hydroxy fatty acids were examined as TMS ethers of methyl esters.[193] Analyses of autoxidized fats by g.c.-m.s. have continued.[194]

Organic matter adsorbed on particles in seawater[195] was found to be a much more complicated mixture of lipids than was that actually dissolved in the marine environment,[196] although the major components, n-alkanes and fatty-acid esters, were common to both particulate and dissolved phases. Lipids of fungi[197] and *Thiobacillus thioparus*[198] have been studied by use of g.c.-m.s. Two new series of branched dimethylalkanes have been identified in *Solenopsis invicta* and *S. richteri*.[199] A previous and tentative structural assignment to 10,12-dimethyltricosane was reinterpreted as being a mixture of 3,9- and 3,11-dimethyltricosane.[199] Synthetic long-chain esters, ketones, and propanediol diesters have been subjected to g.c.-m.s. analysis using a packed Dexsil 400 column.[200] Their fragmentation behaviour was discussed.

Prostaglandins and Related Eicosanoids.- Workshop papers on measurement of prostaglandins, prostacyclin, and thromboxanes have been published in book form.[201] Another book[202] describes structure elucidation of the leukotrienes and slow-reacting substance-A. Analysis of prostaglandins and thromboxanes by g.c.-m.s. has been reviewed[203] and compared with determination by radioimmunoassay (r.i.a.).[204]

Much work has been carried out on prostaglandin (PG) biosynthesis in this review period. For example, eicosapolyenoic acids have been observed as possible prostaglandin precursors in amphibians.[205] The fate of arachidonic acid in the gastro-intestinal tract,[206] mouse peritoneal marcrophage cells,[207] and carcinoma tissue[208] has been studied. Metabolites include PGE_2, PGD_2, $PGF_{2\alpha}$, 6-keto-$PGF_{1\alpha}$, thromboxane B_2, and several monohydroxylated eicosatetraenoic acids. Novel leukotrienes found in human leukocytes include dihydroxyeicosatetraenoic acids[209] resulting from oxygenation of arachidonic acid at C-15 and a trihydroxy-eicosatetraenoic acid[210] from ω-hydroxylation of leukotriene B_4. Leukotriene A_4 has been reported[211] as an intermediate in the biosynthesis of leukotrienes C_4 and D_4.

Whilst methyl esters of methoxime TMS ethers are the most widely used derivatives of prostaglandins for g.c.-m.s. work, detection limits can be improved by using methyl esters of butaneboronate TMS ethers.[207a] Formation of mixed derivatives (the methyl esters of butaneboronate, pentafluorobenzyloxime, TMS ethers) requires a multi-step procedure but enhances capillary g.c-m.s. of prostaglandins.[212] Various TMS derivatizing agents for PGE methyl ester do not always afford a single product.[213] Improved sample preparation[115] and new derivatives[156,157] of prostaglandins have been discussed above.

Prostacyclin (PGI_2) is unstable, but its hydrolysis product, 6-keto-$PGF_{1\alpha}$, is stable and, after derivatization, amenable to identification and/or quantification by g.c.-m.s. The number of publications dealing with this topic has been large because of clinical interest in prostacyclin. Notable advances have concerned the application of capillary g.c.-c.i.m.s. in the positive-[214] or negative-ion[215] mode, which enables assays in the pg range. G.c.-m.s. has been compared with r.i.a.[216,217] and g.c. (electron-capture detection)[218] for analysis of 6-keto-$PGF_{1\alpha}$ in biological fluids. Twelve urinary metabolites of prostacyclin have been detected in monkey[219] and two 2,3-dinor metabolites assayed in human urine.[220]

A novel procedure for simultaneous quantification of eight prostaglandins against one internal standard is available.[221] The method involves reductive incorporation of deuterium at $\Delta^{5,6}$ and g.c.-c.i.m.s. on a packed column. Quantitative determination of 9-deoxo-16,16-dimethyl-9-methylene-PGE_2 in human plasma can be performed by g.c.-m.s., h.p.l.c., or r.i.a. methods.[222] A trimethyl-PGE_2 analogue in human plasma has been determined as its pentafluorobenzyl derivative by g.c.-n.i.c.i.m.s. with methane as carrier and buffer gas.[223] Metabolites of PGH_2[224] and thromboxane B_2[225] have also been studied by g.c.-m.s.

Isoprenoid Compounds.- Carotenoid and retinoid compounds can be analysed by g.c., g.c.-m.s., and/or h.l.p.c.[226] The inherent instability of polyene chains limits the utility of methods based on g.c. so that carbon skeletons only are identified following reduction to the saturated (perhydro) derivative. The shrub *Euphorbia lathyris* has a remarkably high content of readily extractable and energy-intensive constituents.[227] An overview of the terpenoid components has been published.[227]

There are now four known juvenile hormones in insects (1; JH-0, JH-I, JH-II, and JH-III), and all four can be identified and quantified by the use of g.c.-m.s. The new hormone, JH-0, was isolated from developing embryos of the tobacco hornworm and structurally assigned as methyl (2*E*,6*E*,10-*cis*)-10,11-epoxy-3,7-diethyl-11-methyl-2,6-tridecadienoate (1).[228,229] Following

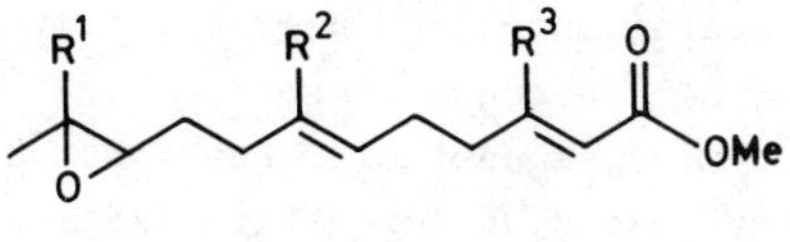

(1) $R^1 = R^2 = R^3 = Et$ (JH-0)
$R^1 = R^2 = Et$, $R^3 = Me$ (JH-I)
$R^1 = Et$, $R^2 = R^3 = Me$ (JH-II)
$R^1 = R^2 = R^3 = Me$ (JH-III)

formation of the 11-($^2H_3$)methoxy-10-hydroxy derivatives, the four juvenile hormones can be assayed down to a level of 10 - 40 pg g^{-1} of insect tissue by selected-ion monitoring.[229] The selectivity of the method appears to be maintained despite monitoring ions of low mass (*m/z* 76 and 90) formed by scission of the C-10-C-11 bond. If the derivatizing agent is changed from methanol to n-1H,1H,2H,2H-perfluorohexanol then the same bond cleavage affords ions at *m/z* 305 or 319 $\{[CF_3(CF_2)_3CH_2CH_2O{-}C(Me)R]^+$ where R = Me or Et$\}$. These ions are better candidates for selected monitoring. When this methodology was employed with a capillary g.c.-m.s. system, a specific and sensitive (femtomolar range) assay for JH-I, JH-II, and JH-III resulted.[230] Two compounds, having strong juvenile-hormone activity and designated juvocimenes 1 and 2, from sweet basil have been characterized by g.c.-m.s. and n.m.r. spectroscopy.[231]

New gibberellins, GA_{57} (*ent*-1β,3α,10β,13-tetrahydroxy-7,19-dioic acid 19-10 lactone)[232] and GA_{59} (Δ^2-gibberellin A_{21}),[233] have been isolated from *Gibberella fujikuroi* and *Canavalia gladiata*, respectively. In developing grains of *Triticum aestivum* (wheat), several gibberellins have been identified as TMS derivatives,[234,235] including GA_{15}, GA_{17}, GA_{19}, GA_{20}, GA_{24}, GA_{44}, and the novel GA_{54}, GA_{55}, GA_{60}, GA_{61}, and GA_{62}. These papers[234,235] constitute the first reports of the natural occurrence of 1β-hydroxygibberellins in plants. Major (GA_1) and minor (GA_3) endogenous gibberellins in germinating barley have been determined as TMS ether methyl esters. Changes in their levels under malting conditions were also examined by g.c.-m.s.[236]

Six diterpenoid lactones, two new and four known, have been isolated from the common Caribbean gorgonian *Briareum asbestinum* and their structures elucidated.[237] Hydrolysis of the pigment bacteriochlorophyll b releases $\Delta^{2,10}$-phytadienol as shown by capillary g.c.-m.s. on OV-1.[238] Terpene-type alcohols can be reduced to the corresponding alkane and subjected to g.c.-m.s. for analysis of their carbon skeletons.[239] Zoapatanol (2) and montanol (3) are novel diterpenes from the Mexican plant zoapatle.[240]

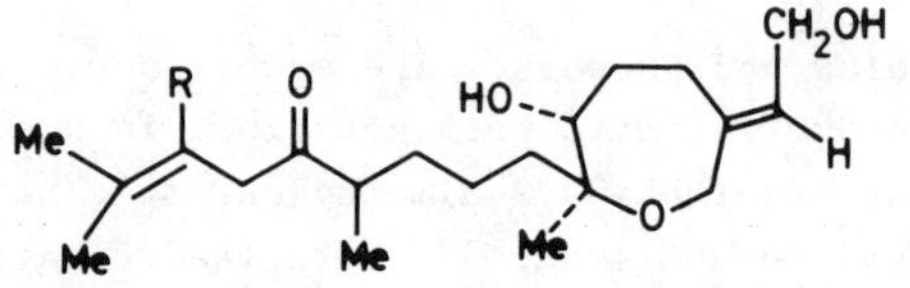

(2) R = H
(3) R = Me

The parent compounds and various derivatives were analysed on a packed-column g.c.-m.s. system. The mass spectra of 55 sesquiterpenes (longifolanes, caryophyllanes, cadinanes, cedranes, longipinanes, and thujopsanes) have been acquired *via* g.c.-m.s.[241]

The most biologically active compound of the vitamin E family is α-tocopherol. As its TMS derivative, α-tocopherol can be measured, against a deuterium-labelled internal standard, in the pg range by selected-ion monitoring using a 3% silar 10C column.[242]

Analysis of essential oils by g.c.-m.s.[243,244] can now be regarded as routine and need not be covered extensively here. A new method for small-scale steam distillation of volatile components of essential oils has been outlined.[245] Distillates were collected in only 1 ml of dichloromethane, ready for analysis by capillary g.c.-m.s. using e.i. and c.i. (CH_4) modes. The essential oil of hops contains cyclic polysulphides (4) and the thiophen (5) derived from myrcene and sulphur.[246] On the other

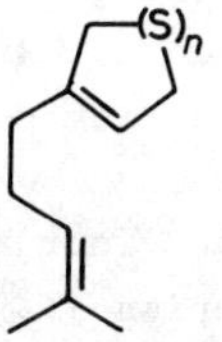

(4) n = 2, 3, or 4

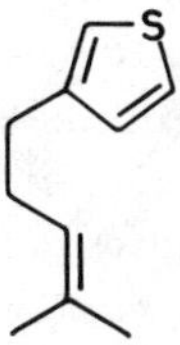

(5)

hand, it is possible to remove terpenes selectively by chromatography on a styrene resin without introducing artefacts from the resin itself. The resulting 'deterpenated' extracts can be examined by g.c.-m.s.[247]

Oxygen Heterocycles and Phenols.- The occurrence and chemistry of the natural coumarins have been described in a book.[248] Five new chromenes and 6-methoxy-2,2-dimethylchromene have been shown to be fungal metabolites.[249] Both the retention indices and mass spectra of the TMS ethers of seven naturally occurring dihydroxymonomethoxyisoflavones can be used for their differentiation.[250] Derivatives and degradation products of melanervin (6) were examined by g.c.-m.s. during its structural determination as the first known member of the triphenyl family of naturally occurring flavanoids.[251] Weak oestrogenic activity in the urine of several animals has been associated with the presence of 7-hydroxy-3-(4'-hydroxyphenyl)chroman (equol). This

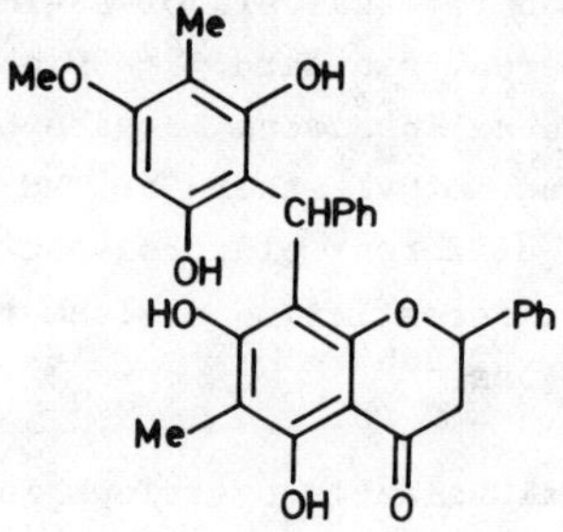

(6)

compound has now been found in human urine, excreted primarily as its monoglucuronide.[252]

Lignans are a new class of compounds containing the 2,3-dibenzylbutane skeleton. They are uncommon constituents of higher plants and have recently been identified for the first time in man and several animals.[253] The two principal animal lignans are *trans*-2,3-bis-(3-hydroxybenzyl)-γ-butyrolactone (enterolactone) and 2,3-bis-(3-hydroxybenzyl)butane-1,4-diol (enterodiol).[254] They may be biosynthesized by intestinal bacteria.[255] The major acidic conjugates of lignans in human urine are the monoglucuronides.[256] The use of TMS ethers and SE-30 WCOT columns is favoured for the g.c.-m.s. analysis of lignans.

Analytical studies on marihuana cannabinoids have been thoroughly reviewed[257,258] and are covered more fully in Chapter 8. Discussion here is correspondingly brief. Metastable-ion methods can be used to quantify Δ^1-tetrahydrocannabinol (Δ^1-THC) in plasma.[259] The technique is discussed in the penultimate section of this chapter. On a column of OV-17 the ubiquitous plasticizer di-iso-octyl phthalate coelutes with Δ^1-THC, so the liquid phase, OV-17, is best avoided at least when underivatized samples are analysed.[260] Neutral and acidic cannabinoids can be separated by chromatography on Amberlite XAD-2 to assist subsequent characterization by g.c.-m.s.[260] Liquid-solid chromatography has been used to separate cannabinoids from unwanted polar compounds in urine.[261] Some workers[262] have preferred pentafluoropropionate, rather than TMS, derivatives for g.c.-m.s. analysis of tetrahydrocannabinol in blood. The enzyme-multiplied immunoassay[263] and r.i.a. techniques[262,263] have been compared with g.c.-m.s. for assay of tetrahydrocannabinols in ng amounts. A detection limit of 400 pg has been claimed for analysis of Δ^9-THC using tetraphenylethylene as internal standard.[264] Finally in this category, the constituents of *Cannabis sativa* L. have been investigated and 61 cannabinoids, as methyl ethers or TMS derivatives, detected. The use of different electron-beam energies (5 - 21 eV) was reported to aid differentiation of isomeric compounds and recognition of homologues.[265]

Carbohydrates.- Several valuable overviews in this field have appeared, including those on analysis of glycoside conjugates[37,38] and structural elucidation of saccharide chains by g.c.-c.i.m.s.[266] and a range of analytical methods.[267,268] Methylation analysis of polysaccharides is now very well established;[269,270] only a limited number of references to its use is given here.

Glucosinolates have been converted into their per-TMS desulpho derivatives for analysis by g.c.-e.i.m.s. and g.c.-c.i.m.s.[271,272] For structure elucidation, the lattter was preferred.[271]

For g.c.-m.s. analysis of monosaccharides, the most widely used derivatives are alditol acetates and TMS ethers, but improved reaction conditions for formation of aldonitrile acetates should make these derivatives a more attractive alternative, particularly for structure determination.[273] Mixtures containing both aldoses and ketoses are best analysed following application of the reaction for preparing aldonitrile acetates because, under these conditions, ketoses afford the chemically distinct peracetylated oxime derivatives. Thus, differentiation of aldoses and ketoses is facilitated, their different derivatives exhibiting different g.c. and fragmentation behaviour.[274] Pertrifluoroacetylated alkoxime derivatives of aldoses have also been prepared.[182] Fused-silica capillary columns have been employed for g.c. separation of TMS alditols of monosaccharides,[275] and quantitative analysis of free carbohydrates in nematode tissue has made use of TMS ethers.[276] Chiral monosaccharides can be resolved gas chromatographically as described above.[137] Assay of hexitols in biological fluid can be brought about, even in the presence of large amounts of endogenous glucose, by peracetylation and selected-ion monitoring.[277] Monosaccharides liberated from glycoproteins have been analysed quantitatively at the picomolar level by g.c. and g.c.-m.s. following methanolysis and deamination by nitrous acid.[278]

Gas chromatography and mass spectrometry of amino sugars have been discussed in a review article.[279] *N*-Acetylneuraminic acid (sialic acid) in erythrocyte ghosts has been estimated by g.c.-c.i.m.s. using the TMS derivative of the methyl glycoside methyl ester.[280] The detection limit for selected-ion monitoring was under 400 pg. Acylation[281] and methylation[282] of partially methylated *N*-acetylneuraminic acid have also been investigated by g.c.-m.s. The use of sodium borodeuteride in preparing alditol acetates permits differentiation and simultaneous assay of *N*-acetylgalactosamine and *N*-acetylgalactosaminitol in the 40 - 1000 nmol range.[283] 9-*O*-Acetyl-*N*-acetylneuraminic acid has been identified in mouse erythrocytes,[284] and a new sialic acid, 4-*O*-acetyl-9-*O*-lactyl-*N*-acetylneuraminic acid, was found in horse submandibular gland.[285] The aminoglycosidic portion of the toxin wedeloside from yellow daisy has been shown to be compound (7), which is β-linked to a diterpene aglycone.[286]

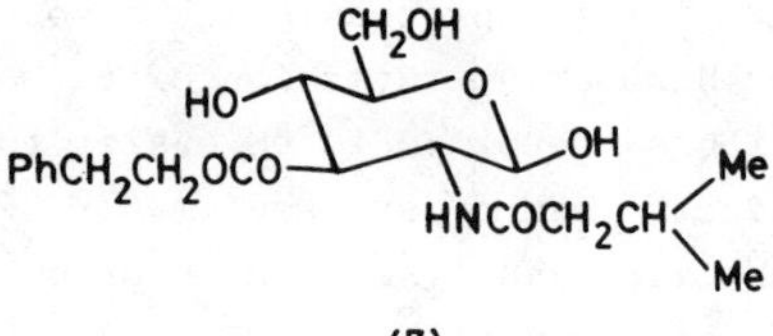

CH_2OH
HO
O
OH
$PhCH_2CH_2OCO$
$HNCOCH_2CH(Me)_2$

(7)

The number and position of sulphate and phosphate groups in carbohydrates have been determined by g.c.-m.s. methodology. Sulphate groups were substituted by formyl to 'label' sulphate sites[287] whilst phosphorylated carbohydrates were subjected to the following reaction sequence: methylation, dephosphorylation, and alkylation with trideuterioiodomethane or iodoethane[288] prior to g.c.-m.s. analysis. The latter method was applied to enterobacterial lipopolysaccharides.

Hydrolysis of the antigenic polysaccharide of *Eubacterium saburreum* strain V5 yields 3,6-dideoxy-3-(L-glyceroylamino)-D-glucose, as determined by g.c.-m.s. of TMS alditols.[289] By use of isobutane c.i., direct analysis of ^{13}C abundance in plant carbohydrates as TMS-oxime derivatives can be effected.[290]

Sequence analysis of polysaccharides requires chemical or enzymatic methods for specific cleavage of the chains to yield smaller units amenable to analysis. Hydrazinolysis-nitrous acid deamination causes selective cleavage of *N*-acetylglucosamine linkages and forms a useful reaction in sequence strategies for aminopolysaccharides.[291,292] The resulting oligosaccharides can be reduced, methylated, and analysed by g.c.-m.s.[292] With current procedures, the largest oligosaccharides amenable to g.c.-m.s. are 4 - 5 residues in length. Partial acetolysis of polysaccharides or glycoproteins affords mixtures of oligosaccharides which, after formation of methylated alditol derivatives, can be analysed by direct-probe mass spectrometry and/or g.c.-m.s.[293]

The positions of methyl and acetyl groups of partially methylated and acetylated methyl glycosides of galactose, mannose, glucose, and *N*-acetylglucosamine can be determined by g.c.-m.s. using WCOT columns of OV-101 or Carbowax 20M.[294] The method has been applied to acetylated methanolysis products of glycoprotein glycans and requires less than 0.5 mg of protein. Methylation analysis of polysaccharides does not normally differentiate between 4-linked aldopyranosyl and 5-linked aldofuranosyl residues, but a modified derivatization procedure makes possible the distinction.[295] For Hakomori methylation of carbohydrates, dimsyl potassium has been reported to be easier to prepare and afford fewer by-products than dimsyl sodium.[296] Further improvements to methylation analysis can be brought about, not surprisingly, by use of fused-silica WCOT columns for resolution of alditol acetates[297] or application of chemical ionization.[298] Methylation analysis has aided studies of the structures of polysaccharides in fungi[299,300] and *Pneumococcus* Type IX,[301] hog submaxillary-gland glycoproteins,[302] and plant mucilages.[303]

Structure elucidation of the acidic polysaccharide secreted by *Rhizobium meliloti* strain 1021 is an example[304] of an interesting approach to polysaccharide analysis using several forms of mass spectrometry, including g.c.-m.s. and l.c.-m.s. to examine complex mixtures of peralkylated oligosaccharide-alditols.[268] Intact, permethylated oligosaccharides that have been identified, at least partly, by g.c.-m.s. include a tetrasaccharide specific to blood group A[305] and two carbohydrates (Manβ1 → 4GlcNAc and Manβ1 → 4GlcNAcβ1 → 4GlcNAc) that accumulate in the kidney of goats with β-mannosidosis.[306]

Phospholipids.- The methaneboronate derivative was employed for g.c.-m.s. analysis of 1-*O*-octadec-*cis*-11-enyl glycerol from *Paramecium* phospholipids.[307] A novel sialoglycolipid from hepatopancreas of starfish[308] and two unusual gangliosides of eggs and embryos of the sea urchin[309] have been identified by a variety of degradative procedures followed by g.c.-m.s. The ganglioside NeuACα2 → 6-neolactotetraosylceramide found in human meconium may be the first foetal antigen of glycolipid type to be characterized.[310] Methylation analysis has been used to identify neutral glycosphingolipids in hairy cell leukaemia.[311] The composition of sphingomyelins in different lipoprotein classes has been examined.[312] The tBDMS ethers of ceramides were analysed by g.c.-m.s. using a short WCOT column of SP-2100. Both field-desorption mass spectrometry and g.c.-m.s. were used in the first report of chemical-structure determination of naturally occurring platelet-activating factor, an acetyl-alkylglyceryl phosphorylcholine.[313]

Pyrimidines, Purines, Nucleosides, and Nucleotides.- Exogenous caffeine often interferes with selected-ion monitoring assays of xanthine and hypoxanthine in biological fluids. Such interference can be avoided by use of butyl rather than methyl derivatives.[314] Extractive alkylation with pentafluorobenzyl bromide and selected-ion monitoring form an effective assay method for 6-mercaptopurine in plasma.[315] The bases uracil and dihydrouracil have been detected by g.c.-m.s. following hydrolysis of streptovirudin, a complex of related nucleoside antibiotics.[316]

The TMS derivatives of *trans*-zeatin and *trans*-ribosylzeatin, representative of a wide range of cytokinin bases and ribonucleosides occurring as natural products, have been examined by capillary g.c.-m.s.[317] The levels of *cis*- and *trans*-zeatin riboside and *cis*- and *trans*-methylthiozeatin riboside in roots and tops of tobacco plants have been determined.[318] Preparation of (^{15}N)-labelled cytokinins and their use as internal standards for quantification of cytokinins in *Vinca rosea* have been described.[319] A novel nucleoside,[320] containing ribose and a base with the same molecular weight as acetyladenine, has been found in RNase T_2.

Adenosine-3':5'-cyclic monophosphate (cAMP) can be determined at the pmol level by selected-ion monitoring of TMS derivatives, using synthetic (2,8-$^2H_2$,6-^{15}N)cAMP as internal standard. The g.c.-m.s. method has been compared to biological assays for cAMP[321] and applied to the analysis of cAMP in various plants including cultured tobacco.[322]

Nucleosides and nucleotides of 6-mercaptopurine extracted from blood have been separated on Amberlite XAD-4 resin prior to g.c.-m.s. analysis. Since quantification of nucleotides by g.c.-m.s. is difficult, they were dephosphorylated enzymatically and the resulting nucleosides assayed as permethylated derivatives using ammonia c.i. and selected-ion monitoring.[323]

Derivatization methods in this field are discussed above.[161,162]

Steroids and Triterpenes.- Reviews have been published on the topics of steroid hormones,[324-326] sterols and bile acids,[327] marine sterols,[328-330] and steroids in insects and algae.[331] G.c.-m.s. is now used routinely for analysis of steroids. The papers cited below are restricted to those contributing significant advances in the field. Applications of metastable-ion techniques are described in a later section.

Fractionation of steroids and steroid conjugates into discrete chemical groups has been accomplished with ion-exchange chromatography. The method complements g.c.-m.s. for determination of urinary-steroid profiles.[332] Computerized g.c.-m.s. has been applied to automatic qualitative and quantitative analysis of steroid metabolites in urine.[333] Both packed and capillary (fused-silica OV-101 WCOT) columns were used. Unconjugated and sulphated steroids in human faeces have been separated on Sephadex LH-20 and analysed by g.c.-m.s.[334]

The first firm mass-spectrometric evidence for free and esterified 26-hydroxycholesterol in blood serum of adult humans has been given.[335] The new plant-growth promoter brassinolide has been analysed as its bis-methaneboronate derivative.[336] Isopropyl and isopropenyl groups in the side chains of marine steroids are well known, but only recently have they been shown also to have a terrestrial source.[337] Nervisterol (22-dehydro-24-isopropenylcholesterol) was found in an orchidaceous plant.[337] Unusual steroid nuclei have also been observed: several 4α-methyl sterols with unsaturation at $\Delta^{8(14)}$ or Δ^{14} positions were identified in cultured marine dinoflagellates.[338] Another novel dinoflagellate sterol to be reported is 4α,23ξ,24ξ-trimethyl-cholestanol.[339] A new steroid of adrenal origin, 3α-hydroxy-

17-methyl-18-nor-5β-androst-13(17)-en-16-one, has been characterized in human urine.[340] Incubation of ($4-^{14}C$)pregnenolone and progesterone in confluent embryonic rat fibroblast cultures results in four novel metabolites, 3α,20α- and 3β,20α-dihydroxy-5α-pregnane, 4-pregnene-3α,20α-diol, and 5α-pregnane-3β,20α,21-triol.[341] The latest in a series of papers on minor sterols in marine invertebrates concerns structure elucidation of 26-methyl-strongylosterol and 28-methylxestosterol from a sponge, a study throwing light on the mechanisms involved in biomethylation of steroid side chains.[342]

Unconjugated oestrogens have been assayed by g.c.-m.s. in a variety of biological fluids using several different procedures. For example, oestrone, oestradiol-17β, and oestriol have been determined as trifluoroacetyl derivatives in blood serum of pregnant women. Selected-ion monitoring of molecular ions was compared to the r.i.a. method.[343] Other workers[344,345] have preferred the use of TMS derivatives and capillary columns for quantification of endogenous oestradiol-17β in rat uterus[344] and breast tumour tissue.[345] A g.c.-m.s. method has been proposed for assessing uptake, retention, and processing of oestrogens in target cell nuclei under *in vivo* conditions. Again, TMS ethers were employed. With selected-ion monitoring and using an SE-30 WCOT column, assays at the pg level were possible.[346] For many laboratories, g.c.-m.s. equipment is prohibitively expensive and complex. Cheaper alternatives for determination of oestrogens and their conjugates are r.i.a.[343,345] and g.c. with flame-ionization detection.[347]

Catecholoestrogens are known metabolites of oestrogens, but 4-hydroxyoestrogens have only recently been isolated and identified in human urine.[348,349] In the first report of 4-hydroxylation, 4-hydroxyoestrone was identified by g.c.-m.s. as its TMS ether.[348] A combination of h.p.l.c. and g.c.-m.s. has been used by other workers to confirm the presence of both 4-hydroxyoestrone and 4-hydroxyoestradiol in pregnancy urine.[349] The mechanism of loss of C-19 as formate during biosynthesis of oestrogens from androstenedione has been investigated by use of (^{18}O)-isotope labelling and g.c.-m.s.[350] As a representative example of analysis of ethynyloestrogens, ethynyloestradiol glucuronides were isolated by h.p.l.c., hydrolysed, separated from non-ethynyl steroids on an SP-Sephadex (Ag^+) column, and then assayed quantitatively by g.c.-m.s. using TMS derivatives.[351]

Quantification of dehydroepiandrosterone sulphate has been effected by selected-ion monitoring of silylated derivatives.[352-354] It has been found that g.c.-m.s. is superior to r.i.a. since the latter technique overestimates the amount of steroid, presumably due to cross-reaction of the antibodies with other substances present in the analytical sample.[352] G.c.-m.s. assay can be made highly specific for dehydroepiandrosterone sulphate by use of selected-ion monitoring at high resolution.[354] The steroid can be identified and quantified in saliva, suggesting a new, convenient, and non-intrusive sampling procedure for biological assays.[354] Simple methods for the synthesis of 20 different, highly enriched deuterium-labelled steroids[355] and of deuteriated testosterone[356] have been described. Differentiation between endogenous and exogenous testosterone in human plasma and urine after oral administration of isotopically labelled testosterone can be effected by selected-ion monitoring.[357]

A reference method for quantification of 17-hydroxyprogesterone and pregnenolone has been proposed.[358] Steroid biosynthesis in the newborn infant affords 5β-pregnane-3α,6α,20α-triol[359] and 15β-hydroxylated C_{21} steroids such as 5β-pregnane-3α,15β,17α-triol-20-one.[360] A new oxidative, microsomal conversion of Δ^5 steroids to 5α,6β-diols, *via* α- and β-epoxides, has been noted in bovine liver.[361]

Two isomeric forms of aldosterone have been observed by g.c.-m.s. and tentatively identified as the 11,18-hemiacetal and 11,18-hemiacetal 18,20-hemiketal,[362] but other workers have refuted these assignments.[363] They argue that, in dilute alkali, aldosterone rearranges to 11β,18:18,21-diepoxy-20,21-dihydroxy-4-pregnen-3-one (apoaldosterone).[363] Such conflicting views illustrate the fallibility of g.c.-m.s. for analysis of isomers which may, at least in principle, interconvert thermally in the injector or ion source, catalytically in the g.c. column or after ionization as mass-spectral rearrangement processes. Urinary acidic metabolites of cortisol have been identified as α- and β-cortolonic acids and α- and β-cortolic acids as TMS ether methyl esters. Resolution of the four metabolites was effected on a Carbowax 20M WCOT column.[364] Monohydroxylated derivatives of cortisol in human urine have been partially characterized by g.c.-m.s. as *O*-methoxime TMS derivatives.[365]

Cholesterol biosynthesis is an active area of research for which g.c.-m.s. is an excellent analytical tool. Demethylation mechanisms for cholesterol formation have been studied in rat,[366] crab,[367] and silkworm.[368] A useful investigation to detect, identify, and quantify cholesterol autoxidation products has been reported.[369] Towards development of a reference method for the clinically important determination of cholesterol in serum, two papers have reported improved precision in isotope dilution assays.[370,371] Metabolic hydroxylations of cholesterol have been extensively researched. Cholesterol 7α-hydroxylase activity can be measured by selected-ion monitoring.[372] Side-chain hydroxylation is a very important process because it leads to cleavage and formation of pregnenolone. Adrenocortex mitochondrial cytochrome P-450, which cleaves the side chain of cholesterol, has been shown to contain the intermediates 20- and 22-hydroxycholesterol and 20,22-dihydroxycholesterol, by g.c.-m.s. of TMS derivatives.[373] Twenty monohydroxylated cholesterol derivatives have been differentiated by h.p.l.c. and c.i. mass spectrometry.[374]

Analysis of the insect moulting hormones ecdysteroids by h.p.l.c. and g.c.-m.s. has been reviewed.[375] A combination of n.m.r. spectroscopy, fast-atom-bombardment mass spectrometry, and g.c.-m.s. has led to the first structural elucidation of ecdysteroid conjugates.[376] Ecdysone-22-phosphate and 2-deoxyecdysone-22-phosphate were found in eggs of the desert locust. Liberated phosphate was analysed by g.c.-m.s. as its TMS derivative.[376] In an independent study, eggs and embryos of the migratory locust were shown to contain 2-deoxyecdysone-22-adenosine-monophosphate and 3-epi-2-deoxyecdysone-3-phosphate.[377]

Application of g.c.-m.s. to the study of bile acids[378] has been prolific. Bile acids can be analysed as a variety of derivatives, including trifluoroacetyl hexafluoroisopropyl esters[379] and acetyl methyl esters.[380,381] The latter derivatives are best analysed by g.c.-c.i.m.s. with ammonia as reactant gas.[380,381] Several investigations of the rare, inherited lipid-storage disease *cerebrotendinous xanthomatosis* (CTX) have been reported.[382-384] The unusual composition of bile acids in CTX patients includes bile alcohol glucuronides[382] and may arise through defective liver mitochondrial hydroxylation of chenodeoxycholic acid precursors.[383] Profiles of serum bile acids could form useful criteria for diagnosis of CTX.[384] Bile acids in human liver have been assayed as dimethylethylsilyl ether ethyl esters.[385] The novel bile alcohol 27-nor-5β-cholestane-3α,7α,12α,24ξ,25ξ-pentol occurs in urine as its glucuronide,[386] and urinary bile acid glucuronides have been analysed as TMS ether methyl esters by g.c.-m.s.[387] Repetitive scanning during g.c.-m.s. provides a method for qualitative and quantitative determinations of faecal bile acids as their permethyl derivatives.[388] The effect of cholesterol feeding on faecal bile acid profiles of male and female rats has been investigated.[389]

Structural and stereochemical analyses of vitamin D and its analogues have been reviewed.[390-392] Major difficulties in g.c.-m.s. analyses in this field are lability of vitamin D and its metabolites and differentiation between hydroxylated vitamins D_2 and D_3. Several different derivatization procedures have overcome these problems.[392-395] Currently, the best method involves isomerization to isotachysterols followed by silylation,[394,395] methylation,[392] or heptafluorobutyrylation.[392] Quantitative analyses are available for vitamin D_3,[395] 25-hydroxyvitamins D_2 and D_3,[393,395,396] and 25,26-dihydroxyvitamins D_2 and D_3.[395] Distinction between the dihydroxyvitamins has been accomplished by g.c.-m.s. for the first time,[395] and Sep-Pak C_{18} cartridges have been used to isolate 25-hydroxyvitamin D_2 and D_3 prior to measurement by h.p.l.c. or g.c.-m.s.[396]

Screening for anabolic steroids can be effected by selected-ion monitoring during g.c.-m.s.[397] Fused-silica capillary columns have been applied to g.c.-m.s. determination of anabolic steroids in horse urine.[398] A splitless injection system was used at elevated temperatures to reduce analysis time, the solvent effect being maintained by employing high-boiling hydrocarbon solvents.[398]

Two pentacyclic triterpenes of the hopane series in a photosynthetic bacterium have been studied. 3β-Hydroxy-17-methylhopane and 29-hydroxy-3,17-dimethylhopane were identified by g.c.-m.s. of their acetates and n.m.r. spectroscopy.[399] Mixtures of triterpenes from sow thistle, containing α- and β-amyrin, lupeol, taraxasterol, and pseudo-taraxasterol, have been examined by acetylation then g.c.-m.s. on a fused-silica OV-1 capillary column.[400] Nineteen triterpene alcohols in various seed oils have been characterized.[401]

Amines.- A series of reviews concerning assays of catecholamines,[402a] biogenic amines,[402b] norepinephrine,[402c] and aromatic amines[402d] has been published (in Japanese). Mass-spectrometric characterization of naturally occurring pyrazines has also been reviewed.[403]

For nitrogen-containing aromatic compounds, h.p.l.c. has proven useful for initial fractionation[404] and capillary g.c.-c.i.m.s. for characterization.[405] When deuteriated ammonia is used as c.i. reactant gas, active hydrogen atoms of the sample exchange with the ND_3 deuterium atoms, and thus amino-substituted polycyclic hydrocarbons are differentiated from aza-arenes.[405] There have been few g.c.-m.s. studies of alkaloids in the review period. The determination of alkaloidal profiles in several New World *Lupinus* species is typical.[406] Nineteen quinolizidine alkaloids were identified that had not previously been observed in the genus, and the alkaloid distribution was found to be diagnostic of taxonomy.[406]

By far the largest field of study in this category is that of biogenic amines. Several publications have dealt with the natural occurrence of 1,2,3,4-tetrahydro-β-carbolines. The parent compound and its 1-methyl homologue have been determined quantitatively, after conversion to heptafluorobutyryl derivatives, in human plasma and platelets.[407,408] It is claimed that 1-methyl-1,2,3,4-tetrahydro-β-carboline (1-methyl-THCB) occurs only after ethanol consumption,[408] yet others[409] have reported significant urinary levels of endogenous 1-methyl-THBC. Moreover, the excretion rate did not increase after acute intake of ethanol.[409] It has been shown by g.c.-m.s. that 6-methoxy-THCB occurs in pineal gland of chickens and cocks[410] and that THCB, 2-methyl-THCB, 6-methoxy-THCB, and 6,7-dihydroxy-THCB are normal constituents of rat brain.[411,412]

Another active area is the analysis of cholinergic quaternary ammonium compounds by pyrolysis g.c.-m.s. A comparison of direct-inlet pyrolysis mass spectrometry and pyrolysis g.c.-m.s. has revealed that highest sensitivity for choline and choline esters is attained using the latter method.[413] The detection limits currently quoted for pyrolysis g.c.-m.s. techniques fall in the 10^{-13} - 10^{-12} mole range for choline and acetylcholine when using packed columns, chemical ionization, and selected-ion monitoring.[413-415] An improved pyrolyser has been introduced and applied to blood samples.[415] An alternative analytical method for choline and acetylcholine involves demethylation to the tertiary amines followed by g.c.-m.s.[416] Using this method, some evidence has been presented for increased metabolism of the neurotransmitter, acetylcholine, in the septum region of rabbit brain during methoxypyridoxine-induced convulsions.[416]

Several biogenic amines (λ-aminobutyric acid, tyramine, norepinephrine, epinephrine) and metabolites have been assayed as *N*-acetyl TMS derivatives on an SE-30 capillary column.[417] High sensitivity was achieved through selected-ion monitoring. A valuable report has appeared concerning practical aspects of analysing for traces of neurochemically important indoleamines and their metabolites.[418] Minimization of sample losses during derivatization and in the g.c.-m.s. interface and the limitations of TMS derivatives were discussed. A further note of discord is sounded over the occurrence (or not) of 5-methoxytryptamine in rats. Metabolic deacetylation of melatonin to 5-methoxytryptamine in rat liver (but not in brain) has been claimed following g.c.-m.s. of the pentafluoropropionyl (PFP) derivative.[419] Other workers,[420] also failing to detect 5-methoxytryptamine in rat brain, have noted that pentafluoropropionylation of melatonin is accompanied by some transacylation, inadvertently affording the PFP derivative of 5-methoxytryptamine. Reports of the natural occurrence of this amine and of the *in vivo* conversion of melatonin to 5-methoxytryptamine are open to question at least until a more reliable derivatization procedure is devised.

Several mass-spectrometric assays of serotonin (5-hydroxytryptamine) and melatonin (*N*-acetyl-5-methoxytryptamine) in biological fluids are available. As an example, a new and efficient extraction procedure has been coupled with g.c.-e.i.m.s. and g.c.-c.i.m.s. (positive- and negative-ion modes) for determination of serotonin in various biological fluids.[421] Selected-ion monitoring during g.c.-m.s. has been compared to fluorimetry[422] and h.p.l.c. with electrochemical detection[423] for assay of serotonin in biological tissues. Quantification of melatonin[22] and its metabolite, 6-hydroxymelatonin,[424] has been described. The latter compound was analysed in urine by g.c.-n.i.c.i.m.s. as its tBDMS/PFP derivative. An *in vivo* measurement of the tryptophan-5-hydroxylase system has been developed for study of phenylketonuria. It consists of dosing the subject with $(^2H_5)$tryptophan and subsequent assay of urinary metabolites, $(^2H_4)$serotonin and $(^2H_5)$tryptamine, as PFP derivatives by g.c.-m.s. and selected-ion monitoring.[425] Assays for dopamine,[423,426] octopamine and tyramine,[427] and norepinephrine[428] have been published. Boric acid gel chromatography provides a simple method for extraction of norepinephrine from plasma.[428] G.c.-m.s. has been employed to measure many catecholamine and related metabolites, as for example phenylethylene glycol metabolites of epinephrine,[429-431] 4-hydroxy-3-methoxyphenylacetic acid,[432] 4-hydroxy-3-methoxymandelic acid,[432] homovanillic acid,[431,433,434] and the monohydroxylated mandelic acids.[435-437]

Histamine and N^{τ}-methylhistamine have been determined simultaneously in human plasma and urine by g.c.-m.s.,[438] and other publications[439-441] have described the analysis of N^{τ}-methylhistamine as perfluoroacyl derivatives. Given the electronegative character of these derivatives, it is surprising that g.c.-n.i.c.i.m.s. has not been tried as a means of increasing sensitivity of the assay. Selected-ion monitoring has been applied to the measurement of pyrrolidine[442] and piperidine,[443] as dinitrophenyl derivatives, in the brain. The latter amine appears to have a role in regulating levels of consciousness.

An analytical procedure for separation and identification of urinary amine metabolites has been described.[444] During the study, *N*-3-hydroxypropyl-1,4-diaminobutane was observed for the first time in urine. Metabolism of spermidine and/or spermine to isoputreanine has been proposed on the grounds of g.c.-m.s. analysis.[445] A fused-silica capillary column inserted directly into the ion source of a mass spectrometer has been used successfully to analyse traces of underivatized and potentially carcinogenic aromatic diamines in urine.[446] With a transfer line of glass-lined tubing, tailing of the g.c. peaks was observed, and this problem was not resolved by standard deactivation techniques.

Amino-acids and Peptides.- An extensive, authoritative review of gas-chromatographic analysis of amino-acids has appeared,[447] and some aspects of g.c.-m.s. of amino-acids released by hydrolysis from marine peptides have been described in a review article.[30] General agreement has been reached on derivatization of amino-acids. *N*-Perfluoroacyl butyl esters, and *N*-trifluoroacetyl n-butyl esters in particular, should be considered optimal for gas chromatography and mass spectrometry in either the positive-ion or negative-ion mode.

Development of chiral stationary phases for capillary g.c. columns has aided study of the optically active amino-acids. For example, D-amino-acids can be used as internal standards for quantification by single-ion monitoring of the naturally occurring L-enantiomers. Such a procedure is cheaper, simpler, and more sensitive than when using isotopically labelled internal standards and multiple-ion monitoring. The technique has been used to estimate exposure of humans and rats to toxic methylating agents such as methyl methanesulphonate.[129,130] Some L-cysteine residues in haemoglobin suffer methylation on exposure, and the extent of this is measured by hydrolysis of the protein and analysis by g.c.-c.i.m.s. of liberated *S*-methyl-L-cysteine against *S*-methyl-D-cysteine as *N*-trifluoroacetyl n-butyl esters on a capillary column of Chirasil-Val.[130] The only disadvantage of the method is that, if some racemization occurred during work-up, a mathematical correction of analytical data would be required. It should be noted here that alkylation of proteins and subsequent g.c.-m.s. analysis of hydrolysates has been accomplished using conventional capillary columns[448] and that alkylated tyrosines and lysines have been examined as *N*-trifluoroacetyl n-butyl esters.[449] Chiral g.c. columns have also been exploited for determination of configuration. The optical purity of amino-acid residues in proteins can be assessed by hydrolysis in $^{2}H_2O/^{2}HCl$ so that any molecule that is inverted thereby becomes labelled with deuterium. Subsequent g.c.-m.s. with enantiomeric resolution of *N*-trifluoroacetyl amino-acid isopropyl esters affords a method for differentiating D-amino-acids initially present in the sample protein from those formed during hydrolysis.[450,451] The technique appears to be useful but the data should be interpreted with caution. The presumed mechanism of racemization in acid is *via* enolization, the direction of approach of the incoming $^{2}H^{+}$ determining whether or not inversion accompanies ^{2}H-labelling. Also, residues undergoing two inversions become labelled but regain their original configuration.

The plant hormone indole-3-acetic acid and its analogues can be detected, identified, and quantified by g.c.-m.s. methods. Indole-3-acetic acid has been found in Scots Pine[452] and in resinous plant material.[453] The latter publication[453] describes the preparation of the pentafluorobenzyl ester and its analysis by g.c.-n.i.c.i.m.s., achieving the impressive detection limit of 5 pg. Three different derivatives of indole-3-acetic acid have been tested for g.c.-m.s. analysis. The methyl ester and TMS derivatives were satisfactory, but anomalous results were obtained for the heptafluorobutyryl methyl ester.[454] Quantification of methyl 4-chloroindole-3-acetate in *Vicieae* species[455] and identification of *N*-(3-indoleacetyl)aspartic acid in Scots Pine[456] have been reported.

Assays for the amino-sulphonic acid taurine in plasma[457] and N^{τ}-methylhistidine in urine and plasma[458] have been developed. As an example of the many papers to describe quantification of γ-aminobutyric acid (GABA) by g.c.-m.s., both GABA and glutamic acid have been determined in Ringer's solution without desalination at the femtomolar level by the use of chemical ionization.[459] Endogenous glutamic acid and glycine have been shown to be released from pigeon optic tectum.[460]

Kinetic aspects of metabolism in the amino-acid field are readily studied by use of (^{15}N)-labelled precursors (tracers). Assimilation of (*amide*-^{15}N)glutamine in human plasma can be evaluated[461] following administration of $^{15}NH_4Cl$, and glycine turnover rates and pool sizes in neonates after dosage with (^{15}N)glycine.[462] Up to eleven heptafluorobutyryl amino-acid isobutyl esters can be measured for ^{15}N abundance in one g.c.-m.s. analysis.[463] The method has been applied to ammonia-ion assimilation in plant tissues. On the other hand, leucine flux in dogs can be assessed by administration of a deuteriated tracer, (6,6,6-$^{2}H_3$)leucine, with ($^{2}H_7$)leucine as an internal standard for g.c.-m.s. assay.[464] The stable ^{34}S is a safe alternative to the radioactive ^{35}S isotope for metabolic studies of cysteine and methionine when g.c.-m.s. is available for analysis.[465]

Two SCOT columns of SE-30 and OV-17 have been coupled together for g.c.-m.s. analysis of free amino-acids in environmental samples.[466] Detection limits were under 1 pg for most *N*-heptafluorobutyryl amino-acid isobutyl esters. Fourteen urinary amino-acids have been quantified, with their (^{13}C) analogues as internal standards, on 0.65% EGA by g.c.-c.i.m.s.[467] The method has a number of disadvantages (asparagine, glutamine, tryptophan, histidine, tyrosine, and arginine could not be determined). For routine amino-acid analysis, the well established and relatively cheap ion-exchange chromatographic method seems to be the most appropriate technique at present. However, g.c.-m.s. is invaluable for analysis of novel or unusual amino-acids. The urine of a baby who died with major physical malformations and failure of growth and development has been shown by g.c.-m.s. to contain *S*-(2-carboxypropyl)cysteine and *S*-(2-carboxypropyl)cysteamine.[468] The new amino-acid β-carboxyaspartic acid has been identified in the ribosomal proteins of *E. coli* as its *N*-acetyl trimethyl ester on an SE-52 WCOT column.[469]

Peptide and protein sequence analysis has been described in a recent book[470] and the role of g.c.-m.s. discussed.[471] Of recent reviews of mass-spectrometric amino-acid sequencing, that of Biemann[472] remains the most reasoned account.

Dipeptidyl aminopeptidases are well known for selective cleavage of polypeptides in sequence studies. The liberated dipeptides can be divided into basic, acidic, and neutral groups by h.p.l.c. and the latter two classes of dipeptides examined by g.c.-m.s. following derivatization.[473] Because of poor g.c. yields of basic dipeptides, these are characterized by other means.[473] Dipeptidyl aminopeptidase methods like this have been reviewed, along with the relatively new and complementary use of dipeptidyl carboxypeptidase for amino-acid sequencing from the C-terminus.[474] Reported dipeptidyl carboxypeptidases[474-476] cleave all peptide bonds except those involving proline and are applicable to the elucidation of primary structure of *N*-blocked peptides. A didactic discussion of techniques available for g.c.-m.s. study of dipeptides has been published,[6] and a method for quantification of methionine-enkephalin has been developed[477] using dipeptidylaminopeptidase I and g.c.-m.s. analysis of the liberated dipeptides.

The most successful g.c.-m.s. strategy for sequencing peptides involves partial hydrolysis, conversion of the resulting oligopeptides (di- to hexa-peptides) to *N*-trifluoroethyl *O*-TMS polyamino-alcohols, and g.c.-m.s. with computerized spectral interpretation. The most recent of a series of papers describing this methodology is concerned with automatic data processing.[478] The approach has been supplemented with data from DNA sequencing of the gene for the protein of interest.[479] This welcome synergic combination of two independent methods of amino-acid sequencing will lead to confident assignments of primary structure. It has been applied to alanine tRNA synthetase.[479]

Decarboxylation of γ-carboxyglutamic acid (Gla) residues in bovine prothrombin[480] and osteocalcin from chicken bone[481] using deuteriated reagent solutions results in formation of γ,γ-dideuterioglutamic acid residues. Subsequent proteolytic or acid digestion of the labelled protein, derivatization of the resulting oligopeptides, and g.c.-m.s. analysis afford a means of locating Gla residues, differentiating them from Glu residues, and estimating the ratio of Gla to Glu at specific sites in the protein.[480,481] Apart from Gla, another 'difficult' amino-acid residue is that of arginine. In peptides, the guanidino function has been reacted with cyclohexane-1,2-dione to form a 1,2-diol adduct and, after enzymic hydrolysis, peptides containing this grouping have been selectively absorbed on an affinity chromatography column of immobilized 3-aminobenzeneboronic acid. G.c.-m.s. analysis can be applied following elution from the column and further derivatization.[482] Peptide sequencing *via* reaction with pyridine-2-carboxaldehyde has been described.[483]

Cyclic[484] and acyclic[485] peptide antibiotics have been characterized, partly by g.c.-m.s. of amino-acids released by hydrolysis. The occurrence of unusual amino-acids in such structures justifies the use of g.c.-m.s. The peptide antibiotic trichotoxin A-40 from a fungus has been subject to partial hydrolysis and the resulting di- to tetra-peptides have been analysed by capillary g.c.-e.i.m.s. and g.c.-c.i.m.s. as *N*-trifluoroacetyl methyl esters.[486] Not surprisingly, the longer peptides were more satisfactorily identified by c.i. owing to their excessive fragmentation in the electron-impact mode. Cyclic peptides to be identified partly by g.c.-m.s. include islanditoxin from *Penicillium islandicum*[487] and dolestatin 3 from Indian Ocean sea hare.[488] The cell-growth inhibitor dolestatin 3 contained two novel thiazole amino-acids (8).

COOH N S RCH NH$_2$

(8) R = H or $H_2NCOCH_2CH_2$

Insect Pheromones and Other Secretions.- Reviews of insect sex pheromones[489] and plant secretions[490] may be noted. Research effort in this category has concentrated on elucidation of new pheromone structures. Investigations of insect juvenile hormones are covered in an earlier category (see above section on isoprenoid compounds).

Alkenyl acetates frequently occur in insect secretions. The double-bond position in these substances can be determined by the methoxymercuration-demercuration technique.[491,492] A field method for collecting sub-microgramme amounts of odour components released by insects has been described and applied to compounds of solitary bees.[493]

The sting of the honey bee contains an important new pheromone, (*Z*)-11-eicosen-1-ol.[494] The phenomenon of 'false queens' among worker honey bees has been investigated by comparing the mandibular-gland contents of workers and queens by g.c.-m.s.[495] Citronellyl citronellate and citronellyl geranate have been identified for the first time in an animal source, the van der Vecht's gland of *Vespa crabro* (European hornet).[496] The low-molecular-weight components of wasp venom have been examined by g.c.-e.i.m.s. and g.c.-c.i.m.s. with an SE-30 capillary column.[497]

Terpenoid substances in defence secretions of nasute termite soldiers have been analysed,[498,499] and the defence secretion of *Armitermes* termites has been shown to contain several macrocyclic latones.[500]

Whereas the trail pheromone of the red imported fire ant has been variously identified as (*Z*,*Z*)- and (*Z*,*E*)-3,4,7,11-tetramethyldodeca-1,3,6,10-tetraene[501] and (*Z*,*Z*,*Z*)-2,6,10-trimethyldodeca-2,4,6,10-tetraene,[502] *Myrmica* ants were said to secrete 7-ethyl-3,11-dimethyldodeca-1,3,6,10-tetraene and 7-ethyl-3,11-dimethyltrideca-1,3,6,10-tetraene on the basis of degradation studies and g.c.-m.s.[503] The venom of *Monomorium* ants has been shown to comprise 2,5-dialkylpyrrolidine alkaloids.[504]

Synthesis and g.c.-m.s. study has confirmed the occurrence of *erythro*-6-acetoxy-5-hexadecanolide as the major component of a mosquito attractant pheromone.[505] Newly discovered in the animal kingdom, (-)-β-fenchol has been shown to be synthesized by the male fruit fly and occurs in its sex pheromones.[506] The major pheromone of the olive fly is 1,7-dioxaspiro(5.5)undecane. Two novel and minor hydroxyspiroacetals, 3- and 4-hydroxy-1,7-dioxaspiro(5.5)-undecane, have also been characterized.[507] An unusual cyclobutanyl compound (9) has been identified as the sex attractant of the citrus mealybug.[508] The female sex pheromones of the avocado pest[509] and the wing-gland pheromone of the male African sugar-cane borer[510] have also been examined by g.c.-m.s.

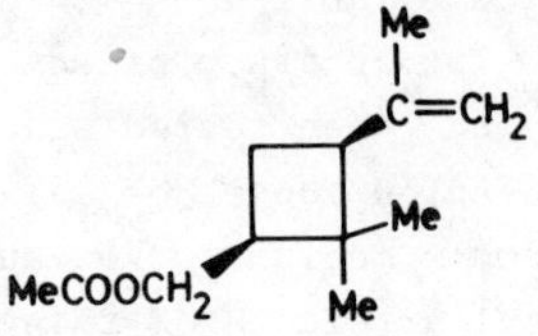

Me
C=CH$_2$
Me
MeCOOCH$_2$ Me

(9)

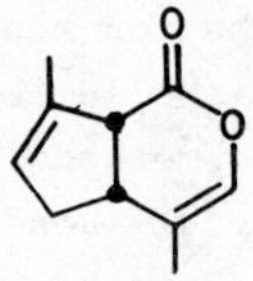

(10)

The first reported unsaturated sec-butyl ester as a lepidopteran sex pheromone is sec-butyl (*Z*)-7-tetradecenoate from the Western grapeleaf skeletonizer.[511] The double-bone position was established by preparation of the TMS derivative of the corresponding diol and g.c.-m.s. analysis. An acetylenic structure has been observed for the first time in the insect pheromone field. (*Z*)-13-Hexadecen-11-ynyl acetate was identified in the processionary moth.[512] G.c.-m.s. has revealed the novel cyclopentenoid monoterpene (10) in the laval secretion of *Gastrophysa cyanea*.[513]

The edible fungus truffle has been shown to contain 5α-androst-16-en-3α-ol.[514] This steroid happens to be a major component of boar pheromone, so its occurrence in the fungus may explain the boar's aptitude for locating truffles growing deep underground.

Clinical and Metabolic Studies.- The proliferation of g.c.-m.s. techniques in this field, noted in the last Report,[1c] has continued. Key papers only are discussed here. The study of drugs and their metabolites is covered in the following chapter. Readers' attention is drawn to several books[42,515-518] and reviews[19,21,46,519-528] that have recently appeared. The role of g.c.-m.s. in clinical diagnosis,[519-521] inherited acidemias,[522] pharmacology,[19,21,46,523-526] microbiology,[527] and barbiturate analysis[528] has been reviewed.

A discussion of the obstacles between a biological sample and a meaningful analytical result is recommended reading.[529] The important point is made that highly selective mass-spectrometric methods of detection and quantification can be useful for contaminated samples but their very selectivity can prevaricate. Impurities that are out of sight tend to be out of mind, despite the fact that the matrix can still markedly affect the analytical result (for instance, by suppression effects upon ionization or by detuning the ion source).[6,529] The accuracy of some routine methods used in clinical chemistry has been tested against g.c.-m.s. reference methods.[530] As indicated above,[354] saliva is becoming a useful, non-invasive sample medium for measurement of metabolites. Possible environmental artefacts of this sampling procecure have been studied.[531] Whilst analysis of drugs is described elsewhere, attention is drawn to the potential of negative-ion techniques in this field. Using electron capture, flurazepam can be quantified at the pg ml^{-1} level by g.c.-n.i.c.i.m.s. (detection limit, 12 pg ml^{-1}).[532] G.c.-m.s. has been applied to the analysis of pharmaceutical packaging materials.[533]

'Profiling' of urinary metabolites, usually organic acids, continues to be the chief application of g.c.-m.s. in this category because of its ability to diagnose and characterize metabolic diseases.[183,516,517,519-522,534-536] New polar and acidic metabolites, such as *N*-acetyl-2-amino-octanoic acid, in human urine have been identified by a combination of chromatographic methods including capillary g.c.-m.s.[537] Excretion of dicarboxylic acids by individuals with normal and increased fatty-acid oxidation has been compared.[538] Metabolic profiles are good indicators of the diabetic condition.[539,540] The absolute configuration of urinary 2-hydroxydicarboxylic acids can be assigned by capillary g.c. and g.c.-m.s.[139] New metabolites of isovaleric acidemia (an inborn error of leucine metabolism) include 4-hydroxyisovaleric acid,[541,542] mesaconic acid,[541] 3-hydroxyisoheptanoic acid,[542] and *N*-isovalerylglutamic acid.[543] Urinary acidic metabolites have been studied in patients with methylmalonic aciduria,[544-546] propionic acidemia,[545] and tyrosinemia.[547] Quantification of 2-ketoacids by g.c.-c.i.m.s can be effected by use of the TMS ethers of quinoxalinol derivatives.[548] As well as organic acids, steroids in urine can be used for clinical diagnosis, as for example in cases of hypertension[549] and 21-hydroxylase deficiency.[550] In virtually all of the referenced papers, high-resolution g.c. columns have been employed. With such a complex mixture as urine, this approach is advisable on the grounds of separation efficiency and sensitivity.[527,534,535] As a word of caution, it has been found that 3-methylglutaconic acid in the urine of patients with 3-hydroxy-3-methylglutaryl-CoA lyase deficiency decomposes during analysis (probably in the g.c. injector) to 3-methyl-3-butenoic acid and 3-methylcrotonic acid, adding artefact peaks to the metabolic profile.[551]

Whilst urine is the most widespread medium for metabolic profiling, the more complex biological fluid plasma has become amenable to metabolic profile studies with the development of a new sample-preparation method for g.c.-m.s.[120,121] Methyl esters of organic acids in uremic haemofiltrates,[552] phenols in uremic serum,[553] and organic acids in renal-tissue biopsy[554] have been analysed. Amniotic fluid is an important medium for analysis since it can lead to prenatal diagnosis of foetal diseases. Methylmalonic acidemia was detectable as early as 12 - 13 weeks of gestation by measuring levels of methylmalonic acid in amniotic fluid (or maternal urine).[546] Organic acids of amniotic fluid of different gestational ages have been analysed, as TMS derivatives, by capillary g.c.-m.s.[555]

Metabolism studies of urea have been reported.[556,557] Both publications describe very sensitive g.c.-m.s. analyses of (^{15}N)urea, capable of detecting about 0.1 atom % excess of (^{15}N)urea. Selected-ion monitoring assays of tuberculostearic ((*R*)-10-methyloctadecanoic) and C_{32} mycocerosic acids have been described.[558] Rapid diagnosis of pulmonary tuberculosis by analysis of sputum specimens appears to be feasible. Organic acids in the hearts of patients with idiopathic cardiomyopathy have been determined by g.c.-m.s.,[559] and abnormal metabolites, long-chain esters, have have been identified in nails of psoriatic patients.[560] Volatile substances in blood plasma and cerebrospinal fluids of humans have been quantitatively analysed. Levels of 3-methylbutanal were elevated in patients with chronic encephalopathics.[561] Selected-ion monitoring with a capillary g.c.-m.s. system has been used to measure endogenous, short-chain alcohols in serum and urine.[562] It was found that, for diabetics, concentrations of ethanol, propanol, and butanol were higher than for healthy individuals.

The biochemical composition of suction blister fluid is similar to that of serum, as determined by capillary g.c.-m.s. and two-dimensional electrophoresis.[563] In another multichromatographic analysis, l.c., t.l.c., and g.c.-m.s. were used to identify guanidino compounds in the urine of patients with hyperargininaemia.[564] The guanidino compounds were converted to dimethylpyrimidyl derivatives and silylated or acylated for g.c.-m.s. analysis. Guanidinoacetic acid, *N*-α-acetylarginine, argininic acid, γ-guanidinobutyric acid, arginine, and α-keto-δ-guanidinovaleric acid were excreted at high levels owing to the deficiency in arginase.[564] A method has been devised for evaluation of 3-hydroxy-3-methylglutaryl-CoA reductase activity based on the amount of mevalonic acid lactone formed in incubations of the reductase with microsomal proteins. The mevalonic acid lactone was measured by selected-ion monitoring of its TSM ether.[565] Finally, components of mouse-urine vapour have been examined by g.c.-m.s. in a study of the mechanisms of ageing. Eighty compounds were identified of which two, 3-hydroxy-3-methylbutene and *trans*-3-hepten-2-one, were strongly correlated with age of mice.[566]

Food and Agricultural Chemistry.- Two types of study are distinguishable in this area: the identification and origin of (*a*) endogenous flavour and/or odour components and (*b*) toxic compounds in food sources. Two books have been published concerning food-flavour analysis.[567,568] Reviews in category (*a*) concern food and agricultural applications of quantitative mass spectrometry[17] and general analysis of foodstuffs with considerable emphasis on mass-spectrometric methods.[569-571] Toxicological aspects have also been reviewed extensively.[569-574] The two areas are dealt with separately below.

Food Flavour and Odour. The explanation for the apparent low incidence of coronary diseases amongst arctic populations does not seem to involve dietary fat. The fat content of major arctic sources of meat was not lower than that of domestic animals used as standard meat sources in western countries.[176] Volatile compounds from boiled beef[575] and a synthetic meat-flavour mixture[576] have been examined by g.c. and g.c.-m.s.

Analysis of cigarette smoke continues to be an active area of research. Two 5-isopropyl-8,12-dimethyl-pentadeca-3,8,12,14-tetraen-2-one isomers[577] and ten nitrogen-containing compounds[578] have been newly reported. After derivatization with *o*-phenylenediamine, α-dicarbonyl compounds in cigarette smoke can be determined on a capillary column of Carbowax 20M.[579] There are substantial qualitative and quantitative differences between neutral constituents of tobacco and marihuana smoke, as shown by capillary g.c.-m.s.[580] The basic fraction of marihuana smoke has been characterized by a combination of h.p.l.c. and capillary g.c.-m.s., nearly 300 substances being tentatively identified.[404,581]

Again, there are considerable differences between the basic components of marihuana and tobacco smoke.[581]

The flavour chemistry of oat groats has been studied.[582,583] In rancid oat groats, 45 compounds were identified, many of which probably arise from autoxidation of lipids.[583] Thirty volatile compounds of baked 'Jewel' sweet potatoes have been characterized by g.c.-m.s. on an SF-96 capillary column.[584] The substances were trapped on porous polymer precolumns prior to analysis. Fractions of flavour components of Idaho Russet Burbank baked potatoes have been collected by preparative g.c., assessed for odour, and then identified by i.r. spectroscopy and g.c.-m.s. Pyrazines, thiazoles, and oxazoles were prevalent.[585] Both packed and SCOT columns have been applied to the analysis of odour components of wood-apple fruit and a canned substitute by g.c.-e.i.m.s. and g.c.-c.i.m.s.[586] The volatile oil of alfalfa leaves and stems includes possible insect attractants.[587] Examples of new and unusual food constituents are the occurrence of 2-hexyl-3-methylmaleic anhydride in raisins and almond hulls,[588] gingerdiones and hexahydrocurcumin analogies in ginger,[589] and 3,4-dimethyl-2,5-dioxo-2,5-dihydrothiophene in onion and leek.[590]

Traces of alkylpyridines in beer have been extracted by steam distillation and ion-exchange chromatography, and subsequently detected by g.c.-m.s.[591] The identification of 38 volatile compounds not previously reported in cider and their origin have been discussed.[592] Several novel constituents of an oxygen-rich fraction of Burgundy red wines[593] and volatile substances in eight varieties of wines[594] have been characterized. In more sobering studies, freshly roasted coffee was shown to contain many *N*-alkylpyrroles and *N*-furfurylpyrroles,[595] and the odour components of high-quality pouchong tea were examined by g.c.-m.s.[596]

Toxicological Aspects. In most studies of adulteration of food by pesticides, petroleum products, and so on, the mass spectrometer is used as a specific and now routine detector for gas chromatography. Much of the work is still carried out with packed columns. A few examples suffice here. Pentachloronitrobenzene and its metabolites in roots of peanut plants have been isolated, identified, and quantified. A pathway for formation of the observed metabolites was proposed.[597] In two studies that have employed capillary columns, residues of oestrogenic anabolic steroids were determined in meat at the parts per billion level[598] and 2,3,7,8-tetrachlorodibenzo-*p*-dioxin (2,3,7,8-TCDD) was observed in vegetables.[599] As part of a long series of papers describing the analysis of pesticide residues in food, g.c.-m.s. revealed the presence of endrin, dicofol, parathion, and ethoprop in some common crops.[600] Procedures for assay of polymethylbiphenyls (fuel contaminants)[601] and pirimiphos-methyl (an insecticide)[602] in various foods have been described. A case of scombroid poisoning was attributed to high levels of histamine in fresh tuna-fish tissues following g.c.-c.i.m.s. analysis of the TMS derivative on a short SP-2100 capillary column.[603] Food-packaging materials have been subject to g.c.-m.s. study.[604,605] Volatile organic residues inside sealed, laminated plastic bags were collected on a Tenax trap prior to g.c.-m.s. identification. In this way benzene was found in commercial potato-crisp bags.[604] Poly(vinyl chloride) resins used for food packaging contain vinyl chloride oligomers, phthalates, and alkanes, amongst other substances.[605]

The problem of nitrosamines in food has been amenable to g.c.-m.s. study for a long time, and analytical methodology, at least for volatile nitrosamines, is sufficiently advanced to be applicable at the 1 μg kg^{-1} level.[606] Analysis of involatile nitrosamines falls well short of this mark. An improved clean-up procedure for *N*-nitrosamines utilizing chromatography on celite and silica gel has been developed and a survey of many market food samples undertaken.[607] The occurrence of *N*-nitrosodimethylamine in beers has been confirmed.[608] A comparative study of chemical and electron-impact ionization for analysis of *N*-nitrosodimethylamine has come out in favour of the former.[609] Food contaminated with mould has been found to contain the novel *N*-nitroso-*N*(1'-methyl-2'-oxopropyl)-3-methylbutylamine.[610] The same group of Chinese workers has identified the organoiron nitroso complex (11) in pickled vegetables popular in an area of North China where oesophageal cancer is common. It is speculated that compound (11) reacts with secondary amines within the vegetable to form carcinogenic nitrosamines.[611]

Isolation of the mycotoxins patulin and penicillic acid has been improved and used for their determination as TMS derivatives in unroasted cocoa beans.[612] Broiling of meat and fish can produce potent mutagens in the food.[613] For instance, 2-amino-

NO NO
Fe
Me—S S—Me
Fe
NO NO

(11)

3-methylimidazo(4,5-*f*)quinoline is present in broiled sardines and beef. Its quantification has been described.[614] Both g.c.-m.s. and h.p.l.c. have been applied to the assay of styrene in wines.[615] As if the health and safety hazards of drinking alcohol and smoking tobacco were not enough, it now transpires that, when one does both concurrently, the mutagenic ethyl nitrite results.[616] This mutagen cannot be detected in breath when one takes one's pleasures singly.

Environmental Science and Toxicology.- In terms of numbers of publications, the interdisciplinary field that comes under the umbrella term of 'the environment' has become the most heavily subscribed category in this chapter. Newcomers to the subject can refer to a large number of recent books and reviews for background information (see Table). The newer techniques of capillary columns, negative-ion chemical ionization, atmospheric-pressure ionization, and metastable-ion techniques have all been applied advantageously for analysis of environmental samples. Discussion of metastable-ion methods is deferred till the penultimate section. Fused-silica capillary-column g.c.-m.s. is the preferred mass-spectral method for analysis of priority pollutants,[617,618] and it is complemented well by g.c.-FT i.r.[618]

Sample Preparation. Progress in collection and analysis of hazardous organic emissions has been summarized.[619] Pollutants in air were trapped on cartridges of polymeric beads prior to capillary g.c.-m.s. Such commercial solid sorbents should be washed well before use to prevent contamination of the air samples

Table *Books and reviews pertaining to environmental science and toxicology*

Topic	*References*
General environmental analysis	*a - e*
Extraction techniques	*f*
Contamination of environmental samples	*g*
Applications of negative-ion chemical ionization	*h*
Air pollution	*i - k*
Water pollution	*l - p*
Polycyclic aromatic hydrocarbons	*q, r*
Forensic science	*s - u*
Drug abuse	*v*
Pesticides	*w - y*
Chlorinated dioxins	*z - cc*

(a) 'Analytical Techniques in Environmental Chemistry', ed. J. Albaiges, Pergamon, Oxford, 1980. *(b)* 'Applications of Mass Spectrometry to Trace Analysis', ed. S. Facchetti, Elsevier, Amsterdam, 1982. *(c)* B.S. Middleditch, S.R. Missler, and H.B. Hines, 'Mass Spectrometry of Priority Pollutants', Plenum Press, New York, 1981. *(d)* W.C. Schnute and D.E. Smith, *Int. Lab.*, 1980, 75. *(e)* Y. Kato, T. Yamamoto, and K. Moriya, *Seikatsu Eisei*, 1981, 25, 135. *(f)* Ref. 105. *(g)* Ref. 49. *(h)* R.C. Dougherty, *Anal. Chem.*, 1981, 53, 625; *Biomed. Mass Spectrom.*, 1981, 8, 283.
(i) D.L. Fox and H.E. Jeffries, *Anal. Chem.*, 1981, 53, 1R.
(j) A.B. Belikov and Yu. S. Drugov, *Zh. Anal. Khim.*, 1981, 36, 1624.
(k) D. Schultzle and C.V. Hampton in 'Mass Spectrometry in the Environmental Sciences', ed. S. Safe, F. Karasek, and O. Hutzinger, Pergamon, Oxford, 1982, Ch. B2. *(l)* 'Hydrocarbons and Halogenated Hydrocarbons in the Aquatic Environment', ed. B.K. Afghan and D. MacKay, Plenum Press, New York, 1980. *(m)* Q.V. Thomas, J.R. Stork, and S.L. Lammert, *J. Chromatogr. Sci.*, 1980, 18, 583.
(n) J.M. McGuire and R.G. Webb in 'Water Quality Measurement', ed. H. Mark and J. Mattson, Marcel Dekker, New York, 1981, p. 1.
(o) M.J. Fishman, D.E. Erdmann, and T.R. Steinheimer, *Anal. Chem.*, 1981, 53, 182R. *(p)* D. Meek, *Water Res. Top.*, 1981, 1, 164.
(q) M.L. Lee, M. Novotony, and K.D. Bartle, 'Analytical Chemistry of Polycyclic Aromatic Compounds', Academic Press, New York, 1981.
(r) K.D. Bartle, M.L. Lee, and S.A. Wise, *Chem. Soc. Rev.*, 1981, 10, 113. *(s)* 'Forensic Toxicology, Proceedings of the European Meeting of the International Association of Forensic Toxicology', ed. J.S. Oliver, Croom Helm, London, 1980. *(t)* A. Maehly and L. Strömberg, 'Chemical Criminalistics', Springer, Berlin, 1981.
(u) B.S. Finkle, *Anal. Chem.*, 1982, 54, 433A. *(v)* Ref. 515.
(w) H.A. Moye, 'Analysis of Pesticide Residues', Wiley, Chichester, 1981. *(x)* S.H. Safe and R.K. Boyd, *Pestic. Anal.*, 1981, 329.
(y) H.-J. Stan, *Pestic. Anal.*, 1981, 369. *(z)* T. Cairns, L. Fishbein, and R.K. Mitchum, *Biomed. Mass Spectrom.*, 1980, 7, 484.
(aa) 'Chlorinated Dioxins and Related Compounds: Impact on the Environment', ed. O. Hutzinger, R.W. Frei, E. Merian, and F. Pocchiari, Pergamon, Oxford, 1982. *(bb)* F.W. Karasek and

F.I. Onuska, *Anal. Chem.*, 1982, 54, 309A. *(cc)* N.H. Mahle and L.A. Shadoff, *Biomed. Mass Spectrom.*, 1982, 9, 45.

with residual organic compounds.[620] The limitations and applications of a simple liquid-liquid extraction procedure for water pollutants[621] and a general approach to the fractionation of complex mixtures of chlorinated aromatic compounds[622] have been discussed. In an interlaboratory exercise, the efficiencies of six different clean-up procedures for 2,3,7,8-TCDD in fish were evaluated and ranked in order of preference.[623] Simple shaking or ultrasonic treatment of fly ash in solvent is reported to be unsuitable for extraction of adsorbed polychlorinated residues. Soxhlet extraction with toluene was shown to be much more efficient.[624]

A combination of silica gel and Sephadex LH-20 chromatography has been employed in the isolation of polycyclic aromatic hydrocarbons (PAHs) from marine sediments and biological samples.[625] Extraction of organic pollutants from drinking water by use of Amberlite XAD-2 resin requires considerable care if impurities from the resin are to be avoided.[108] Workers in the PAH field should beware of introducing PAHs into samples accidentally during flame sealing of glass ampules. Such adulteration arises through thermal reactions of hydrocarbon solvents.[626] Analysis of synthetic fuel samples is affected also by interactions between any trace-metal ions present and the organic compounds.[627] Adsorption of pollutants on glass can lead to sample losses during storage in glass containers. Additionally, phthalates and polychlorinated biphenyls (PCBs) adsorb reversibly so that contamination of the next solution to be placed in the container is probable.[628] Impurities in cyclohexane, dichloromethane, and methanol (commonly used for environmental samples) include traces of phthalate esters, hydrocarbons, and chlorinated compounds. Solvents distilled in glass are recommended to minimize contamination of samples.[629]

Air and Airborne Particulate Pollution. Hydrocarbon pollutants in the atmosphere have been shown to react with chlorine with or without exposure to u.v. radiation to yield organochlorine compounds.[630] A specific and sensitive method has been devised for assay of tetramethyl-lead and tetraethyl-lead in air.[631] Using a packed EGSS-X column, the detection limit for tetramethyl-lead was 20 pg m^{-3} by selected-ion monitoring. Other compounds determined in air include phthalate ester plasticizers,[632] monoterpenes,[633] and aldehydes.[181] Atmospheric chlorofluorocarbons over the Red Sea, Indian Ocean, and Italian locations were concentrated on Carbopack B prior to qualitative and quantitative analysis by g.c.-m.s. with medium mass-spectrometric resolving power.[634] Some advantages accrue from the use of h.p.l.c. with on-line scanning u.v. spectrometry for analysis of airborne PAHs.[635] The method is complementary to g.c.-m.s.

Some sound cautionary notes are to be found in a critical discussion of the interpretation of analytical data for atmospheric aerosols with respect to the health hazards of aliphatic and polycyclic aromatic hydrocarbons.[636] G.c.-m.s. with an SP-2100 capillary column has revealed the presence of mutagenic alkylated phenanthrenes, fluorenes, fluorenones, and other PAHs, aldehydes, quinones, and a nitroarene in diesel-engine soot.[637] Capillary g.c.-m.s. has also been utilized by others to examine organic adsorbates on diesel particulate matter.[638] Fused-silica capillary columns inserted right into the ion source and a modified split/splitless injector have enabled partial characterization of 190 PAHs in airborne particulate matter to be made. The lack of reference substances precludes complete structural analysis.[639] Diesel-exhaust particles have been shown to contain toxic nitroaromatic compounds by a variety of methods including g.c.-n.i.c.i.m.s.[640] and h.p.l.c., capillary g.c.-m.s., and metastable-ion techniques.[641]

Effluents and Water Pollution. The range of compounds that have been observed in polluted water is very wide indeed. It includes pesticides such as permethrin[74] and carbofuran,[642] organochlorine compounds,[643,644] odorous fungal metabolites,[645] nonylphenols,[646] amino-acid derivatives,[647] and faecal steroids.[648] Identification and quantification of traces of organic substances (alkylbenzenes, plasticizers, *etc.*) in tap water have been effected by computerized g.c.-m.s. and selected-ion monitoring.[649] The insecticide and nematicide carbofuran was quantified at levels of μg l^{-1} of water by selected-ion monitoring.[642] Several new halogenated substances were identified in extracts of superchlorinated septage, a study utilizing a computer program to search for specific Cl/Br isotope clusters in g.c.-m.s. data.[644]

Effluents from the leather industry entering the highly

polluted Hayashida River have been shown to contain large amounts of ethanediol monoalkyl ethers, fatty acids, and poly-(ethylene glycols) by g.c.-m.s. and field-desorption mass spectrometry.[650] Papermill effluents can have damaging effects on aquatic life.[651] The determination of resin and fatty acids at the 10 p.p.b. level in such effluents has been reported.[652] Organic compounds adsorbed on activated carbons used in water treatment can be desorbed thermally and analysed directly by g.c.-m.s., affording a method for monitoring water-treatment processes.[653]

Fuel Spills and Polycyclic Aromatic Hydrocarbons. Despite the finding of the Royal Commission on Environmental Pollution that "oil pollution entails no permanent threat to the environment", research in this area continues unabated. A considerable amount of oil remains in the sediments[654] of the Brittany coast following the wreck of the 'Amoco Cadiz', and long-term pollutants of alkylated dibenzothiophenes have been noted.[655] The chemical complexity of oil necessitates the use of capillary columns for its analysis. Information on molecular weight of hydrocarbon eluants is most easily obtained by use of chemical ionization. In analyses of oil spills and source oils, g.c.-c.i.m.s. with SCOT columns of SE-30 and OV-17 has been employed.[656] Selected-ion retrieval of $[M + H]^+$ ions facilitated location and identification of oil components such as naphthalene and its alkylated homologues. A possible case of a pollutant becoming a nutrient has been reported.[657] Discharge of elemental sulphur from the Buccaneer oil field and its occurrence in water effluent and sediment samples were investigated by g.c.-m.s. with $[^{34}S]$-enriched sulphur as internal standard. It seems likely that the discharged sulphur supports a population of bacteria which in turn supports fish and shrimp populations in the region of the production platform.[657]

The occurrence of PAHs in airborne particles is discussed above. A number of PAHs can be determined at p.p.b. levels in fractions from an activated-charcoal column used for separation of methylnaphthalenes and methylphenanthrenes in crude oil and marine organisms.[658] An inventory of carcinogenic PAHs and S- and O-containing polycyclic aromatic compounds in the environment continues to expand with profiles of PAHs observed in used engine oil, as determined by capillary g.c.-m.s.[659] Characterization of organic sulphur compounds and PAHs in shellfish[660] and PAHs in vertebrate fish tissue[661] has been accomplished by capillary g.c. techniques.

Pesticides and Halogenated Residues. Truly innovative work, from the mass spectrometrist's point of view, forms a low proportion of the total literature on this topic because of the abundance of reports of routine g.c.-m.s. methods.

A rapid procedure for screening of the insecticide mirex at the 500 fg level with selected-ion monitoring and an OV-101 glass capillary column[662] has been criticized by some workers and defended by its originators.[663] Organochlorine pesticides (including DDT) and PCB residues have been identified in fish from Austrian rivers by capillary g.c.-m.s.[664] Derivatization and g.c.-m.s. analysis of triorganotin hydroxide pesticide residues have been described.[665] Metabolic studies of the organophosphorus pesticide Diazinon in a human poisoning case[666] and of lindane in rats[667] are good examples of their genre. The first[666] used a wide range of g.c. and mass-spectrometric techniques and the second[667] was based on deuterium-isotope effects for elucidating mechanisms of formation of mercapturic acids from lindane.

Negative-ion mass spectrometry is a particularly useful technique for polyhalogenated compounds. Using an SE-52 capillary column and several n.i.c.i. reactant-gas systems, assays were developed for various polyhalogenated hydrocarbons. Oxygen-enhanced methane n.i.c.i. was the most sensitive method but suffered from undesirable cluster ions and shortened filament lifetime.[668] Determination of polybrominated biphenyls (PBBs) from Firemaster BP-6 in serum has been effected by selected-ion monitoring of $^{79}Br^-$ ions obtained by electron-capture n.i.c.i. mass spectrometry.[669] Detection limits were of the order of 10 pg cm^{-3}. Fused-silica capillary columns have been coupled to an atmospheric-pressure-ionization mass spectrometer.[670-672] When this equipment was used for assay of 2,3,7,8-TCDD in the negative-ion mode, chlorinated methoxybiphenyls, tetrachlorodibenzofurans (TCDFs), several chlorinated pesticides, and PCBs were shown not to interfere with the analysis at low p.p.t. levels.[670-672] Three TCDF isomers[671] and all 22 isomeric TCDDs[672] were studied by use of a 50 m SP-2100 column. Extremely toxic polychlorinated dibenzofuran isomers in turtle and seal fats have been identified at the pg g^{-1} level with g.c.-n.i.c.i.m.s. and capillary columns of OV-17 and

Silar 10C.[673]

Contamination of the food chain in Michigan with PBBs was attributed to Firemaster FF-1. G.c.-m.s. analysis revealed the occurrence not only of toxic PBBs but also of the minor contaminants, mixed polybromochlorobiphenyls.[674] A procedure for identifying and quantifying any of the 209 possible PCBs in 10 - 200 ng cm^{-3} amounts involves two separate analyses with a packed-column g.c.-m.s. system.[675] Chemical ionization with methane reactant gas has also been used for detection of PCBs.[676]

Polychlorinated dibenzo-*p*-dioxins (PCDDs) and dibenzofurans (PCDFs) in solid environmental samples and water condensate from urban incinerators have been determined simultaneously at the p.p.t. level. Both fused-silica OV-1 capillary and packed SP-2250 columns were applied, with mass-spectrometric resolving powers of 3000 - 5000.[677] Reverberations of the Seveso incident, which resulted in accidental release to the environment of 2,3,7,8-TCDD, continue with the result that 2,3,7,8-TCDD is probably the compound most widely studied by mass spectrometry. Despite this, analysis of 2,3,7,8-TCDD still gives rise to controversy.[1c] Its detection and quantification at p.p.t. levels in vegetation[678,679] and soil around Seveso[678] have been reported, using low-resolution mass spectrometry with low-[678] or high-resolution[679] gas chromatography. Generally, low-resolution techniques are applicable only when potentially interfering compounds such as chlorinated methoxybiphenyls[680] are known to be absent. This necessitates highly specific clean-up procedures for 2,3,7,8-TCDD. Even so, it is prudent to confirm positive results using g.c.-high-resolution m.s. (g.c.-h.r.m.s.), especially at levels under about 20 p.p.t. A statistical study of precision and accuracy of 2,3,7,8-TCDD quantification in reference solutions and beef adipose tissues using g.c.-h.r.m.s. has been published.[681] Quantification of 2,3,7,8-TCDD at the 2 - 10 pg level with a precision of 20% is possible with g.c.-h.r.m.s., utilizing a 30 m SE-30 capillary column.[682] Methods suitable for preparing tissue, milk, water, soil, and sediment samples and the possibility of interference from other TCDD isomers were also discussed. Capillary g.c.-m.s. at medium mass-spectrometric resolution has been reported to be applicable to goats' milk and tissue samples for determination of 2,3,7,8-TCDD.[683] The selected-ion monitoring technique used to determine variations in concentrations of PCDDs in fly ash from a municipal incinerator[684] suffers from a systematic error, producing unreliable quantitative data for PCDDs other than TCDDs.[685] Traces of PCDDs in a variety of particulate samples[686,687] and in 2,4-D products[688] have been determined by g.c.-m.s. Complete resolution of the 22 TCDD isomers by capillary g.c. alone has not yet been achieved. However, a combination of h.p.l.c. and capillary g.c.-h.r.m.s. has led to separation of all isomers,[689] a narrow-bore Silar 10C capillary column has been reported to resolve 2,3,7,8-TCDD from the other 21 isomers,[690] and a 60 m OV-101 WCOT column has enabled identification of 12 of the possible 22 isomers.[691] The ten isomers of hexachlorodibenzo-*p*-dioxin have been synthesized, separated by multiple h.p.l.c. fractionations, and characterized by g.c.-m.s.[692]

Forensic Science and Toxicology. A survey of some 29 compounds identified by g.c.-m.s. in a veterinary toxicology laboratory has appeared.[693] A typical application of g.c.-m.s. in this area has involved the determination of cantharidin (12) in urine and stomachs of cows due to ingestion of blister beetles.[694] Analysis of T-2 toxin (13) in maize has been accomplished by selected-ion

(12)

(13)

monitoring of the TMS ether.[695] The effect of oil pollution on clams and eels has been simulated *in vitro* by controlled dosage of aquarium water with crude oil.[696] The flesh of the fish and shell-fish was subsequently examined by capillary g.c.-m.s. for organosulphur compounds as possible indicators of oil pollution. *In vivo* epoxidation of styrene affords styrene-7,8-epoxide, considered to be the metabolite responsible for styrene toxicity. The epoxide has been determined in biological specimens at the ng level by selected-ion monitoring.[697] Another mutagenic monomer widely used in the polymer industry, epichlorohydrin, can also be assayed in blood by selected-ion monitoring.[698] Hydrocarbons, and octachlorostyrene and related compounds, have been quantified in fish from the Great Lakes.[699] *N*-Nitrosamines in environmental samples at the ng level have been analysed by g.c.-m.s. following denitrosation.[700,701] Detection of volatile *N*-nitrosamines in waste water and effluents from amine-production plants has been effected with g.c.-m.s. utilizing a glass WG11 capillary column.[702]

Applications of g.c.-m.s. in the field of forensic science have been concerned mostly with proving manufacture and/or use of illicit drugs as, for example, with amphetamines,[131,703] cocaine,[704,705] and mescaline and dimethyltryptamine.[706] A fatal exposure to tetrachloroethylene at a dry cleaners has been investigated by g.c.-m.s.[707] Headspace vapour from a portion of lung was found to contain tetrachloroethylene. A series of explosives has been identified by a variety of mass-spectrometric methods, most notably by pulsed positive-ion/negative-ion chemical ionization with ammonia as reactant gas.[708]

Organic Geochemistry and Fuel.- Porphyrins are an important class of compounds to the geochemist. For example, petroporphyrins, which could be described as 'fossilized' chlorophyll originating from photosynthetic micro-organisms, are present in crude oils and their source rocks as red pigments (usually V or Ni complexes) and can give an indication of the amount of organic degradation that has occurred in rocks. Hence, their analysis aids oil exploration. Petroporphyrins with extended alkyl substituents have been characterized by g.c.-m.s. following appropriate derivatization.[709,710] Formation of vanadyl petroporphyrins in Boscan crude oil was discussed.[709] Intact metallated porphyrin complexes can now be determined by g.c.-m.s.[143]

New methods for isolation and separation[711] and identification[72,711] of sulphur heterocycles in coal liquids and oils have been devised. Fractionation and characterization of nitrogen-containing compounds in fuels have been the subject of a number of publications.[71,404,712-716] The use of h.p.l.c. to locate the bases is particularly promising,[714] and alkyl anilines are readily distinguished from alkyl pyridines by acetylation followed by capillary g.c.-m.s. since only the former compounds can afford acetyl derivatives.[716]

Capillary g.c.-m.s. should be regarded as an essential tool for determining oil-oil, oil-source rock, and oil-condensate correlations. Chemical 'profiles' or 'fingerprints' have been reported of Alaskan crude oils,[717] aromatic hydrocarbons by low-voltage electron-impact ionization,[718] C_{15} - C_{40} isoalkanes,[719] crude oils and source rocks in the German Molasse Basin,[720] and branched and cyclic alkanes.[721] A comparison of g.c.-m.s., n.m.r. spectroscopy, and fluorescent indicator analysis for determining the hydrocarbon content of petroleum fractions has concluded that light fractions free of alkenes and heteroatomic compounds are best characterized by g.c.-m.s.[722] Individual organic compounds in shale oil can be analysed by g.c., g.c.-m.s., and/or h.p.l.c.,[723] and carboxylic acids in oil samples have been examined as methyl esters by g.c.-m.s. on an SP-1000 WCOT column.[724] Degradation of pentacyclic triterpenoid acids in Nigerian crude oil involves ring-A cleavage and affords tetracyclic carboxylic acids.[725]

New compounds recently characterized include: dicyclic and tricyclic hydrocarbons in a tertiary crude oil[726] and botryococcane in petroleum;[727] triaromatic steroids,[728] tetracyclic terpene hydrocarbons,[729] and tricyclic terpanes[730] in petroleum and sediments; ring-A monoaromatic steroids[731] and 25,28,30-trisnormoretane (a C_{27} triterpane)[732] in shales; and 27-nor-24-methyl-5α-cholestan-3β-ol,[733] long-chain alkane-1,15-diols and alkan-15-one-1-ols,[734] and 2,6,10-trimethyl-7-(3-methylbutyl)-dodecane[735] in sediments. The final compound of this list accounts for several previous observations of an unknown branched alkane in surface sediments.[735] On standard capillary columns, it co-elutes with the more widely occurring pristane, so care should be taken during the common practice of measuring pristane/phytane ratios for verifying oil origin.

From soil and sediment samples suspended in water, volatile compounds can be extracted and analysed by a conventional purging device interfaced to a computerized g.c.-m.s. instrument.[736] The polyunsaturated fatty acids found in sediment sections of the upper 20 m of a lacustrine core appear to be correlated with paleoclimate (as recognized by pollen analysis).[737] More than 50 compounds have been identified by capillary g.c.-m.s. in a river-bed sediment. The role of the sediment as a 'sink' for organic pollutants was also discussed.[738] The lipid content of sediments trapped at different depths in equatorial Atlantic Ocean, as determined by capillary g.c.-e.i.m.s. and g.c.-c.i.m.s., reflects the changing particulate composition during transport through the water column.[739] Sediments from the deep-sea drilling project[740] and carboxylic acids in oil shale retort and process waters[741] have also been examined.

Characterization and mechanisms of formation of C_{36} - C_{40} pentacyclic triterpanes of the hopane family in bitumen have been discussed.[742] Microbial and geothermal degradation of carotenes has been proposed to account for the occurrence of seven alkylbenzene homologues in coals.[743] Crude coal tar of different geographical origins[744] and from the gasification of low-ranked coals[745] has been studied by g.c.-m.s. Sulphur heterocycles in coal tar and pitch and nitrogen-sulphur heterocycles in anthracene oil have been analysed with an SP-2250 SCOT column and at medium mass-spectrometric resolving power.[746]

Pyrolysis g.c.-m.s. is exploited to advantage in geochemistry, particularly to probe the precise molecular structure of kerogen.[747-749] Other geochemical applications include studies of samples of the *Tasmanites* species,[750-752] vitrinite and Appalachian Devonian black shales,[750] and tricyclic hydroaromatic compounds from coal hydrogenates.[753] The ratio of aromatic compounds to n-alkanes in the pyrolysis products of Appalachian shale was found to correlate with its uranium content.[750]

Pyrolysis G.C.-M.S.- Geochemical applications of pyrolysis g.c.-m.s. are discussed in the previous category. The other two principal uses for this technique are in the polymer industry and microbiology.

The efficiency of vulcanization of polyisoprene (the crosslink density)[754] and the diol sequence distribution in vinyl copolymers[755] have been determined by pyrolysis g.c.-m.s. Traces of polydimethylsiloxanes can be analysed quantitatively by selected-ion monitoring of the major pyrolysis product, hexamethylsiloxane.[756] The use of capillary columns in pyrolysis g.c.-m.s. experiments has enabled characterization of 82 volatile products of cellulose pyrolysis.[757]

The origins of acetamide and propionamine in pyrolysates of bacterial cell walls have been elucidated. Acetamide is a pyrolysis product of *N*-acetyl groups on the glycan backbone in cell-wall peptidoglycan, and propionamide is formed from the lactyl-peptide bridge portion of the peptidoglycan.[758] Different *Klebsiella pneumoniae* strains can be distinguished clearly by pyrolysis g.c.-m.s. The potential application of the method to diagnoses of epidemic diseases was discussed.[759]

Finally, vitamins B_5, B_6, and C, separated by t.l.c., have been characterized by pyrolysis g.c.-m.s.[760]

Miscellaneous Applications.- Some archaeological artefacts to be dated by ^{14}C-counting have been shown by g.c.-m.s. to be contaminated with fossil fuels. These impurities dilute the ^{14}C content and thus account for the falsely high ages recorded. Molecular analysis should accompany ^{14}C-dating for verification of the source of carbon.[761] Neutral products of gaseous ion/molecule reactions may be trapped on polymer-coated glass wool and subsequently flushed into a g.c.-m.s. instrument for analysis.[762]

Organometallic species have been subjected to g.c.-m.s. Organomercury compounds disproportionated during chromatography,[763] and isotopes of Ca, Cr, Cu, Fe, Mg, Ni, and Zn were determined by g.c.-m.s. of their volatile chelates.[764]

Several studies have been concerned with silicon compounds. The TMS derivatives of several silicate anions have been identified by g.c.-m.s.[765,766] In one report,[765] the silicate derivatives were separated first by gel-permeation chromatography. Methylethyloctasilsesquioxanes have also been characterized by g.c.-m.s.[767]

Inorganic phosphate can be quantified, against an internal standard of $(^{18}O_4)$phosphoric acid, as its TMS derivative by the use of g.c.-m.s.[768] The method was utilized by other workers to confirm the occurrence of phosphate in steroid conjugates.[376] An internal standard of ^{82}Se has been employed for assay of selenium in biological samples.[769] Thiosulphonates with additional functional groups, separated from the methane thiosulphonate moiety by alkyl chains, are reported to be amenable to g.c.-m.s. analysis.[770]

A mixture of metallic oxides and superoxides, Hopcalite, is a catalyst for oxidation of carbon monoxide to carbon dioxide. This process has been utilized in a g.c.-m.s. method for quantification of (^{13}C)carbon monoxide in air. After oxidation, the carbon dioxide formed can be measured by selected-ion monitoring.[771] Assay of carbon monoxide at the 10 p.p.m. level can be effected, but the technique is not suitable for analyses of (^{18}O)carbon monoxide because the oxygen atom exchanges with the oxygen pool in the catalyst.[772]

Product and quality control of synthetic-ester lubricating oils has been based on capillary g.c.-c.i.m.s. methods with ammonia as reactant gas,[773] and analysis of additives and process residues in the plastics industry relies on several instrumental methods including g.c.-m.s. and pyrolysis g.c.-m.s.[774] Plastic medical devices contain abundant phthalate plasticizers. G.c.-m.s. has revealed that such plasticizers are to be found in critically ill surgical patients.[775] Metabolites of di-(2-ethylhexyl)phthalate have been identified in urine of the African green monkey.[776] The antioxidant 3-t-butyl-4-hydroxyanisole has been observed in rat plasma[777] and the antioxidant and heat stabilizer octadecyl 3-(3,5-di-t-butyl-4-hydroxyphenyl)propionate in calf thymus.[778]

4 High-performance Liquid Chromatography-Mass Spectrometry

A great many applications, particularly in the life sciences, await the coming-of-age of the h.p.l.c.-m.s. technique. Yet it is with some optimism that it is regularly stated that 'in a few years' technical improvements in l.c.-m.s. methodology will precipitate its widespread application. Several factors account for its reluctant integration into the arsenal of analytical techniques. (*a*) Any substance amenable to gas chromatography can be analysed by g.c.-m.s., but the equivalent statement does not apply to l.c. and l.c.-m.s. No one h.p.l.c.-m.s. interface is sufficiently advanced to cope with every l.c. problem that a single user may encounter. (*b*) No one interface is obviously superior, making choice of purchase difficult. Potential users of a rapidly developing technique tend to wait for a definitive advance before buying. (*c*) It is easy (relative to g.c.) to collect l.c. fractions for subsequent mass-spectrometry analysis, as long as the chromatogram is not too complex and peaks are well resolved. The off-line combination does not compromise either analytical technique, *e.g.* any method of ionization can be applied to eluants, and the transfer efficiency in the off-line mode (50 - 100% in the pg - μg range)[779] is better than that quoted for many current on-line systems. (*d*) To date, much l.c.-m.s. literature has been devoted to test mixtures and contrived investigations. To potential users, applications to 'real' problems, *not readily solved by g.c.-m.s.*, are much more persuasive. Fortunately, the literary tide seems to be turning in favour of genuine applications. (*e*) Most l.c.-m.s. interfaces are complex and expensive.

Progress towards a routine and universal l.c.-m.s. instrument and current applications are discussed in this section. There is a plethora of relevant reviews: general l.c.-m.s.,[780-783] mechanical transport interfaces,[784] microcolumns in l.c.[785-787] and l.c.-m.s.,[788] detectors for liquid chromatography,[789] ionization techniques used in l.c.-m.s.,[28,29] off-line combination of l.c. and field-desorption mass spectrometry,[790] and lipid analysis by h.p.l.c.-m.s.[791,792] Off-line and on-line l.c.-m.s. methods had been compared for polar lipids.[792]

Regrettably, fifteen references were omitted inadvertently from the previous report.[1c] With the agreement of the previous Reporter and the Senior Reporter, these publications are cited here as if they fell within the current review period.

Chromatographic Aspects.- The value of microcolumns for l.c.-m.s. is now generally appreciated. Compatibility with mass spectrometry is increased owing to the low flow rates employed and, whilst the resolving powers obtained at first were disappointing (*ca.* 1000 theoretical plates[784a]), resolution is now improving with 6000 - 13500 theoretical plates having been achieved with a moving-belt interface utilizing a procedure akin to 'heart-cutting' in g.c. work.[788a] Almost all reports of capillary h.p.l.c.-m.s. have concerned the commercial JASCO system and, unfortunately, few data on column efficiency have been published.

Highly efficient packed microbore columns have been described[793] and their use for l.c.-m.s. is recommended. Moderate yet encouraging resolving powers have been attained with similar columns incorporated into a direct inlet probe of a c.i. mass spectrometer.[794] The system is cheap and simple, but problems with involatile compounds and long elution times remain to be solved. Reasonable efficiency and sensitivity were also obtained with a micro l.c. column attached to a modified direct-liquid-introduction (DLI) interface.[795] Capillary l.c. columns have been utilized in conjunction with a vacuum nebulizing interface,[796] an unmodified g.c.-m.s. jet separator,[797] and various DLI systems.[798-801] The use of a splitter with microbore columns is not normally necessary, but it can lead to versatility: the operator can opt for high sensitivity (low flow rate, low split ratio) or for rapid analysis (high flow rate, high split ratio, lower sensitivity).[800] Very promising results have been obtained with fused-silica open tubular columns with diameters of 10 - 50 μm.[801] Long, narrow l.c. columns displayed theoretical plate numbers up to 5 x 10^6. Fundamental aspects of column efficiency and direct coupling to a c.i. mass spectrometer were considered.[801] The development of chiral l.c. phases for separation of enantiomers should be noted by l.c.-m.s. users.[802] Advantages of chiral columns for l.c. mirror those already described for g.c. and g.c.-m.s. (see above).

Thermospray Interface.- Of the new l.c.-m.s. interfaces, the thermospray technique[803,804] shows the most promise. The l.c. eluent is heated extremely rapidly to vaporization as it emerges from a nozzle, and the majority of the solvent is removed by molecular diffusion through a jet separator. It had been noted in an earlier interface that pyrolysis of the sample was avoided when the nozzle of the separator was heated to incandescence by a focussed laser beam, so the laser was replaced by four oxy-hydrogen burners. It was found that the droplets passing through the modified molecular separator had been charged by thermal and frictional ionization *in the interface*. These droplets carry eluents into the mass spectrometer and impinge on a heated metallic probe whereupon they vaporize and some of the resulting molecules appear as ionized species. Ion/molecule reactions occur to produce $(M + H)^+$ and $(M - H)^-$ ions. Before it was realized that a new method of ionization, thermospray ionization, was operating, the particles hitting the probe were bombarded with electrons from a heated filament and c.i. mass spectra resulted with the vaporized solvent acting as reactant gas.[805] But best results were obtained with the filament off. Thermospray ionization has substantially extended the range of samples to which l.c.-m.s. is applicable. For instance, dinucleotides afford large $(M + H)^+$ peaks, and the base peak for the pentapeptide leucine-enkephalin is due to quasi-molecular ions.[803] Transfer efficiencies and detection limits are not far short of those specified for g.c.-m.s., and the system tolerates ammonium formate and alkali salts (but not phosphate buffers) in the l.c. effluent.[804] Gradient elution poses no special problems. Commercial exploitation of the thermospray interface/ionization method is strongly encouraged. Useful advances would include replacement of the rather cumbersome torches by some form of electrical-resistance heating of the nozzle and a greater understanding of the ionization process(es) involved. With development, this inexpensive and simple interface may provide a routine l.c.-m.s. tool for study of involatile and thermally labile substances.

Direct Liquid Introduction.- Usually referred to as DLI, this interface is commercially available and being applied in several laboratories. Direct introduction of h.p.l.c. effluents into the ion source of a mass spectrometer requires some form of flow restrictor to prevent ion-source pressure rising to unacceptable levels (unless capillary l.c. columns are utilized - see above). Viscous restrictions are not recommended.[806] A pinhole of diameter 1 - 3 μm is a suitable non-viscous device producing an axial jet of small droplets. Independent addition of hot gas to the ion source aids droplet vaporization and provides a reactant gas for chemical ionization.[806] The main drawback of the system is that partial blockage of the pinhole causes erratic deviations in the direction of the jet. These can result in collisions of the sample molecules with the hot walls of the ion source, thereby transferring vibrational energy and resulting in decomposition of thermally labile substances.[806,807] Minimization of blockage difficulties has been discussed and the use of a 0.5 μm porosity filter proposed.[807] Alternatively, larger diaphragms can be used that produce very stable jet streams due to reduced plugging.[808] The jet comprises droplets of relatively large diameter and the gaseous mass obtained is used to achieve nebulization of droplets.[808]

Complete removal of the solvent in an l.c.-m.s. interface is neither necessary nor desirable.[809] This conclusion was reached after drawing a parallel between l.c.-m.s. and mass-spectrometric techniques in which protonation or cationization of polar compounds can occur in solution. The method of 'ionization' then serves only as a means of supplying sufficient energy to overcome electrostatic attraction between pre-formed ions and to desorb (evaporate) them. Finally in this section, a variable splitter has been described for coupling of a micro l.c. column and a magnetic-sector mass spectrometer.[810]

Moving-belt Interfaces.- Mechanical transfer devices, based on an endless moving belt, are commercial competitors to DLI methods. Currently, moving-belt interfaces enable l.c.-m.s. study of underivatized mono- and di-saccharides, but trisaccharides are not adequately analysed.[811] (Note that simple derivatization and g.c.-m.s. are capable of characterizing mono- to tetra-saccharides - see above.) High proportions of water in the l.c. solvent have always caused difficulties because of beading on the belt. However, deposition of a water-miscible solvent like ethanol onto the belt just before application of the l.c. column effluent improves wetting characteristics. The belt can then cope with the whole eluent of acetonitrile-water (20:80) from a microbore column.[788] A more complex solution to this problem requires a segmented-flow extractor between the liquid chromatograph and the l.c.-m.s. interface.[812-814] On-line continuous extraction of organic compounds from the aqueous (mobile) phase into an immiscible organic (extracting) phase such as dichloromethane accommodates flow rates up to 1 ml min^{-1} in reversed-phase l.c. applications. Objections that highly polar (water-miscible) samples were not amenable to the technique have been overcome by use of on-line chemical procedures. Extraction of cationic species into the organic solvent is best effected by use of aromatic counter-ions such as picrate, and anionic substances are subjected to on-line methylation with trimethylanilium hydroxide.[814] Whilst compatibility of l.c. (particularly in the reversed-phase mode) and mass spectrometry is significantly increased by these methods, the technique has become too complex for most potential users of h.p.l.c.-m.s.

It is well known that ionization of samples, as solids or solutions, can be brought about by bombarding them with ions, rapidly moving atoms, or laser beams. These ionization methods, like l.c., are particularly suited to analysis of involatile materials. Therefore, it is not surprising that l.c. eluents on a belt have been used as moving targets in bombardment ionization techniques. Direct ionization from the surface of a silver belt has been accomplished by bombardment with argon ions.[815] Inhomogeneous coverage of the belt with effluent gives rise to erratic ion signals, but this problem can be overcome by aerosol deposition, which improves wetting characteristics.[816] Once eluents have been applied to a high-purity nickel belt and the solvent has evaporated, the belt can be used as a storage system. Analysis by secondary-ion mass spectrometry can be effected any time up to an hour later. Detection limits for several amino-acids were measured to be less than 10^{-11} g with such a system.[816]

Continuous sample introduction by electrospraying onto a moving stainless-steel belt has been described for a laser ionization system. Abundant cationized molecules and little fragmentation were observed for amino-acids, saccharides, peptides, nucleosides, and nucleotides using this method.[817] Spraying of the l.c. effluent onto the belt has two major advantages: (*a*) uniform deposition and (*b*) facilitation of solvent evaporation without heating. The latter effect results in more efficient transfer of the more volatile components, less decomposition of thermally labile ones, and enables use of higher flow rates. The combination of fast-atom-bombardment mass spectrometry[818] (see Chapter 5) and l.c. *via* a moving belt has not yet been reported but may well increase considerably the range of compounds amenable to l.c.-m.s. Yet no matter how good the method of ionization, the limitations of the moving-belt system (mechanical complexity, cost, restricted flow rates, and so on) remain.

As a final note in this section, it is recommended that systems based on a moving belt are tested by true (on-line) h.p.l.c.-m.s. rather than by 'spotting' solutions directly onto the belt since analysis by the former is notoriously more difficult than by the latter.[811]

Other Interfaces.- Concentration of liquid chromatographic effluent prior to introduction into a quadrupole mass spectrometer has been accomplished by a clever device for evaporating solvents.[819] The effluent was allowed to flow down an electrically heated and stationary wire, the current being controlled by feedback from a volume-sensing photocell. After concentration by up to 20 times, the solution was sprayed into the ion source through a small needle valve to regulate the flow to the mass spectrometer. Poor wetting of the wire by aqueous solutions restricted maximum flow rates to 0.7 ml min^{-1} for methanol/water (50:50). The interface caused some peak broadening, and mass spectra are of mixed e.i./c.i. character. It remains to be seen how this relatively simple interface copes with gradient elution and buffers in the effluent.

A nebulizing inlet for an atmospheric-pressure-ionization instrument has been described.[820] Mass spectra of non-volatile compounds in solution were obtained even in the presence of salts. These findings may well be important for directly coupled l.c.-m.s. Also pertinent to l.c.-m.s. are two reports of ionization of dissolved polar compounds at atmospheric pressure.[821,822] Ionization of sample molecules was effected either by corona discharge with subsequent charge transfer from carrier-gas ions (and some Penning ionization)[821] or by a cooled inductively coupled plasma.[822] With a better understanding of the ion chemistry involved and improvements in detection limits, both methods could prove to be useful for h.p.l.c.-m.s.

Two easily constructed and inexpensive interfaces have been described.[823] One method involves splitting of the total effluent so that a small percentage only enters the mass spectrometer, the other utilizes vacuum nebulization to vaporize solvent and sample. Finally, h.p.l.c. has been combined with a ^{252}Cf plasma-desorption mass spectrometer.[824] The effluent was fed continuously through a capillary tube to the surface of the sample foil using a vacuum drying process to remove solvent. While one sample was being analysed, the next one was collecting. The current acquisition time for each mass spectrum (over 10 min) limits the usefulness of the method for on-line h.p.l.c.-m.s.

Applications.- A wide range of compounds has been used to test l.c-m.s. interfaces. In this section, preference is given to those reports in which l.c.-m.s. has been utilized to solve actual analytical problems.

The moving-belt interface has been applied with various degrees of success to PAHs[825] and heteroaromatics[826] in coal-liquefaction products, saccharides,[783,784,811,827,828] lipids,[829-831] petroporphyrins,[831] steroids,[827,828,830] alkaloids,[783,784,832] drugs and metabolites,[783,784,833] pesticides,[783,784] peptides,[783,784] nucleic acid components,[783,784,811] coumarin natural products,[784a,830] phenoxy-acid herbicides and simple phenols,[788b] glycoside conjugates,[811] liquid crystals,[834] PCBs,[835] and herbicide photolysis products.[836] The application of l.c.-c.i.m.s. to sequencing studies on mixtures of small peptides has shown some potential.[784a,837] The perchloro cage compounds kelevan and kepone (pesticides) degrade in the injection port of a gas chromatograph, but analysis by h.p.l.c.-m.s. avoids this problem.[838] Detection at the 1 p.p.m. level was possible. Underivatized chlorophenols in human urine can be assayed by selected-ion monitoring during l.c.-m.s. With detection limits around 2 - 10 ng, the mass-spectral method was less sensitive than electrochemical detection but provided a more specific analysis.[839] Normal-phase l.c.-m.s. with a moving-belt interface for analysis of carbamate pesticides is somewhat less sensitive than l.c. with u.v. detection (at optimum wavelengths for each compound).[840] The anti-ulcer drug ranitidine and its metabolites have been characterized and quantified in urine by direct injection of up to 10 μl of urine onto a normal-phase l.c. column. Sensitivity was of the order of 1 μl ml^{-1} of ranitidine.[841] Other studies have concerned analysis of effluents from the tannery and leather-finishing industry,[842] synthetic corticosteroids administered to racehorses (by l.c.-n.i.c.i.m.s.),[843] and explosives.[844] Studies of lipids by the on-line hydrogenation technique have continued.[791,845] Thermally labile bases have been studied by on-line extraction from aqueous l.c. effluents by the use of ion-paring techniques.[813] Saccharides, amino-acids, peptides, and nucleic acid components can be determined by laser desorption from a moving belt,[817] and a secondary-ion mass spectrometer has been used as a l.c. detector for the analysis of amino-acids.[816]

The following classes of compounds have been examined by l.c.-m.s. with DLI interfaces: long-chain acids,[846] alkaloidal drugs,[847,848] cannabinoid compounds,[849] steroids,[795] steroid hormones and vitamins,[810] aromatic amines,[796] PAHs,[797,799] drugs and metabolites,[798] phenols,[799] and saccharides.[806] The detection limit for juvenile hormones by selected-ion monitoring during reversed-phase l.c.-m.s. was determined to be 10 pg.[800,850] The method was employed to detect JH-I in *Pieris brassicae*.[850] When deuteriated solvents were used for l.c. and functioned as c.i. reactant gases, it was possible to determine the number of acidic (exchangeable) hydrogen atoms in eluents and to study ion/molecule reactions.[798] The common reversed-phase l.c. solvent acetonitrile/water (60:40) when vaporized in a c.i. ion source can provide conditions of electron capture and dissociative electron capture for negative-ion studies of organophosphorus pesticides. Sensitivity of the l.c.-n.i.m.s. system was 100 - 200 ng of pesticide applied to the column (1 - 2 ng entering the mass spectrometer after splitting the effluent).[851] These workers have used the same technique in both positive-ion and negative-ion mode to analyse triazine herbicides.[852] H.p.l.c.-m.s. with positive-ion and negative-ion chemical ionization has been applied also to pharmacological and toxicological problems.[853a]

Moving-belt and DLI interfaces have been compared and contrasted for the needs of thc pharmaceutical industry.[853b] H.p.l.c.-m.s. has played a significant part in a total strategy for sequence analysis of polysaccharides.[268,304] Study of sulpha drugs by DLI l.c.-m.s.[854] is postponed till the following section since metastable-ion techniques were utilized. Analysis of flavonoids and fucosides in plant extracts and of vitamin C in beer has been effected by use of l.c. with either a u.v./visible or a mass spectrometer as detector.[855]

The thermospray l.c.-m.s. instrument has been applied to a range of compounds including amino-acids, peptides, and nucleic acid components,[803,804] and saccharides, fatty acids, vitamins, and antibiotics.[805] The combination of l.c. with ^{252}Cf plasma-desorption mass spectrometry was utilized to study drugs.[824]

5 Metastable-ion Techniques

If the l.c.-m.s. combination were sufficiently advanced, it would be an answer to very many questions. On the other hand, the combination of metastable-ion techniques and chromatography is an answer awaiting questions. The methodology for utilizing methods based on metastable ions during g.c.-m.s. or l.c.-m.s. is established, but many applications remain to be found. The term 'metastable-ion techniques' refers to the whole range of scanning methods available on a variety of instruments to study metastable ions in the absence of normal ions. They include ion kinetic-energy spectroscopies and linked scanning with conventional magnetic-sector or triple-quadrupole mass spectrometers, with or without collisional activation to induce fragmentation.[6] The unhelpful term m.s.-m.s. has been applied to the same range of methods,[856] but the g.c.-m.s.-m.s. and l.c.-m.s.-m.s. terminology is not used here. Combined chromatography-metastable-ion methods have been used for two distinct types of study: (i) quantification by selected-metastable-peak monitoring and (ii) structure elucidation, particularly of isomers in mixtures, by recording complete metastable-ion mass spectra during g.c.-m.s. or l.c.-m.s.

Selected-metastable-peak Monitoring.- Because of its lesser demands on scanning techniques, selected-metastable-peak monitoring was developed first. It is analogous to the well known selected-ion monitoring, the mass spectrometer being tuned to detect only ions of a single mass or a small number of masses. However, the ions monitored result from the decomposition of metastable ions during flight through the mass spectrometer. Since metastable peaks are almost always smaller than normal peaks, development of the technique was not motivated initially by the prospect of improved sensitivity but on the grounds of specificity.[857] The great specificity of the method arises because (i) normal ions are 'defocused' and (ii) a single *reaction* occurring in a given field-free region of the analyser is monitored. To elicit a response at the detector, a substance must afford ions at both of two specified integer m/z values, a precursor/product relationship must exist between those ions, and a significant metastable peak must occur for the fragmentation. Additionally, identification of an analyte depends upon its eluting at the correct chromatographic retention time. These stringent requirements endow selected-metastable-peak monitoring with specificity similar to that of high-resolution selected-ion monitoring,[858] but without the technical difficulties of working at high resolving powers. The high selectivity also makes for low noise levels, leading to high signal-to-noise ratios and hence *effective sensitivities* that can be much improved with respect to low-resolution selected-ion monitoring. Detection limits of about 20 and 30 pg have been recorded for 5α-dihydrotestosterone[857] and testosterone[858] by monitoring the metastable peaks for the transitions $(M - Bu^{\cdot})^+ \rightarrow (M - Bu^{\cdot} - HMe_2SiOH)^+$ and $M^{+\cdot} \rightarrow (M - Bu^{\cdot})^+$, respectively, of tBDMS derivatives.

In common with conventional selected-ion monitoring, an internal standard is required for accurate quantification. The very specificity of selected-metastable-peak monitoring imposes severe restraints on suitable internal standards. For the simplest and most sensitive technique of single-metastable-peak monitoring, isotope-labelled analogues of the analyte are not suitable internal standards, their metastable peaks falling at different m/z values. Since the mass-spectral behaviour of isomeric compounds is frequently very similar and sometimes identical, unlabelled isomers of the analyte usually satisfy the above criteria for causing a signal at the detector and make good internal standards as long as (i) they can be separated chromatographically from the analyte and (ii) they do not occur as significant impurities in samples under test. For example, epitestosterone meets these requirements as an internal standard for assay of testosterone in hamster prostrate tissue[858] and human plasma and saliva.[859] This approach would reach its most sophisticated using unnatural enantiomers of chiral analytes as internal standards and a chiral g.c. (or l.c.) column to effect their separation. This has not yet been reported despite the common occurrence of asymmetric natural products and drugs and the guaranteed identical mass-spectrometric behaviour of internal standard and analyte.

An alternative approach is to use an isotopically labelled analogue as internal standard, but this entails multiple-metastable-peak monitoring, which is less sensitive and more instrumentally demanding than single-reaction monitoring. Only two reports of multiple-metastable-peak monitoring have appeared.[860,861] Both articles describe switching between two metastable peaks with a conventional geometry, double-focusing mass spectrometer operating in the linked-scan mode under data system control. $(^2H_3)$-Testosterone was utilized as internal standard for testosterone[860] and $(^2H_2)$hydroxyphenylacetic acids for their naturally occurring, unlabelled analogues.[861] The problem of finding a suitable internal standard for measurement of Δ^1-THC in plasma was elegantly solved by Harvey and co-workers.[259] Isomeric THCs not being suitable, these workers synthesized cannabinol labelled with four deuterium atoms in the side chain ($(^2H_4)$CBN). The TMS derivatives of Δ^1-THC and $(^2H_4)$CBN can be separated by g.c., and both afford molecular ions at *m/z* 386 and $(M - Me^{\bullet})^+$ ions at *m/z* 371. The corresponding metastable ion (*m/z* 386 → 371) could be monitored to detect both compounds without the need for complex instrumentation.[259] The method was capable of measuring Δ^1-THC in 1 ml of plasma at the 5 pg ml^{-1} level and was so specific that no clean-up of the sample was required prior to g.c.-m.s.

The range of compounds examined by selected-metastable-peak monitoring has been extremely narrow. Reports have concerned testosterone and dehydroepiandrosterone in various biological samples,[858-860] 5α-dihydrotestosterone,[857] androstanolones,[862,863] oestradiol and dehydroepiandrosterone sulphate in saliva,[864] *m*- and *p*-hydroxyphenylacetic acids in rat brain,[861] and Δ^1-THC.[259] One reason why the method is not generally applied is that research has concerned almost exclusively tBDMS derivatives with monitoring of $M^{+\bullet} \rightarrow (M - Bu^{\bullet})^+$ or $(M - Bu^{\bullet})^+ \rightarrow (M - Bu^{\bullet} - H(Me)_2SiOH)^+$ fragmentations. Tert-butyldimethylsilylation is not always possible or appropriate. Wider applicability requires research effort being expended to discover different derivatives that tend to fragment by ejection of small, stable neutral molecules frequently associated with abundant metastable ions. Further developments for the future include selected-metastable-peak monitoring during l.c.-m.s. and the use of collisional activation. Whilst sensitivity should be increased by use of a collision cell, owing to the enhanced number of decompositions, reproducibility of metastable-peaks' heights must be assessed carefully for the effects of slight pressure changes in the given field-free region and of other day-to-day variations in operating conditions.

Acquisition of Metastable-ion Spectra During Chromatography.- So-called m.s.-m.s. involves allowing or inducing ions in a mass-selected beam to fragment and recording the resulting product ions as a 'metastable-ion spectrum'. Such spectra are frequently more sensitive to fine differences in structure than are normal mass spectra, enabling differentiation of isomeric compounds. Also, metastable-ion techniques with non-chromatographic inlets are growing in importance for rapid screening of samples for the presence of selected compounds.[856] However, this methodology has several drawbacks.[6] Because components of mixtures are resolved by mass selection only, mixtures of isomers cannot be readily handled. Other disadvantages include possible suppression effects, contamination of the ion source by 'raw' samples, the occurrence of artefact peaks particularly when examining low-mass ions in the presence of high-mass ones, and interpretational ambiguities.[865] The desirability of combining chromatography and metastable-ion techniques has been clearly expounded:[168,832,854] g.c. or l.c. can be used to separate isomers prior to their examination by a method that is diagnostic of subtle structural differences. Also, the specificity of metastable-ion methods offers the advantage over conventional g.c.-m.s. or l.c.-m.s. that any impurities co-eluting with the compounds of interest are not detected (assuming they do not afford ions at the m/z value selected as the primary ion beam).

The feasibility of linked scanning or mass-analysed ion kinetic-energy spectroscopy (m.i.k.e.s.) using chromatographic inlets came with increased scan speeds commensurate with chromatographic peak width (10 - 60 s).[866] The first reports of linked scanning[168,865,867,868] and m.i.k.e.s.[869,870] during g.c.-m.s.

and triple-quadrupole mass filters linked to liquid chromatographs[816*c*,854,871] have now appeared. Initial reports of linked scanning during g.c.-m.s. utilized scans of 3 s duration and recording of spectra on u.v.-sensitive paper.[867,868] Mixtures of isomeric PAHs[867] and chlorinated biphenyls[868] were successfully analysed. Latterly, computerized acquisition of full linked-scan mass spectra during g.c.-m.s. has been reported[168,865] for cyclic esters of aromatic boronic acids. Positional isomers in mixtures were readily differentiated.[168] Elucidation of the different side-chain structures of di-(2-ethylhexyl)phthalate metabolites has been effected during elution from a g.c. column by recording m.i.k.e. spectra of the ions corresponding to those side chains,[869] and confirmation of the presence of 2,3,7,8-TCDD in air filter samples has been brought about by m.i.k.e.s. during g.c.-c.i.m.s.[870]

In all these g.c.-m.s. studies, magnetic-sector mass spectrometers were employed. In fact, triple-quadrupole systems have several considerable advantages for chromatography-metastable-ion studies: extremely fast scanning capability, ease of automation, and fewer artefact peaks. The coupling of a triple-quadrupole system to one moving-belt l.c.-m.s. interface has been described, but no metastable-ion work has yet been reported.[816*c*] In l.c.-m.s. experiments, interference by ions from the l.c. solvent can be eliminated by filtering out all ions but the required $(M + H)^+$ ions with one quadrupole analyser, then recording the mass spectra of those $(M + H)^+$ ions by collisional activation and mass analysis with another quadrupole mass filter.[871] The system was used to monitor aromatic acids in a complex mixture, the detection limit for full scans being 5 μg on the l.c. column (50 ng into the ion source). A triple-quadrupole instrument employing atmospheric-pressure ionization has been linked to an l.c. column *via* a DLI interface for analysis of sulpha drugs in racehorse urine.[854] The system was used to record normal and metastable-ion mass spectra, and perform selected-metastable-peak monitoring. The method required little or no sample clean-up yet protected the mass spectrometer from exposure to high levels of endogenous materials that would have resulted from a non-chromatographic inlet. As with all chromatography-metastable-ion techniques, it provided retention-time information not available with the direct-insertion (m.s.-m.s.) method. Again, detection limits for full metastable-ion spectra were in the low μg range after splitting of the l.c. effluent.[854]

The relative merits of direct-probe metastable-ion techniques and conventional h.p.l.c.-m.s. for analysing mixtures have been compared,[52,832] the two groups of workers reaching opposite conclusions. Since unified chromatography-metastable-ion methods combine the advantages of both techniques, their development and application should be pursued. Examples of studies in which a combined method could have been applied to advantage are many. For instance, capillary g.c.-m.s. was required when non-chromatographic metastable-ion analysis with a triple-quadrupole instrument failed to determine specific nitrated PAH isomers in mixtures.[641] A combined g.c.-triple-quadrupole mass spectrometer would have accomplished the desired analysis in one experiment. It is suggested here that major advances in this field are likely to come from combinations of g.c. or l.c. with triple-quadrupole or 'mixed' magnetic-sector/quadrupole instruments because of the speed, versatility, and simplicity of their scanning modes. In particular, the fast scan speed would enable combination with the most efficient capillary g.c. columns. This would be especially beneficial in environmental science because identification of individual PAH, PCB, and PCDD isomers in complex mixtures necessitates high-resolution chromatographic separations.

6 Concluding Remarks

Gas chromatography-mass spectrometry occupies an established and prime position in analytical chemistry owing to its sensitivity, wide applicability, and versatility. The use of capillary g.c. columns (especially those manufactured from fused silica) for g.c.-m.s. has increased considerably, with workers in the fields of clinical chemistry, food flavour research, the environment, and fuel analysis leading the way. In the next two years, capillary g.c.-m.s. will go through a period of consolidation. What else does the future hold? In two years' time, it is hoped that structural features not normally elucidated by current mass-spectrometric techniques will become more amenable to study. The tools to accomplish this are, at the isomeric level, metastable-ion techniques in conjunction with g.c.-m.s. or l.c.-m.s. and, at the level of molecular configuration, chiral stationary phases for g.c. and l.c. Extension of the sensitive and specific method selected-metastable-peak monitoring to the assay of a wider range of compounds is also encouraged.

The time for great strides forward in g.c.-m.s. has passed; progress is now made in small steps. The same is not true for l.c.-m.s. with new interfaces still being announced. Further development of the thermospray method[803-805] and the inevitable marriage of l.c. and fast-atom-bombardment (f.a.b.) mass spectrometry *via* a moving belt are eagerly awaited. However, the l.c./f.a.b./m.s. combination may not necessarily require a moving belt with its expense and mechanical complexity. The effluent from a capillary l.c. column (possibly supplemented with a suitable solvent such as glycerol) could be bombarded directly with fast atoms through a 'window' or small hole in the exit tubing from the chromatograph. Despite the sophistication of many present coupling devices, it may well be that the key to successful l.c.-m.s. is simplicity.

(References begin overleaf)

References

1 (*a*) C.J.W. Brooks and B.S. Middleditch in 'Mass Spectrometry', ed. R.A.W. Johnstone (Specialist Periodical Reports), The Royal Society of Chemistry, London, 1977, Vol. 4, p. 146; (*b*) C.J.W. Brooks and B.S. Middleditch in ref. 1*a*, 1979, Vol. 5, p. 142; (*c*) F.A. Mellon in ref. 1*a*, 1981, Vol. 6, p. 196.
2 'Gas Chromatography-Mass Spectrometry Abstracts', P.R.M. Science and Technology Agency Ltd., London, 1980 - 1982, Vols. 11 - 13.
3 'Liquid Chromatography-Mass Spectrometry Abstracts', P.R.M. Science and Technology Agency Ltd., London, 1981 - 1982, Vol. 1.
4 'Advances in Mass Spectrometry', ed. A. Quayle, Heyden, London, 1980, Vol. 8.
5 I. Howe, D.H. Williams, and R.D. Bowen, 'Mass Spectrometry: Principles and Applications', 2nd Ed., McGraw-Hill, New York, 1981.
6 M.E. Rose and R.A.W. Johnstone, 'Mass Spectrometry for Chemists and Biochemists', Cambridge University Press, Cambridge, 1982.
7 H. Willard, L. Merritt, J. Dean, and F. Settle, 'Instrumental Methods of Analysis', 6th Ed., Van Nostrand Company, New York, 1981.
8 'Mass Spectrometry, Part B', ed. C. Merritt, jun. and C.N. McEwen, Marcel Dekker, New York, 1980.
9 P. Longevialle, 'Principles of Mass Spectrometry of Organic Substances', Masson, Paris, 1981.
10 A.L. Burlingame, A. Dell, and D.H. Russell, *Anal Chem.*, 1982, 54, 363R.
11 W.G. Jennings, 'Comparisons of Fused Silica and Other Glass Columns in Gas Chromatography', Huethig, Heidelberg, 1981.
12 J.A. Perry, 'Introduction to Analytical Gas Chromatography: History, Principles and Practice', Marcel Dekker, New York, 1981.
13 'Applications of Glass Capillary Gas Chromatography', ed. W.G. Jennings, Marcel Dekker, Basel, 1981.
14 'Chromatographic Methods: Recent Advances in Capillary Gas Chromatography', ed. W. Bertsch, W.G. Jennings, and R.E. Kaiser, Huethig, Heidelberg, 1981, Vol. 1.
15 K.S. Brenner and W. Huber, *J. Chromatogr.*, 1981, 220, 95.
16 T.H. Risby, L.R. Field, F.J. Yang, and S.P. Cram, *Anal. Chem.*, 1982, 54, 410R.
17 R. Self, *Ann. Chim. (Rome)*, 1980, 70, 15.
18 S.H. Koslow, *Phys.-Chem. Methodol. Psychiatr. Res.*, 1980, 155.
19 M. McCamish, *Eur. J. Mass Spectrom. Biochem., Med. Environ. Res.*, 1980, 1, 7,
20 D. Hazelby and K.T. Taylor, *Ann. Chim. (Rome)*, 1980, 70, 201.
21 F. Cattanbeni, *Ann. Chim. (Rome)*, 1981, 71, 77.
22 S.P. Markey, *Biomed. Mass Spectrom.*, 1981, 8, 426.
23 S.J. Gaskell, *Trends Anal. Chem.*, 1982, 1, 110.
24 A.M. Greenway and C.F. Simpson, *J. Phys. E*, 1980, 13, 1131.
25 I. Hanin and T.-M. Shih, *Phys.-Chem. Methodol. Psychiatr. Res.*, 1980, 111.
26 T. Tsuchiya, *Kagaku, Zokan (Kyoto)*, 1980, 3 (*Chem. Abstr.*, 1981, 94, 219058).
27 T. Murada, *Bokin Bobai*, 1980, 8, 218.
28 K. Levsen in ref. 4, p. 897.
29 N.M.M. Nibbering, *J. Chromatogr.*, 1982, 251, 93.
30 K.L. Rinehart, jun. *et al.*, *Pure Appl. Chem.*, 1981, 53, 795.
31 'Isotopes: Essential Chemistry and Applications', ed. J.A. Elvidge and J.R. Jones, Special Publication No. 35, The Royal Society of Chemistry, London, 1980.
32 J. Sjövall in ref. 4, p. 1069.

33 A.M. Lawson, C.K. Lim, W. Richmond, D.M. Samson, K.D.R. Setchell, and A.C.S. Thomas in ref. 42, p. 135.
34 S. Baba, *Radioisotopes*, 1982, 31, 119.
35 'Stable Isotopes', ed. H.-L. Schmidt, H. Förstel, and K. Heinzinger, Elsevier, Amsterdam, 1982.
36 L.A. Currie, *Pure Appl. Chem.*, 1982, 54, 715.
37 C. Fenselau and L.P. Johnson, *Drug Metab. Dispos.*, 1980, 8, 274.
38 E.M. Martinelli, *Eur. J. Mass Spectrom. Biochem., Med. Environ. Res.*, 1980, 1, 33.
39 A. Tatematsu *et al.*, 'Practical Mass Spectrometry for the Medical and Pharmaceutical Sciences', Kodansha, Tokyo, 1979.
40 'Recent Developments in Mass Spectrometry in Biochemistry and Medicine', ed. A. Frigerio and M. McCamish, Elsevier, Amsterdam, 1980, Vol. 6.
41 'Recent Developments in Mass Spectrometry in Biochemistry, Medicine and Environmental Research', ed. A. Frigerio, Elsevier, Amsterdam, 1980, Vol. 7.
42 'Clinical Research Center Symposium, No. 1: Current Developments in the Clinical Applications of HPLC, GC and MS', ed. A.M. Lawson, C.K. Lim, and W. Richmond, Academic Press, London, 1980.
43 Y. Seyama, *Bunseki*, 1980, 151.
44 T. Murata, S. Takahashi, T. Takeda, and S. Ohnishi, *Shimadzu Hyoron*, 1980, 37, 29.
45 M. Suzuki, *Hakko to Kogyo*, 1980, 38, 197.
46 B. Halpern, *CRC Crit. Rev. Anal. Chem.*, 1981, 11, 49.
47 S. Ando and Y. Tanaka, *Shitsuryo Bunseki*, 1981, 29, 113.
48 'Biochemical Applications of Mass Spectrometry, First Supplementary Volume', ed. G.R. Waller and O.C. Dermer, Wiley-Interscience, New York, 1980.
49 M. Ende and G. Spiteller, *Mass Spectrom. Rev.*, 1982, 1, 29.
50 'Mass Spectrometry Reviews', ed. G.R. Waller and O.C. Dermer, Wiley, New York, first issue, 1982.
51 R. Kraft, A. Otto, A. Makower, and G. Etzold, *Anal. Biochem.*, 1981, 113, 193.
52 S.E. Unger, A. Vincze, R.G. Cooks, R. Chrisman, and L.D. Rothman, *Anal. Chem.*, 1981, 53, 976.
53 J.F.J. Todd, *Dyn. Mass Spectrom.*, 1981, 6, 3.
54 C.L. Wilkins, G.N. Giss, G.M. Brissey, and S. Steiner, *Anal. Chem.*, 1981, 53, 113.
55 R.W. Crawford, T. Hirschfeld, R.H. Sanborn, and C.M. Wong, *Anal. Chem.*, 1982, 54, 817.
56 K.H. Shafer, T.L. Hayes, and J.E. Tabor, *Proc. SPIE-Int. Soc. Opt. Eng.*, 1981, 289, 160.
57 P.A. Barnes, G. Stevenson, and S.B. Warrington, *Proc. Eur. Symp. Therm. Anal.*, 1981, 2nd, 47.
58 D.L. Doerfler, G.T. Emmons, and I.M. Campbell, *Anal. Chem.*, 1982, 54, 832.
59 D.L. Doerfler, E.R. Rosenblum, J.M. Malloy, J.D. Naworal, I.R. McManus, and I.M. Campbell, *Biomed. Mass Spectrom.*, 1980, 7, 259.
60 B. Munson, *Int. Lab.*, 1981, 16.
61 C.W. Polley, jun. and B. Munson, *Anal. Chem.*, 1981, 53, 308.
62 V.A. Bogdanov, Yu.I. Savel'ev, R.N. Shchelokov, and K.I. Sakodynskii, *Zh. Fiz. Khim.*, 1981, 55, 1315.
63 V.A. Bogdanov, Yu.I. Savel'ev, and K.I. Sakodynskii, *Dokl. Akad. Nauk SSSR*, 1981, 256, 616.
64 H. Takagi, N. Washida, H. Akimoto, and M. Okuda, *Anal. Chem.*, 1981, 53, 175.

65 D.I. Carroll, I. Dzidic, E.C. Horning, and R.N. Stillwell, *Appl. Spectrosc. Rev.*, 1981, 17, 337.
66 F.W. Karasek and G.E. Spangler, *J. Chromatogr. Libr.*, 1981, 20, 377.
67 S.H. Kim and F.W. Karasek, *J. Chromatogr.*, 1982, 234, 13.
68 C.L. Wilkins and M.L. Gross, *Anal. Chem.*, 1981, 53, 1661A.
69 E.B. Ledford, jun., R.L. White, S. Ghaderi, C.L. Wilkins, and M.L. Gross, *Anal. Chem.*, 1980, 52, 2450.
70 M.T. Nguyen, J. Wronka, S. Starry, and D.P. Ridge, *Int. J. Mass Spectrom. Ion Phys.*, 1981, 40, 195.
71 E.J. Gallegos, *Anal. Chem.*, 1981, 53, 187.
72 J.L. Selves, *Analusis*, 1980, 8, 410.
73 D. Picart, F. Jacolet, F. Berthou, and H.H. Floch, *Biomed. Mass Spectrom.*, 1980, 7, 464.
74 M.M. Siegel, B.E. Hildebrand, and D.R. Hall, *Int. J. Environ. Anal. Chem.*, 1980, 8, 107.
75 B. Hedfjäll and R. Ryhage, *Anal. Chem.*, 1981, 53, 1641.
76 R.W. Silverman and D.J. Jenden, *IEEE Trans. Instrum. Meas.*, 1980, 1M-29, 7.
77 H. Eustache and G. Histi, *J. Membr. Sci.*, 1981, 8, 105.
78 W.D. Koller and G. Tressl, *J. High Resolut. Chromatogr. Chromatogr. Commun.*, 1980, 3, 359.
79 F. Friedli, *J. High Resolut. Chromatogr. Chromatogr. Commun.*, 1981, 4, 495.
80 C.H.L. Shackleton, *GC-MS News*, 1981, 9, 58, 72.
81 E. Wetzel, Th. Kuster, and H.-Ch. Curtius, *J. Chromatogr.*, 1982, 239, 107.
82 B.D. Andresen, K.J. Ng, J. Wu, and J.R. Bianchime, *Biomed. Mass Spectrom.*, 1981, 8, 237.
83 C.A. Cramers, G.J. Scherpenzeel, and P.A. Leclercq, *J. Chromatogr.*, 1981, 203, 207.
84 P.A. Leclercq, G.J. Scherpenzeel, E.A.A. Vermeer, and C.A. Cramers, *J. Chromatogr.*, 1982, 241, 61.
85 L.G. Randall and A.L. Wahrhaftig, *Rev. Sci. Instrum.*, 1981, 52, 1283; L.G. Randall, *Sep. Sci. Technol.*, 1982, 17, 1.
86 B.E. Blaisdell and C.C. Sweeley, *Anal. Chim. Acta*, 1980, 117, 1.
87 W.F. Hargrove and D. Rosenthal, *Anal. Chem.*, 1981, 53, 538.
88 F.J. Knorr, H.R. Thorsheim, and J.M. Harris, *Anal. Chem.*, 1981, 53, 821.
89 M.S. Khots and N.M. Bogel'fer, *Zh. Anal. Khim.*, 1980, 35, 2203.
90 P.C. Tway, L.J.C. Love, and H.B. Woodruff, *Anal. Chim. Acta*, 1980, 117, 45.
91 M.A. Sharaf and B.R. Kowalski, *Anal. Chem.*, 1981, 53, 518.
92 J.-H. Chen and L.-P. Hwang, *Anal. Chim. Acta*, 1981, 133, 271.
93 H.B. Woodruff, P.C. Tway, and L.J.C. Love, *Anal. Chem.*, 1981, 53, 81.
94 R.J. Anderegg, *Anal. Chem.*, 1981, 53, 2169.
95 B.E. Blaisdell and C.C. Sweeley, *Anal. Chim. Acta*, 1980, 117, 17.
96 B.E. Blaisdell, S.C. Gates, F.E. Martin, and C.C. Sweeley, *Anal. Chim. Acta*, 1980, 117, 35.
97 I.K. Mun, D.R. Bartholomew, D.B. Stauffer, and F.W. McLafferty, *Anal. Chem.*, 1981, 53, 1938.
98 F.C. Falkner, *Biomed. Mass Spectrom.*, 1981, 8, 43.
99 A.D. Sauter, P.E. Mills, W.L. Fitch, and R. Dyer, *J. High Resolut. Chromatogr. Chromatogr. Commun.*, 1982, 5, 27.
100 D.A. Schoeller, *Biomed. Mass Spectrom.*, 1980, 7, 457.

101 L.D. Gruenke, J.C. Craig, and D.M. Bier, *Biomed. Mass Spectrom.*, 1980, 7, 381.
102 P. Olynyk, W.L. Budde, and J.W. Eichelberger, *J. Chromatogr. Sci.*, 1981, 19, 377.
103 B.N. Colby, A.E. Rosecrance, and M.E. Colby, *Anal. Chem.*, 1981, 53, 1907.
104 W. Schramm, T. Louton, W. Schill, and H.O. Hoppen, *Biomed. Mass Spectrom.*, 1980, 7, 273.
105 F.W. Karasek, R.E. Clement, and J.A. Sweetmen, *Anal. Chem.*, 1981, 53, 1050A.
106 H.G. Eaton, *J. Chromatogr. Sci.*, 1980, 18, 580.
107 R.L. Hanson, C.R. Clark, R.L. Carpenter, and C.H. Hobbs, *Environ. Sci. Technol.*, 1981, 15, 701.
108 H.A. James, C.P. Stell, and I. Wilson, *J. Chromatogr.*, 1981, 208, 89.
109 M.D. Erickson, M.K. Alsup, and I.A. Hyldburg, *Anal. Chem.*, 1981, 53, 1265.
110 M.H. Hiatt, *Anal. Chem.*, 1981, 53, 1541.
111 J.G. Moncur, T.E. Sharp, and E.R. Byrd, *J. High Resolut. Chromatogr. Chromatogr. Commun.*, 1981, 4, 603.
112 A.R. Trussell, J.G. Moncur, F.-Y. Lieu, and L.Y.C. Leong, *J. High Resolut. Chromatogr. Chromatogr. Commun.*, 1981, 4, 156.
113 C.H.L. Shackleton and J.O. Witney, *Clin. Chim. Acta*, 1980, 107, 231.
114 L.C. Ramirez, C. Millot, and B.F. Maume, *J. Chromatogr.*, 1982, 229, 267.
115 H. Müller, R. Mrongovius, and H.W. Seyberth, *J. Chromatogr.*, 1981, 226, 450.
116 N. Narasimhachari, *J. Chromatogr.*, 1981, 225, 189.
117 E. Dahlberg, M. Snochowski, and J.-Å. Gustafsonn, *Anal. Biochem.*, 1980, 106, 380.
118 S.G. Colgrove and H.J. Svec, *Anal. Chem.*, 1981, 53, 1737.
119 D.W. Later, M.L. Lee, K.D. Bartle, R.C. Kong, and D.L. Vassilaros, *Anal. Chem.*, 1981, 53, 1612.
120 D. Issachar and C.C. Sweeley, *Anal. Biochem.*, 1981, 113, 43.
121 D. Issachar, J.F. Holland, and C.C. Sweeley, *Anal. Chem.*, 1982, 54, 29.
122 G.D. Veith and D.W. Kuehl, *Anal. Chem.*, 1981, 53, 1132.
123 K.E. MacLeod, R.C. Hanisch, and R.G. Lewis, *J. Anal. Toxicol.*, 1982, 6, 38.
124 K. Grob and G. Grob, *J. Chromatogr.*, 1981, 213, 211.
125 K. Grob and G. Grob, *J. High Resolut. Chromatogr. Chromatogr. Commun.*, 1981, 4, 491.
126 K. Grob, G. Grob, and K. Grob, jun., *J. Chromatogr.*, 1981, 219, 13.
127 D. Rosenthal, *Anal. Chem.*, 1982, 54, 63.
128 H.G. Nowicki, C.A. Kieda, and A.S. Nakagawa, *J. High Resolut. Chromatogr. Chromatogr. Commun.*, 1981, 4, 236.
129 P.B. Farmer, E. Bailey, and T.A. Connors in ref. 4, p. 1227.
130 E. Bailey, P.B. Farmer, and J.H. Lamb, *J. Chromatogr.*, 1980, 200, 145.
131 J.H. Liu, W.W. Ku, J.T. Tsay, M.P. Fitzgerald, and S. Kim, *J. Forensic Sci.*, 1982, 27, 39.
132 J.H. Liu and W.W. Ku, *Anal. Chem.*, 1981, 53, 2180.
133 N. Ôi, M. Hariba, H. Kitahara, T. Doi, T. Tani, and T. Sakakibara, *J. Chromatogr.*, 1980, 202, 305.
134 N. Ôi, H. Kitahara, Y. Inda, and T. Doi, *J. Chromatogr.*, 1982, 237, 297.
135 N. Ôi, T. Doi, H. Kitahara, and Y. Inda, *J. Chromatogr.*, 1981, 208, 404.

136 W.A. König, I. Benecke, and S. Sievers, *J. Chromatogr.*, 1982, 238, 427.
137 W.A. Koenig, I. Benecke, and H. Bretting, *Angew. Chem., Int. Ed. Engl.*, 1981, 20, 693.
138 J.A. Thompson, J.L. Holtzman, M. Tsuru, C.L. Lerman, and J.L. Holtzman, *J. Chromatogr.*, 1982, 238, 470.
139 J.P. Kamerling, M. Duran, G.J. Gerwig, D. Ketting, L. Bruinvis, J.F.G. Vliegenthart, and S.K. Wadman, *J. Chromatogr.*, 1981, 222, 276.
140 G. Schomburg, H. Husmann, and R. Rittmann, *J. Chromatogr.*, 1981, 204, 85.
141 M. Gaili and S. Trestiannu, *J. Chromatogr.*, 1981, 203, 193.
142 A.P.J.M. De Jong, *J. High Resolut. Chromatogr. Chromatogr. Commun.*, 1981, 4, 125.
143 P.J. Marriott, J.P. Gill, and G. Eglinton, *J. Chromatogr.*, 1982, 236, 395.
144 J.E. Picker and R.E. Sievers, *J. Chromatogr.*, 1981, 217, 275.
145 W.V. Ligon and R.J. May, *Anal. Chem.*, 1980, 52, 901.
146 P. Vouros in ref. 8, p. 129,
147 W.P. Cochrane in 'Chemical Derivatization in Analytical Chemistry', ed. R.W. Frei and J.F. Lawrence, Plenum Press, New York, 1981, Vol. 1, p. 1.
148 C.F. Poole and A. Zlatkis, *Anal. Chem.*, 1980, 52, 1002A.
149 W.C. Kossa in 'Chemical Derivatization in Analytical Chemistry', ed. R.W. Frei and J.F. Lawrence, Plenum Press, New York, 1981, Vol. 1, p. 99.
150 T. Veysoglu and L.A. Mitscher, *Tetrahedron Lett.*, 1981, 22, 1303.
151 T. Veysoglu and L.A. Mitscher, *Tetrahedron Lett.*, 1981, 22, 1299.
152 B.M. Trost and C.G. Caldwell, *Tetrahedron Lett.*, 1981, 22, 4999.
153 D.J. Harvey, *J. Chromatogr.*, 1980, 196, 156.
154 D.J. Harvey, *Biomed. Mass Spectrom.*, 1980, 7, 278.
155 D.J. Harvey, *Biomed. Mass Spectrom.*, 1980, 7, 211.
156 H. Miyazaki, M. Ishibashi, K. Yamashita, Y. Nishikawa, and M. Katori, *Biomed. Mass Spectrom.*, 1981, 8, 521.
157 A.C. Bazan and D.R. Knapp, *J. Chromatogr.*, 1982, 236, 201.
158 A.P.J.M. De Jong, J. Elema, and B.J.T. Van de Berg, *Biomed. Mass Spectrom.*, 1980, 7, 359.
159 M. Donike and J. Zimmermann, *J. Chromatogr.*, 1980, 202, 483.
160 H. Miyazaki, M. Ishibashi, K. Yamashita, and M. Yakushiji, *Chem. Pharm. Bull.*, 1981, 29, 796.
161 M.A. Quilliam. K.K. Ogilvie, and J.B. Westmore, *Org. Mass Spectrom.*, 1981, 16, 129.
162 C.F. Gelijkens, D.L. Smith, and J.A. McCloskey, *J. Chromatogr.*, 1981, 225, 291.
163 A.S. Christopherson, E. Houland, and K.E. Rasmussen, *J. Chromatogr.*, 1982, 234, 107.
164 C.J.W. Brooks, W.J. Cole, J.H. Borthwick, and G.M. Brown, *J. Chromatogr.*, 1982, 239, 191.
165 C.J.W. Brooks, W.J. Cole, H.B. McIntyre, and A.G. Smith, *Lipids*, 1980, 15, 745.
166 D.W. Robinson and D.S. Reive, *J. Anal. Toxicol.* 1981, 5, 69.
167 S.G. Batrakov, B.V. Rozynov, and A.N. Ushakov, *Khim. Prir. Soedin.*, 1981, 283.
168 C. Longstaff and M.E. Rose, *Org. Mass Spectrom.*, 1982, 17, 508.
169 M.T. Bush and E. Sanders-Bush, *Anal. Biochem.*, 1980, 106, 351.
170 S. Barnes, D.G. Pritchard, R.L. Settine, and M. Geckle, *J. Chromatogr.*, 1980, 183, 269.

171 L. Rivier, P. Gaskin, K.S. Albone, and J. MacMillan, *Phytochemistry*, 1981, 20, 687.
172 J. Rosenfeld, T.L. Ting, and A. Phatak, *Prostaglandins*, 1981, 21, 41.
173 D.P. Schwartz and C. Allen, *J. Chromatogr.*, 1981, 208, 55.
174 G.W. Francis and K. Veland, *J. Chromatogr.*, 1981, 219, 379; G.W. Francis, *Chem. Phys. Lipids*, 1981, 29, 369.
175 B.A. Bieri-Leonhardt, E.D. DeVilbiss, and J.R. Plimmer, *J. Chromatogr. Sci.*, 1980. 18, 364.
176 V.K.S. Shukla, J. Clausen, H. Egsgaard, and E. Larsen, *Fette, Seifen, Anstrichm.*, 1980, 82, 193.
177 W. Vetter and W. Meister, *Org. Mass Spectrom.*, 1981, 16, 118.
178 E.A. Bergner and R.C. Dougherty, *Biomed. Mass Spectrom.*, 1981, 8, 204.
179 L.R. Hogge and D.J.H. Olson, *J. Chromatogr. Sci.*, 1982, 20, 109.
180 V.P. Uralets, J.A. Rijks, and P.A. Leclercq, *J. Chromatogr.*, 1980, 194, 135.
181 S.P. Levine, T.M. Harvey, T.J. Waeghe, and R.H. Shapiro, *Anal. Chem.*, 1981, 53, 805.
182 H. Schweer, *J. Chromatogr.*, 1982, 236, 355, 361.
183 J. Greter, S. Lindstedt, and G. Steen in ref. 4, p. 1362.
184 L.R. Phillips and B.A. Fraser, *Biomed. Mass Spectrom.*, 1981, 8, 327.
185 (*a*) P. Beaune, B. Pileire, F. Rocchiccoli, M. Hardy, J.P. Leroux, and P. Cartier in ref. 40, p. 1; (*b*) B. Pileire, P. Beaune, P. Cartier, and M.H. Laudat, *J. Chromatogr.*, 1980, 182, 269.
186 J. Vine, *J. Chromatogr.*, 1980, 196, 415.
187 A.G. Rehfeldt, E. Schulte, and F. Spencer, *Phytochemistry*, 1980, 19, 1685.
188 R.K. Christopher and A.M. Duffield, *Biomed. Mass Spectrom.*, 1980, 7, 429.
189 M. Suzuki, T. Ariga, M. Sekine, E. Araki, and T. Miyatake, *Anal. Chem.*, 1981, 53, 985.
190 G.P. Hazlewood. A.J. Northrop. and R.M.C. Dawson, *Br. J. Nutr.*, 1981, 45, 159.
191 F. Warnaar, *Phytochemistry*, 1981, 20, 89.
192 H.-J. Stan and M. Scheutwinkel-Reich, *Lipids*, 1980, 15, 1044.
193 H. Wada, H. Okada, H. Suginaka, I. Tomiyasu, and I. Yano, *FEMS Microbiol. Lett.*, 1981, 11, 187.
194 E.N. Frankel, W.E. Neff, and E. Selke, *Lipids*, 1981, 16, 279.
195 M.C. Kennicutt and L.M. Jeffrey, *Mar. Chem.*, 1981, 10, 389.
196 M.C. Kennicutt and L.M. Jeffrey, *Mar. Chem.*, 1981, 10, 367.
197 R.K. Christopher, A.M. Duffield, B.J. Ralph, and J.J.H. Simes, *Aust. J. Biol. Sci.*, 1981, 34, 115.
198 R.K. Christopher, A.M. Duffield, and B.J. Ralph, *Aust. J. Biol. Sci.*, 1980, 33, 737,
199 D.R. Nelson, C.L. Fatland, R.W. Howard, C.A. McDaniel, and G.J. Blomquist, *Insect Biochem.*, 1980, 10, 409.
200 M. Vajdi, W.W. Nawar, and C. Merritt, jun., *J. Am. Oil Chem. Soc.*, 1981, 58, 106
201 'Prostaglandins, Prostacyclin, and Thromboxanes Measurement', ed. J.M. Boeynaems and A.G. Herman, Kluwer Academic, Dordrecht, 1980.
202 'SRA-A and the Leukotrienes', ed. P.J. Piper, Wiley, Chichester, 1981.
203 K. Yamashita and H. Miyazaki, *Kagaku no Ryoiki, Zokan*, 1981, 95.
204 E. Granstroem, *NATO Adv. Study Inst. Ser., Ser. A*, 1981, 36, 67.

205 G.A. Eiceman, V.A. Fuavao, K.D. Doolittle, and C.A. Herman, *J. Chromatogr.*, 1982, 236, 97.
206 K. Green, A. Aly, and C. Johansson, *Prostaglandins*, 1981, 21, (Suppl.), 1.
207 (*a*) J. Rosello, E. Gelpi, M. Rigaud, J. Durand, and J.C. Breton, *Biomed. Mass Spectrom.*, 1981, 8, 149; (*b*) H. Rabinovitch, J. Durand, M. Rigaud, F. Mendy, and J.C. Breton, *Lipids*, 1981, 16, 518.
208 W.C. Hubbard, A.J. Hough, A.R. Brash, J.T. Watson, and J.A. Oates, *Prostaglandins*, 1980, 20, 431.
209 W. Lubiz, O. Radmark, J.A. Lindgren, C. Malmsten, and B. Samuelsson, *Biochem. Biophys. Res. Commun.*, 1981, 99, 976.
210 G. Hansson, J.A. Lindgren, S. Dahlén, P. Hedqvist, and B. Samuelsson, *FEBS Lett.*, 1981, 130, 107.
211 S. Hammarström and B. Samuelsson, *FEBS Lett.*, 1980, 122, 83.
212 J. Rosello, E. Gelpi, M. Rigaud, and J.C. Breton, *J. High Resolut. Chromatogr. Chromatogr. Commun.*, 1981, 4, 437.
213 K. Uobe, R. Takeda, M. Wato, T. Nishikawa, S. Yamaguchi, T. Koshimura, Y. Kawaguchi, and M. Tsutsui, *J. Chromatogr.*, 1981, 214, 177.
214 M. Suzuki, I. Morita, M. Kawamura, S.-I. Murota, M. Nishizawa T. Miyatake, H. Nagase, K. Ohno, and H. Shimizu, *J. Chromatogr.*, 1980, 221, 361.
215 S.E. Barrow, K.A. Waddell, M. Ennis, C.T. Dollery, and I.A. Blair, *J. Chromatogr.*, 1982, 239, 71.
216 C.N. Hensby, M. Jogee, M.G. Elder, and L. Myatt, *Biomed. Mass Spectrom.*, 1981, 8, 111.
217 S. Fischer, B. Scherer, and P.C. Weber, *Biochim. Biophys. Acta*, 1982, 710, 493.
218 C. Chiabrando, A. Noseda, M.A. Noe, and R. Fanelli, *Prostaglandins*, 1980, 20, 747.
219 F.F. Sun and B.M. Taylor, *Prostaglandins*, 1981, 21, 307.
220 P. Falardeau, J.A. Oates, and A.R. Brash, *Anal. Biochem.*, 1981, 115, 359.
221 R.G. Megargle, L.E. Slivon, J.E. Grass, and A.H. Andrist, *Anal. Chim. Acta*, 1980, 120, 193.
222 K. Green, F.A. Kimball, B.A. Thornburgh, and A.J. Wickremasinha, *Prostaglandins*, 1980, 20, 767.
223 B.H. Min, J. Pao, W.A. Garland, J.A.F. De Silva, and M. Parsonnet, *J. Chromatogr.*, 1980, 183, 411.
224 M.J. Dembele-Duchesne, H. Thaler-Dao, C. Chavis, and A. Crastes de Paulet, *Prostaglandins*, 1981, 22, 979.
225 L.J. Roberts, B.J. Sweetman, and J.A. Oates, *J. Biol. Chem.*, 1981, 256, 8384.
226 R.F. Taylor and M. Ikawa, *Methods Enzymol.*, 1980, 67, 233.
227 E.K. Nemethy, J.W. Otvos, and M. Calvin, *Pure Appl. Chem.*, 1981, 53, 1101.
228 B.J. Bergot, G.C. Jamieson, M.A. Ratcliff, and D.A. Schooley, *Science*, 1980, 210, 336.
229 B.J. Bergot, M. Ratcliff, and D.A. Schooley, *J. Chromatogr.*, 1981, 204, 231.
230 H. Rembold, H. Hagenguth, and J. Rascher, *Anal. Biochem.*, 1980, 101, 356.
231 W.S. Bowers and R. Nishida, *Science*, 1980, 209, 1030.
232 N. Murofushi, N. Takahashi, M. Sugimoto, and K. Itoh, *Agric. Biol. Chem.*, 1980, 44, 1583.
233 T. Yokota and N. Takahashi, *Agric. Biol. Chem.*, 1981, 45, 125
234 P. Gaskin, P.S. Kirkwood, J.R. Lenton, J. MacMillan, and M.E. Radley, *Agric. Biol. Chem.*, 1980, 44, 1589.

302 H. Van Halbeek, L. Dorland, J. Haverkamp, G.A. Veldink, J.F.G. Vliegenthart, B. Fournet, G. Ricart, J. Montreuil, W.D. Gathmann, and D. Aminoff, *Eur. J. Biochem.*, 1981, 118, 487.
303 M. Tomoda, K. Ishikawa, and M. Yokoi, *Chem. Pharm. Bull.*, 1981, 29, 3256.
304 P. Aman, M. McNeil, L.-E. Franzen, A.G. Darvill, and P. Albersheim, *Carbohydr. Res.*, 1981, 95, 263.
305 C. Derappe, A. Lundblad, L. Messeter, and S. Svensson, *FEBS Lett.*, 1980, 119, 177.
306 F. Matsuura, R.A. Laine, and M.Z. Jones, *Arch. Biochem. Biophys.*, 1981, 211, 485.
307 D.E. Rhoads, K.B. Meyer, and E.S. Kaneshiro, *Biochem. Biophys. Res. Commun.*, 1981, 98, 858.
308 G.P. Smirnova and N.K. Kochetkov, *Biochem. Biophys. Acta*, 1980, 618, 486.
309 N.V. Prokazova, A.T. Mikhailov, S.L. Kocharov, L.A. Malchenko, N.D. Zvezdina, G. Buzinkov, and L.D. Bergelson, *Eur. J. Biochem.*, 1981, 115, 671.
310 O. Nilsson, J.E. Månsson, E. Tibblin, and L. Svennerholm, *FEBS Lett.*, 1981, 133, 197.
311 W.M.F. Lee, J.C. Klock, and B.A. Macher, *Biochemistry*, 1981, 20, 6505.
312 J.J. Myher, A.A. Kuksis, W.C. Breckenridge, and J.A. Little, *Can. J. Biochem.*, 1981, 59, 626.
313 D.J. Hanahan, C.A. Demopoulos, J. Liehr, and R.N. Pinckhard, *J. Biol. Chem.*, 1980, 255, 5514.
314 J.L. Chabard, C. Lartigue-Mattei, F. Vedrine, J. Petit, and J.A. Berger, *J. Chromatogr.*, 1980, 221, 9.
315 S. Floberg, P. Hartvig, B. Lindström, G. Lönnerholm, and B. Odlind, *J. Chromatogr.*, 1981, 225, 73.
316 W. Ihn, D. Krebs, K. Eckardt, and D. Tresselt, *Tetrahedron*, 1982, 38, 1781.
317 T.R. Kemp and R.A. Andersen, *J. Chromatogr.*, 1981, 209, 467.
318 J.A. McCloskey, B. Basile, K. Kimura, and T. Hashizume, *Proc. Jpn. Acad. B*, 1981, 57, 276.
319 I.M. Scott and R. Horgan, *Biomed. Mass Spectrom.*, 1980, 7, 446; I.M. Scott, G.C. Martin, R. Horgan, and J.K. Heald, *Planta*, 1982, 154, 273.
320 Z. Yamaizumi, Y. Kuchino, F. Harada, S. Nishimura, and J.A. McCloskey, *J. Biol. Chem.*, 1980, 255, 2220.
321 L.P. Johnson, J.K. Macleod, R.E. Summons, and N. Hunt, *Anal. Biochem.*, 1980, 106, 285.
322 L.P. Johnson, J.K. Macleod, C.W. Parker, D.S. Letham, and N.H. Hunt, *Planta*, 1981, 152, 195.
323 I. Jardine and M.M. Weidner, *J. Chromatogr.*, 1980, 182, 395.
324 H. Adlercreutz in ref. 4, p. 1165.
325 P. Vestergaard, *Lipids*, 1980, 15, 710.
326 D.W. Johnson, G. Phillipou, and R.F. Seamark, *J. Steroid Biochem.*, 1981, 14, 793.
327 W.H. Elliott, *Lipids*, 1980, 15, 764.
328 J.A. Ballantine, *Anal. Proc.*, 1980, 17, 326.
329 C. Djerassi, *Pure Appl. Chem.*, 1981, 53, 873.
330 L.J. Goad, *Pure Appl. Chem.*, 1981, 53, 837.
331 M.J. Thompson, G.W. Patterson, S.R. Dutky, J.A. Svoboda, and J.N. Kaplanis, *Lipids*, 1980, 15, 719.
332 M. Axelson, B.-L. Sahlberg, and J. Sjövall, *J. Chromatogr.*, 1981, 224, 355.
333 J.J. Vrbanac, W.E. Braselton, jun., J.F. Holland, and C.C. Sweeley, *J. Chromatogr.*, 1982, 239, 265.

334 M.A. Islam, R.F. Raicht, and B.I. Cohen, *Anal. Biochem.*, 1981, 112, 371.
335 N.B. Javitt, E. Kok, S. Burstein, B. Cohen, and J. Kutscher, *J. Biol. Chem.*, 1981, 256, 12644.
336 S. Takatsuto, B. Ying, M. Morisaki, and N. Ikekawa, *J. Chromatogr.*, 1982, 239, 233.
337 T. Kikuchi, S. Kadota, H. Suehara, and T. Namba, *Chem. Pharm. Bull.*, 1982, 30, 370.
338 W.C.M.C. Kokke, W. Fenical, and C. Djerassi, *Phytochemistry*, 1981, 20, 127.
339 M. Alam, T.B. Sansing, J.R. Guerra, and A.D. Harmon, *Steroids*, 1981, 38, 375.
340 C. Popp-Snijders and A.P.J.M. De Jong, *J. Steroid Biochem.*, 1981, 14, 1209.
341 M. Damon and C. Chavis, *J. Steroid Biochem.*, 1982, 16, 771.
342 L.N. Li and C. Djerassi, *Tetrahedron Lett.*, 1981, 22, 4639.
343 M. Iwai, H. Kanno, M. Hashino, J. Suzuki, T. Yanaihara, T. Nakayama, and H. Mori, *J. Chromatogr.*, 1981, 225, 275.
344 M. Tetsuo, H. Eriksson, and J. Sjövall, *J. Chromatogr.*, 1982, 239, 287.
345 M. Edery, J. Goussard, L. Dehennin, R. Scholler, J. Reiffsteck, and M.A. Drosdowsky, *Eur. J. Cancer*, 1981, 17, 115.
346 M. Axelson, J.H. Clark, H.A. Eriksson, and J. Sjövall, *J. Steroid Biochem.*, 1981, 14, 1253.
347 T. Cains, E.G. Siegmund, and B. Rader, *Anal. Chem.*, 1981, 53, 1217.
348 G. Emons, H.-O. Hoppen, P. Ball, and R. Knuppen, *Steroids*, 1980, 36, 73.
349 K. Shimada, T. Tanaka, and T. Nambara, *J. Chromatogr.*, 1981, 223, 33.
350 M. Akhtar, M.R. Calder, D.L. Corina, and J.N. Wright, *J. Chem. Soc., Chem. Commun.*, 1981, 129.
351 B.-L. Sahlberg, M. Axelson, D.J. Collins, and J. Sjövall, *J. Chromatogr.*, 1981, 217, 453.
352 D.W. Johnson, G. Phillipou, S.K. James, C.J. Seaborn, and M.M. Ralph, *Clin. Chim. Acta*, 1980, 106, 99.
353 C. Corpéchot, P. Robel, M. Axelson, J. Sjövall, and E.-E. Baulieu, *Proc. Natl. Acad. Sci., U.S.A.*, 1981, 78, 4704.
354 E.M.H. Finlay, M.S. Morton, and S.J. Gaskell, *Steroids*, 1982, 39, 63.
355 L. Dehennin. A. Reiffsteck, and R. Scholler, *Biomed. Mass Spectrom.*, 1980, 7, 493.
356 S.J. Gaskell and E.M.H. Finlay, *J. Labelled Comp. Radiopharm.*, 1980, 17, 861.
357 S. Baba, Y. Shinohara, and Y. Kasuya, *J. Clin. Endocrinol. Metab.*, 1980, 50, 889.
358 L. Dehennin, A. Reiffsteck, and R. Scholler, *Pathol. Biol.*, 1981, 29, 329.
359 R.-J. Begue, M. Moriniere, C. Nivois, D. Sandre, and P. Padieu, *J. Steroid Biochem.*, 1981, 14, 489.
360 G.E. Joannou, *J. Steroid Biochem.*, 1981, 14, 901.
361 T. Watabe, M. Kanal, M. Isobe, and N. Ozawa, *J. Biol. Chem.*, 1981, 256, 2900.
362 C.P. De Vries and A.P.J.M. De Jong, *J. Steroid Biochem.*, 1980, 13, 387.
363 V.R. Mattox, P.C. Carpenter, and E. Graf, *J. Steroid Biochem.*, 1981, 14, 19.
364 C.H.L. Shackleton, E. Roitman, C. Monder, and H.L. Bradlow, *Steroids*, 1980, 36, 289.

365 P. Bournot, N. Pitoizet, M. Zachmann, and B.F. Maums, *J. Steroid Biochem.*, 1982, 16, 467.
366 M. Galli-Kienle, M. Anastasia, G. Cighetti, G. Galli, and A. Fiecchi, *Eur. J. Biochem.*, 1980, 110, 93.
367 T.S. Douglass, W.E. Connor, and D.S. Lin, *J. Lipid Res.*, 1981, 22, 961.
368 S. Maruyama, Y. Fujimoto, M. Morisaki, and N. Ikekawa, *Tetrahedron Lett.*, 1982, 23, 1701.
369 V. Karahani, J. Bascoul, and A.C. De Paulet, *J. Chromatogr.*, 1981, 211, 392.
370 J. Freudenthal, H.J.G.M. Derks, L.G. Gramberg, G.J. Ten Hove, and R. Klaassen, *Biomed. Mass Spectrom.*, 1981, 8, 5.
371 R. Schaffer, L.T. Sniegoski, M.J. Welsh, E. White, A. Cohen, H.S. Hertz, J. Mandel, R.C. Paule, L. Svensson, I. Björkhem, and R. Blomstrand, *Clin. Chem.*, 1982, 28, 5.
372 A. Sanghvi, E. Grassi, C. Bartman, R. Lester, M.G. Kienle, and G. Galli, *J. Lipid Res.*, 1981, 22, 720.
373 C. Larroque, J. Rousseau, and J.E. Van Lier, *Biochemistry*, 1981, 20, 925.
374 Y.Y. Lin, C.-E. Low, and L.L. Smith, *J. Steroid Biochem.*, 1981, 14, 563.
375 R. Lafont, G. Somme-Martin, B. Mauchamp, B.F. Maume, and J.P. Delbecque, *Dev. Endocrinol.*, 1980, 7, 45.
376 R.E. Isaac, M.E. Rose, H.H. Rees, and T.W. Goodwin, *J. Chem. Soc., Chem. Commun.*, 1982, 249.
377 G. Tsouporas, C. Hetru, B. Luu, M. Lagueux, E. Constantin, and J.A. Hoffmann, *Tetrahedron Lett.*, 1982, 23, 2045.
378 S. Ikawa, *Shitsuryo Bunseki*, 1981, 29, 141.
379 R. Edenharder and J. Slemr, *J. Chromatogr.*, 1981, 222, 1.
380 B.R. DeMark and P.D. Klein, *J. Lipid Res.*, 1981, 22, 166.
381 T. Murata, S. Takahashi, S. Ohnishi, K. Hosoi, T. Nakashima, Y. Ban, and K. Kuriyama, *J. Chromatogr.*, 1982, 239, 571.
382 T. Hoshita, M. Yasuhara, M. Une, A. Kibe, E. Itoga, S. Kito, and T. Kuramoto, *J. Lipid, Res.*, 1980, 21, 1015.
383 H. Oftebro, I. Björkhem, F.C. Stormer, and J.I. Pedersen, *J. Lipid Res.*, 1981, 22, 632.
384 T. Beppu, Y. Seyama, T, Kasama, S. Serizawa, and T. Yamakawa, *Clin. Chim. Acta*, 1982, 118, 167.
385 J. Yanagisawa, M. Itoh, M. Ishibashi, H. Miyazaki, and F. Nakayama, *Anal. Biochem.*, 1980, 104, 75.
386 G. Karlaganis, B. Almé, V. Karlaganis, and J. Sjövall, *J. Steroid Biochem.*, 1981, 14, 341.
387 B. Almé and J. Sjövall, *J. Steroid Biochem.*, 1980, 13, 907.
388 G.A. De Weerdt, R. Beke, H. Verdievel, and F. Barbier, *Biomed. Mass Spectrom.*, 1980, 7, 515.
389 G.G. Parmentier, L.M.J. Smets, G.A. Jannsen, and H.J. Eyssen, *Eur. J. Biochem.*, 1981, 116, 365.
390 Z.V.I. Zaretskii in ref. 41, p. 227.
391 D.A. Seamark, D.J.H. Trafford, and H.L.J. Makin, *J. Steroid Biochem.*, 1981, 14, 111.
392 A.P. DeLeenheer and A.A. Cruyl, *Methods Enzymol.*, 1980, 67, 335.
393 I. Björkhem and I. Holmberg, *Methods Enzymol.*, 1980, 67, 385.
394 D.A. Seamark, D.J.H. Trafford, and H.L.J. Makin, *J. Steroid Biochem.*, 1980, 13, 1057.
395 D.A. Seamark, D.J.H. Trafford, and H.L.J. Makin, *Clin. Chim. Acta*, 1980, 106, 51.
396 H. Turnbull, D.J.H. Trafford, and H.L.J. Makin, *Clin. Chim. Acta*, 1982, 120, 65.

397 C. Clausnitzer, D. Behrendt, and A. Storch, *Med. Sport*, 1980, 20, 180.
398 E. Houghton and P. Teale, *Biomed. Mass Spectrom.*, 1981, 8, 358.
399 D.L. Howard and D.J. Chapman, *J. Chem. Soc., Chem. Commun.*, 1981, 468.
400 S.N. Hooper, R.F. Chandler, E. Lewis, and W.D. Jamieson, *Lipids*, 1981, 17, 60.
401 T. Itoh, T. Uetsuki, T. Tamura, and T. Matsumoto, *Lipids*, 1980, 15, 407.
402 (*a*) I. Matsumoto, *Tanpakushitsu Kakusan Koso*, 1981, 26, 1151; (*b*) Y. Hashimoto and H. Miyazaki, *ibid.*, p. 1159; (*c*) A. Hayashi, *ibid.*, p. 1172; (*d*) M. Nakagawara, A. Wanatabe, and T. Kariya, *ibid.*, p. 1259.
403 J.J. Brophy and G.W.K. Cavill, *Heterocycles*, 1980, 14, 477.
404 F. Merli, M. Novotny, and M.L. Lee, *J. Chromatogr.*, 1980, 199, 371.
405 M.V. Buchanan, *Anal. Chem.*, 1982, 54, 570.
406 A.D. Kinghorn, M.A. Selim, and S.J. Smolenski, *Phytochemistry*, 1980, 19, 1705.
407 I. Kari, P. Peura, and M.M. Airaksinen, *Biomed. Mass Spectrom.*, 1980, 7, 549.
408 P. Peura, I. Kari, and M.M. Airaksinen, *Biomed. Mass Spectrom.*, 1980, 7, 553.
409 J.R.F. Allen, O. Beck, S. Borg, and R. Skroeder, *Eur. J. Mass Spectrom. Biochem., Med. Environ. Res.*, 1980, 1, 171.
410 I. Kari, *FEBS Lett.*, 1981, 127, 277.
411 S.A. Barker, R.E.W. Harrison, J.A. Monti, G.B. Brown, and S.T. Christian, *Biochem. Pharmacol.*, 1981, 30, 9.
412 S.A. Barker, J.A. Monti, L.C. Tolbert, G.B. Brown, and S.T. Christian, *Biochem. Pharmacol.*, 1981, 30, 2461.
413 F. Mikes, G. Boshart, and P.G. Waser in ref. 40, p. 21.
414 J.K. Khandelwal, P.I.A. Szilagyi, L.A. Barker, and J.P. Green, *Eur. J. Pharmacol.*, 1981, 76, 145.
415 Y. Hasegawa, M. Kunihara, and Y. Maruyama, *J. Chromatogr.*, 1982, 239, 335.
416 M. Suzuki, C. Nitsch, W. Wunn, B. Schmude, and P. Haug, *Biomed. Mass Spectrom.*, 1980, 7, 537.
417 R. Kobelt, G. Wiesener, and R. Kuehn, *J. High Resolut. Chromatogr. Chromatogr. Commun.*, 1981, 4, 520.
418 F. Artigas, E. Martinez, and E. Gelpi, *J. Chromatogr. Sci.*, 1982, 20, 75.
419 O. Beck and G. Jonsson, *J. Neurochem.*, 1981, 36, 2013.
420 N. Narasimhachari, E. Kempster, and M. Anbar, *Biomed. Mass Spectrom.*, 1980, 7, 231.
421 S.P. Markey, R.W. Colburn, and J.N. Johannessen, *Biomed. Mass Spectrom.*, 1981, 8, 301.
422 M. Mumtaz, N. Narasimhachari, R.O. Friedel, G.N. Pandey, and J.M. Davis, *Res. Commun., Chem. Pathol. Pharmacol.*, 1982, 36, 45.
423 J.J. Warsh, A. Chiu, P.P. Li, and D.D. Godse, *J. Chromatogr.*, 1980, 183, 483.
424 M. Tetsuo, S.P. Markey, R.W. Colburn, and I.J. Kopin, *Anal. Biochem.*, 1981, 110, 208.
425 H.-Ch. Curtius, H. Farner, and F. Rey, *J. Chromatogr.*, 1980, 199, 171.
426 M.R. Holdiness, M.T. Rosen, J.B. Justice, and D.B. Neill, *J. Chromatogr.*, 1980, 198, 329.
427 P.H. Duffield, D.F.H. Dougan, D.N. Wade, and A.M. Duffield, *Biomed. Mass Spectrom.*, 1981, 8, 170.

428 J.-I. Yoshida, K. Soshino, T. Matsunaga, S. Higa, T. Suzuki, A. Hayashi, and Y. Yumamura, *Biomed. Mass Spectrom.*, 1980, 7, 396.

429 J.J. Warsh, D.D. Godse, S.W. Cheung, and P.P. Li, *J. Neurochem.*, 1981, 36, 893.

430 D.C. Jimerson, S.P. Markey, J.A. Oliver, and I.J. Kopin, *Biomed. Mass Spectrom.*, 1981, 8, 256.

431 F.A.J. Muskiet, G.T. Nagel, and B.G. Wolthers, *Anal. Biochem.*, 1980, 109, 130.

432 J. Girault, M.A. Lefebvre, J.B. Fourtillan, P. Courtois, and J. Gombert, *Ann. Pharm. Fr.*, 1980, 38, 439.

433 W. Vogt, K. Jacob, A.-B. Ohnesorge, and G. Schwertfeger, *J. Chromatogr.*, 1980, 199, 191.

434 T. Matsumoto, H. Uchimura, M. Hirano, J.S. Kim, T. Nakahara, and K. Tanaka, *J. Neurochem.*, 1982, 38, 285.

435 B.A. Davis and A.A. Boulton, *J. Chromatogr.*, 1981, 222, 271.

436 J.R. Crowley, M.W. Couch, C.M. Williams, R.M. Threatte, and M.J. Fregly, *Clin. Chim. Acta*, 1981, 109, 125.

437 D.A. Durden and A.A. Boulton, *J. Neurochem.*, 1981, 36, 129.

438 H. Mita, H. Yasueda, and T. Shida, *J. Chromatogr.*, 1980, 221, 1.

439 J.J. Keyzer, B.G. Wolthers, F.A.J. Muskiet, H.F. Kauffman, and A. Groen, *Clin. Chim. Acta*, 1981, 113, 165.

440 L.B. Hough, J.K. Khandelwal, A.M. Morrishow, and J.P. Green, *J. Pharmacol. Methods*, 1981, 5, 143.

441 C.G. Swahn and G. Sedvall, *J. Neurochem.*, 1981, 37, 461.

442 Y. Okano, T. Miyata, K. Iwasaki, K. Takahama, T. Hitoshi, Y, Kasé, I. Matsumoto, and T. Shinka, *Anal. Biochem.*, 1981, 115, 254.

443 T. Miyata, Y. Okano, K. Iwasaki, K. Takahama, T. Hitoshi, and Y. Kase, *Eur. J. Pharmacol.*, 1982, 78, 457.

444 T. Ohki, A. Saito, N. Yamanaka, K. Ohta, J. Sakakibara, T. Niwa, and K. Maeda, *J. Chromatogr.*, 1982, 228, 51.

445 F.A.J. Muskiet, C.M. Stratingh. D.C. Fremouw-Ottevangers, and M.R. Halie, *J. Chromatogr.*, 1982, 230, 142.

446 R.E. Hurst, R.L. Settine, F. Fish, and E.C. Roberts, *Anal. Chem.*, 1981, 53, 2175.

447 S.L. MacKenzie, *Methods Biochem. Analysis*, 1981, 27, 1.

448 L. Vollner, *GSF-Ber. O*, 1980, 599, 44.

449 M. Sakamoto, N. Tsuji, F. Nakyama, and K.-I. Kajiyama, *J. Chromatogr.*, 1982, 235, 75.

450 R. Liardon, S. Lederman, and U. Ott, *J. Chromatogr.*, 1981, 203, 385.

451 S. Kusumoto, M. Matsukura, and T. Shiba, *Chem. Lett.*, 1981, 1017.

452 G. Sandberg, B. Andersson, and A. Dunberg, *J. Chromatogr.*, 1981, 205, 125.

453 E. Epstein and J.D. Cohen, *J. Chromatogr.*, 1981, 209, 413.

454 J. McDougall and J.R. Hillman, *Z. Pflanzenphysiol.*, 1980, 98, 89.

455 K.C. Engvild and H. Egsgaard, *Physiol. Plant.*, 1981, 53, 79.

456 B. Andersson and G. Sandberg, *J. Chromatogr.*, 1982, 238, 151.

457 C.S. Irving and P.D. Klein, *Anal. Biochem.*, 1980, 107, 251.

458 D.E. Mathews, J.B. Starren, A.J. Drexler, D.M. Kipnis, and D.M. Bier, *Anal. Biochem.*, 1981, 110, 308.

459 K. Murayama, N. Shindo, R. Mineki, and K. Ohta, *Biomed. Mass Spectrom.*, 1981, 8, 165.

460 M. Wolfensberger, J.C. Reubi, V. Canzek, U. Redweik, H.C. Curtius, and M. Cuenod, *Brain Res.*, 1981, 224, 327.

461 M. Yudkoff, I. Nissim, and S. Segal, *Clin. Chim. Acta*, 1982, 118, 159.

462 J. Amir, S.H. Reisner, and A. Lapidot, *Pediatr. Res.*, 1980, 14, 1238.
463 D. Rhodes, A.C. Myers, and G. Jamieson, *Plant Physiol.*, 1981, 68, 1197.
464 M.W. Haymond, C.P. Howard, J.M. Miles, and J.E. Gerich, *J. Chromatogr.*, 1980, 183, 403.
465 R.H. White, *Anal. Biochem.*, 1981, 114, 349.
466 G. Bengtsson, G. Odham, and G. Westerdahl, *Anal. Biochem.*, 1981. 111, 163.
467 P.J. Finlayson, R.K. Christopher, and A.M. Duffield, *Biomed. Mass Spectrom.*, 1980, 7, 450.
468 R.J.W. Truscott, D. Malegan, E. McCairns, B. Halpern, J. Hammond, R.G.H. Cotton, J.F.B. Mercer, S. Hunt, J.G. Rogers, and D.M. Danks, *Biomed. Mass Spectrom.*, 1981, 8, 99.
469 M.R. Christy, R.M. Barkley, T.H. Koch, J.J. Van Buskirk, and W.M. Kirsch, *J. Am. Chem. Soc.*, 1981, 103, 3935.
470 L.R. Croft, 'Introduction to Protein Sequence Analysis', Wiley, Chichester, 1980.
471 K. Biemann, *Dev. Biochem.*, 1981, 131.
472 K. Biemann in ref. 48, p.469.
473 S.-N. Lin, L.A. Smith, and R.M. Caprioli, *J. Chromatogr.*, 1980, 197, 31.
474 H.C. Krutzsch, *Dev. Biochem.*, 1981, 149.
475 H.C. Krutzsch, *Biochemistry*, 1980, 19, 5290.
476 C.F. Beckner and R.M. Caprioli, *Biochem. Biophys. Res. Commun.*, 1980, 93, 1290.
477 E. Peralta, H.Y.T. Yang, J. Hong, and E. Costa, *J. Chromatogr.*, 1980, 190, 43.
478 W.C. Herlihy and K. Biemann, *Biomed. Mass Spectrom.*, 1981, 8, 70.
479 S.D. Putney, N.J. Royal, H. Neuman de Vegvar, W.C. Herlihy, K. Biemann, and P. Schimmel, *Science*, 1981, 213, 1497.
480 K. Rose, J.D. Priddle, R.E. Offord, and M.P. Esnouf, *Biochem. J.*, 1980, 187, 239.
481 S.A. Carr, P.V. Hauschka, and K. Biemann, *J. Biol. Chim.*, 1981, 256, 9944.
482 K. Rose, J.D. Priddle, and R.E. Offord, *J. Chromatogr.*, 1981, 210, 301.
483 R.A. Day, K. Jayasimhulu, J.V. Evans, G. Bhat, L.C. Lin, and M.J. Wieser, *J. Heterocycl. Chem.*, 1980, 17, 1651.
484 K.L. Rinehart, jun., J.B. Gloer, J.C. Cook, jun., S.A. Mizsak, and T.A. Scahill, *J. Am. Chem. Soc.*, 1981, 103, 1857.
485 K.L. Rinehart, jun., L.A. Gaudioso, M.L. Moore, R.C. Pandey, J.C. Cook, jun., M. Barber, R.D. Sedgwick, R.S. Bordoli, A.N. Tyler, and B.N. Green, *J. Am. Chem. Soc.*, 1981, 103, 6517.
486 W.A. König, H. Krohn, M. Greiner, H. Brückner, and G. Jung in ref. 4, p. 1109; M. Aydin, D.H. Bloss, W.A. König, H. Brückner, and G. Jung, *Biomed. Mass Spectrom.*, 1982, 9, 39.
487 A.C. Ghosh and M. Rampopal, *J. Heterocycl. Chem.*, 1980, 17, 1809.
488 G.R. Pettit, Y. Kamano, P. Brown, D. Gust, M. Inoue, and C.L. Herald, *J. Am. Chem. Soc.*, 1982, 104, 905.
489 Y. Tamaki, *JEOL News (Ser.) Anal. Instrum.*, 1982, 18A, 43.
490 H. Schildknecht, *Angew. Chem., Int. Ed. Engl.*, 1981, 20, 164.
491 O. Vostrowsky and K. Michaelis, *Z. Naturforsch., Teil B*, 1981, 36, 402.
492 R. Baker, J.W.S. Bradshaw, and W. Speed, *Experientia*, 1982, 38, 233.
493 J.H. Cane and T. Jonsson, *J. Chem. Ecol.*, 1982, 8, 15.

494 J.A. Pickett, I.H. Williams, and A.P. Martin, *J. Chem. Ecol.*, 1982, 8, 163.
495 R.M. Crewe and H.H.W. Velthius, *Naturwissenschaften*, 1980, 67, 467.
496 J.W. Wheeler, F.O. Ayorinde, A. Greene, and R.M. Duffield, *Tetrahedron Lett.*, 1980, 23, 2071.
497 H. Klein, W. Francke, and W.A. König, *Z. Naturforsch., Teil B*, 1981, 36, 757.
498 S.H. Goh, S.L. Tong, and Y.P. Tho, *Mikrochim. Acta*, 1982, 1, 219.
499 G.D. Prestwick, S.H. Goh, and Y.P. Tho, *Experientia*, 1981, 37, 11.
500 G.D. Prestwick and M.S. Collins, *Tetrahedron Lett.*, 1981, 22, 4587.
501 R.K. Vander Meer, F.D. Williams, and C.S. Lofaren, *Tetrahedron Lett.*, 1981, 22, 1651.
502 H.J. Williams, M.R. Strand, and S.B. Vinson, *Experientia*, 1981, 37, 1159.
503 A.B. Attygalle and E.D. Morgan, *J. Chem. Soc., Perkin Trans. 1*, 1982, 949.
504 T.H. Jones, M.S. Blum, R.W. Howard, C.A. McDaniel, H.M. Fales, M.B. BuBois, and J. Torres, *J. Chem. Ecol.*, 1982, 8, 285.
505 B.R. Laurence and J.A. Pickett, *J. Chem. Soc., Chem. Commun.*, 1982, 59.
506 M. Jacobson and K. Ohinata, *Experientia*, 1980, 36, 629.
507 R. Baker, R.H. Herbert, and A.H. Parton, *J. Chem. Soc., Chem. Commun.*, 1982, 601.
508 B.A. Bierl-Leonhardt, D.S. Moreno, M. Scharz, J. Fargerlund, and J.R. Plimmer, *Tetrahedron Lett.*, 1981, 22, 389.
509 L.M. McDonough, M.P. Hoffmann, B.A. Bierl-Leonhardt, C.L. Smithhisler, J.B. Bailey, and H.G. Davis, *J. Chem. Ecol.*, 1982, 8, 255.
510 G. Kunesch, P. Zagatti, J.Y. Lallemand, A. Debal, and J.P. Vigneron, *Tetrahedron Lett.*, 1981, 22, 5271.
511 J. Myerson, W.F. Haddon, and E.L. Soderstrom, *Tetrahedron Lett.*, 1982, 23, 2757.
512 A. Guerrero, F. Camps, J. Coll, M. Riba, J. Einhorn, Ch. Descoins, and Y.L. Lallemand, *Tetrahedron Lett.*, 1981, 22, 2013.
513 T.H. Jones and M.S. Blum, *Tetrahedron Lett.*, 1981, 22, 4373.
514 R. Claus, H.O. Hoppen, and H. Karg, *Experientia*, 1981, 37, 1178.
515 R.L. Foltz, A.F. Fentiman, jun., and R.B. Foltz, 'NIDA Research Monograph, No. 32: GC/MS Assays for Abused Drugs in Body Fluids', GPO, Washington, 1980.
516 S.I. Goodman and S.P. Markey, 'Laboratory and Research Methods in Biology and Medicine, Vol. 6: Diagnosis of Organic Acidemias by Gas Chromatography-Mass Spectrometry', Alan R. Liss, New York, 1981.
517 R.A. Chalmers and A.M. Lawson, 'Organic Acids in Man', Chapman and Hall, London, 1982.
518 'Analysis of Drugs and Metabolites By Gas Chromatography-Mass Spectrometry', ed. B.J. Gudzinowicz and M.J. Gudzinowicz, Marcel Dekker, New York, 1980: (*a*) Vol. 6: 'Cardiovascular, Antihypertensive, Hypoglycemic, and Thyroid-Related Agents'; (*b*) Vol. 7: 'Natural, Pyrolytic and Metabolic Products of Tobacco and Marijuana'.
519 T. Mizuno, N. Abe, H. Teshima, E. Yamauchi, Y. Itagaki, I. Matsumoto, T. Kuhara, and T. Shinka, *Biomed. Mass Spectrom.*, 1981, 8, 593.

520 A. Zlatkis, R.S. Brazell, and C.F. Poole, *Clin. Chem.*, 1981, 27, 789.
521 E. Jellum, *Trends Anal. Chem.*, 1981, 1, 12.
522 S.I. Goodman, *Am. J. Hum. Genet.*, 1980, 32, 781.
523 V.N. Reinhold and C.E. Costello in ref. 8, p. 1.
524 W.A. Garland and M.L. Powell, *J. Chromatogr. Sci.*, 1981, 19, 392.
525 E.J. Cone, *Drugs Pharm. Sci.*, 1981, 11, 143.
526 T.A. Baillie, *Pharmacol. Rev.*, 1981, 33, 81.
527 C.W. Moss, *J. Chromatogr.*, 1981, 203, 337.
528 D.N. Pillai and S. Dilli, *J. Chromatogr.*, 1981, 220, 253.
529 J. Vink, H.J.M. van Hal, and P.C.J.M. Koppens in ref. 4, p. 1251.
530 I. Bjoerkhem, A. Bergman, O. Falk, A. Kallner, O. Lantto, L. Svensson, E. Aekerloef, and R. Blomstrand, *Clin. Chem.*, 1981, 27, 733.
531 J.G. Kostelc, G. Preti, P.R. Zelson, J. Tonzetich, and G.R. Huggins, *J. Chromatogr.*, 1981, 226, 315.
532 B.J. Miwa, W.A. Garland, and P. Blumenthal, *Anal. Chem.*, 1981, 53, 793.
533 P.H. Kiang, *J. Parenter. Sci. Technol.*, 1981, 35, 152.
534 C.C. Sweeley, J. Vrbanac, D. Pinkston, and D. Issachar, *Biomed. Mass Spectrom.*, 1981, 8, 436.
535 E. Jellum, *J. Chromatogr.*, 1982, 239, 29.
536 I. Matsumoto, T. Kuhara, T. Shinka, T. Mizuno, H. Teshima, and N. Abe, *CODATA Bull.*, 1981, 41, 29.
537 A. Grupe and G. Spiteller, *J. Chromatogr.*, 1981, 226, 301.
538 H.M. Liebich, A. Pickert, U. Stierle, and J. Wöll, *J. Chromatogr.*, 1980, 199, 181.
539 G. Rhodes, M.L. Holland, D. Wiesler, M. Novotny, S.A. Moore, R.G. Peterson, and D.L. Felten, *J. Chromatogr.*, 1982, 228, 33.
540 T. Niwa, K. Maeda, T. Ohki, A. Saito, and I. Tsuchida, *J. Chromatogr.*, 1981, 225, 1.
541 R.J.W. Truscott, D. Malegan, D. Burke, L. Hick, P. Sims, B. Halpern, K. Tanaka, L. Sweetman, W.L. Nyhan, J. Hammond, C. Bumack, E.A. Haan, and D.H. Danks, *Clin. Chim. Acta*, 1981, 110, 187.
542 W. Lehnert, *Clin. Chim. Acta*, 1981, 113, 101.
543 W. Lehnert, *Clin. Chim. Acta*, 1981, 116, 249.
544 T. Kuhara and I. Matsumoto, *Biomed. Mass Spectrom.*, 1980, 7, 424.
545 M. Duran, L. Bruinvis, D. Ketting, J.P. Kamerling, S.K. Wadman, and R.B.H. Schutgens, *Biomed. Mass Spectrom.*, 1982, 9, 1.
546 F.K. Trefz, H. Schmidt, B. Tauscher, E. Depene, R. Baumgartner, G. Hammersen, and W. Kochen, *Eur. J. Pediatr.*, 1981, 137, 261.
547 E. Christensen, B.B. Jacobsen, N. Gregersen, H. Hjeds, J.B. Pedersen, N.J. Brandt, and U.B. Baekmark, *Clin. Chim. Acta*, 1981, 116, 331.
548 F. Rocchiccioli, J.P. Leroux, and P. Cartier, *Biomed. Mass Spectrom.*, 1981, 8, 160.
549 C.H.L. Shackleton, J.W. Honour, M.J. Dillon, C. Chander, and R.W.A. Jones, *J. Clin. Endocrinol. Metab.*, 1980, 50, 786.
550 K.R. van der Ploeg, B.G. Wolthers, G.T. Nagel, M. Volmer, and N.M. Drayer, *Clin. Chim. Acta*, 1982, 120, 341.
551 C. Jacobs, M. Bojasch, M. Duran, D. Ketting, S.K. Wadman, and D. Leupold, *Clin. Chim. Acta*, 1980, 106, 85.
552 D. Pinkston, G. Spiteller, H. von Henning, and D. Matthaei, *J. Chromatogr.*, 1981, 223, 1.
553 T. Niwa, K. Maeda, T. Ohki, A. Saito, and K. Kobayashi, *Clin. Chim. Acta*, 1981, 110, 51.

554 T. Niwa, K. Maeda, H. Asada, M. Shibata, T. Ohki, A. Saito, and H. Furukawa, *J. Chromatogr.*, 1982, 230, 1.

555 K.J. Ng, B.D. Andresen, J.R. Bianchine, J.D. Iams, R.W. O'Shaugnessy, L.E. Stempel, and F.P. Zuspan, *J. Chromatogr.*, 1982, 228, 43.

556 I. Nissim, M. Yudkoff, W. Yang, T. Terwilliger, and S. Segal, *Clin. Chim. Acta*, 1981, 109, 295.

557 K.Y. Tserng and S.C. Kalhan, *Anal. Chem.*, 1982, 54, 489.

558 L. Larsson, P.A. Mardh, G. Odham, and G. Westerdahl, *J. Chromatogr.*, 1980, 182, 402; *Acta Pathol. Microbiol. Scand., Sect. B*, 1981, 89B, 245.

559 S.-I Haraguchi, M. Terasawa, H. Toshima, I. Matsumoto, T. Kuhara, and T. Shinka, *J. Chromatogr.*, 1982, 230, 7.

560 K. Maeda, S, Kawaguchi, T. Niwa, T. Ohki, and K. Kobayashi, *J. Chromatogr.*, 1980, 221, 199.

561 E.M. Goldberg, L.M. Blendis, and S. Sandler, *J. Chromatogr.*, 1981, 226, 291.

562 H.M. Liebich, H.J. Buelow, and R. Kallmayer, *J. Chromatogr.*, 1982, 239, 343.

563 G. Volden, A.K. Thorsrud, I. Bjoernson, and E. Jellum, *J. Invest. Dermatol.*, 1980, 75, 421.

564 B. Marescau, A. Lowenthal, E. Esmans, Y. Luyten, F. Alderweireldt, and H.G. Terheggen, *J. Chromatogr.*, 1981, 224, 185.

565 G. Cighetti, E. Santaniello, and G. Galli, *Anal. Biochem.*, 1981, 110, 153.

566 K. Miyashita and A.B. Robinson, *Mech. Ageing Dev.*, 1980, 13, 177.

567 W. Jennings and T. Shibamoto, 'Qualitative Analysis of Flavour and Fragrance Volatiles by Glass Capillary Gas Chromatography', Academic Press, New York, 1980.

568 'Food Flavours Part A - Introduction', ed. I.D. Morton and A.J. MacLeod, Elsevier, Amsterdam, 1981.

569 J. Gilbert and R. Self, *Chem. Soc. Rev.*, 1981, 10, 255.

570 K.G. Sloman, A.K. Foltz, and J.A. Yeransian, *Anal. Chem.*, 1981, 53, 242R.

571 I. Horman, *Biomed. Mass Spectrom.*, 1981, 8, 384.

572 J.R. Giacin, *Package Eng.*, 1980, 25, 70.

573 G.W. Bowes, *Biomed. Mass Spectrom.*, 1981, 8, 419.

574 W.C. Brumley and J.A. Sphon, *Biomed. Mass Spectrom.*, 1981, 8, 390.

575 V.P. Uralets and R.V. Golovnja, *Nahrung*, 1980, 24, 155.

576 Y.P.C. Hsieh, A.M. Pearson, C.C. Sweeley, and F.E. Martin, *J. Food Sci.*, 1980, 45, 1078.

577 T. Katz, T.P. Pitner, R.D. Kinser, R.N. Ferguson, and W.N. Einolf, *Tetrahedron Lett.*, 1981, 22, 4771.

578 Y. Saint-Jalm and P. Moree-Testa, *J. Chromatogr.*, 1980, 198, 188.

579 P. Moree-Testa and Y. Saint-Jalm, *J. Chromatogr.*, 1981, 217, 197.

580 M. Novotny, F. Merli, D. Wiesler, M. Fencl, and T. Saeed, *J. Chromatogr.*, 1982, 238, 141.

581 F. Merli, D. Wiesler, M.P. Maskarinec, M. Novotny, D.L. Vassilaros, and M.L. Lee, *Anal. Chem.*, 1981, 53, 1929.

582 M.G. Heydanek and R.J. McGorrin, *J. Agric. Food Chem.*, 1981, 29, 950.

583 M.G. Heydanek and R.J. McGorrin, *J. Agric. Food Chem.*, 1981, 29, 1093.

584 A.E. Purcell, D.W. Later, and M.L. Lee, *J. Agric. Food Chem.*, 1980, 28, 939.

585 E.C. Coleman, C.-T. Ho, and S.S. Chang, *J. Agric. Food Chem.*, 1981, 29, 42.
586 A.J. MacLeod and N.M. Pieris, *J. Agric. Food Chem.*, 1981, 29, 49.
587 R.G. Butterey and J.A. Kamm, *J. Agric. Food Chem.*, 1980, 28, 978.
588 R.G. Butterey, R.M. Seifert, W.F. Haddon, and R.E. Lundin, *J. Agric. Food Chem.*, 1980, 28, 1336.
589 D.J. Harvey, *J. Chromatogr.*, 1981, 212, 75.
590 M. Albrand, P. Dubois, P. Etievant, R. Gelin, and B. Tokarska, *J. Agric. Food Chem.*, 1980, 28, 1037.
591 T.L. Peppard and S.A. Halsey, *J. Chromatogr.*, 1980, 202, 271.
592 A.A. Williams and H.V. May, *J. Inst. Brew.*, 1981, 87, 372.
593 P. Schreier, *J. Agric. Food Chem.*, 1980, 28, 926.
594 S.P. Avakyants, E.G. Rastyannikov, B.S. Chernyaga, and V.I. Navrotskii, *Vinodel. Vinograd. SSSR*, 1981, 50.
595 R. Tressl, K.G. Gruenewald, and R. Silwar, *Chem., Mikrobiol., Technol. Lebensm.*, 1981, 7, 28.
596 T. Yamanishi, M. Kosuge, Y. Tokitomo, and R. Maeda, *Agric. Biol. Chem.*, 1980, 44, 2139.
597 D.G. Rusness and G.L. Lamoureaux, *J. Agric. Food Chem.*, 1980, 28, 1070.
598 H.-J. Stan and B. Abraham, *J. Chromatogr.*, 1980, 195, 231.
599 A. Cavallaro, G. Bartolozzi, D. Carreri, G. Bandi, L. Luciani, G. Villa, A. Gorni, and G. Invernizzi, *Chemosphere*, 1980, 9, 623.
600 H. Sekita, K. Sasaki, Y. Kawamura, M. Takeda, Y. Saito, and M. Uchiyama, *Eisei Shikensho Hokoku*, 1981, 89.
601 K. Adachi, *Bull. Environ. Contam. Toxicol.*, 1981, 26, 737,
602 T.L. Barry, G. Petzinger, and F.M. Gretch, *Bull. Environ. Contam. Toxicol.*, 1981, 27, 524.
603 J.D. Henion, J.S. Nosanchuk, and B.M. Bilder, *J. Chromatogr.*, 1981, 213, 475.
604 G.A. Eiceman and F.W. Karasek, *J. Chromatogr.*, 1981, 210, 93.
605 J. Gilbert, M.J. Shepherd, J.R. Startin, and M.A. Wallwork, *J. Chromatogr.*, 1982, 237, 249.
606 J.H. Hotchkis, *J. Assoc. Off. Anal. Chem.*, 1981, 64, 1037.
607 T. Fazio, D.C. Havery, and J.W. Howard, *IARC Sci. Publ.*, 1980, 31, 419.
608 D. Andrzejewski, D.C. Havery, and T. Fazio, *J. Assoc. Off. Anal. Chem.*, 1981, 64, 1457.
609 Y. Fang, J.H. Ding, and S.L. Liu, *Anal. Lett.*, 1981, 14, 1165.
610 G.-H. Wang, W.-X. Zhang, Y.-W. Fang, and Z.-L Bian, *Hua Hsueh Hsueh Pao*, 1980, 38, 231.
611 G.-H. Wang, W.-X. Zhang, and W.-G. Chai in ref. 4, p. 1369.
612 J.P. Chaytor and M.J. Saxby, *J. Chromatogr.*, 1981, 214, 135.
613 S. Nishimura and Z. Yamaizumi, *Shitsuryo Bunseki*, 1981, 29, 151.
614 Z. Yamaizumi, T. Shiomi, H, Kasai, K. Wakabayashi, M. Nagao, T. Sugimura, and S. Nishimura, *Koenshu-Iyo Masu Kenkyukai*, 1980, 5, 245.
615 G. Modi, M. Piccinni, P. Fiorentino, and G. Simiani, *Boll. Chim. Unione Ital. Lab. Prov., Parte Sci.*, 1981, 32, 19.
616 J.C. MacDonald and A. Jonsson, *Acta Chem. Scand., Ser. B*, 1981, 35, 485.
617 A.D. Sauter, L.D. Betowski, T.R. Smith, V.A. Strickler, R.G. Beimer, B.N. Colby, and J.E. Wilkinson, *J. High Resolut. Chromatogr. Chromatogr. Commun.*, 1981, 4, 366.
618 K.H. Shafer, M. Cooke, F. DeRoos, R.J. Jakobson, O. Rosario, and J.D. Mulik, *Appl. Spectrosc.*, 1981, 35, 469.

619 K.J. Krost, E.D. Pellizzari, S.G. Walburn, and S.A. Hubbard, *Anal. Chem.*, 1982, 54, 810.
620 G. Hunt and N. Pangaro, *Anal. Chem.*, 1982, 54, 369.
621 R. Otson and D.T. Williams, *J. Chromatogr.*, 1981, 212, 187.
622 P.W. Albro and C.E. Parker, *J. Chromatogr.*, 1980, 197, 155.
623 W.C. Brumley, J.A.G. Roach, J.A. Sphon, P.A. Dreifuss, D. Andrzejewski, R.A. Niemann, and D. Firestone, *J. Agric. Food Chem.*, 1981, 29, 1040.
624 R.M.M. Kooke, J.W.A. Lustenhouwer, K. Olie, and O. Hutzinger, *Anal. Chem.*, 1981, 53, 461.
625 L.S. Ramos and P.G. Prohaska, *J. Chromatogr.*, 1981, 211, 284.
626 G.R. Sirota, J.F. Uthe, C.J. Musial, and V. Zitko, *J. Chromatogr.*, 1980, 202, 294.
627 A.D. Jorgensen and J.R. Stetter, *Anal. Chem.*, 1982, 54, 381.
628 K.F. Sullivan, E.L. Atlas, and C.-G. Giam, *Anal. Chem.*, 1981, 53, 1718.
629 W.D. Bowers, M.L. Parsons, R.E. Clement, G.A. Eiceman, and F.W. Karasek, *J. Chromatogr.*, 1981, 206, 279.
630 R. Massot, P. Foster, M. Laffond, and C. Nicotra, *Actual Chim.*, 1980, 15.
631 T. Nielsen, H. Egsgaard, E. Larsen, and G. Schroll, *Anal. Chim. Acta*, 1981, 124, 1.
632 J.L. Bove and P. Dalven, *Int. J. Environ. Anal. Chem.*, 1981, 10, 189.
633 Y. Yokouchi, T. Fujii, Y. Ambe, and K. Fuwa, *J. Chromatogr.*, 1981, 209, 293.
634 F. Bruner, G. Crescentini, F. Mangani, E. Brancaleoni, A. Cappiello, and P. Ciccioli, *Anal. Chem.*, 1981, 53, 798.
635 D.R. Choudhury and B. Bush, *Anal. Chem.*, 1981, 53. 1351.
636 L. Van Vaeck, G. Broddin, and K. Van Cauwenberghe, *Biomed. Mass Spectrom.*, 1980, 7, 473.
637 M.-L. Yu and R.A. Hites, *Anal. Chem.*, 1981, 53, 951.
638 J.A. Yergey, T.H. Risby, and S.S. Lestz, *Anal. Chem.*, 1982, 54, 354.
639 T. Romanowski, W. Funcke, J. Koenig, and E. Balfanz, *J. High Resolut. Chromatogr. Chromatogr. Commun.*, 1981, 4, 209.
640 D.L. Newton, M.D. Erickson, K.B. Tomer, E.D. Pellizzari, P. Gentry, and R.B. Zweidinger, *Environ. Sci. Technol.*, 1982, 16, 206.
641 D. Schuetzle, T.L. Riley, T.J. Prater, T.M. Harvey, and D.F. Hunt, *Anal. Chem.*, 1982, 54, 265.
642 T.R. Nelsen and M.H. Gruenauer, *J. Chromatogr.*, 1981, 212, 366.
643 J.L. Lake, P.F. Rogerson, and C.B. Norwood, *Environ. Sci. Technol.*, 1981, 15, 549.
644 J.E. Henderson and W.G. Glaze, *Water Res.*, 1982, 16, 211.
645 T. Kikuchi, S. Kadota, H. Suehara, A. Nishi, and K. Tsubaki, *Chem. Pharm. Bull.*, 1981, 29, 1782.
646 W. Giger, E. Stephanou, and C. Schaffner, *Chemosphere*, 1981, 10, 1253.
647 J.L. Burleson, G.R. Peyton, and W.H. Glaze, *Environ. Sci. Technol.*, 1980, 14, 1354.
648 D.V. McCalley, M. Cooke, and G. Nickless, *Water Res.*, 1981, 15, 1019.
649 R. Shinohara, A. Kido, S. Eto, T. Hori, M. Koga, and T. Akiyama, *Water Res.*, 1981, 15, 535.
650 A. Yasuhara, H. Shiraishi, M. Tsuji, and T. Okuno, *Environ. Sci. Technol.*, 1981, 15, 570.
651 N. Kinae, T. Hashizume, T. Makita, I. Tomita, I. Kumura, and H. Kanamori, *Water Res.*, 1981, 15, 17.

652 V.E. Turoski, M.E. Kuehnl, and B.F. Vincent, *Tappi*, 1981, 64, 117.
653 K. Alben, *Anal. Chem.*, 1980, 52, 1821.
654 R.J. Law, *Sci. Total Environ.*, 1980, 15, 37.
655 F. Berthou, Y. Gourmelun, Y. Dreano, and M.P. Friocourt, *J. Chromatogr.*, 1981, 203, 279.
656 M.H. Carter, J. Chromatogr., 1982, 235, 165.
657 B.S. Middleditch and B. Basile, *J. Chromatogr.*, 1980, 199, 161.
658 K. Adachi, *Bull. Environ. Contam. Toxicol.*, 1980, 25, 416.
659 G. Grimmer, J. Jacob, and K.-W. Naujack, *Fresenius' Z. Anal. Chem.*, 1981, 306, 347; *ibid.*, 1981, 309, 13.
660 M. Ogata and Y. Miyake, *Water Res.*, 1981, 15, 257.
661 D.L. Vassilaros, P.W. Stoker, G.M. Booth, and M.L. Lee, *Anal. Chem.*, 1982, 54, 106.
662 F.I. Onuska, M.E. Comba, and J.A. Coburn, *Anal. Chem.*, 1980, 52, 2272.
663 J.L. Laseter and I.R. Deleon, *Anal. Chem.*, 1982, 54, 594; F.I. Onuska, M.E. Comba, and J.A. Coburn, *ibid.*, p. 595.
664 H. Tausch, G. Stehlik, and H. Wihlidal, *Chromatographia*, 1981, 14, 403.
665 T.E. Stewart and R.D. Cannizzaro, *ACS Symp. Ser.*, 1980, 136, 367.
666 G.W. Sovocool, R.L. Harless, D.E. Bradway, L.H. Wright, E.M. Lores, and L.E. Feige, *J. Anal. Toxicol.*, 1981, 5, 73.
667 N. Kurihara, J. Suzuki, and M. Nakajima, *Pestic. Biochem. Physiol.*, 1980, 14, 41.
668 F.W. Crow, A. Bjorseth, K.T. Knapp, and R. Bennett, *Anal. Chem.*, 1981, 53, 619.
669 J. Roboz, J. Greaves, J.F. Holland, and J.G. Bekesi, *Anal. Chem.*, 1982, 54, 1104.
670 R.K. Mitchum, G.F. Moler, and W.A. Korfmacher, *Anal. Chem.*, 1980, 52, 2278.
671 W.A. Korfmacher and R.K. Mitchum, *J. High Resolut. Chromatogr. Chromatogr. Commun.*, 1981, 4, 294.
672 R.K. Mitchum, W.A. Korfmacher, G.F. Moler, and D.L. Stalling, *Anal. Chem.*, 1982, 54, 719.
673 C. Rappe, H.R. Buser, D.L. Stalling. L.M. Smith, and R.C. Dougherty, *Nature (London)*, 1981, 292, 524.
674 E.F. Domino and S.E. Domino, *J. Chromatogr.*, 1980, 197, 258.
675 G.W. Tindal and P.E. Wininger, *J. Chromatogr.*, 1980, 196, 109.
676 T. Cairns and E.G. Siegmund, *Anal. Chem.*, 1981, 53, 1599.
677 A. Liberti, P. Ciccioli, E. Brancaleoni, and A. Cecinato, *J. Chromatogr.*, 1982, 242, 111.
678 A. Di Domenico, V. Silano, G. Viviano, and G. Zapponi, *Ecotoxicol. Environ. Saf.*, 1980, 4, 283.
679 A. Cavallaro, G. Bartolozzi, D. Carreri, G. Bandi, L. Luciani, G. Villa, A. Gorni, and G. Invernizzi, *Chemosphere*, 1980, 9, 623.
680 D.W. Phillipson and B.J. Puma, *Anal. Chem.*, 1980, 52, 2328.
681 M.L. Gross, T. Sun, P.A. Lyon, S.F. Wojinski, D.R. Hilker, A.E. Dupuy, jun., and R.G. Heath, *Anal. Chem.*, 1981, 53, 1902.
682 R.L. Harless, E.O. Oswald, M.K. Wilkinson, A.E. Dupuy, jun., D.D. McDaniel, and H. Tai, *Anal. Chem.*, 1980, 52, 1239.
683 S. Räisänen, R. Hiltunen, A.U. Arstila, and T. Sipiläinen, *J. Chromatogr.*, 1981, 208, 323.
684 G.A. Eiceman, R.E. Clement, and F.W. Karasek, *Anal. Chem.*, 1981, 53, 955.
685 T.J. Nestrick, L.L. Lamparski, W.B. Crummett, and L.A. Shadoff, *Anal. Chem.*, 1982, 54, 823.

686 L.L. Lamparski and T.J. Nestrick, *Anal. Chem.*, 1980, 52, 2045.
687 F.W. Karasek, R.E. Clement, and A.C. Viau, *J. Chromatogr.*, 1982, 239, 173.
688 W.P. Cochrane, J. Singh, W. Miles, and B. Wakeford, *J. Chromatogr.*, 1981, 217, 289.
689 P.W. O'Keefe, R. Smith, C. Meyer, D. Hilker, K. Aldous, and B. Jelus-Tyror, *J. Chromatogr.*, 1982, 242, 305.
690 H.R. Buser and C. Rappe, *Anal. Chem.*, 1980, 52, 2257.
691 G.F. Van Ness, J.G. Solch, M.L. Taylor, and T.O. Tiernan, *Chemosphere*, 1980, 9, 553.
692 L.L. Lamparski and T.J. Nestrick, *Chemosphere*, 1981, 10, 3.
693 A.C. Ray, L.O. Post, T.P. Hewlett, and J.C. Reagor, *Vet. Hum. Toxicol.*, 1981, 23, 418.
694 A.C. Ray, L.O. Post, and J.C. Reagor, *Vet. Hum. Toxicol.*, 1980, 22, 398.
695 J.P. Chaytor and M.J. Saxby, *J. Chromatogr.*, 1982, 237, 107.
696 M. Ogata and Y. Miyake, *J. Chromatogr. Sci.*, 1980, 18, 594.
697 F. Bidoli, L. Airoldi, and C. Pantarotto, *J. Chromatogr.*, 1980, 196, 314.
698 L. De Petrocellis, M. Tortoreto, S. Paglialunga, R. Paesani, L. Airoldi, E. Ramos Castaneda, and C. Pantarotto, *J. Chromatogr.*, 1982, 240, 218.
699 K.J. Welch, D.W. Kuehl, E.N. Leonard, G.D. Veith, and N.D. Schoenthal, *Bull. Environ. Contam. Toxicol.*, 1981, 26, 724; D.W. Kuehl, K.L. Johnson, B.C. Butterworth, E.N. Leonard, and G.D. Veith, *J. Great Lakes Res.*, 1981, 7, 330.
700 F.A. Medvedev, D.B. Melamed, and Ya.L. Kostyukovskii, *Biomed. Mass Spectrom.*, 1980, 7, 354.
701 Ya.L. Kostyukovskii, F.A. Medvedev, and D.B. Melamed, *Biomed. Mass Spectrom.*, 1981, 8, 480.
702 G. Hartmetz and J. Slemrova, *Bull. Environ. Contam. Toxicol.*, 1980, 25, 106.
703 T. Yamamoto, T. Yoshida, K. Aoki, Y, Kuroiwa, M. Terada, S. Yoshimura, T. Sato, and H. Kitagawa, *Eisei Kagaku*, 1981, 27, 331.
704 T. Lukaszewski and W.K. Jeffery, *J. Forensic Sci.*, 1980, 25, 499.
705 C.C. Clark, *J. Assoc. Off. Anal. Chem.*, 1981, 64, 884.
706 W. Gielsdorf, K. Schubert, and K. Allin, *Arch. Kriminol.*, 1980, 166, 21.
707 B. Levine, M.F. Fierro, S.W. Goza, and J.C. Valentour, *J. Forensic Sci.*, 1981, 26, 206.
708 W. Gielsdorf, *Fresenius' Z. Anal. Chem.*, 1981, 308, 123.
709 J. Martin, E. Quirke, G.J. Shaw, P.D. Soper, and J.R. Maxwell, *Tetrahedron*, 1980, 36, 3261.
710 R. Alexander, G. Eglinton, J.P. Gill, and J.K. Volkman, *J. High Resolut. Chromatogr. Chromatogr. Commun.*, 1980, 3, 521.
711 C. Willey, M. Iwao, R.N. Castle, and M.L. Lee, *Anal. Chem.*, 1981, 53, 400.
712 F.P. Di Sanzo, *J. High Resolut. Chromatogr. Chromatogr. Commun.*, 1981, 4, 649.
713 B.A. Tomkins and C.-h. Ho, *Anal. Chem.*, 1982, 54, 91.
714 J.-M. Schmitter, H. Colin, J.-L. Excoffier, P. Arpino, and G. Guiochon, *Anal. Chem.*, 1982, 54, 769.
715 D.W. Later, M.L. Lee, and B.W. Wilson, *Anal. Chem.*, 1982, 54, 117.
716 L.J. Felice, *Anal. Chem.*, 1982, 54, 869.
717 L.V.S. Hood and C.M. Erikson, *J. High Resolut. Chromatogr. Chromatogr. Commun.*, 1980, 3, 516.
718 J. Rullkoetter and D.H. Wette, *Phys. Chem. Earth*, 1980, 12, 93.

719 J. Albaigés, J. Borbón, and M. Gassiot, *J. Chromatogr.*, 1981, 204, 491.
720 H. Wehner and M. Teschner, *J. Chromatogr.*, 1981, 204, 481.
721 M.E. Hohn, N.W. Jones, and R. Patience, *Geochim. Cosmochim. Acta*, 1981, 45, 1131.
722 R.S. Ozubko, D.M. Clugston, and E. Furimsky, *Anal. Chem.*, 1981, 53, 183.
723 H.S. Hertz, J.M. Brown, S.N. Chesler, F.R. Guenther, L.R. Hilpert, W.E. May, R.M. Parris, and S.A. Wise, *Anal. Chem.*, 1980, 52, 1650.
724 J. Shen, *Anal. Chem.*, 1981, 53, 475.
725 J.M. Schmitter, P.J. Arpino, and G. Guiochon, *Geochim. Cosmochim. Acta*, 1981, 45, 1951.
726 J.S. Richardson and D.E. Miiller, *Anal. Chem.*, 1982, 54, 765.
727 J.M. Moldowan and W.K. Seifert, *J. Chem. Soc., Chem. Commun.*, 1980, 912.
728 B. Ludwig, G. Hussler, P. Wehrung, and P. Albrecht, *Tetrahedron Lett.*, 1981, 22, 3313.
729 J.-M. Trendel, A. Restle, J. Connan, and P. Albrecht, *J. Chem. Soc., Chem. Commun.*, 1982, 304.
730 F.R. Aquino Neto, A. Restle, J. Connan, P. Albrecht, and G. Ourisson, *Tetrahedron Lett.*, 1982, 23, 2027.
731 G. Hussler, B. Chappe, P. Wehrung, and P. Albrecht, *Nature (London)*, 1981, 294, 556.
732 M. Bjorøy and J. Rullkötter, *Chem. Geol.*, 1980, 30, 27.
733 S.C. Brassell and G. Eglinton, *Nature (London)*, 1981, 290, 579.
734 J.W. De Leeuw, W.I.C. Rijpstra, and P.A. Schenck, *Geochim. Cosmochim. Acta*, 1981, 45, 2281.
735 D.A. Yon, J.R. Maxwell, and G. Ryback, *Tetrahedron Lett.*, 1982, 23, 2143.
736 H.G. Nowicki and R.F. Devine, *J. High Resolut. Chromatogr. Chromatogr. Commun.*, 1980, 3, 360.
737 K. Kawamura and R. Ishiwatari, *Geochim. Cosmochim. Acta*, 1981, 45, 149.
738 T.R. Steinheimer, W.E. Pereira, and S.M. Johnson, *Anal. Chim. Acta*, 1981, 129, 57.
739 S.G. Wakeham, J.W. Farrington, R.B. Gagosian, C. Lee, H. DeBaar, G.E. Nigrelli, B.W. Tripp, S.O. Smith, and N.M. Frew, *Nature (London)*, 1980, 286, 798.
740 S.C. Brassell, A.P. Gowar, and G. Eglinton, *Phys. Chem. Earth*, 1980, 12, 421.
741 R.H. Fish, A.S. Newton, and P.C. Babbitt, *Fuel*, 1982, 61, 227.
742 J. Rullkötter and P. Philip, *Nature (London)*, 1981, 292, 616.
743 E.J. Gallegos, *J. Chromatogr. Sci.*, 1981, 19, 177.
744 M. Novotny, J.W. Strand, S.L. Smith, D. Wiesler, and F.J. Schwende, *Fuel*, 1981, 60, 213.
745 D.J. Miller, J.K. Olson, and H.H. Schobert, *Fuel*, 1981, 60, 370.
746 P. Burchill, A.A. Herod, and E. Pritchard, *J. Chromatogr.*, 1982, 242, 51, 65.
747 D. Van de Meent, S.C. Brown, R.P. Philp, and B.R.T. Simoneit, *Geochim. Cosmochim. Acta*, 1980, 44, 999.
748 H. Solli, S.R. Larter, and A.G. Douglas, *Phys. Chem. Earth*, 1980, 12, 591.
749 G. Van Graas, J.W. De Leeuw, P.A. Schenck, and J. Haverkamp, *Geochim. Cosmochim. Acta*, 1981, 45, 2465.
750 J.S. Leventhal, *Geochim. Cosmochim. Acta*, 1981, 45, 883.
751 R.P. Philp, N.J. Russell, and T.D. Gilbert, *Fuel*, 1981, 60, 937.

752 R.P. Philp. T.D. Gilbert, and N.J. Russell, *Fuel*, 1982, 61, 221.
753 C. Braekman-Danheux, *J. Anal. Appl. Pyrolysis*, 1981, 3, 173.
754 X.-G. Jin and H.-M. Li, *J. Anal. Appl. Pyrolysis*, 1981, 3, 49; P.-J. Wu, *Tzu Jan Tsa Chih*, 1981, 4, 557.
755 J. Kubat and J. Zachoval, *Makrotest*, 1980, 96.
756 J.C. Kleinert and C.J. Weschler, *Anal. Chem.*, 1980, 52, 1245.
757 H. Sakuma, S. Munakata, and S. Sugawara, *Agric. Biol. Chem.*, 1981, 45, 443.
758 J.R. Hudson, S.L. Morgan, and A. Fox, *Anal. Biochem.*, 1982, 120, 59.
759 L.E. Abbey, A.K. Highsmith, T.F. Moran, and E.J. Reiner, *J. Clin. Microbiol.*, 1981, 13, 313.
760 S.J. Lyle and M.S. Tehrani, *J. Chromatogr.*, 1982, 236, 31.
761 M.I. Venkatesan, T.W. Linick, H.E. Suess, and G. Buccellati, *Nature (London)*, 1982, 295, 517.
762 M.A. Smith, R.M. Barkley, and G.B. Ellison, *J. Am. Chem. Soc.*, 1980, 102, 6851.
763 A. Touabet, K. Abdeddaim, and M.-H. Guermouche, *J. High Resolut. Chromatogr. Chromatogr. Commun.*, 1981, 4, 525.
764 D.L. Hachey, J.-C. Blais, and P.D. Klein, *Anal. Chem.*, 1980, 52, 1131.
765 T. Shimono, H. Takagi, T. Isobe, and T. Tarutani, *J. Chromatogr.*, 1980, 197, 59.
766 K. Kuroda, T. Koike, and C. Kato, *J. Chem. Soc., Dalton Trans.*, 1981, 1957.
767 V.I. Lavrentev, V.M. Kovrigin, and G.G. Treer, *Zh. Obshch. Khim.*, 1981, 51, 124.
768 G. Graff, T.P, Krick, T.F. Walseth, and N.D. Goldberg, *Anal. Biochem.*, 1980, 107, 324.
769 D.C. Reamer and C. Veillon, *Anal. Chem.*, 1981, 53, 2166.
770 D.L. Corina, D.P. Bloxham, and G.K. Cooper, *J. Chromatogr.*, 1980, 198, 287.
771 A.R. Swanson and M.W. Anders, *J. Chromatogr.*, 1981, 207, 365.
772 A.R. Swanson and M.W. Anders, *J. Chromatogr.*, 1982, 234, 268.
773 A. Zeman, *Fresenius' Z. Anal. Chem.*, 1982, 310, 243.
774 D.C.M. Squirrell, *Analyst (London)*, 1981, 106, 1042.
775 N.P.H. Ching, G.N. Jham, C. Subbarayan, D.V. Bowen, A.L.C. Smit, jun., C.E. Grossi, R.G. Hicks, F.H. Field, and T.F. Nealon, jun., *J. Chromatogr.*, 1981, 222, 171.
776 P.W. Albro, J.R. Hass, C.C. Peck, D.G. Odam, J.T. Corbett, F.J. Bailey, H.E. Blatt, and B.B. Barrett, *Drug Metab. Dispos.*, 1981, 9, 223.
777 E. Bailey, L.D. Corte, P.B. Farmer, and A.J. Gray, *J. Chromatogr.*, 1981, 225, 83.
778 J. Reisch and A.S. El-Sharaky, *Fresenius' Z. Anal. Chem.*, 1981, 307, 287.
779 E.R. Schmid, I. Fogy, F. Heresch, E. Kenndler, and J.F.K. Huber, *Mikrochim. Acta*, 1979, 2, 207.
780 S. Tsuge, *Kagaku no Ryoiki, Zokan*, 1981, 155.
781 P.J. Arpino, *Trends Anal. Chem.*, 1982, 1, 154.
782 R.E. Majors, H.G. Barth, and C.H. Lochmüller, *Anal. Chem.*, 1982, 54, 323R.
783 (*a*) D.E. Games, C. Eckers, J.L. Gower, P. Hirter, M.E. Knight, E. Lewis, K.R.N. Rao, and N.C. Weerasinghe in ref. 42, p. 97; (*b*) C. Eckers, D.E. Games, M.L. Games, W. Kuhnz, E. Lewis, N.C.A. Weerasinghe, and S.A. Westwood in ref. 41, p. 169; (*c*) D.E. Games, *Biomed. Mass Spectrom.*, 1981, 8, 454.

784 (*a*) D.E. Games, P. Hirter, W. Kuhnz, E. Lewis, N.C.A. Weerasinghe, and S.A. Westwood, *J. Chromatogr.*, 1981, 203, 131; (*b*) N.J. Alcock, C. Eckers, D.E. Games, M.P.L. Games, M.S. Lant, M.A. McDowall, M. Rossiter, R.W. Smith, S.A. Westwood, and H.-Y. Wong, *J. Chromatogr.*, 1982, 251, 165.
785 M. Novotny, *Anal. Chem.*, 1981, 53, 1294A.
786 G. Guiochon, *Anal. Chem.*, 1981, 53, 1318.
787 F.J. Yang, *J. Chromatogr. Sci.*, 1982, 20, 241.
788 (*a*) S.A. Westwood, D.E. Games, M.S. Lant, and B.J. Woodhall, *Anal. Proc.*, 1982, 19, 121; (*b*) D.E. Games, S.A. Westwood, M.J. Cocksedge, N. Evans, J. Williamson, and B.J. Woodhall, *Biomed. Mass Spectrom.*, 1982, 9, 215.
789 S.A. Borman, *Anal. Chem.*, 1982, 54, 327A.
790 H.-R. Schulten, *J. Chromatogr.*, 1982, 251, 105.
791 O.S. Privett and W.L. Erdahl, *Methods Enzymol.*, 1981, 72, 56.
792 F.R. Sugnaux and C. Djerassi, *J. Chromatogr.*, 1982, 251, 189.
793 C.E. Reese and R.P.W. Scott, *J. Chromatogr. Sci.*, 1980, 18, 479.
794 J.J. Brophy, D. Nelson, and M.K. Withers, *Int. J. Mass Spectrom. Ion Phys.*, 1980, 36, 205.
795 J.D. Henion and T. Wachs, *Anal. Chem.*, 1981, 53, 1963.
796 S. Tsuge, Y. Yoshida, T. Takeuchi, K. Mochizuki, N. Kokubun, and K. Hibi, *Chem., Biomed., Environ. Instrum.*, 1980, 10, 405.
797 T. Takeuchi, D. Ishii, A. Saito, and T. Ohki, *J. High Resolut. Chromatogr. Chromatogr. Commun.*, 1982, 5, 91.
798 J.D. Henion, *J. Chromatogr. Sci.*, 1981, 19, 57.
799 K.H. Schäfer and K. Levsen, *J. Chromatogr.*, 1981, 206, 245.
800 P. Krien, G. Devant, and M. Hardy, *J. Chromatogr.*, 1982, 251, 129.
801 R. Tijssen, J.P.A. Bleumer, A.L.C. Smit, and M.E. Van Kreveld, *J. Chromatogr.*, 1981, 218, 137.
802 W.H. Pirkle, J.M. Finn, J.L. Schreiner, and B.C. Hamper, *J. Am. Chem. Soc.*, 1981, 103, 3964.
803 C.R. Blakley, J.J. Carmody, and M.L. Vestal, *J. Am. Chem. Soc.*, 1980, 102, 5931.
804 C.R. Blakley, J.J. Carmody, and M.L. Vestal, *Anal. Chem.*, 1980, 52, 1636.
805 C.R. Blakley, J.J. Carmody, and M.L. Vestal, *Clin. Chem.*, 1980, 26, 1467.
806 P.J. Arpino, P, Krien, S. Vajta, and G. Devant, *J. Chromatogr.*, 1981, 203, 117.
807 B. Mauchamp and P. Krien, *J. Chromatogr.*, 1982, 236, 17.
808 M. Dedieu, C. Juin, P.J. Arpino, J.P. Bounine, and G. Guiochon, *J. Chromatogr.*, 1982, 251, 203.
809 P.J. Arpino and G. Guiochon, *J. Chromatogr.*, 1982, 251, 153.
810 E. Yamauchi, T. Mizuno, and K. Azuma, *Shitsuryo Bunseki*, 1980, 28, 227.
811 D.E. Games and E. Lewis, *Biomed. Mass Spectrom.*, 1980, 7, 433.
812 D.P. Kirby, P. Vouros, and B.L. Karger, *Science*, 1980, 209, 495.
813 D.P. Kirby, P. Vouros, B.L. Karger, B. Hidy, and B. Petersen, *J. Chromatogr.*, 1981, 203, 139.
814 P. Vouros, E.P. Lankmayr, M.J. Hayes, B.L. Karger, and J.M. McGuire, *J. Chromatogr.*, 1982, 251, 175.
815 A. Benninghoven, A. Eicke, M. Junack, W. Sichtermann, J. Krizek, and H. Peters, *Org. Mass Spectrom.*, 1980, 15, 459.
816 (*a*) R.D. Smith and A.L. Johnson, *Anal. Chem.*, 1981, 53, 739; (*b*) *ibid.*, p. 1120; (*c*) R.D. Smith, J.E. Burger, and A.L. Johnson, *ibid.*, p. 1603.
817 E.D. Hardin and M.L. Vestal, *Anal. Chem.*, 1981, 53, 1492.

818 M. Barber, R.S. Bordoli, R.D. Sedgwick, and A.N. Tyler, *Science*, 1981, 293, 270; F.M. Devienne and J.-C. Rouston, *Org. Mass Spectrom.*, 1982, 17, 173.
819 R.G. Christensen, H.S. Hertz, S. Meiselman, and E. White, *Anal. Chem.*, 1981, 53, 171.
820 H. Kambara, *Anal. Chem.*, 1982, 54, 143.
821 M. Tsuchiya and T. Taira, *Int. J. Mass Spectrom. Ion Phys.*, 1980, 34, 351.
822 R.S. Houk, V.A. Fassel, and H.J. Svec, *Org. Mass Spectrom.*, 1982, 17, 240.
823 N. Evans and J.E. Williamson, *Biomed. Mass Spectrom.*, 1981, 8, 316.
824 H. Jungclas, H. Danigel, and L. Schmidt, *Org. Mass Spectrom.*, 1982, 17, 86.
825 W.A. Dark, W.H. McFadden, and D.C. Bradford, *J. Chromatogr. Sci.*, 1977, 15, 454.
826 W.A. Dark and W.H. McFadden, *J. Chromatogr. Sci.*, 1978, 16, 289.
827 D.E. Games, *Chem. Phys. Lipids*, 1978, 21, 389.
828 D.E. Games, J.L. Gower, M.G. Lee, I.A.S. Lewis, M.E. Pugh, and M. Rossiter, *Methodol. Surv. Biochem.*, 1978, 7, 185.
829 W.H. McFadden, D.C. Bradford, D.E. Games, and M.L. Gower, *Am. Lab.*, 1977, 9, 55.
830 D.E. Games, J.L. Gower, M.G. Lee, I.A.S. Lewis, M.E. Pugh, and M. Rossiter, *Proc. Anal. Div. Chem. Soc.*, 1978, 15, 101.
831 W.H. McFadden, D.C. Bradford, G. Eglinton, S.K. Hajlbrahim, and N. Nicolaides, *J. Chromatogr. Sci.*, 1979, 17, 518.
832 C. Eckers, D.E. Games, D.N.B. Mallen, and B.P. Swann, *Anal. Proc.*, 1982, 19, 133; *Biomed. Mass Spectrom.*, 1982, 9, 162.
833 D.E. Games and N.C.A. Weerasinghe, *J. Chromatogr. Sci.*, 1980, 18, 106.
834 T.I. Martin, *J. Chromatogr. Sci.*, 1980, 18, 104.
835 D.L. Stalling, J.D. Petty, G.R. Dubay, and R.A. Smith, *J. Chromatogr. Sci.*, 1980, 18, 107.
836 R.F. Skinner, O. Thomas, J. Giles, and D.G. Crosby, *J. Chromatogr. Sci.*, 1980, 18, 108.
837 T.J. Yu, H. Schwartz, R.W. Giese, B.L. Karger, and P. Vouros, *J. Chromatogr.*, 1981, 218, 519.
838 T. Cairns, E.G. Siegmund, and G.M. Doose, *Anal. Chem.*, 1982, 54, 953.
839 L.H. Wright, T.R. Edgerton, S.J. Arbes, jun., and E.M. Lores, *Biomed. Mass Spectrom.*, 1981, 8, 475.
840 L.H. Wright, *J. Chromatogr. Sci.*, 1982, 20, 1.
841 L.E. Martin, J. Oxford, and R.J.N. Tanner, *Xenobiotica*, 1981, 11, 831; *J. Chromatogr.*, 1982, 251, 215.
842 A.D. Thruston, jun. and J.M. McGuire, *Biomed. Mass Spectrom.*, 1981, 8, 47.
843 E. Houghton, M.C. Dumasia, and J.K. Welby, *Biomed. Mass Spectrom.*, 1981, 8, 558.
844 I.S. Krull and P. Vouros, *Altex Chromatogram*, 1980, 3, 3.
845 W.L. Erdahl, W. Beck, C. Jones, D.E. Jarvis, and O.S. Privett, *Lipids*, 1981, 16, 614.
846 J.M. Schmitter, P.J. Arpino, and G. Guiochon, *J. Chromatogr.*, 1978, 167, 149.
847 J.D. Henion, *J. Chromatogr. Sci.*, 1980, 18, 101.
848 C.N. Kenyon, A. Melera, and F. Erni, *J. Chromatogr. Sci.*, 1980, 18, 103.
849 P.J. Arpino and P. Krien, *J. Chromatogr. Sci.*, 1980, 18, 112.
850 B. Mauchamp, R. Lafont, and P. Krien, *Dev. Endocrinol.*, 1981, 15, 21.

851 C.E. Parker, C.A. Haney, and J.R. Hass, *J. Chromatogr.*, 1982, 237, 233.
852 C.A. Parker, C.A. Haney, D.J. Harvan, and J.R. Hass, *J. Chromatogr.*, 1982, 242, 77.
853 (*a*) C.N. Kenyon, A. Melera, and F. Erni, *J. Anal. Toxicol.*, 1981, 5, 216; (*b*) F. Erni, *J. Chromatogr.*, 1982, 251, 141.
854 J.D. Henion, B.A. Thomson, and P.H. Dawson, *Anal. Chem.*, 1982, 54, 451.
855 R. Schuster, *Chromatographia*, 1980, 13, 379.
856 F.W. McLafferty, *Acc. Chem. Res.*, 1980, 13, 33; F.W. McLafferty, P.J. Todd, D.C. McGilvery, M.A. Baldwin, F.M. Bockhoff, G.J. Wendell, M.R. Wixom, and T.E. Niemi in ref. 4, p. 1589; J.H. Beynon and R.M. Caprioli in ref. 48, p. 89; D.H. Russell, E.H. McBay, and T.R. Mueller, *Int. Lab.*, 1980, 49; S.E. Ungar, R.G. Cooks, R. Mata, and J.L. McLaughlin, *J. Nat. Prod.*, 1980, 43, 288; F.W. McLafferty, *Science*, 1981, 214, 280; F.W. McLafferty and E.R. Lory, *J. Chromatogr.*, 1981, 203, 109; R.G. Cooks and G.L. Glish, *Chem. Eng. News*, 1981, 59, 40.
857 S.J. Gaskell and D.S. Millington, *Biomed. Mass Spectrom.*, 1978, 5, 557.
858 S.J. Gaskell, R.W. Finney, and M.E. Harper, *Biomed. Mass Spectrom.*, 1979, 6, 113.
859 S.J. Gaskell, A.W. Pike, and K. Griffiths, *Steroids*, 1980, 36, 219; E.M.H. Finlay and S.J. Gaskell, *Clin. Chem.*, 1981, 27, 1165.
860 S.J. Gaskell, E.M.H. Finley, and D.S. Millington in ref. 4, p. 1908.
861 D.A. Durden, *Anal. Chem.*, 1982, 54, 666.
862 S.J. Gaskell and A.W. Pike in ref. 4, p. 279.
863 S.J. Gaskell and A.W. Pike, *Biomed. Mass Spectrom.*, 1981, 8, 125.
864 S.J. Gaskell, E.M.H. Finlay, and A.W. Pike, *Biomed. Mass Spectrom.*, 1980, 7, 500.
865 M.E. Rose, *Org. Mass Spectrom.*, 1981, 16, 323.
866 W.F. Haddon, *Anal. Chem.*, 1979, 51, 983; M.H. Bozorgzadeh, J.H. Beynon, R.P. Morgan, and A.G. Brenton, *Int. J. Mass Spectrom. Ion Phys.*, 1979, 29, 191.
867 B. Shushan, S.H. Safe, and R.K. Boyd, *Anal. Chem.*, 1979, 51, 156; B. Shushan and R.K. Boyd, *Org. Mass Spectrom.*, 1980, 15, 445.
868 B. Shushan, N.J. Bunce, R.K. Boyd, and C.T. Corke, *Biomed. Mass Spectrom.*, 1981, 8, 225.
869 D.J. Harvan, J.R. Hass, P.W. Albro, and M.D. Friesen, *Biomed. Mass Spectrom.*, 1980, 7, 242.
870 D.J. Harvan, J.R. Hass, J.L. Schroeder, and B.J. Corbett, *Anal. Chem.*, 1981, 53, 1755.
871 R.D. Voyksner, J.R. Hass, and M.M. Bursey, *Anal. Lett.*, 1982, 15, 1.

8

The Use of Mass Spectrometry in Pharmacokinetic and Drug-metabolism Studies

BY D. J. HARVEY

1 General Considerations

Introduction.- This review covers the period from early 1980 until mid-June 1982 and is concerned with the application of mass spectrometry to drug detection in biological matrices, drug metabolism and pharmacokinetics, and methodology for drug quantification. As a major proportion of the reported work involves g.c.-m.s., references to this technique applied to drug research are included in this review rather than in the chapter on g.c.-m.s. and h.p.l.c.-m.s. methods, as has been the practice in earlier volumes of this series. Although the term 'drug' has been interpreted fairly broadly, this review does not include studies on compounds such as insecticides and toxins of no therapeutic value, even though their metabolism has been studied extensively by mass spectrometry. Included, however, are a few reports of work performed with model compounds where these have been used to obtain information on specific pathways of drug biotransformation. Within this framework over 900 relevant papers have been found and some selection has had to be made. It is hoped, however, that this review will give a broad overview of the subject with particular reference to the different ways in which mass spectrometry can be used in drug research.

Books of interest appearing during the review period include a comprehensive set of reviews on drugs of abuse,[1] two additional volumes of Florey's 'Analytical Profiles of Drug Substances',[2a,b] the seventh in the series on the analysis of drugs by g.c.-m.s.,[3] and two books on drug quantification.[4,5] Several recent meetings also contained papers relating to drug research. Thus the proceedings of the 6th and 7th International Symposia on Mass

Spectrometry, held in Venice (1979) and Milan (1980), respectively, and sponsored by the Italian Group for Mass Spectrometry in Biochemistry and Medicine,[6,7] and of the 4th International Conference on Stable Isotopes[8] held in Jülich, in March 1981, have been published in *Anal. Chem. Symp. Ser.* (Elsevier). Many of the papers presented at the first in a series of international symposia organized by the Clinical Research Centre, Harrow, U.K., and published by Academic Press are also of interest.[9] A special edition of *Biomed. Mass Spectrom.*[10] published the papers from the 3rd International Symposium on Quantitative Mass Spectrometry in the Life Sciences (Ghent, June 1980), and abstracts from the GAMS Conference on Fundamental and Applied Mass Spectrometry have appeared in the *Eur. J. Mass Spectrom.*[11] General reviews have been published on quantitative aspects of mass spectrometry,[12-15] biomedical applications,[16] the use of liquid-chromatography (l.c.) and field-desorption (f.d.) techniques,[17] drug metabolism,[18] and the applications of stable isotopes in biomedical and drug-metabolism studies.[19-21]

Current Trends.- The past two years have seen major advances in several aspects of mass-spectrometric technology particularly those such as fast atom bombardment (f.a.b.) and h.p.l.c.-m.s. interfacing relating to the analysis of involatile compounds of high molecular weight. Although most drugs are relatively small molecules that can be handled by g.c.-m.s. techniques, these developments will be of considerable significance in studies of drug conjugation and the binding of active drug metabolites to tissues. Investigations of the structures and molecular interactions of the antibiotics are also possible with this technique, and a study on the application of f.a.b. to some penicillins has recently been published.[22] Developments in l.c.-m.s. interfacing have been mainly concerned with direct coupling especially to microbore l.c. columns. It is significant that most of the papers discussing aspects of l.c.-m.s. presented at the 1982 ASMS meeting in Honolulu were concerned with this method of sample introduction. A study of the use of belt interfaces and direct coupling of microbore columns has recently been made;[23] microbore columns gave increased sensitivity and spectral quality with both interfaces. The main advantage offered by the belt system appears to be the capability of using electron-impact (e.i.) ionization as well as chemical ionization (c.i.). A study of several drugs by micro-l.c.-m.s. with acetonitrile:water (9:1) containing 0.1% trimethylamine has given very promising results with a detection limit of 20 pg being reported for phenothiazines.[24] The use of deuteriated solvents has yielded structural information by exchange reactions.[25] A new technique in which the column effluent is vaporized with an oxyhydrogen flame and the sample introduced into the mass spectrometer by molecular- and particle-beam techniques has been applied to various antibiotics,[26] and a method using ^{252}californium plasma desorption[27] from l.c. effluents deposited on a rotating metal foil has also been used for pharmaceutical analysis. Post-column derivatization techniques for on-line h.p.l.c.-m.s. analyses have recently been investigated.[28]

Another recent development that shows considerable promise for the detection and measurement of small amounts of drugs in biological matrices is m.s.-m.s. using multi-analyser instruments. Ions from the drug can be selected, fragmented by collision-induced decomposition, and a fragment ion monitored to give greatly increased selectivity and hence sensitivity. An application of this technique using a triple-quadrupole instrument for the detection of sulphonamides in race-horse urine has recently been published;[29] clean-up stages were reduced to a minimum, although separation by t.l.c. was recommended. Related linked-scan techniques for metastable-ion analysis have been used for the identification of metabolites of 5-ethyl-3-methyl-5-phenylhydantoin (1)[30] and phenacetin (6),[31] and fragmentation of the antibiotic maleimycin has been studied using B/E and B^2/E linked scans.[32]

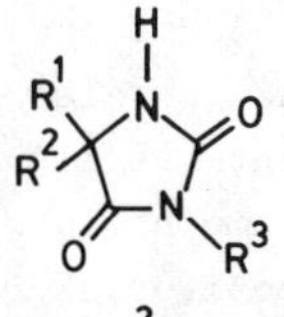

(1) $R^1 = Ph, R^2 = Et, R^3 = Me$
(2) $R^1 = R^2 = Ph, R^3 = H$
(3) $R^1 = Ph, R^2 = H, R^3 = Et$
(4) $R^1 =$ 5,6-dihydroxycyclohexa-1,3-dienyl (OH, H, H, OH), $R^2 = Et, R^3 = H$
(5) $R^1 = Ph, R^2 = Et, R^3 = H$

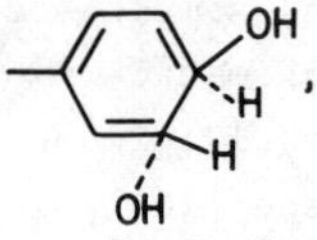

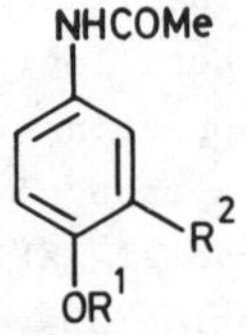

(6) $R^1 = Et, R^2 = H$
(7) $R^1 = R^2 = H$
(8) $R^1 = H, R^2 = OH$
(9) $R^1 = H, R^2 = SMe$
(10) $R^1 = H, R^2 = SCH_2CH(NH_2)COOH$
(11) $R^1 = PO_3H_2, R^2 = H$

Decomposition and Contamination of Samples.- In any drug-metabolism study, recognition of impurities and artefacts is of vital importance. Recent studies of commercial drugs have shown that many are impure; samples of 5-iodo-2'-deoxyuridine have been found to contain the 5-bromo analogue,[33] and phenylbutazone (1-butyl-1,2-diphenyl-3,5-pyrazolidinedione) stored at 60 ^{o}C for 203 days has been shown to undergo partial decomposition to α-(*N*-phenylcarbamoyl)-*N*-caproylhydrazobenzene and its α-hydroxy analogue together with aniline.[34] The anti-tumour agent 2,2-diaziridinyl-3,6-bis-(carboethoxyamino)-1,4-benzoquinone has been found to contain products produced by hydrolysis of the aziridine rings,[35] and samples of imipramine (12) have been found to contain the metabolite desipramine (13).[36] Pain following injection of aged solutions of sodium cephalothin (15) has been traced to acetic acid and deacyl- and deacetyl-cephalothin,[37] and

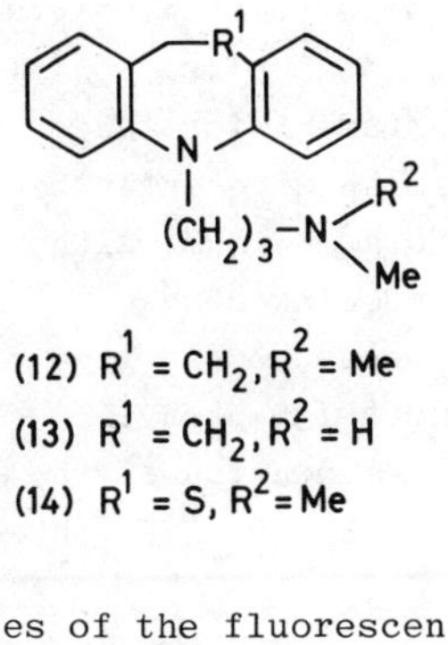

(12) $R^1 = CH_2, R^2 = Me$
(13) $R^1 = CH_2, R^2 = H$
(14) $R^1 = S, R^2 = Me$

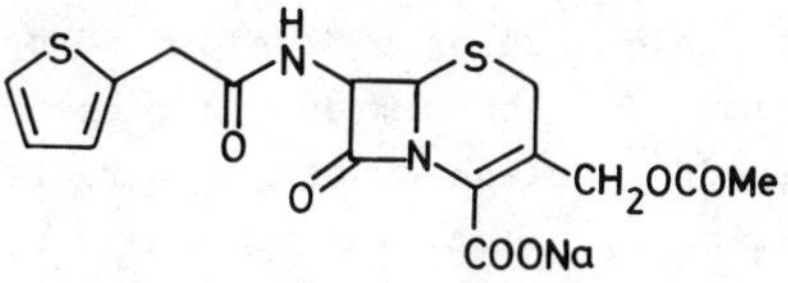

(15)

studies of the fluorescent degradation products of ampicillin have implicated penilloaldehyde as an intermediate.[38] Additional compounds appearing in pharmaceutical preparations can also arise from the reaction of the drug with other ingredients as illustrated by the characterization of glyceryl esters of indomethacin formed by the reaction of the drug with glycerol in suppositories.[39] In some cases the presence of impurities can act as a fingerprint to enable the source of an illicit drug to be identified, as has recently been shown with phencyclidine[40,41] and cocaine.[42]

Recent examples of artefacts arising during sample storage and work-up include large amounts of the photochemical-decomposition product 2-chlorothioxanthen-9-one (17) found in the gastric aspirate in a case of chlorprothixene (16) poisoning[43] (Scheme 1) and the products of side reactions of various drugs

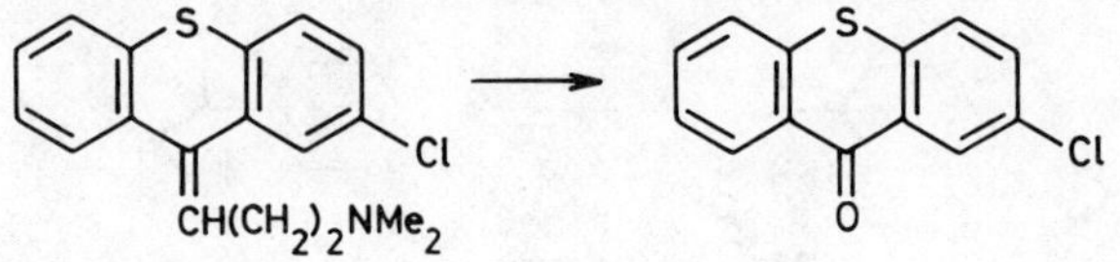

(16) (17)

Scheme 1

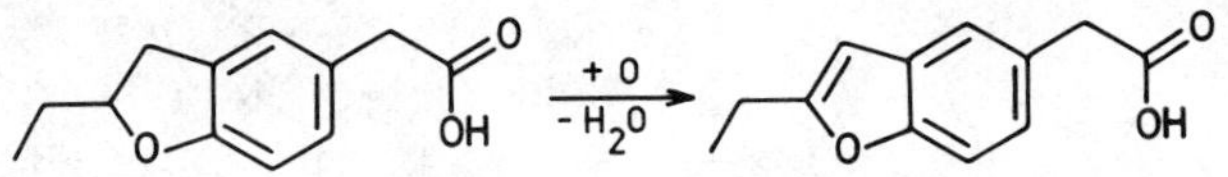

(18) (19)

Scheme 2

occurring during extraction with chloroform and other halogenated solvents. Such side reactions are shown by imipramine (12), which is methylated and chloromethylated to quaternary ammonium products by chloroform, dichloromethane, and 1,2-dichloroethane,[44] by the related octriptyline, which is attacked by phosgene in unstabilized chloroform to give alkyl carbamates,[45] and by clofibrate [2-(4'-chlorophenoxy)-2-methylpropionic acid, ethyl ester], which forms *N*-formyl analogues through a suspected reaction with chloroform.[46] A urinary metabolite of the latter drug, 2-(4'-chlorophenoxy)-2-methylpropionic acid, has been reported to form an amide if the urine is made alkaline with ammonium salts.[47] Dehydration of hydroxy metabolites has been found for furofenac (18)[48] (Scheme 2) and ketamine (20)[49] (Scheme 3) to give compounds (19) and (21), respectively. Acid-catalysed ring opening with the formation of substituted benzophenones has been reported from the benzodiazepines midazolam (22)[50] and triazolam (23)[51] (Scheme 4) and their 4-hydroxy metabolites. Hydrazino drugs such as hydralazine [(26), Scheme 5] can form hydrazones with carbonyl compounds both *in vivo* and *in vitro*; the hydrazone (27) formed with pyruvic acid *in vivo* has been shown to decompose to

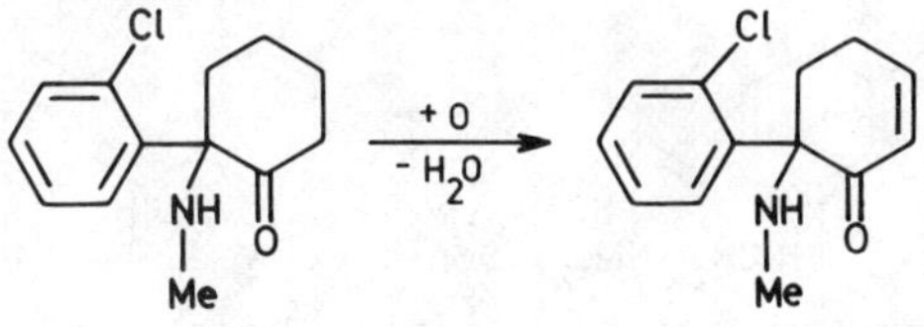

(20) (21)

Scheme 3

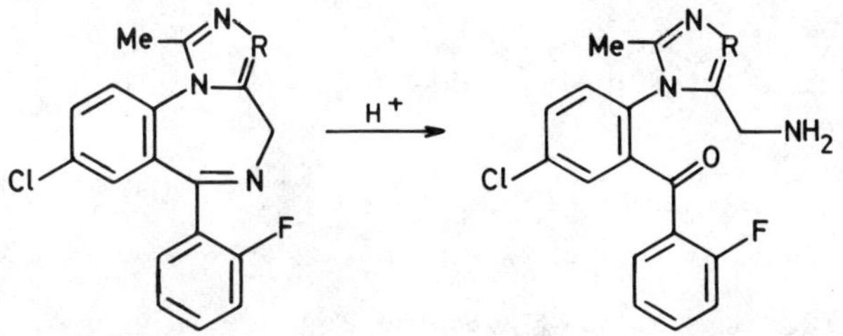

(22) R = CH_2
(23) R = N

(24) R = CH_2
(25) R = N

Scheme 4

3-methyl-S-triazolo(3,4-*a*)phthalazine (28), a compound previously thought to be a product arising exclusively from hydralazine acetylation.[52] Thermal decomposition can also give rise to artifacts at various stages of a study; for example, phencyclidine (29) forms phenylcyclohexene (30) when heated, and this must be allowed for if the drug is administered by smoking.[53,54] Many underivatized drugs such as propoxyphene (4-dimethylamino-3-methyl-1,2-diphenyl-2-butanol propionate) decompose thermally during gas chromatography,[55] and several drugs such as the steroid dexamethasone undergo thermal decomposition in the mass spectrometer.[56] Thermal decomposition of *N*-oxide metabolites during gas chromatography is well documented;[57-60] consequently analysis by direct insertion with f.d.[61] or e.i. ionization is usually used. G.c.-m.s. analysis has been reported, but the products analysed are not of the oxide itself.[62,63] Detection by g.c.-m.s. of *N*-oxide formation from oxybutynin (32) has been achieved using analogues of the drug deuteriated on the carbon α to the nitrogen

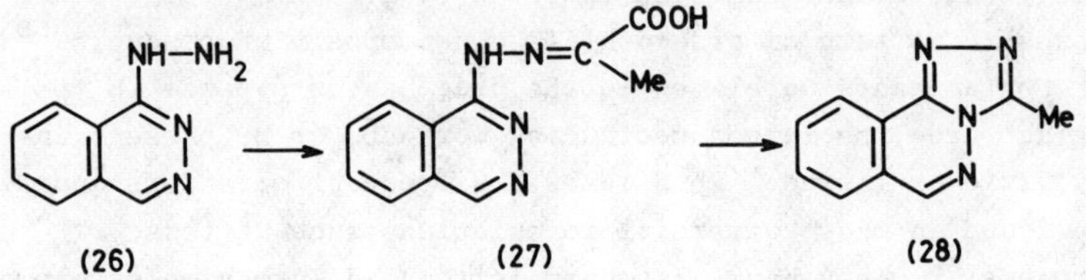

NH—NH2 (26) → NH—N=C(COOH)(Me) (27) → N—N Me (28)

Scheme 5

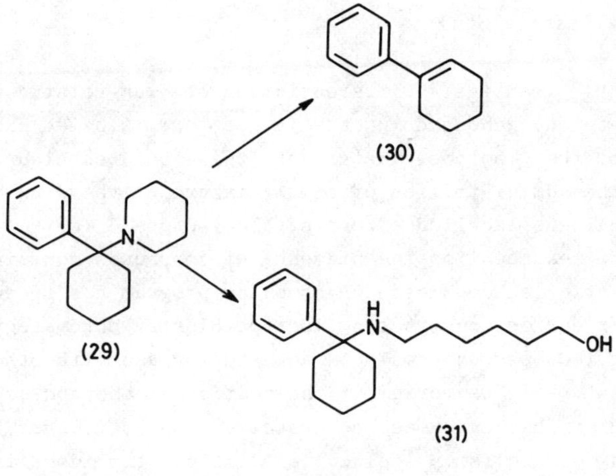

(29) → (30)

(29) → H N OH (31)

Scheme 6

C(OH)(Ph)—COO—CH2—C≡C—CH2—NEt2

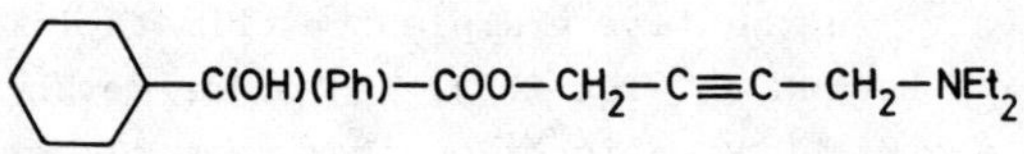

(32)

to preclude attack at this site.[64] *N*-oxide formation from *N*-acetylenic amines has been detected by g.c.-m.s. using a diagnostic pattern of pre-identified decomposition products.[65]

Contaminants unrelated to the drug can be present in the original drug, be introduced during work-up, or be present in the body fluid or tissue. Phthalates are frequent offenders and have been found in many commercial formulations such as those of estrogens.[66] They have also been introduced into samples blown down with nitrogen led through a PVC tube.[67] Mass spectra of phthalates have been reported[68] and their metabolism has been studied.[69,70] Other recently identified contaminants include 2-(2-hydroxymethylmercapto)benzothiazole from disposable syringes[71] and the antioxidant 2,5-di-tert-amyl-*p*-benzoquinone found in adhesive tape.[72]

2 Qualitative Studies

Use of Stable Isotopes.- Differentiation of drug-related peaks from peaks due to endogenous compounds in g.c.-m.s. studies has been achieved by the 'isotope-cluster' or 'twin-ion' technique, which involves the administration of a 1:1 mixture of the drug together with an analogue labelled with a stable isotope. The resulting spectra are searched for the presence of compounds containing the characteristic ion doublets. A computer program has recently been developed for constructing isotope-cluster chromatograms[73] of halogen isotopes and could be adapted for use with other isotope mixtures. Deuterium is the most usual isotope used in these studies, but care must be exercised to ensure that it is not lost during metabolism. Applications include the identification of 3-hydroxyparacetamol (8) as a metabolite of paracetamol (7) by the use of a $(^{2}H_{3})$ analogue,[74] hydroxy and desmethyl metabolites of aminopyrine (33) by the use of the $(N\text{-}^{2}H_{3}C)$ analogue,[75] a ring-opened metabolite (31) of phencyclidine (29) by the use of $(^{2}H_{5})$phencyclidine,[76,77] and hydroxylated metabolites of Δ^{1}-tetrahydrocannabinol (Δ^{1}-THC (37)) by the use of the $(1'',1'',2'',2''\text{-}^{2}H_{4})$ analogue.[78] In the last example the trimethylsilyl (TMS) derivative of 7-hydroxy-Δ^{1}-THC (38), the major metabolite, had a spectrum and a g.l.c. retention time that were almost identical to those of the corresponding derivative of endogenous 1-monopalmitin. Other stable isotopes do not suffer from a potential metabolic loss to the same extent as deuterium. ^{13}C

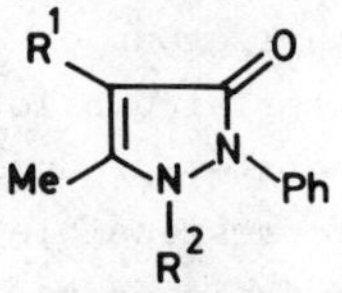

(33) $R^1 = N(CH_2)_2, R^2 = Me$
(34) $R^1 = H, R^2 = Me$
(35) $R^1 = CHMe_2, R^2 = H$
(36) $R^1 = R^2 = H$

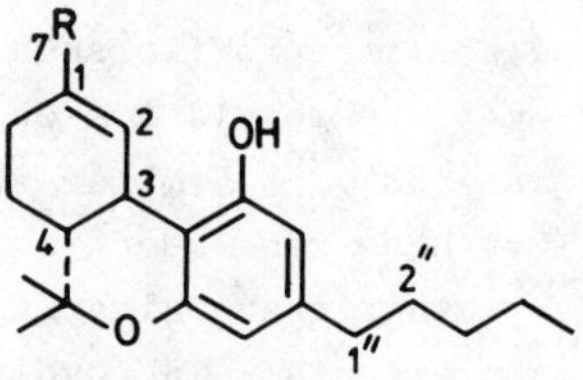

(37) R = Me
(38) $R = CH_2OH$
(39) R = COOH

has been used in studies with cambendazole conjugates,[79] and a radioactive ^{14}C-label has been used in a study with the synthetic prostaglandin cloprostenol.[80] The use of stable isotopes as tracers for t.l.c. separations[81] and for the investigation of drug disposition by isotope ratio measurements[82] has also been reported.

For differentiation of metabolites from compounds of the same structure occurring naturally, administration of the labelled drug alone is more appropriate. Phenylacetic acid and related compounds have been identified as metabolites of the flavone rutin following administration of $(^2H_3)$rutin,[83] dimethyl-$(\alpha,\alpha,\beta,\beta-{}^2H_4)$-tryptamine has been used to study the formation of indole-acetic acid, *N*-methyltryptamine, and related compounds for *NN*-dimethyl-tryptamine,[84] and hippuric acid formed as a conjugate of benzoic acid has been differentiated from endogenous hippuric acid by the use of $(^2H_5)$benzoic acid.[85,86]

Replacement of deuterium in studies of this type can indicate the site of metabolic attack as demonstrated by hydroxylation of $(N-{}^2H_3)$- and $(3-{}^2H_3)$-aminopyrine (33)[75] and two deuteriomethyl analogues of tofizopam.[87] Lack of replacement has been used to exclude aromatic hydroxylation in certain metabolites of ketamine (20)[49] and phencyclidine (29).[77] In another study, the structure of the clucuronide conjugate of the neuroleptic bromperidol has been shown to involve coupling to the tertiary alcohol group by the metabolite's inability to undergo hydrogen: deuterium exchange.[88]

Pronounced isotope effects associated with deuterium replacement in the rate-determining step of a metabolic reaction can result in 'metabolic switching'. This produces an alteration of the metabolite profile such as has been observed with the inhibition of hydroxylation on the (3-^2H_3)methyl group of aminopyrine (33).[89] Similar studies with deuteriated halothane ($CF_3CHClBr$) have been used to show that metabolites produced by reductive rather than oxidative metabolism are responsible for hepatotoxic effects; in conditions of oxygen deficiency, concentrations of metabolites of deuteriated halothane resulting from oxidative pathways were low, whereas hepatotoxicity was not.[90] Metabolic inhibition can be used to advantage to prolong the action of a drug; thus the anti-tumour properties of 6-mercaptopurine are potentiated by deuterium incorporation at the 2 and 8 positions, which inhibits the action of xanthine oxidase.[91] The absence of an isotope effect as seen with many drugs such as diphenylhydantoin (2)[92] with aromatic deuterium substitution has been interpreted as indicating the intermediacy of arene oxides rather than metabolite production by direct substitution. Comparison of the inter- and intra-molecular deuterium-isotope effects during demethylation of *NN*-dimethylphentermine (40) containing deuterium in one or both *N*-methyl groups has indicated that cleavage of the C–H bond does not contribute significantly to V_{max} and that the methyl groups are free to exchange at the enzyme-active site.[93]

Incubations of paracetamol (7) with microsomal preparations in an ^{18}O-enriched atmosphere have been used to demonstrate air as the source of oxygen in the 3-hydroxy metabolite (8) probably *via* an arene oxide.[74] Similar incubations of (+)-1-(4'-ethylphenoxy)-3,7-dimethyl-6,7-epoxyoctane with either ^{16}O or ^{18}O in the epoxide ring and isotopic ratio measurements of the products have demonstrated that enzymic attack is at the secondary carbon (C-6) whereas acid-catalysed ring opening involves attack at the tertiary centre (C-7).[94] In a related study involving mass spectrometry, the formation of a dihydroxy metabolite from dimethylclondidine [(54), Scheme 7] was shown to involve direct substitution of both hydroxy groups (route a) rather than the formation of an intermediary epoxide (58).[95] In another study, $(^2H_3)$acetyl groups have been used to monitor the enzymic deacetylation of *N*-acetylprocainamide and to overcome the problem of re-acetylation.[96] By using a mixture of $(^2H_8)$ and unlabelled

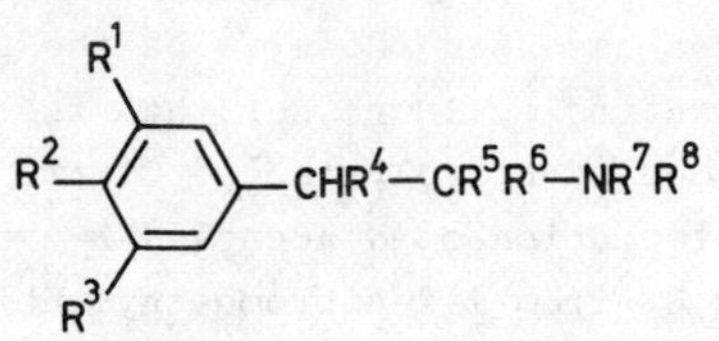

$R^1R^2R^3C_6H_2\text{-}CHR^4\text{—}CR^5R^6\text{—}NR^7R^8$

(40) $R^1 = R^2 = R^3 = R^4 = H, R^5 = R^6 = R^7 = R^8 = Me$

(41) $R^1 = R^3 = R^5 = R^7 = H, R^2 = NO_2, R^4 = OH, R^6 = CH_2OH, R^8 = COCHCl_2$

(42) $R^1 = R^3 = R^5 = R^7 = H, R^2 = NO_2, R^4 = OH, R^6 = CH_2OH, R^8 = COCHO$

(43) $R^1 = R^2 = R^3 = R^4 = R^5 = R^7 = R^8 = H, R^6 = Me$

(44) $R^1 = OMe, R^2 = R^3 = R^4 = R^5 = R^7 = R^8 = H, R^6 = Me$

(45) $R^1 = R^3 = R^4 = R^5 = R^7 = R^8 = H, R^2 = OMe, R^6 = Me$

(46) $R^1 = R^2 = R^3 = R^4 = R^5 = R^7 = H, R^6 = Me, R^8 = Pr$

(47) $R^1 = R^2 = R^3 = R^4 = R^5 = H, R^6 = R^7 = Me, R^8 = CHC{\equiv}CH$

(48) $R^1 = R^3 = R^5 = R^7 = H, R^2 = NO_2, R^4 = OH, R^6 = CH_2OH, R^8 = COCOOH$

(49) $R^1 = R^2 = R^3 = OMe, R^4 = R^5 = R^6 = R^7 = R^8 = H$

(50) $R^1 = R^3 = R^4 = OH, R^2 = R^5 = R^6 = R^7 = H, R^8 = Bu^t$

(51) $R^1 = CH_2OH, R^2 = R^4 = OH, R^3 = R^5 = R^6 = R^7 = H, R^8 = Bu^t$

(52) $R^1 = R^3 = R^4 = OH, R^2 = R^5 = R^6 = R^7 = H, R^8 = p\text{-}CH_2PhOH$

(53) $R^1 = Cl, R^2 = R^3 = R^5 = R^6 = R^7 = H, R^4 = OH, R^8 = Bu^t$

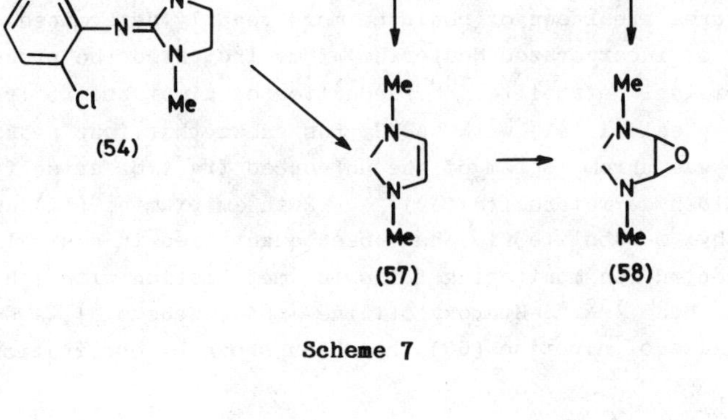

(54) (55) (56) (57) (58)

Scheme 7

1,3-bis(2-chloroethyl)-1-nitrosourea (BCNU), the metabolite 1,3-bis(2-chloroethyl)urea has been shown to be the product of denitrosation of the intact molecule rather than the recombination product of molecules formed by hydrolysis.[97] Of two pathways (a, b) proposed for the formation of *N*-acetyl-*S*-2-hydroxyethyl-*l*-cysteine [(61), Scheme 8] from 1,2-dibromoethylene (59)[98] the oxidative route was shown to contribute most following administration of the tetradeuterio analogue and observing exchange of the α-hydrogens.[99]

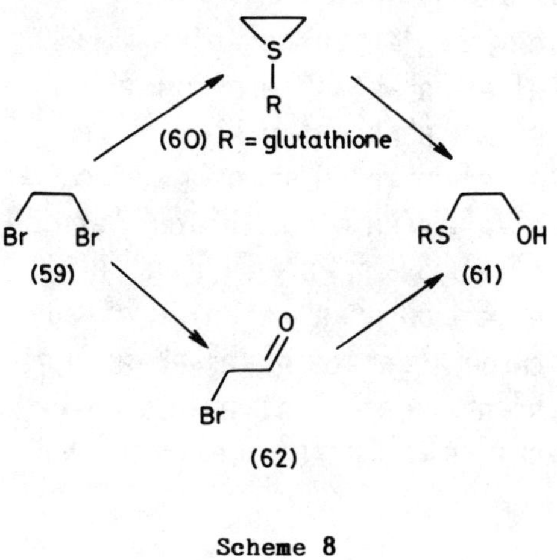

Scheme 8

As an alternative to the administration of labelled drugs, deuterium can be introduced chemically into metabolites during work-up to yield structural information. Thus ketone and acid metabolites of THC have been reduced with $LiAl^2H_4$ to alcohols whose structures are known or could be more readily determined. The number of incorporated deuterium atoms indicated the structure of the original metabolite.[100] Reduction of a metabolite fraction from chloramphenicol (41) with NaB^2H_4 has shown that four peaks present in the gas chromatogram of the unreduced fraction arise from the same aldehyde metabolite (42).[101] Both imipramine (12) and its desmethyl metabolite (13) have been quantified in a single peak by selected-ion monitoring following methylation with ($^2H_2$)-formaldehyde.[102] 5-Hydroxycotinine [(64), Scheme 9], a secondary metabolite of nicotine (63), has been shown by observation of the

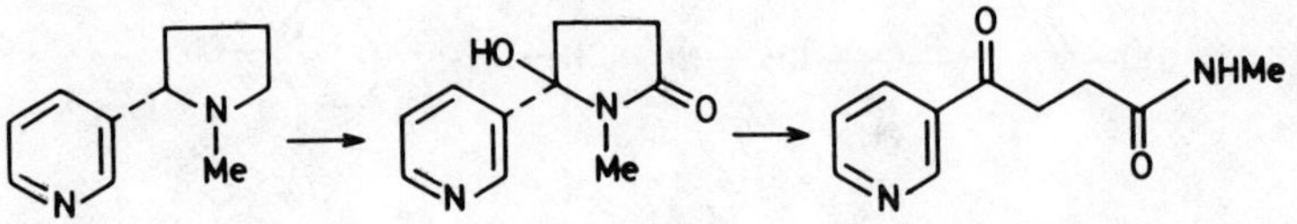

(63) (64) (65)

Scheme 9

umber of exchangeable hydrogens to exist as an open-chain tructure (65) in aqueous solution.[103] Deuteriated ethanol has een used to label NADH pools[104,105] and to trace the incorporation f metabolically removed hydrogen into bile acids.[105]

erivatization.- Chemical derivatization in g.c.-m.s. has been eviewed.[106] TMS derivatives continue to be the most widely used or oxygen-containing compounds and generally give single peaks ith no decomposition. However, dehydration of the pyrrolizidine lkaloids indicine and heliotrine *N*-oxides has been reported to ield dehydro products with BSTFA and TMS-imidazole but not when STFA is omitted.[107] Chlorotrimethylsilane or chloro-tert-utyldimethylsilane has been used to trap reactive epoxide etabolites of the allyl-substituted drug alclofenac[108] and of llylbenzene analogues[109] as the silyl ethers of the derived hlorohydrins. The very abundant $(M - 57)^+$ ion from tert-utyldimethylsilyl derivatives is useful in several highly sensitive ssays of, for example, valproic acid (di-n-propylacetic acid)[110] nd steroid anti-arrhythmic drugs.[111] Numbers of TMS groups in molecule can be estimated by using perdeuteriated TMS erivatives.[100,112-114]

Fluoroacyl analogues are the derivatives of choice for amines n account of their high volatility. G.c.-m.s. properties of the rifluoroacyl (TFA) derivatives of apomorphines have been tudied,[115] and an on-column method has been reported for ntroducing both TMS and TFA groups into phenylalkylamines.[116] ixed TMS/pentafluoropropionyl (PFP) derivatives of trimetoquinol ave been prepared from the tris-TMS derivative by replacement of he *N*-TMS group by a PFP moiety.[117] Recently reported decompositions f fluoroacyl derivatives include elimination of derivatized ydroxy groups from formoterol [(66), Scheme 10] during preparation

NHCHO

HO–C6H3–CH(OH)–CH2–NH–CH(Me)–CH2–C6H4–OMe

(66)

↓

NHCHO, PFP

MeO–C6H3–CH=CH–N(PFP)–CH(Me)–CH2–C6H4–OMe

(67)

Scheme 10

of a Me:PFP derivative[118] and from the 17α position of 17α-ethylnyl oestradiol.[119] In addition, the reduction of the derivatized amide group of atenolol (68) to a nitrile (69)[120] and the removal of the nitroso group from nitrosoureas[121] have been observed. Several of these reactions are quantitative and finally yield satisfactory derivatives.

R^1, R^2, R^3, R^4 substituted benzene–O–CH_2–CH(OH)–CH_2–$NHCR^5Me_2$

(68) $R^1 = R^2 = R^4 = R^5 = H, R^3 = CH_2CONH_2$
(69) $R^1 = R^2 = R^4 = R^5 = H, R^3 = CH_2CN$
(70) $R^1 = CH_2CH_2{=}CH_2, R^2 = R^3 = R^4 = R^5 = H$
(71) $R^1 = R^2 = R^4 = R^5 = H, R^3 = CH_2CH_2\text{-}O\text{-}Me$
(72) $R^1 = O\text{-}CH_2CH_2{=}CH_2, R^2 = R^3 = R^4 = R^5 = H$
(73) $R^1 = R^2 = R^4 = R^5 = H, R^3 = NHCOMe$
(74) $R^1 = Cl, R^2 = R^3 = H, R^4 = R^5 = Me$
(75) $R^1 = R^2 = R^4 = R^5 = H, R^3 = OH$
(76) $R^1 = R^2 = Me, R^3 = OCOMe, R^4 = R^5 = H$
(77) $R^1 = R^4 = Cl, R^2 = R^3 = H, R^5 = Me$

The structures of the products formed by alkylation with dimethylformamide dialkylacetals have received some attention. Thus barbiturates have been shown to yield the 2-dimethyl acetal derivative rather than the *NN*-dimethyl derivative on flash-heater methylation with DMF-dimethyl acetal,[122] and secondary amines, as illustrated by desipramine (13), give *N*-formyl derivatives.[123-125]

Derivatives recently used for other functional groups include *S*-acetyl analogues produced from glutathione conjugates,[126] *N*-ethylmaleimide derivatives of the thiol group of captopril (78),[127,128] a propylene oxide adduct of the pyrazole ring in the heterocyclic steroid stanozolol (pyrazolo-17β-hydroxy-17α-methylandrostane),[129] and the hydrazones formed from the

(78)

reactions of the hydrazine-containing drugs phenalzine ($PhCH_2CH_2NHNH_2$)[130] and isoniazid[131,132] with pentafluorobenzaldehyde and benzaldehyde, respectively. The homologous n-propyloxime rather than the methyloxime of the steroid anaesthetic alphaxalone has been used to produce selective g.l.c. shifts in the ketone metabolite 3α,11-dihydroxy-5α-pregnan-20-one.[133] In contrast to their use for increasing volatility, derivatives have also been used to reduce volatility of small molecules, an example being the ethanolamine amide of acid metabolites of the anaesthetic enflurane (CF_2HOCF_2CFHCl).[134]

Reported derivatives of bifunctional compounds include cyclic alkane boronates of diols, diamines, and hydroxyamines,[135-137] and the hexafluoroacetylacetone condensation product (80) of the amidine group in debrisoquine ((79), Scheme 11).[138]

Chiral derivatization for the chromatographic resolution of several racemic drugs has been reported. Thus *R*-(+)- and *S*-(-)-propranolol (81) have been separated as diastereoisomeric amides formed by reaction with *N*-trifluoroacetyl-1-prolylchloride[139] or as the urea derivatives formed with *R*-(+)- or *S*-(-)-1-phenylethyl isocyanate.[140] (*R*)- and (*S*)-alprenolol (70) and metoprolol (71) have been resolved as derivatives with L-leucine.[141]

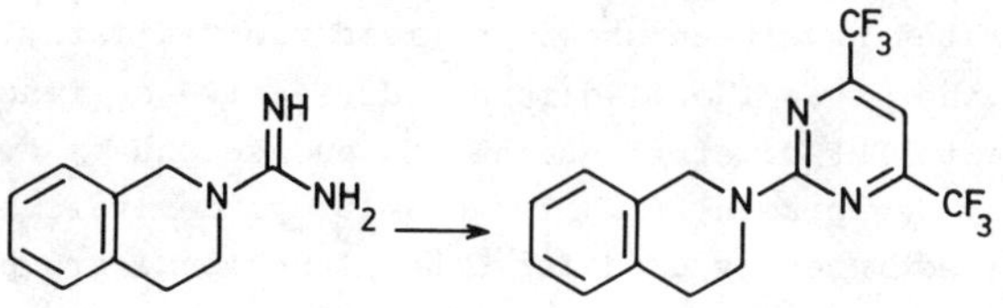

(79) → (80)

Scheme 11

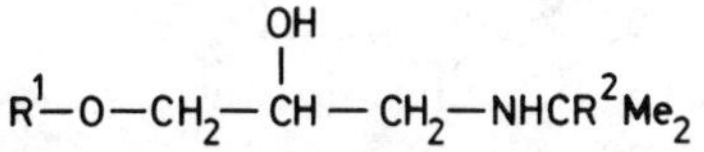

$R^1{-}O{-}CH_2{-}CH(OH){-}CH_2{-}NHCR^2Me_2$

(81) R^1 = 1-naphthyl, R^2=H

(82) R^1 = 3-morpholino-tetrahydrothien-4-yl, R^2 = Me

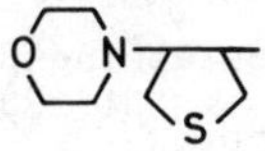

Derivatives have also been exploited for the trapping of reactive intermediates in metabolic reactions. In addition to the epoxide example cited above, labile aldehyde intermediates of propranolol (81) and oxprenolol (72) have been trapped in incubates as methoximes[142] and propriolaldehyde from pargyline, $PhCH_2N(Me)CH_2C{\equiv}CH$, has been trapped as its semicarbazone.[143] Cyanide complexes have been used to detect iminium ions from 1-benzylpyrrolidine,[144] methapyrilene,[145] and phencyclidine (29).[146]

Metabolite Identification.- *General.* Fused-silica capillary columns are becoming widely used in g.c.-m.s. studies and can be introduced directly into the ion source of the mass spectrometer. The main disadvantage lies in the difficulty in coating them with polar phases and thus most reported work has been performed with low polarity phases such as OV-1. For packed-column work the silicon phases OV-1 and OV-17 are the most widely used. A chiral Chirasil-Val column has been employed to separate four optical isomers of the *N*-trifluoroacetyl-L-prolylamide derivative of amphetamine (43).[147]

Although e.i. continues to be the most extensively used ionization method, other forms of ionization are receiving increasing attention. Chemical ionization with a variety of reagent gases yields complementary information, particularly with respect to molecular weight. Pulsed positive-ion/negative-ion mass spectrometry with methane as the reagent gas has yielded more informative spectra than e.i. for a series of sulphonamides.[148] Negative-ion mass spectra of benzodiazepines and related benzophenones have been studied,[149,150] and methane negative-ion c.i., unlike e.i., has been reported to give abundant high-mass ions from metabolites of probucol.[151] Nitrate esters also give abundant ions.[152] Negative-ion mass spectrometry under high-pressure conditions has been reported for several drugs.[153]

Model Compounds. A study of the *in vitro* metabolism of 1-(4-acetylnaphthyl)ethers with chain lengths from C_1 to C_4 has shown an increasing tendency for side-chain hydroxylation in the longer-chain compounds with ω-1-hydroxylation predominating.[154] Aromatic hydroxylation following metabolic induction with several metabolic inducers such as phenobarbitone has been studied with biphenyl,[155,156] and β-naphthoflavone has been shown to induce 4-hydroxylation in mice.

Conjugation of carboxylic acids such as benzoic and salicylic acids with amino-acids has been studied.[85,86] A novel reaction found in this study was that the horse was able to produce β-hydroxyphenylpropionic acid from benzoic acid by chain elongation.[86] Conjugation of 3-phenoxybenzoic acid by several species has been compared[157,158] and shown to exhibit considerable species variation; a dipeptide conjugate was identified in the mallard duck.[159] Incorporation of 3-phenoxybenzoic acid into triglycerides of rat skin has also been reported.[160] Related conjugations of various molecules with lipids having considerable significance for toxicity and drug retention by tissues include fatty-acid conjugates of ethanol,[161] trichloroethanol,[162] etofenamate,[163] and DDT[164] and phospholipid conjugates of halothane metabolites.[165]

Glucuronide Conjugates. The analysis of intact glucuronide conjugates by mass spectrometry has been reviewed[166] with emphasis on derivatization and determination of the linkage with the aglycone. Pyridinium ions have been used for glucuronide identification by c.i.-mass spectrometry,[167] and the utility of NH_3-c.i.[168] and negative-ion desorption c.i.[169] has been evaluated for the study of underivatized glucuronides. Glucuronyl transferase has been immobilized on Sepharose and used for glucuronide synthesis,[170] and modified Sephadex LH-20 has been used to separate A- and D-ring glucuronides of steroids.[171] G.c.-m.s. of the TMS derivatives was used in both of these studies. Other glucuronides identified by similar techniques include the acid glucuronide of Δ^1-tetrahydrocannabinol-7-oic acid (39)[172] and conjugates of nomifensine,[173] dezocine,[174] and propranolol (81).[175] Four isomeric (α- and β-pyranose and -furanose) glucuronides of clofibrate have been characterized by g.c.-m.s. using a 30m SE-30 capillary column.[176] A *N*-glucuronide of sulfamethazine (83) has been identified as its methyl derivative by ammonia chemical ionization.[177] Novel glucuronides include a quaternary ammonium glucuronide (85) of

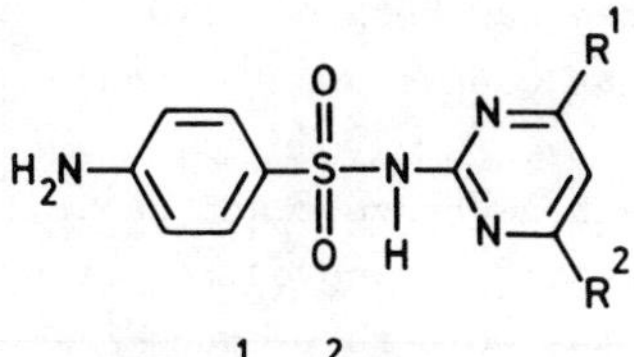

(83) $R^1 = R^2 = Me$

(84) $R^1 = R^2 = H$

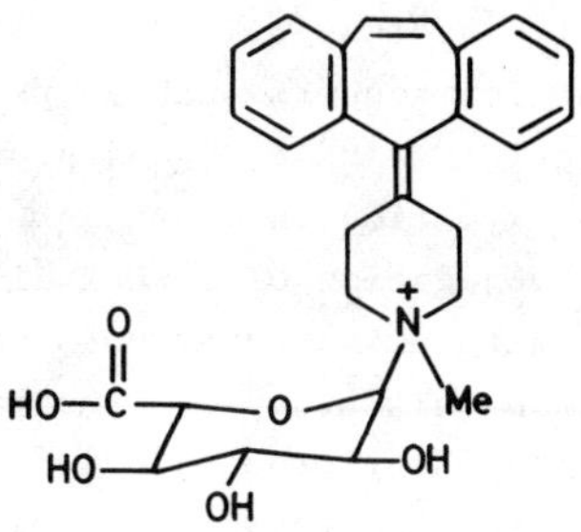

(85)

cyproheptadine[178] and a glucuronide (87) of tocainide (86)[179] involving incorporation of an additional molecule of carbon monoxide by the primary amine group (Scheme 12). Cyclization of the unconjugated CO_2 adduct has been shown to give 3-(2,6-xylyl)-5-methylhydantoin (88).[180]

Paracetamol. Studies on the toxic metabolites produced from paracetamol (7) have continued. Deacetylation to *p*-aminophenol by the kidney has been studied following conversion of the metabolite to *p*-hydroxybutyranilide and examination by direct-probe mass

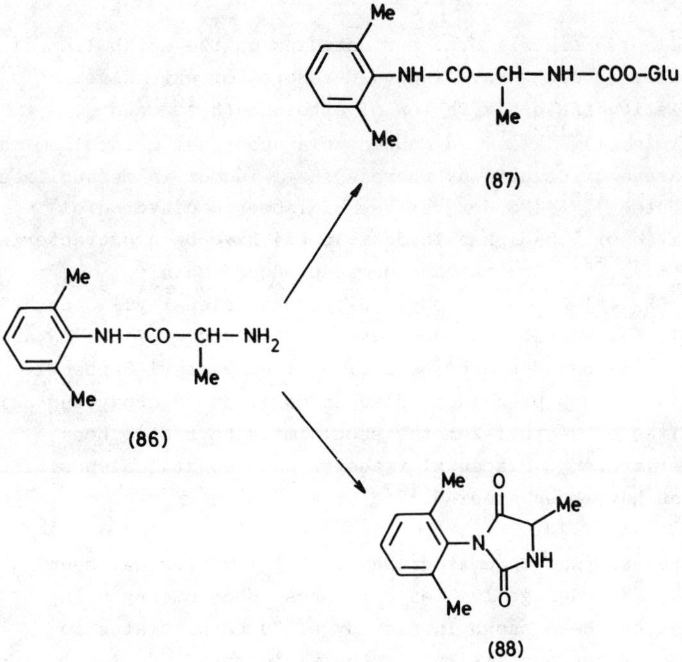

(86) NH—CO—CH(Me)—NH₂ → (87) NH—CO—CH(Me)—NH—COO-Glu; (86) → (88)

Scheme 12

spectrometry.[181] 3-Hydroxyparacetamol (8) derived from an arene oxide has been identified but not implicated in toxicity[74] and, similarly, *N*-hydroxy metabolites[31] are thought not to be involved.[182,183] Current theories implicate direct oxidation to *N*-acetyl-*p*-benzoquinone imine.[183] *O*-De-ethylation of phenacetin (6) has been shown to display biphasic kinetics indicating multiple enzyme involvement.[184] Thioether [*e.g.* (9) and (10)] conjugates have been identified[185] and their e.i., c.i., and f.d. spectra examined.[186] Acetylation of phenacetin and paracetamol by aspirin has been observed,[187] and a paracetamol pro-drug, 4-acetamidophenyl phosphate (11), has been demonstrated by g.c.-m.s. to undergo rapid hydrolysis by alkaline phosphatase[188] to paracetamol (7).

Diethylstilbestrol. Toxic epoxide-derived metabolites of diethylstilbestrol have been studied.[189,190] The compound forms alcohols, phenols, and quinones with attack on both the aromatic rings and ethyl groups.

Anti-epileptic Drugs. G.c.-m.s. studies on the metabolism of anti-epileptic drugs have included reports of extensive hydroxylation and dealkylation of ethotoin (3)[191] and mephenytoin (1).[192,193] d-Mephenytoin undergoes metabolism mainly by the arene oxide pathway whereas the 1-isomer is metabolized by other routes.[193] TMS derivatives of isomeric dihydrodiol metabolites of 5,5-diphenylhydantoin (2) have been characterized by g.c.-m.s.;[194] the rat and human produced mainly (5*S*)-5-[(3*R*,4*R*)-3,4-dihydroxy-1,5-cyclohexadien-1-yl]-5-phenylhydantoin (4) whereas the dog gave a 2:1 mixture of the 5*R* and 5*S* isomers. The (*S*)-dihydrodiol derived from 5-ethyl-5-phenylhydantoin (5) has been identified in dogs.[195] Hydroxy and acid metabolites of 2-ethyl-2-methylsuccinimide have also been characterized.[196] Placental transfer and neonatal disposition of primidone have been studied.[197]

Barbiturates. Gas chromatography of barbiturates has been reviewed,[198] and negative-ion c.i. mass spectrometry using isobutane has been shown in a study of 30 barbiturates to discriminate between all but two butyl isomers.[199] Alkylation is necessary for g.l.c. separations to prevent excessive tailing, but, as many barbiturates contain an *N*-methyl substituent, higher alkyl groups such as ethyl[200] or n-propyl[201-203] have been used to prevent loss of information following metabolic *N*-demethylation reactions. Barbiturate metabolites of primidone have been treated similarly.[204] A 3',4'-diol metabolite of amylobarbitone (89) has been identified in dog, mouse, hamster, and guinea-pig, but not in man.[205] The unsaturated barbiturates (90) - (92) are metabolized by the epoxide-diol pathway,[200,206-208] and a 4'-hydroxy metabolite has been identified and quantified from the unsaturated barbiturate methohexital (93).[209]

Xanthines. Methylation and demethylation of the xanthines have been studied by similar alkylation techniques. Thus, metabolism of theophylline (99), a drug used to treat apnea in premature infants, has been studied with the aid of ethyl,[210] n-propyl,[211]

(89) $R^1 = CH_2CH_2CHMe_2, R^2 = Et, R^3 = H$
(90) $R^1 = CH_2CH{=}CH_2, R^2 = CH_2CHMe_2, R^3 = H$
(91) $R^1 = CH{=}CH_2, R^2 = CHMePr, R^3 = H$
(92) R^1 = *cyclo*-hexen-1-yl, $R^2 = R^3 = Me$
(93) $R^1 = CH_2CH{=}CH_2, R^2 = CHMeC{\equiv}CCH_2Me, R^3 = Me$
(94) $R^1 = Et, R^2 = Ph, R^3 = H$
(95) R^1 = *cyclo*-hexen-1-yl, $R^2 = Et, R^3 = Me$
(96) $R^1 = CH_2CH{=}CH_2, R^2 = CHMe(CH_2)_2Me, R^3 = H$
(97) $R^1 = Et, R^2 = CHMe(CH_2)_2Me, R^3 = H$
(98) $R^1 = R^2 = Et, R^3 = H$

and n-pentyl[212,213] derivatives, and among the many metabolites identified was caffeine (100), formed by *N*-methylation; this metabolite is not formed by adults. Placental transfer of

(99) R = H
(100) R = Me
(101) $R = CH_2SH$
(102) $R = CH_2{-}SO{-}Me$
(103) $R = CH_2$-S-glutathione
(104) $R = CH_2CH(OH)Me$

theophylline has been demonstrated.[214] Deuteriomethylation can be used as an alternative to alkylation by larger groups in structural analysis of isomeric dimethylxanthines; the position of the deuteriomethyl group in the resulting caffeine yields diagnostic fragment ions from which its position can be deduced.[215] The rate of caffeine demethylation has been measured in humans using ^{13}C-labelled caffeine and respiratory gas analysis,[216] and thioether (101) and (102) and glutathione (103) conjugates have been identified.[217] The metabolism of the 7β-hydroxypropyl analogue of caffeine (proxyphylline (104)) has also been examined by the use of direct-probe[218] and g.c.-m.s. techniques.[219] Methylxanthine mass spectra have been discussed.[220]

Cannabinoids. G.c.-m.s. techniques have proved ideal for cannabinoid analysis in view of the low concentrations of both drugs and their metabolites in body fluids and tissues. The subject has been reviewed,[3,221] and the spectra of the TMS derivatives of Δ^1- (37) and Δ^6-THC[222] and a large number of their metabolites[223,224] have been studied. On-column methylation with DMF dimethyl acetal gives good yields of methyl derivatives of both neutral and acidic natural cannabinoids.[225] Methyl and TMS derivatives have also been used in an identification technique for naturally occurring cannabinoids involving ion-intensity plots at low electron energy (5 - 23 eV).[226] Metabolism of the major psychoactive cannabinoid Δ^1-THC (37) has been studied in man,[227] guinea-pig,[228] rat,[229] mouse,[230] and various micro-organisms,[231] and a large number of hydroxy and acid metabolites has been characterized. Epoxides are produced by the rat,[229] and the acid glucuronide of Δ^1-THC-7-oic acid (39) has been identified as the major urinary metabolite in humans.[172] Metabolic pathways involved in diol formation have been shown to be specific in mouse following administration of major monohydroxy metabolites,[230] and metabolic switching towards side-chain β-oxidation by inhibition of 7-hydroxylation has been demonstrated by administration of 6β-hydroxy-Δ^1-THC.[232] Δ^6-THC metabolism has been studied in the mouse[100] and guinea-pig[233] and shown to be similar to Δ^1-THC metabolism by these species. The epoxide-diol pathway is more prominent when the double bond is exocyclic as in Δ^7-THC; in the studies on the metabolism of this compound, 39 metabolites were identified by g.c.-m.s.[234,235] Stereospecific removal of deuterium during TMSOH elimination from TMS derivatives of $(3-{}^{2}H_{1})-\Delta^{7}$-THC metabolites yielded information on the stereochemistry of several of these metabolites.[234] Stereospecific production of alcohols by ketone reduction in the synthetic cannabinoid nabilone has also been reported.[236]

Other Hallucinogens. Direct-probe analysis of metabolites of LSD[237,238] and lisuride[239] have shown dealkylation, and hydroxylation reactions and the production of the ethyl-vinyl amide from LSD. Phencyclidine (29) undergoes hydroxylation in the cyclohexane ring to *cis* and *trans* isomers[240] and is metabolized by cleavage of the piperidine ring to give (31)[76,77,146,241] following hydroxylation α to the nitrogen.[146]

Opiates. Methylecgonine[242] and an epoxide metabolite have been identified from codeine;[243] hydroxy, dealkyl, and conjugated metabolites of naltrexone have been characterized in rat,[244] rabbit,[245] and man,[246] and a major metabolite of butorphanol in man has been identified as a diol having both hydroxy groups in the cyclobutyl ring.[247] Metabolites of codorphone have been derivatized as TMS oximes to prevent enolization during subsequent derivatization.[248]

Steroids. Studies with steroids have demonstrated interactions between norethindrone and the hydrazine-containing drug isoniazid resulting in the formation of a norethindrone hydrazone in rat stomach; the compound was trapped as its *p*-methoxybenzaldehyde hydrazone and identified by direct-insertion mass spectrometry.[249] In rat liver this steroid is metabolized to an epoxide.[250] G.c.-m.s. using ammonia c.i. has yielded information on seven hydroxylated, reduced, and deacetylated metabolites of ethynodiol diacetate,[251] and negative-ion ammonia c.i. has been used in studies with anabolic steroids.[252] An interesting rearrangement found in studies of the anabolic steroid dianabol (105) was migration of the 18-methyl group from C-13 to C-17 (Scheme 13) to produce the major urinary excretion product 18-nor-17,17-dimethyl-1,4,13(14)-androstatrien-3-one (106).[253] Catalysis by gastric acid was proposed for its formation. Detection of anabolic steroids of the 19-nortestosterone type has been performed by monitoring metabolic estrane-3,17α-diol as its TMS derivative.[254] Defluorination of flunisolide[255] [(107), Scheme 14] is thought to involve a ketone intermediate (109),[256] and spirolactone has

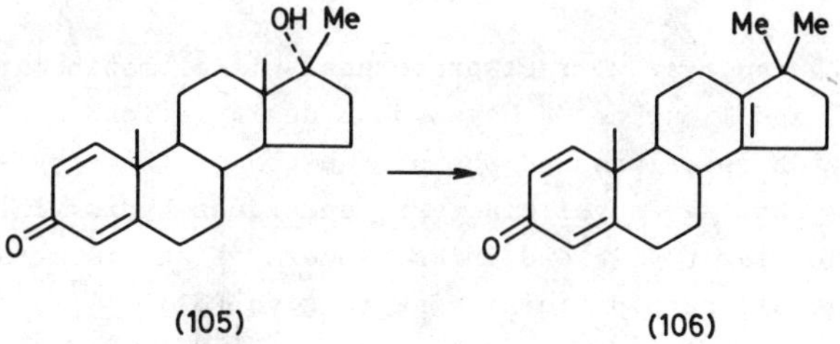

OH Me
Me Me

(105) (106)

Scheme 13

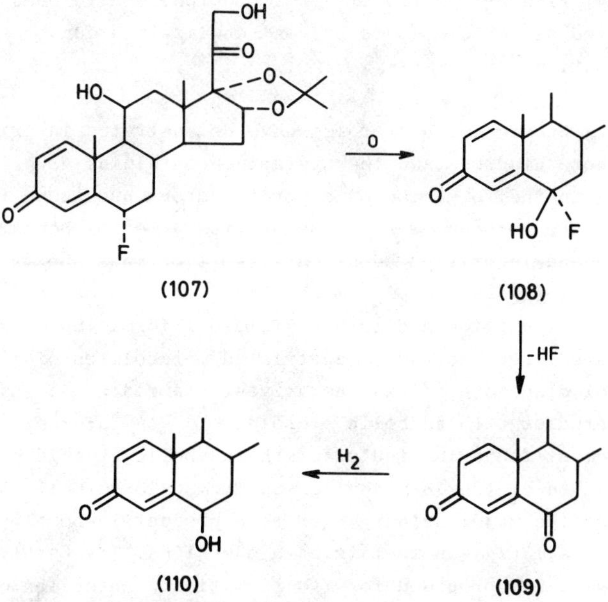

HO OH O O O F
O
HO F
-HF
H_2
OH O

(107) (108) (109) (110)

Scheme 14

been shown to form a 7α-thiol metabolite.[257] Hydroxy and dehydro metabolites have been reported for althesin (a mixture of alphaxalone and alphadalone acetate),[258] trilostane,[259,260] ethylestrenol,[261] and 19-nortestosterone;[262] plasma alphaxalone levels have been measured by g.c.-m.s.[263]

Phenothiazines. The phenothiazines promethazine (111)[264,265] and promazine (112)[266] are extensively metabolized by *N*-demethylation and oxidation on both nitrogen and sulphur atoms. Mass-spectral properties of phenothiazine *S*-oxides[267] and *SS*-dioxides[268] have been reported. Ring hydroxylation of the phenothiazine methotrimeprazine (113) has been demonstrated,[269] but the positions of the hydroxy groups were not determined.

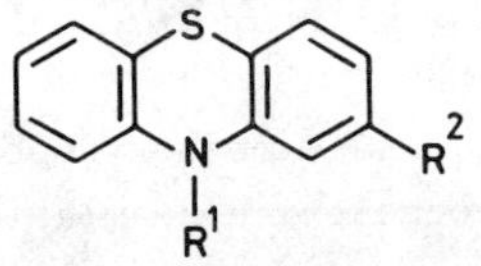

(111) $R^1 = CH_2CHMeNMe_2, R^2 = H$
(112) $R^1 = (CH_2)_3NMe_2, R^2 = H$
(113) $R^1 = CH_2CHMeCH_2NMe_2, R^2 = OMe$

Benzodiazepines. Studies on benzodiazepine metabolism have shown *N*-demethylation of lormetazepam (114) in dogs[270] and a sex difference favouring males in the production of the 3-hydroxy metabolite of diazepam (115) in rats.[271] Peptido-aminobenzophenones form benzodiazepines *in vivo*,[272] and ethyl loflazepate (116) rapidly decarboxylates *in vivo* following hydrolysis of the ethyl ester group.[273]

β-Blockers. G.c.-m.s. studies on the metabolism of the β-blockers practolol (73)[274] and propranolol (81)[275] have shown deamination to glycols and removal of the entire alkyl chain to give phenols. In the case of practolol the resulting phenol is the analgesic paracetamol (7). Timolol (82) is metabolized by oxidation and opening of the morpholine ring,[114,276] pranolium chloride, the *NN*-dimethylammonium quaternary ammonium derivative of propranolol (81), yields 1-naphthol,[277] and the aromatic methyl group of bupranolol (74) is oxidized to a carboxylic acid.[278]

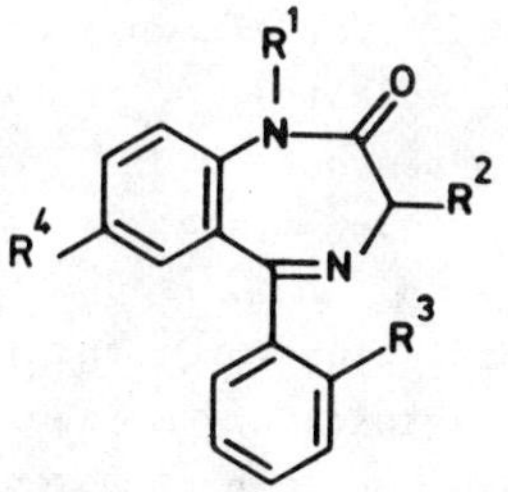

(114) $R^1 = Me, R^2 = OH, R^3 = R^4 = Cl$
(115) $R^1 = Me, R^2 = R^3 = H, R^4 = Cl$
(116) $R^1 = H, R^2 = COOEt, R^3 = F, R^4 = Cl$
(117) $R^1 = (CH_2)_2NEt_2, R^2 = H, R^3 = F, R^4 = Cl$
(118) $R^1 = R^2 = H, R^3 = Cl, R^4 = NO_2$

Amphetamines. Both 3- (44)[279] and 4-methoxyamphetamine (45)[280] are metabolized to methylated catechols and deaminated; *N*-propylamphetamine (46) also forms a catechol metabolite.[281] Metabolism of intermediates containing *N*-hydroxy[282] and nitro[283] groups implicated in the deamination reaction has been examined. 1-Methyl-1,2,3,4-tetrahydro-β-carboline has been detected[284] and quantified[285] in human platelets after ethanol intake; cyclization of the phenylethylamine Schiff base formed from metabolically produced acetaldehyde is implicated in its formation. Related measurements of its 5,6-dihydroxy (Salsolinol) and 5-hydroxy-7-methoxy (Salsoline) derivatives have also been reported.[286,287] Etilefrine is also metabolized to a tetrahydroisoquinoline.[288] *Cunninghamella echinolata* has been proposed as an excellent model of mammalian metabolism of (-)-deprenyl (47) and pargyline.[289]

Hydralazine. Studies on polymorphic acetylation of hydralazine (26) using the derived methyltriazolophthalazine (28)[138,290,291] have indicated a correlation with acetylator phenotype in the production of some metabolites.[138,290-293] However, methyltriazolophthalazine thought to arise only as a decomposition product of acetylhydralazine, is now known to be derived from hydralazone pyruvic acid hydrazone as well (27).[52,294,295]

Other Drugs Studied Mainly by G.C.-M.S. In other metabolic studies by g.c.-m.s., *N*-benzylaniline and *p*-benzylaminophenol have been found as metabolites of antazoline, 2-(*N*-phenyl-*N*-benzylaminomethyl)-imidazoline, in rabbit liver,[296] and the phenol has been shown to be a metabolite of *N*-benzylaniline.[297] Procarbazine, *N*-isopropyl-α-(2-methylhydrazino)-*p*-toluamide, is extensively metabolized[298] with the azo compound, *N*-isopropyl-α-(2-methylazo)-*p*-toluamide, the major circulating metabolite in the rat.[299] Eprazinone,[300] timperone,[301] clebropride,[61,302-304] and fentanyl[305] are *N*-dealkylated at their piperidine or pyrazine[300] rings. Clebropride is also *N*-oxidized and fentanyl is oxidized at the ethyl group to a carboxylic acid.[306] Drugs such as ninaprine, containing morpholine rings, undergo oxidative cleavage of this ring.[307] Econazole undergoes oxidation in the imidazole ring.[308] Sulphones are major metabolites of the sulphur-containing drugs tiflorex[309] and tolmesoxide.[310] An aldehyde adduct that presumably cyclizes to an observed hydantoin derivative (88) has been found in the rat[311] as a metabolite of tocainide (86). β-Oxidation is a major metabolic route for chlorambucil in rats,[312] and the product, *p*-(*NN*-bis(2-chloroethyl))aminophenylacetic acid, has been quantified by g.c.-m.s.[313] Epoxides have been detected as metabolites of quinine,[314] alclofenac,[315] and carbamezepine.[316,317] Plasma-desorption mass spectrometry has been reported to give good $M^{+\cdot}$ ions for carbamazepine oxide; during g.c.-m.s. studies of its TMS derivative, decomposition to acridine-9-carboxaldehyde and iminostilbene oxide occurred.[317] Desorption c.i. with ammonia has assisted g.c.-m.s. identification of three metabolites of procainamide in a study relating metabolite production with toxicity,[318] and f.d. has aided a study of suxibutazone and phenylbutazone metabolism in humans.[319] Oxidative dechlorination gives 2-difluoromethoxy-2,2-difluoroacetic acid from enflurane (1,1,2-trifluoro-2-chloro-difluoromethyl ether),[320] and a metabolite of haloperidol (involving reduction of a carboxyl group to a hydroxyl moiety) has been reported.[321] Extensive metabolism involving hydroxylation, dealkylation, and conjugation has also been reported for tramadol,[322] bromoprid,[323] verapamil,[324] and the anti-viral agent arildone.[325] Metabolic dechlorination of chloramphenicol (41) to an aldehyde (42)[101] and an oxamic acid (48)[326] has been shown to involve several metabolic steps.[327] Hydroxylation and conjugation reactions have been studied for the non-steroid anti-inflammatory agents fenclofenac,[328] tolfenamic acid (*N*-(3-chloro-2-methylphenyl)anthranilic acid),[329,330] indobufen,[331] and carprofen,[332] and unchanged carprofen has been shown to be the major circulating species of the latter drug in man.[333] The short plasma half-life of less than 1 minute reported for thymoxamine has been shown to be a first-order reaction involving deacylation by plasma esterases.[334,335] The anaesthetic chloroprocaine is also hydrolysed by plasma esterases, and the product, 2-chloro-4-aminobenzoic acid, is then *N*-acetylated.[336] Warfarin alcohols and hydroxy derivatives are the major metabolites of warfarin in rabbits,[337] and 1-(3-hydroxycyclohexyl)-1-phenyl-3-(1-pyrrolidinyl)-1-propanol has been identified as a metabolite of procyclidine in man.[338] Aromatic hydroxy metabolites of (-)-3-phenoxy-*N*-methyl-morphinan have been characterized in dogs.[339] Other drugs studied include the synthetic terpenoid epomedol[340] and the hypnotic ethchlorvynol.[341]

Both direct-probe and g.c.-m.s. methods have been used in metabolic studies of anti-tumour and anti-viral nucleoside analogues. These include 5-fluorouricil (5-FU,119),[342,343] 5-azacytidine,[112] 1-(tetrahydro-2-furanyl)-5-FU (120),[344] 1-(2'-deoxy-β-D-ribofuranosyl)-5-ethyluracil (121),[345] and

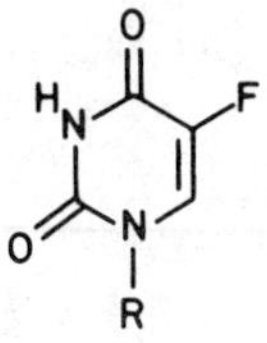

(119) R = H

(120) R = 2-tetrahydrofuranyl

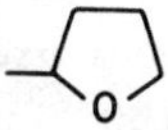

(121) R = $CONH(CH_2)_5Me$

1-hexylcarbamoyl-5-FU, a prodrug for 5-FU.[346] Extensive cleavage of the uricil and cytidine rings has been found in some of these studies. Assays for 5-FU have been reported[347-349] and used for the measurement of the drug after administration of various prodrugs.[348,349] Mass spectra of 5-FU derivatives have been described.[350,351]

Studies Using Direct-introduction Techniques. Analysis by direct-probe introduction of metabolites separated by other techniques has not been as widely used as g.c.-m.s. but has yielded much valuable information with compounds behaving poorly on gas-chromatographic columns. 4-Hydroxyphenazone has been identified as a metabolite of phenazone (antipyrine, 34) in man;[352] aminopyrine (33) yields a dihydro and a catechol metabolite when incubated with *Phenylobacterium immobilis*,[353] and isopropylphenazone (35) is hydroxylated and oxidized on the isopropyl group by the rat.[354] Twelve metabolites of the non-steroid anti-inflammatory drug fenbufen have been identified in six animal species[355] and man.[356] In a study on the metabolism of 2-acetamido-4-(chloromethyl)thiazole in germ-free and conventional rats, methylthio ester, sulphoxide, and sulphone metabolites were only found in the conventional animals, indicating their production by intestinal microflora.[357] The first study on the metabolism of a branched heterocycle, oxapadol, has been reported in man, dog, and rat,[358] and a unique deaminated metabolite of sulfamethazine (83) has been found in swine;[359] isolation was by h.p.l.c. and identification by both e.i. and methane-c.i. mass spectrometry. H.p.l.c. separations have also preceded mass-spectrometric identification of metabolites of BCNU, related nitrosoureas,[360,361] and tamoxifen.[362]

T.l.c. and Sephadex column chromatography have been used in the identification of thirteen *N*-oxide and deaminated metabolites of the antidepressant zimelidine, (*Z*)-3-(4-bromophenyl)-*NN*-dimethyl-3-(3-pyridyl)allylamine.[363] Metabolites of sulfinalol,[364] separated by t.l.c., have been shown to include the sulphone, sulphide, and desmethyl sulphide by direct-probe mass spectrometry. T.l.c. separation has also been used in the identification of desmethyl and hydroxy metabolites of guaifenesine[365] and piroxicam[366] and *N*-desalkyl metabolites of ziperol.[367] Metabolism of enclomiphene,[368] tizoprolique,[369] cartazolate,[370] and tripamide[371] has also been studied. Direct-probe analysis has also been used to identify *N*-acetyl metabolites of pipemidic acid,[372] benzophenones formed from the antihistamine bromazine,[373] and thirteen oxidized and hydrolysed metabolites from the CNS stimulant methylphenidate (*threo*-dl-methyl-α-phenyl-2-piperidine acetate).[374]

Metabolism of antibiotics has been studied by a variety of techniques. Thus f.d. has been used to identify penamaldic acid metabolites formed by opening of the thiazole ring of ampicillin and amoxycillin,[375] and g.c.-m.s. of the TMS derivative has yielded information on the presence of a 5-hydroxymethyl metabolite (isoxazole ring) of oxacillin by selected-ion monitoring.[376] Pulsed positive and negative c.i. spectra of several penicillins have been described.[377] G.c.-m.s. studies on cephaloglycin metabolism in man have shown deacylation and amide hydrolysis,[378] and deacylation has also been reported with 9,3"-diacetylmidecamycin.[379]

F.d. mass spectrometry has, until the advent of f.a.b. ionization, been the most useful technique for characterizing metabolites of ionic or involatile drugs and underivatized conjugates. Reports on the metabolism of ethidium bromide have appeared;[380,381] acylation and conjugation reactions predominate. Several halogenated hydrocarbons such as chloroform have been shown to form diglutathionyl dithiocarbonate,[382] and glutathione conjugates have also been characterized from acridine produced from the anti-cancer drug 4'-(9-acridinylamino)methanesulfon-*m*-anisidide.[383,384] Lamprene,[385] pinaverium bromide,[386] and nitrofurantoin[387] metabolisms have also been studied by f.d.-m.s.

Drug Screeening.-Reliability of drug screening by g.c.-m.s. in overdose cases has recently been investigated[388] in a comparative survey involving three commercial and one academic laboratory. Results from the analysis of samples from 20 patients suggest that laboratories do not reliably identify drugs in serum. Drugs responsible for the overdose were correctly identified in only 50 - 70% of cases with reported concentrations varying over a 10-fold range. Ethanol and salicylates were not detected. Screening methods for 1,4- and 1,5-benzodiazepines in urine using a computerized g.c.-m.s. technique have been reported,[389] and a selected-ion-monitoring technique for laxatives enabled these drugs to be monitored for at least four days.[390] Drug screening of motorists by g.c.-m.s. has been discussed;[391] barbiturates and methaqualone are the drugs most frequently encountered by the technique. Benzodiazepines, although often present, are detected by other methods.

3 Quantitative Studies

General Aspects.- Mass spectrometry continues to provide the most selective and sensitive methods for measurement of a large number of drugs and endogenous compounds and is frequently used as the standard for the evaluation of less selective techniques. Thus, radioimmunoassays (r.i.a.)[392,393] and homologous enzyme immunoassays[394] (e.m.i.t. technique) for Δ^1-THC (37) and an r.i.a. method for Δ^1-THC-7-oic acid (39)[395] have been validated by g.c.-m.s., and good agreement has been found between g.c.-m.s. and r.i.a. methods for benzodiazepines,[396] ethynyloestradiol,[397] and cyclazocine.[398] In comparisons between g.c.-m.s., g.l.c., and r.i.a. methods for imipramine (12)[399] and between g.l.c., h.p.l.c., and g.c.-m.s. methods for the benzodiazepine midazolam (22),[400] g.c.-m.s. was generally the most sensitive but r.i.a. methods were cheaper and more convenient. H.p.l.c. offered the advantage that derivatization was not usually required but could be subject to more interference than g.c.-m.s.[401] G.c.-m.s. also gave more reliable data than g.l.c. or colorimetric assays for sulfamethazine in tissues.[402] Confirmation of peak identity and purity in g.l.c.[403-406] and h.p.l.c.[407-410] is a frequent use of the technique, and many examples have been published of its use in determining the structure of derivatives used in these assays[411-413] and in confirming the absence of decomposition in the case of underivatized drugs.[414] The recent identification by g.c.-m.s. of deaminated metabolites of sulfadiazine (84) has cast doubt on the accuracy of quantitative techniques such as the Bratton-Marshall procedure that rely on the presence of amino groups.[415]

Many parameters such as choice of derivative and standard, ions monitored, and extraction procedures have to be optimized in any g.c.-m.s. assay, and several discussions of these topics have been published.[416-418] Interference at the mass of the ions measured in both standard and drug must be minimized by judicious choice of derivative and mass-monitored. In general, ions of high mass are less subject to interference, but an abundant ion at a lower mass may often give superior results if ions at high mass are not abundant. Models to predict the optimum mass,[419] to overcome problems associated with ion overlap,[420] and to correct for long-term changes in instrument parameters[421] in isotope-dilution assays have been published. The calculation of calibration graphs has also been discussed.[422] In contrast to the situation existing a few years ago, there has been little debate on the use of carriers for the measurement of very low concentrations of drugs, and only a few recent assays such as those reported for pethidine[423] and norphenazone (36)[424] include carriers. In both cases, structural rather than isotopically labelled analogues of the drug were used; this overcomes the problem of the inadvertent addition of unlabelled drug with the labelled carrier. An unknown compound was suspected as acting as a carrier in a recent assay for ethynyloestradiol;[425] its effect seemed to be greatest in the ion source and separator.

The sensitivity of conventional assays using selected-ion monitoring particularly for lipophilic drugs is frequently severely limited by interference by coextracted endogenous compounds. Extensive pre-g.c.-m.s. clean-up stages are inadvisable for a variety of reasons. A better approach is to improve the selectivity of the g.c.-m.s. system. Chemical ionization is often employed such as in a recent assay for the diuretics ethacrynic and tienilic acid,[426] or alternatively a deuteriated[427] or homologous[428] derivative can be prepared to produce an ion not present in the background. Another approach is to use a high-resolution mass spectrometer to filter out interfering ions of the same nominal mass but of different composition. Selected-ion monitoring assays for tamoxifen,[429] ketoprofen,[430] pentoxyfylline,[431] clomipramine,[432] and trioxsalen[433] using this technique have been reported, the trioxsalen assay being usable at the 25 pg cm^{-3} level with the spectrometer operated at a resolution of 2200. Related methods using direct-probe introduction and repetitive scanning of spectra have been used to measure urapidil in serum with isobutane chemical ionization[434] and morphine in liver using a resolution of 8000 - 10000.[435] Alternatively, high-resolution capillary g.l.c. can be used to separate interfering g.l.c. peaks.[433] The m.s.-m.s. technique of monitoring a metastable ion can also give much improved signal:noise ratios as exemplified by a recent assay for Δ^1-THC (37) in plasma.[436] The drug was measured with a double-focusing mass spectrometer as its TMS derivative by single-ion monitoring of the $M^{+\cdot} \rightarrow (M-Me)^+$ transition (m/z 386 → 371) from the drug and $(1'',1'',2'',2''-{}^2H_4)$cannabinol used as the internal standard. Additional selectivity was achieved by selective methylation of contaminating fatty acids, using diazomethane. The lower limit for measurement was 5 pg cm^{-3}, and the method has been used for pharmacokinetic studies.[437,438] A method for adapting the instrument for monitoring metastable ions at different masses has been published.[439]

Very few assays have been published in which the standard is not added directly to the biological matrix before extraction. Corrections for extraction losses are included in most assays. Several types of standard are in use together with both single- and multiple-ion detection or the use of scanned spectra. Analogues of the drug labelled with stable isotopes are the most satisfactory in correcting for extraction deficiencies, but care must be taken, in the case of deuteriated drugs in particular, that the label is not lost during the assay. Several deuterium atoms should be incorporated to prefent interference between the ions being monitored and, in this respect, synthetic methods of deuterium incorporation that do not rely on exchange techniques should preferably be used as more complete labelling is possible.

Assays Using Stable-isotope Standards.- *Deuterium.* An assay for chlorpheniramine in serum with the $[^2H_4]$ analogue as standard has been described[440] and used for pharmacokinetic analysis of the drug in children,[441] and two assays for trifluoperazine have appeared,[442,443] one with a 70 pg cm^{-3} detection limit.[443] In these assays and assays for pethidine,[444] mainserin,[445] and benztropine[446] no derivatization was necessary. Fluoracyl derivatization has been used for analyses of prenalterol (75),[447] mescaline (49),[448] and atropine[449] with $[^2H_5]$, $[^2H_2]$, and $[^2H_3]$ analogues, respectively, as standards; the latter two assays were used for pharmacokinetic studies in humans. Atrophine has also been measured in mouse brain to 1 ng cm^{-1} as its TMS derivative,[450] and the same derivative has been used for pharmacokinetic studies on Δ^1-THC (37),[451-454] bromhexine,[455] and naproxen;[456] $[^2H_3]$ analogues were employed in the latter two assays. Aspirin has been measured following extractive alkylation and conversion to its benzyl derivative.[457] The $[^2H_7]$ benzyl analogue was used as the internal standard. Other drugs quantified by the use of g.c.-m.s. with deuteriated standards include dianabol (105),[458] flurbiprofen,[459] diphenylhydantoin (2),[460] chlorambucil,[461] and ethambutol $[(CH_2NHCHEtCH_2OH)_2]$.[462] With ethambutol, a $[^2H_4]$ label was incorporated, but the drug was monitored with the use of the $M/2^+$ ions, giving a mass-unit difference of only two between drug and standard. A similar assay using methane c.i. for detection has been reported for this drug, but the sensitivity, 36 ng cm^{-3}, was poor compared with electron ionization.[463]

Detection using c.i. usually results in improved signal-to-noise ratios but requires more accurate control of the ion-source conditions, particularly temperature and reagent-gas pressure. Methane, isobutane, and ammonia account for the gases used in most reported assays. Frequently the gas is used as both reagent and g.l.c. carrier. Deacetylmetipranolol, a metabolite of the β-blocker metipranolol (76), has been measured using methane c.i. against a $(^{2}H_{6})$ standard to 1 - 2 ng cm^{-3} in serum and urine,[464] and the method has been used for pharmacokinetic analysis of the parent drug.[465] Other assays employing methane as the reagent gas include those for dibucaine,[466] dothiepin (14),[467] and fentanyl;[468] 500 pg cm^{-3} was the detection limit in the latter two assays. Isobutane has been used in an assay for phenylalanine mustard,[469] and ammonia for assays of terbutaline (50)[470] and clonidine,[471] both of which had lower detection limits of 100 pg cm^{-1} of plasma.

By monitoring more than two ions it is possible to include measurements of metabolites together with the parent drug. Thus, dothiepin (14) and its desmethyl metabolite have been measured[472] and their pharmacokinetics studied[473] by e.i. mass spectrometry with a $(^{2}H_{3})$ standard; the method has been extended to the related drug mianserin.[474] Maprotiline and desmethyl maprotiline have also been measured to 2 ng cm^{-3} with a trideuteriated standard.[475]

More usual for multi-component monitoring is the use of several standards. An assay for metoprolol (71) and two metabolites using e.i., with a 300 pg cm^{-3} detection limit,[476] has been used to study the pharmacokinetics of the drug in patients with hepatic[477] and renal[478,479] insufficiency. Amitriptyline and seven metabolites have been studied in depressive patients.[480] Simultaneous assay of five antidepressant drugs involving extraction with a mixed solvent (toluene, heptane, and isoamyl alcohol) and direct injection of the extract onto the g.l.c. column followed by methane c.i. has been described.[481] Methanol c.i. has been used in an assay for several tricyclic antidepressants;[482,483] good $(M + 1)^{+}$ ions were produced, enabling picogram sensitivity to be achieved. Ammonia c.i. has been used in a simultaneous assay for the β-sympathomimetics terbutaline (50), salbutamol (51), and fenoterol (52) following extraction with a Sep-Pak C-18 cartridge and chromatography on a 60m fused-silica capillary column.[484] The method was not sensitive enough to measure fenoterol in the therapeutic range. Cocaine and its metabolites have also been measured in blood and tissues by ammonia c.i.[485]

Other Stable Isotopes. Use of other stable isotopes is much less common. However, an assay using a ^{15}N standard and involving e.i. detection has been reported for allopurinol,[486] and ammonia c.i. has been used in an assay for 5-fluorouracil (119) and pyrimidine bases.[487] Phenobarbitone (94) and its *p*-hydroxy metabolite have been assayed with a mixed $^{13}C^{15}N$ standard,[488] and the method has been used in a study of phenobarbitone:valproic acid interaction.[489]

Assays Using Analogue Standards.- *Single-ion Monitoring.* Analogue standards offer the advantage of cheapness and availability. If g.l.c. separation can be obtained, a common ion can be monitored for both drug and standard, and this minimizes instrumental variables. The first assay for the β-blocker tobanum (77) in plasma has been reported using this technique,[490] with propranolol (81) as the internal standard; both compounds were monitored at m/z 266. Captopril (78) and its disulphide have also been measured by single-ion monitoring at m/z 366, using the *N*-ethylmaleimide derivatives; the standard was the corresponding *N*-butylmaleimide derivative.[491] A sensitive assay for *erythro*-α(*p*-piperidyl)-2,8-bis(trifluoromethyl)-4-quinoline has appeared;[492] the 2,7,8-trifluoromethyl analogue was used as the internal standard, the *N*-TFA derivatives being monitored at m/z 180. The pyrrolidine analogue of the *NN*-diethyl-containing anti-cholinergic drug oxybutynin was used in an assay sensitive to 500 pg cm^{-3}. The common base peak m/z 189 was monitored.[493] A rapid assay sensitive to 1.5 ng cm^{-3} has been reported for 2-chloroprocaine using procaine as the standard; m/z 86 was monitored.[494] Ammonia c.i. has been used in related assays for tolobuterol (53)[495] and 5-fluoro-2'-deoxyuridine.[496] In these cases the standards were the 4-methoxy analogue and the 3'-epi analogue, respectively. Homologues of the drug are also frequently used as standards when single-ion monitoring is used, as with the nitrosourea anti-cancer agent chloroethylcyclohexyl-nitrosourea (CCNU) and its 4-methyl derivative[497] and methyl- and ethyl-phenidate.[498] Methazolamide, 5-acetylimino-4-methyl-Δ^2-1,3,4-thiadiazoline-2-sulphonamide, has been measured in plasma and erythrocytes[499] against its propionyl analogue. The use of homologous standards and multiple-ion monitoring have been reported for assays of apomorphine in plasma[500] and brain,[501] valproic acid and various metabolites,[502,503] bromhexine,[504] oxepinac,[505] acetazolamide,[506] and clebropride.[507]

Multiple-ion Monitoring. Methods involving the use of analogue standards and multiple-ion monitoring appear to be the most popular for drug quantification. Thus timolol (82) has been measured to 500 pg cm^{-3} with propranolol (81) as the standard, ions at *m/z* 86 and 72 being monitored; the method was claimed to be more sensitive than existing g.l.c. assays.[508] The pharmacokinetics of cyclobarbitone (95) and quinalbarbitone (secobarbitone, 96) have been studied simultaneously with pentobarbitone (97) as the standard,[509] and an assay for phencyclidine (29) and its hydroxy metabolite that uses methane c.i. with 1-(1-phenylcyclohexyl)-morpholine as the standard has been used to study the pharmacokinetics of the drug in patients on methadone maintenance.[510] Trimethobenzamide, an anti-emetic, has been quantified by the use of reconstructed selected-ion chromatogram plots. The detection limit was not determined, but 100 ng cm^{-3} could be measured with a signal-to-noise ratio of 63.[511] The rapidly metabolized benzodiazepine ethyl loflazepate (116) has been studied by an assay involving the monitoring of its metabolites against a butylated standard,[512] and a method has been developed for quantifying two metabolites of sulphinpyrazone[513] and for investigating their pharmacokinetics.[514] The disposition and side effects of pentazocine, 6,11-dimethyl-2,6-methano-3-(3-methyl-2-butenyl)-1,2,3,4,5,6-hexahydro-3-*b* enzazocine-8-ol, have been studied in humans[515] and its chronopharmacokinetics studied in the beagle dog,[516] using the *N*-cyclopropylmethyl analogue as internal standard.

Other Quantitative Methods.- *Negative-ion C.I.* Negative chemical ionization has been used for the quantification of glyceryl trinitrate[517] and its metabolites[518] by selected-ion monitoring with deuteriated standards. $(M + Cl)^-$ ions were produced with dichloromethane as reagent gas; methane, isobutane, and ammonia gave NO_3^- as the base peak. Picogram sensitivity has also been reported for isosorbide dinitrate and pentaerythritol nitrates.[152] Positive-ion ammonia c.i. gave comparable sensitivity.[519]

Both positive and negative c.i. have been investigated for the measurement of 5,5-diphenylhydantoin (2)[520] and the quinidine analogue 7-trifluoromethyldihydrocinchonidine.[521] Sensitivity was comparable with the hydantoin but higher in the positive mode for the quinidine derivative. A very sensitive negative-ion g.c.-m.s. method using methane as both carrier and reagent gas has been reported for the benzodiazepine flurazepam (117) in human plasma.[522] The detection limit was 12 pg cm^{-3}; at 135 pg cm^{-1} the precision was 4%. This is eight times as sensitive as existing r.i.a. assays and over one hundred times as sensitive as electron-capture g.l.c.

H.P.L.C.-M.S. Very few combined h.p.l.c.-m.s. assays have been reported. The histamine H_2 receptor antagonist ranitidine and its metabolites have been measured with the use of $(^2H_3)$ranitidine as the standard.[523,524] About 1 μg cm^{-3} of each compound was required, and sample introduction to the spectrometer employed a belt interface. The dilactonic antibiotic mixture Antimycin-A has also been examined.[525]

Field Desorption. Quantification using field desorption has found application for involatile compounds. Perisoxal and its glucuronide have been measured with a (phenyl-$^2H_5$) standard and repetitive scanning over a narrow range, but difficulty was experienced in achieving constant ion current.[526] Cyclophosphamide and its metabolites have been measured by a similar technique[527-530] with a precision of better than 1% being reported.[529] A f.d. method for methotrexate quantification compared favourably with assays by h.p.l.c. and e.m.i.t., although the latter technique gave higher levels at the top end of the dose range.[531]

Direct Insertion. Direct-probe introduction with isobutane c.i. and deuteriated or ^{13}C internal standards has been used for assays of sodium valproate,[532] methadone,[533] and procarbazine.[534] Although the methods were rapid, they were generally less sensitive than g.c.-m.s. techniques.

Pharmacokinetic Studies.- Papers describing pharmacokinetic studies using previously reported or unreported methods include studies on propranolol (81),[535-538] its 4-hydroxy metabolite,[539] and other β-blockers.[540,541] The antihypertensive drug guanfacine has been studied during long-term treatment and withdrawal and shown to be eliminated by non-renal routes in patients with renal failure.[542-545] A useful correlation between plasma desipramine levels 24 hours after a single dose and the steady-state level has been found,[546] but no correlation was found between the plasma level of the drug and heart rate.[547] Pethidine metabolism[548] and disposition[549] have been studied, ketobemidone has been shown to have a low bioavailability and to be rapidly excreted,[550-552] and polymorphic hydroxylation of nortriptyline[553,554] and phenazone (34),[554] paralleling that of debrisoquine (79), has been demonstrated in man. Small amounts of atenolol (68), metoprolol (71),[541] terbutaline (50),[555] captopril (78),[556] and paracetamol (7) [557] have been measured in breast milk, but in all cases amounts were insufficient to merit discontinuation of breast feeding. The effects of chronic anticonvulsant therapy with phenobarbital (94 or diphenylhydantoin (2) on the pharmacokinetics of *d*-propoxyphene,[558] clonazepam (118),[559] and paracetamol (7)[560] were to increase rates of excretion, probably as the result of metabolic induction. Pharmacokinetics of diphenylhydantoin (2) itself have been studied in children[561] and adults;[460] its half-life increased with increasing doses. Ibuprofen has been found to partition freely into the synovial fluid of arthritic patients;[562] levels of cyclophosphamide partitioning into saliva correlate well with plasma levels and can be used in non-invasive studies of the drug's pharmacokinetics.[563] On the other hand, poor correlation between blood and saliva levels of nortriptyline has been found.[564] Studies on the pharmacokinetics and excretion of the anti-malarials mefloquine,[565] primaquine,[566,567] and chloroquine[567] have been studied. Mefloquine was monitored to 1 ng cm^{-3} by g.c.-m.s. for 84 days and shown to have a half-life of 15 - 33 days;[565] its metabolism has been studied in the rat.[568] Pharmacokinetics and oral bioavailability of pyridostigmine[569] and metoclopramide[570] have been reported, and terbutaline (50) kinetics have been studied in both healthy and asthmatic patients.[571] The disposition of bupivacaine in mother, foetus, and neonate following epidural anaesthesia for caesarean section has been reported,[572] with the drug detectable for at least three days in the neonate. Tricyclic antidepressant monitoring by several methods has been reviewed,[573,574] and extraction of these drugs from urine and plasma by Sep-Pak cartridges was evaluated by g.c.-m.s.[575]

Use of Stable Isotopes in Pharmacokinetic Studies.- *Bioavailability.* Drug-bioavailability studies involving coadministration of the drug and a pharmacokinetically equivalent labelled variant by two routes or as two formulations eliminate errors introduced by sequential administration. Thus, verapamil has been shown to have a low oral bioavailability;[576,577] $(^{2}H_{3})$verapamil was given orally and measured against intravenous unlabelled varapamil by selected-ion monitoring, using $(^{2}H_{7})$verapamil as the internal standard. Similar studies have been reported for pethidine[578] and methadone.[579] The use of a ^{13}C label is exemplified in a study with timolol (82) and its $(^{13}C_{3})$ analogue, which was administered orally; measurement was by g.c.-m.s. using $(^{2}H_{9})$timolol as the standard.[580] A mixed $^{13}C^{15}N$ label has been used in a bioavailability study with disopyramide[581-583] that involved the development of a matrix-calibration technique. The fractional first-pass metabolism due to gut contents has been detected for clonazepam (118) by the use of a ^{15}N analogue; the drug was quantified by negative-ion methane c.i. using an $(^{18}O_{3}{}^{15}N)$ analogue of clonazepam (118) as the internal standard.[584] To study the effect of congestive heart failure on drug disposition, the amount of drug arising from orally administered barbital (98) was measured against intravenous $(^{15}N_{2}{}^{13}C)$barbital with butalbital (90) as the internal standard.[585] Studies on the relative bioavailability of different formulations of verapamil[586] and maprotiline[587] and of a sustained release formulation of clovoxamine[588] have been reported. Sequential oral administration resulting in altered pharmacokinetics gives information on metabolism. Thus, saturable first-pass effects have been demonstrated for both methoxalen[589] and nicardipine.[590]

Use of Pseudo-racemates. Administration of 'pseudo-racemates' consisting of enantiomeric labelled and unlabelled drug yields comparative information on each enantiomer in the presence of the other. A study with 2*S*-(-)-propranolol (81) and $(isopropyl\text{-}^{2}H_{6})$-2*R*-(+)-propranolol has shown that the 2*S*-(-) enantiomer is hydroxylated in the 4' position to a greater extent than the other isomer *in vivo*, whereas *in vitro* there was no significant difference.[591] Similar results have been reported for oxprenolol (72) where, in addition, greater 4'- and 5'-hydroxylation was found with the 2*R* isomer.[592] *N*-Methylation of propranolol (81) enantiomers in the dog has also been reported.[593] The *S* enantiomer of warfarin has been shown to be excreted more rapidly than the *R* form in studies with $(^{2}H_{5})$-*R* warfarin in a pseudo-racemate; 4'-chlorowarfarin was used as the internal standard.[594] Quinalbarbitone (96) increases the plasma clearance of both isomers as studied with a $(2-^{13}C)$-*S*-(-)-warfarin-containing pseudo-racemate.[595] Stereoselective demethylation of chlorpheniramine[596] and a longer half-life of the analgesically active 1-enantiomer of methadone have also been found.[597,598]

Pulse Labelling. Pharmacokinetics of drugs in patients receiving long-term treatment have been studied by the 'pulse-labelling' technique in which one dose of a pharmacokinetically equivalent labelled analogue is substituted for the unlabelled drug during the course of treatment and its levels are measured by mass spectrometry. Methodology for the study of $(2-^{13}C-1,3-^{15}N_{2})$diphenyl-hydantoin (2) and its 4'-hydroxy metabolite using deuteriated internal standards has been reported [599,600] and the pharmacokinetic equivalence of the labelled drug demonstrated.[601] Similar studies have been reported with phenobarbitone (94),[602,603] and its plasma levels have been observed to rise during valproic acid therapy.[604] Pulse-labelling experiments with $(^{2}H_{4})$-,[605,606] $(^{2}H_{7})$-,[607] and $(1,2-^{13}C_{2})$-valproic acid[608] have demonstrated a fall in the plasma half-life in several groups of patients on maintenance therapy. Autoinduction of carbamazepine metabolism after one month has been shown in children;[609] amitriptyline and nortriptyline interactions[610] and lidocaine kinetics have also been studied.[611] Finally, pulse doses of $(5'',5'',5''-^{2}H_{3})\Delta^{1}$-THC have enabled Δ^{1}-THC (37) kinetics to be studied in the presence of the unlabelled drug in both light and heavy users of the drug.[612]

References

1 L. Fishbein, 'Chromatography of Environmental Hazards', Elsevier, Amsterdam, 1982, Vol. 4.
2 (*a*) 'Analytical Profiles of Drug Substances', ed. K. Florey, Academic Press, New York, 1980, Vol. 9; (*b*) *ibid.*, 1981, ed. K. Florey, Vol. 10.
3 B.J. Gudzinowicz and M.J. Gudzinowicz, 'Analysis of Drugs and Metabolites by Gas Chromatography-Mass Spectrometry', Marcel Dekker, New York, 1980, Vol. 7.
4 R.L. Foltz, A.F. Fentiman, and R.B. Foltz, 'GC/MS Assays for Abused Drugs in Body Fluids', NIDA Monograph No. 32, U.S. Govt. Printing Office, Washington, DC, 1980.
5 W. Sadée and G.C.M. Beelen, 'Drug Level Monitoring - Analytical Techniques, Metabolism and Pharmacokinetics', Wiley-Interscience, New York, 1980.
6 'Recent Developments in Mass Spectrometry in Biochemistry and Medicine', Vol. 6, *Anal. Chem. Symp. Ser.*, 1980, 4.
7 'Recent Developments in Mass Spectrometry in Biochemistry, Medicine and Environmental Research', Vol. 7, *Anal. Chem. Symp. Ser.*, 1981, 7.
8 'Stable Isotopes', ed. H.-L. Schmidt, H. Forsted, and K. Heinzinger, *Anal. Chem. Symp. Ser.*, 1982, 11.
9 'Current Developments in the Clinical Applications of HPLC, GC and MS', ed. A.M. Lawson, C.K. Lim, and W. Richmond, Academic Press, London, 1980.
10 *Biomed. Mass Spectrom.*, 1980, 7 (11 - 12), 457.
11 *Eur. J. Mass Spectrom.*, 1980, 1, 69.
12 M. McCamish, *Eur. J. Mass Spectrom.*, 1980, 1, 7.
13 W.A. Garland and M.L. Powell, *J. Chromatogr. Sci.*, 1981, 19, 392.
14 S. Agurell and J.-E. Lindgren in 'Therapeutic Drug Monitoring', ed. A. Richens and V. Marks, Churchill-Livingstone, 1981, 110.
15 A. Callieux, A. Turcant, A. Premel-Cabic, and P. Allain, *J. Chromatogr. Sci.*, 1981, 19, 163.
16 B. Halpern, *Crit. Rev. Anal. Chem.*, 1981, 11, 49.
17 M.-R. Schulten, *J. Chromatogr.*, 1982, 251, 105.
18 V.N. Reinhold and C.E. Costello in 'Practical Spectroscopy Series', ed. C. Merritt and C.N. McEwen, Marcel Dekker, New York, 1980, Vol. 3, Pt. B, p. 1.
19 P.D. Klein, D.A. Schoeller, and E.R. Klein in ref. 9. p. 119.
20 P.J. Murphy and H.R. Sullivan, *Annu. Rev. Pharmacol. Toxicol.*, 1980, 20, 609.
21 D.J. Harvey in ref. 7, p.1.
22 M. Barber, R.S. Bordoli, R.D. Sedgwick, A.N. Tyler, B.N. Green, V.C. Parr, and J.L. Gower, *Biomed. Mass Spectrom.*, 1982, 9, 11.
23 D.E. Games, M.S. Lant, S.A. Westwood, M.J. Cocksedge, N. Evans, J. Williamson, and B.J. Woodhall, *Biomed. Mass Spectrom.*, 1982, 9, 215.
24 J.D. Henion and G.A. Maylin, *Biomed. Mass Spectrom.*, 1980, 7, 115.
25 J.D. Henion, *J. Chromatogr. Sci.*, 1981, 19, 57.
26 C.R. Blakley, J.C. Carmody, and M.L. Vestal, *Clin. Chem. (Winston-Salem, N.C.)*, 1980, 26, 1467.
27 H. Jungclas, H. Danigel, and L. Schmidt, *Org. Mass Spectrom.*, 1982, 17, 86.
28 P. Vouros, E.P. Lankmayr, M.J. Hayes, B.L. Karger, and J.M. McGuire, *J. Chromatogr.*, 1982, 251, 175.
29 J.D. Henion, B.A. Thomson, and P.H. Dawson, *Anal. Chem.*, 1982, 54, 451.

30 G. Karlaganis, A. Küpfer, J. Bircher, U.P. Schlunegger, H. Gfeller, and P. Bigler, *Drug Metab. Dispos.*, 1980, 8, 173.
31 S. McLean, N.W. Davies, H. Watson, W.A. Favretto, and J.C. Bignall, *Drug Metab. Dispos.*, 1981, 9, 255.
32 D. Ghiringhelli, A. Griffini, and P. Traldi, *Biomed. Mass Spectrom.*, 1981, 8, 155.
33 S.H. Hansen and M. Thomsen, *J. Chromatogr.*, 1981, 209, 77.
34 F. Matsui, D.L. Robertson, M.A. Poirier, and E.G. Lovering, *J. Pharm. Sci.*, 1980, 69, 469.
35 G.K. Poochikian and J.A. Kelley, *J. Pharm. Sci.*, 1981, 70, 162.
36 J.J. Saady, N. Narasimhachari, and R.O. Friedel, *Clin. Chem. (Winston-Salem, N.C.)*, 1981, 27, 343.
37 Y. Adachi, C. Nakamura, N. Yohkoh, M. Ikeda, A. Kato, and K. Shimada, *Yakugaku Zasshi*, 1980, 100, 1104.
38 T. Uno, M. Masada, Y. Kuroda, and T. Nakagawa, *Chem. Pharm. Bull.*, 1981, 29, 1344.
39 N.M. Curran, E.G. Lovering, K.M. McErlane, and J.R. Watson, *J. Pharm. Sci.*, 1980, 69, 187.
40 W.H. Soine, W.C. Vincek, D.T. Agee, J. Boni, G.C. Burleigh, T.H. Casey, M. Christian, A. Jones, D.T. King, D. Martin, D.J. O'Neil, P.E. Quinn, and P. Strother, *J. Anal. Toxicol.*, 1980, 4, 217.
41 L.A. Jones, R.W. Beaver, T.L. Schmoeger, J.F. Ort, and J.D. Leander, *J. Org. Chem.*, 1981, 46, 3330.
42 T. Lukaszewski and W.K. Jeffery, *J. Forensic Sci.*, 1980, 25, 499.
43 J.E. Pettersen, G. Teien, and G.A. Ulsaker, *J. Pharm. Sci.*, 1981, 70, 812.
44 S. Honoré and L. Nordholm, *J. Chromatogr.*, 1981, 204, 97.
45 R. Wester, P. Noonan, C. Markos, R. Bible, jun., W. Aksamit, and J. Hribar, *J. Chromatogr.*, 1981, 209, 463.
46 G.D. De Jongh, H.M. van den Wildenberg, H. Nieuwenhuyse, and F. van der Veen, *Drug Metab. Dispos.*, 1981, 9, 48.
47 T.J. Siek and E.F. Rieders, *J. Anal. Toxicol.*, 1981, 5, 194.
48 C. Casalini, G. Mascellani, G. Tamagnone, G. Cesarano, and A. Giumanini, *J. Pharm. Sci.*, 1980, 69, 164.
49 J.D. Adams, jun., T.A. Baillie, A.J. Trevor, and N. Castagnoli, jun., *Biomed. Mass Spectrom.*, 1981, 8, 527.
50 G.K. Woo, T.H. Williams, S.J. Kolis, D. Warinsky, G.J. Sasso, and M.A. Schwartz, *Xenobiotica*, 1981, 11, 373.
51 F.S. Eberts, jun., Y. Philopoulos, L.M. Reineke, and R.W. Vliek, *Clin. Pharmacol. Ther.*, 1981, 29, 81.
52 M. Nakano, T. Tomitsuka, and K. Juni, *Chem. Pharm. Bull.*, 1980, 28, 3407.
53 A.S. Freeman and B.R. Martin, *J. Pharm. Sci.*, 1981, 70, 1002.
54 C.E. Cook, D.R. Brine, G.D. Quin, M.E. Wall, M. Perez-Reyes, and S.R. Di Guiseppi, *Life Sci.*, 1981, 29, 1967.
55 B.J. Millard, E.B. Sheinin, and W.R. Benson, *J. Pharm. Sci.*, 1980, 69, 1177.
56 M. Attinà, *J. Pharm. Sci.*, 1980, 69, 991.
57 G. Hallström, B. Lindeke, and E. Anderson, *Xenobiotica*, 1981, 11, 459.
58 L.A. Damani, P.A. Crooks, and D.A. Cowan, *Biomed. Mass Spectrom.* 1981, 8, 270.
59 P.A. Crooks, L.A. Damani, and D.A. Cowan, *J. Pharm. Pharmacol.*, 1981, 33, 309.
60 L.A. Damani, P.A. Crooks, and D.A. Cowan, *Drug Metab. Dispos.*, 1981, 9, 270.
61 G. Huizing and A.H. Beckett, *Xenobiotica*, 1980, 10, 593.
62 M. Acara, T. Gessner, H. Breizerstein, and R. Trudnowski, *Drug Metab. Dispos.*, 1981, 9, 75.

63 C. Lindberg, H. Greizerstein, T. Gessner, and M. Acara, *Drug Metab. Dispos.*, 1981, 9, 492.
64 B. Lindeke, G. Hallström, C. Johansson, Ö. Ericsson, L.-I. Olsson, and S. Strömberg, *Biomed. Mass Spectrom.*, 1981, 8, 506.
65 G. Hallström and B. Lindeke, *Biomed. Mass Spectrom.*, 1981, 8, 514.
66 G.K. Pillai and K.M. McErlane, *J. Pharm. Sci.*, 1981, 70, 1072.
67 W.D. Hooper and M.T. Smith, *J. Pharm. Sci.*, 1981, 70, 346.
68 M.P. Friocourt, D. Picart, and H.H. Floch, *Biomed. Mass Spectrom.*, 1980, 7, 193.
69 D.J. Harvan, J.R. Hass, P.W. Albro, and M.D. Friesen, *Biomed. Mass Spectrom.*, 1980, 7, 242.
70 P.W. Albro, J.R. Hass, C.C. Peck, D.G. Odam, J.T. Corbett, F.J. Bailey, H.E. Blatt, and B.B. Barrett, *Drug Metab. Dispos.*, 1981, 9, 223.
71 M.C. McPetersen, J. Vine, J.J. Ashley, and R.L. Nation, *J. Pharm. Sci.*, 1981, 70, 1139.
72 B.S. Middleditch, H.B. Hines, F.O. Bastian, and P.R. Middleditch, *Biomed. Mass Spectrom.*, 1980, 7, 105.
73 R.J. Anderegg, *Anal. Chem.*, 1981, 53, 2169.
74 J.A. Hinson, L.R. Pohl, T.J. Monks, J.R. Gillette, and F.P. Guengerich, *Drug Metab. Dispos.*, 1980, 8, 289.
75 T. Goromaru, T. Furuta, S. Baba, A. Noda, and S. Iguchi, *Chem. Pharm. Bull.*, 1981, 29, 1724.
76 R.C. Kammerer, D.A. Schmitz, E.W. Distefano, and A.K. Cho, *Drug Metab. Dispos.*, 1981, 9, 274.
77 A.K. Cho, R.C. Kammerer, and L. Abe, *Life Sci.*, 1981, 28, 1075.
78 D.J. Harvey and W.D.M. Paton, *Adv. Mass Spectrom.*, 1980, 8, 1194.
79 D.E. Wolf, W.J.A. VandenHeuvel, T.R. Tyler, R.W. Walker, F.R. Koniuszy, V. Gruber, B.H. Arison, A. Rosegay, T.A. Jacob, and F.J. Wolf, *Drug. Metab. Dispos.*, 1980, 8, 131.
80 G.R. Bourne, S.R. Moss, P.H. Phillips, and B. Shuker, *Biomed. Mass Spectrom.*, 1980, 7, 226.
81 S. Higuchi and S. Kawamura, *J. Chromatogr.*, 1981, 223, 341.
82 G.J. Yakatan, W.J. Poynor, R.L. Talbert, B.F. Floyd, C.L. Slough, R.S. Ampulski, and J.J. Benedict, *Clin. Pharmacol. Ther.*, 1982, 31, 402.
83 S. Baba, T. Furata, M. Horie, and H. Nakagawa, *J. Pharm. Sci.*, 1981, 70, 780.
84 S.A. Barker, J.A. Monti, and S.T. Christian, *Biochem. Pharmacol.*, 1980, 29, 1049.
85 M.V. Marsh, J. Caldwell, R.L. Smith, M.W. Horner, E. Houghton, and M.S. Moss, *Xenobiotica*, 1981, 11, 655.
86 M.V. Marsh, A.J. Hutt, J. Caldwell, and R.L. Smith, *Biochem. Pharmacol.*, 1981, 30, 1879.
87 É. Tomori, G. Horváth, I. Elekes, T. Láng, and J. Körösi, *J. Chromatogr.*, 1982, 241, 89.
88 D.R. Hawkins, S.R. Biggs, R.R. Brodie, L.F. Chasseaud, and I. Midgley, *J. Pharm. Pharmacol.*, 1982, 34, 299.
89 T. Goromura, T. Furuta, S. Baba, A. Noda, and S. Iguchi, *Yakugaku Zasshi*, 1981, 101, 544.
90 I.G. Sipes, A.J. Gandolfi, L.R. Pohl, G. Krishna, and B.R. Brown, jun., *J. Pharmacol. Exp. Ther.*, 1980, 214, 716.
91 M. Jarman, J.H. Kiburis, G.B. Elion, V.C. Knick, G. Lambe, D.J. Nelson, and R.L. Tuttle, *Anal. Chem. Symp. Ser.*, 1982, 11, 217.
92 J.A. Hoskins and P.B. Farmer, *Anal. Chem. Symp. Ser.*, 1982, 11, 223.
93 G.T. Miwa, W.A. Garland, B.J. Hodshon, A.Y.H. Lu, and D.B. Northrop, *J. Biol. Chem.*, 1980, 255, 6049.

94 B.D. Hammock, M. Ratcliff, and D.A. Schooley, *Life Sci.*, 1980, 27, 1635.
95 T.A. Baillie, H. Hughes, and D.S. Davies, *Anal. Chem. Symp. Ser.*, 1982, 11, 187.
96 G.P. Stec, T.I. Ruo, J.-P. Thenot, A.J. Atkinson, jun., Y. Morita, and J.J.L. Lertora, *Clin. Pharmacol. Ther.*, 1980, 28, 659.
97 H.-S. Lin and R.J. Weinkam, *J. Med. Chem.*, 1981, 24, 761.
98 P.J. van Bladeren, D.D. Breimer, G.M.T. Rotteveel-Smijs, R.A.W. de Jong, W. Buijs, A. van der Gen, and G.R. Mohn, *Biochem. Pharmacol.*, 1980, 29, 2975.
99 P.J. van Bladeren, D.D. Breimer, J.A.T.C.M. van Huijgevoort, N.P.E. Vermeulen, and A. van der Gen, *Biochem. Pharmacol.*, 1981, 30, 2499.
100 D.J. Harvey and W.D.M. Paton, *Drug. Metab. Dispos.*, 1980, 8, 178.
101 J.L. Martin, J.W. George, and L.R. Pohl, *Drug Metab. Dispos.*, 1980, 8, 93.
102 J.C. Craig, L.D. Gruenke, and T.-L. Nguyen, *J. Chromatogr.*, 1982, 239, 81.
103 T.-L. Nguyen, E. Dagne, L. Gruenke, H. Bhargava, and N. Castagnoli, jun., *J. Org. Chem.*, 1981, 46, 758.
104 J. Sjövall, *Adv. Mass Spectrom.*, 1980, 8, 1069.
105 Z.R. Vlahcevic, T. Cronholm, T. Curstedt, and J. Sjövall, *Biochim. Biophys. Acta*, 1980, 618, 369.
106 P. Vouros in 'Practical Spectroscopy Series', ed. C. Merritt and C.N. McEwen, Marcel Dekker, New York, 1980, Vol. 3, Pt. B, p. 129.
107 J.V. Evans, S.K. Daley, G.A. McClusky, and C.J. Nielsen, *Biomed. Mass Spectrom.*, 1980, 7, 65.
108 J.A. Slack and A.W. Ford-Hutchinson, *Drug Metab. Dispos.*, 1980, 8, 84.
109 M. Delaforge, P. Janiaud, P. Levi, and J.P. Morizot, *Xenobiotica*, 1980, 10, 737.
110 F.S. Abbott, R. Burton, J. Orr, D. Wladichuk, S. Ferguson, and T.-H. Sun, *J. Chromatogr.*, 1982, 227, 433.
111 J. Vink, H.J.M. Van Hal, and C.J. Timmer, *Biomed. Mass Spectrom.*, 1980, 7, 592.
112 C.J. Kelly, E. Coles, L. Gaudio, and D.W. Yesair, *Biochem. Pharmacol.*, 1980, 29, 609.
113 P.D. Schweinsberg, R.G. Smith, and T.L. Loo, *Biochem. Pharmacol.*, 1981, 30, 2521.
114 D.J. Tocco, A.E.W. Duncan, F.A. De Luna, J.L. Smith, R.W. Walker, and W.J.A. VandenHeuvel, *Drug Metab. Dispos.*, 1980, 8, 236.
115 J.F. Green, G.N. Jham, J.L. Neumeyer, and P. Vouros, *J. Pharm. Sci.*, 1980, 69, 936.
116 A.S. Christophersen, E. Hovland, and K.E. Rasmussen, *J. Chromatogr.*, 1982, 234, 107.
117 T. Suzuki, K. Tsuzurahara, T. Murata, and S. Takeyama, *Biomed. Mass Spectrom.*, 1982, 9, 94.
118 H. Kamimura, H. Sasaki, S. Higuchi, and Y. Shiobara, *J. Chromatogr.*, 1982, 229, 337.
119 L. Siekmann, A. Siekmann, and H. Breuer, *Biomed. Mass Spectrom.*, 1980, 7, 511.
120 M. Ervik, K. Kylberg-Hanssen, and P.-O. Lagerström, *J. Chromatogr.*, 1980, 182, 341.
121 R.G. Smith, S.C. Blackstock, L.K. Cheung, and T.L. Loo, *Anal. Chem.*, 1981, 53, 1205.
122 A.S. Christophersen and K.E. Rasmussen, *J. Chromatogr.*, 1980, 192, 363.

123 N. Narasimhachari and R.O. Friedel, *Anal. Lett.*, 1980, **13**, 203.
124 M.S.B. Nayar and P.S. Callery, *Anal. Lett.*, 1980, **13**, 625.
125 J.-P. Thenot, T.I. Ruo, and O.J. Bouwsma, *Anal. Lett.*, 1980, **13**, 759.
126 J.E. Bakke, *Biomed. Mass Spectrom.*, 1982, **9**, 74.
127 P.T. Funke, E. Ivashkiv, M.F. Malley, and A.I. Cohen, *Anal. Chem.*, 1980, **52**, 1086.
128 Y. Kawahara, M. Hisaoka, Y. Yamazaki, A. Inage, and T. Morioka, *Chem. Pharm. Bull.*, 1981, **29**, 150.
129 O. Lantto, I. Björkhem, H. Ek, and D. Johnston, *J. Steroid Biochem.*, 1981, **14**, 721.
130 S.P. Jindal, T. Lutz, and T.B. Cooper, *J. Chromatogr.*, 1980, **221**, 301.
131 B.H. Lauterburg, C.V. Smith, and J.R. Mitchell, *J. Chromatogr.*, 1981, **224**, 431.
132 Y. Horai, T. Ishizaki, T. Sasaki, G. Koya, K. Matsuyama, and S. Iguchi, *Br. J. Clin. Pharmacol.*, 1982, **13**, 361.
133 T.E. Nicholas, M.T. Jones, D.W. Johson, and G. Phillipou, *J. Steroid Biochem.*, 1981, **14**, 45.
134 T.R. Burke, jun., R.V. Branchflower, D.E. Lees, and L.R. Pohl, *Drug Metab. Dispos.*, 1981, **9**, 19.
135 C.F. Poole, L. Johansson, and J. Vessman, *J. Chromatogr.*, 1980, **194**, 365.
136 S. Singhawangcha, C.F. Poole, and A. Zlatkis, *J. Chromatogr.*, 1980, **183**, 433.
137 W.J. Irwin, L.W. Po, and R.R. Wadhwani, *J. Clin. Hospital Pharm.*, 1980, **5**, 55.
138 G.C. Khan, A.R. Boobis, S. Murray, M.J. Brodie, and D.S. Davies, *Br. J. Clin. Pharmacol.*, 1982, **13**, 637.
139 J. Hermansson, *J. Chromatogr.*, 1980, **221**, 109.
140 J.A. Thompson, J.L. Holtzman, M. Tsuru, C.L. Lerman, and J.L. Holtzman, *J. Chromatogr.*, 1982, **238**, 470.
141 J. Hermansson, *J. Chromatogr.*, 1982, **227**, 113.
142 F. Goldszer, G.L. Tindell, U.K. Walle, and T. Walle, *Res. Commun. Chem. Pathol. Pharmacol.*, 1981, **34**, 193.
143 F.N. Shirota, E.G. Demaster, J.A. Elberling, and H.T. Nagasawa, *J. Med. Chem.*, 1980, **23**, 669.
144 B. Ho and N. Castagnoli, jun., *J. Med. Chem.*, 1980, **23**, 133.
145 R. Ziegler, B. Ho, and N. Castagnoli, jun., *J. Med. Chem.*, 1981, **24**, 1133.
146 D. Ward, A. Kalir, A. Trevor, J. Adams, T. Baillie, and N. Castagnoli, jun., *J. Med. Chem.*, 1982, **25**, 491.
147 J.H. Liu and W.W. Ku, *Anal. Chem.*, 1981, **53**, 2180.
148 J.A.G. Roach, J.A. Sphon, D.F. Hunt, and F.W. Crow, *J. Assoc. Off. Anal. Chem.*, 1980, **63**, 452.
149 H. Brandenberger and K. Larsson, *GC/MS News*, 1980, **8**, 46.
150 H. Brandenberger and K. Larsson, *GC/MS News*, 1980, **8**, 62.
151 J. Coutant, E.M. Bargar, and R. Barbuch, *Anal. Chem. Symp. Ser.*, 1981, **7**, 35.
152 J.C. Bignall, N.W. Davies, M. Power, M.S. Roberts, P.A. Cossum, and G.W. Boyd, *Anal. Chem. Symp. Ser.*, 1981, **7**, 111.
153 E.C. Horning, D.I. Carroll, R.N. Stillwell, and I. Dzidic, *Anal. Chem. Symp. Ser.*, 1980, **4**, 453.
154 W.H. Hunter and P. Wilson, *Xenobiotica*, 1981, **11**, 179.
155 K. Halpaap-Wood, E.C. Horning, and M.G. Horning, *Drug. Metab. Dispos.*, 1981, **9**, 97.
156 K. Halpaap-Wood, E.C. Horning, and M.G. Horning, *Drug Metab. Dispos.*, 1981, **9**, 103.
157 K.R. Huckle, D.H. Hutson, and P. Millburn, *Drug. Metab. Dispos.*, 1981, **9**, 352.

158 K.R. Huckle, J.K. Chipman, D.H. Hutson, and P. Millburn, *Drug Metab. Dispos.*, 1981, 9, 360.
159 K.R. Huckle, I.J.G. Climie, D.H. Hutson, and P. Millburn, *Drug Metab. Dispos.*, 1981, 9, 147.
160 J.V. Crayford and D.H. Hutson, *Xenobiotica*, 1980, 10, 349.
161 L.G. Lange, S.R. Bergmann, and B.E. Sobel, *J. Biol. Chem.*, 1981, 256, 12968.
162 E.G. Leighty and A.F. Fentiman, jun., *Res. Commun. Chem. Pathol. Pharmacol.*, 1981, 32, 569.
163 H.-D. Dell, J. Fiedler, R. Kamp, W. Gau, J. Kurz, B. Weber, and C. Wuensche, *Drug Metab. Dispos.*, 1982, 10, 55.
164 E.G. Leighty, A.F. Fentiman, jun., and R.M. Thompson, *Toxicology*, 1980, 15, 77.
165 J.R. Trudell, B. Bösterling, and A. Trevor, *Biochem. Biophys. Res. Commun.*, 1981, 102, 372.
166 C. Fenselau and L.P. Johnson, *Drug Metab. Dispos.*, 1980, 8, 274.
167 C. Fenselau, R. Cotter, and L. Johnson, *Adv. Mass Spectrom.*, 1980, 8, 1159.
168 D.E. Games and E. Lewis, *Biomed. Mass Spectrom.*, 1980, 7, 433.
169 A.P. Bruins, *Biomed. Mass Spectrom.*, 1981, 8, 31.
170 J.P. Lehman, L. Ferrin, C. Fenselau, and G.S. Yost, *Drug Metab. Dispos.*, 1981, 9, 15.
171 B.-L. Sahlberg, M. Axelson, D.J. Collins, and J. Sjövall, *J. Chromatogr.*, 1981, 217, 453.
172 P.L. Williams and A.C. Moffat, *J. Pharm. Pharmacol.*, 1980, 32, 445.
173 I. Hornke, H.-W. Fehlhaber, M. Girg, and H. Jantz, *Br. J. Clin. Pharmacol.*, 1980, 9, 255.
174 S.F. Sisenwine and C.O. Tio, *Drug Metab. Dispos.*, 1981, 9, 37.
175 J.A. Thomson, J.E. Hull, and K.J. Norris, *Drug Metab. Dispos.*, 1981, 9, 466.
176 C.E. Hignite, C. Tschanz, S. Lemons, H. Wiese, D.L. Azarnoff, and D.H. Huffman, *Life Sci.*, 1981, 28, 2077.
177 G.D. Paulson, J.M. Giddings, C.H. Lamoureux, E.R. Mansager, and C.B. Struble, *Drug Metab. Dispos.*, 1981, 9, 142.
178 L.J. Fischer, R.L. Thies, D. Charkowski, and K.J. Donham, *Drug Metab. Dispos.*, 1980, 8, 422.
179 A.T. Elvin, J.B. Keenaghan, E.W. Byrnes, P.A. Tenthorey, P.D. McMaster, B.H. Takman, D. Lalka, C.V. Manion, D.T. Baer, E.M. Wolshin, M.B. Meyer, and R.A. Ronfeld, *J. Pharm. Sci.*, 1980, 69, 47.
180 R. Venkataramanan and J.E. Axelson, *Xenobiotica*, 1981, 11, 259.
181 H.M. Carpenter and G.H. Mudge, *J. Pharmacol. Exp. Ther.*, 1981, 218, 161.
182 S.D. Nelson, A.J. Forte, and D.C. Dahlin, *Biochem. Pharmacol.*, 1980, 29, 1617.
183 I.C. Calder, S.J. Hart, and K. Healey, *J. Med. Chem.*, 1981, 24, 988.
184 A.R. Boobis, G.C. Kahn, C. Whyte, M.J. Brodie, and D.S. Davies, *Biochem. Pharmacol.*, 1981, 30, 2451.
185 M.W. Gemborys and G.H. Mudge, *Drug Metab. Dispos.*, 1981, 9, 340.
186 S.D. Nelson, Y. Vaishnav, H. Kambara, and T.A. Baillie, *Biomed. Mass Spectrom.*, 1981, 8, 244.
187 Y.-N. Kyo, Y. Muranaka, N. Hikichi, and H. Niwa, *Yakugaku Zasshi*, 1980, 100, 1043.
188 M. Taniguchi and M. Nakano, *Chem. Pharm. Bull.*, 1981, 29, 577.
189 R. Gotschlich and M. Metzler, *Xenobiotica*, 1980, 10, 317.
190 R.K. Miller, M.E. Heckmann, and R.C. McKenzie, *J. Pharmacol. Exp. Ther.*, 1982, 220, 358.

191 D.L. Bius, W.D. Yonekawa, H.J. Kupferberg, F. Cantor, and K.H. Dudley, *Drug Metab. Dispos.*, 1980, 8, 223.
192 A. Küpfer, G.M. Brilis, J.T. Watson, and T.M. Harris, *Drug Metab. Dispos.*, 1980, 8, 1.
193 P.G. Wells, A. Küpfer, J.A. Lawson, and R.D. Harbison, *J. Pharmacol. Exp. Ther.*, 1982, 221, 228.
194 J.H. Maguire, T.C. Butler, and K.H. Dudley, *Drug Metab. Dispos.*, 1980, 8, 325.
195 J.H. Maguire, B.L. Kraus, T.C. Butler, and K.H. Dudley, *Drug Metab. Dispos.*, 1981, 9, 393.
196 J.E. Pettersen, *Adv. Mass Spectrom.*, 1980, 8, 1291.
197 H. Nau, E. Jager, and H. Helge, *Anal. Chem. Symp. Ser.*, 1980, 4, 331.
198 D.N. Pallai and S. Dilli, *J. Chromatogr.*, 1981, 220, 253.
199 L.V. Jones and M.J. Whitehouse, *Biomed. Mass Spectrom.*, 1981, 8, 231.
200 J.G. Dain, S.I. Bhuta, R.A. Coombs, K.C. Talbot, and H.A. Dugger, *Drug Metab. Dispos.*, 1980, 8, 247.
201 W.D. Hooper, H.E. Kunze, and M.J. Eadie, *Drug Metab. Dispos.*, 1981, 9, 381.
202 W.D. Hooper, H.E. Kunze, and M.J. Eadie, *J. Chromatogr.*, 1981, 223, 426.
203 I.M. Kapetanović and H.J. Kupferberg, *J. Pharm. Sci.*, 1981, 70, 1218.
204 H. Nau, D. Rating, I. Häuser, E. Jäger, S. Koch, and H. Helge, *Eur. J. Clin. Pharmacol.*, 1980, 18, 31.
205 B.K. Tang, A.A. Grey, P.A.J. Reilly, and W. Kalow, *Can. J. Physiol. Pharmacol.*, 1980, 58, 1169.
206 N.P.E. Vermeulen, B.H. Bakker, D. Eylers, and D.D. Breimer, *Xenobiotica*, 1980, 10, 159.
207 N.P.E. Vermeulen, D.D. Breimer, J. Holthuis, C. Mol, B.H. Bakker, and A. Van Der Gen, *Xenobiotica*, 1981, 11, 547.
208 N.P.E. Vermeulen, D.D. Breimer, B.H. Bakker, and A. Van Der Gen, *Anal. Chem. Symp. Ser.*, 1980, 4, 283.
209 H. Heusler, J. Epping, S. Heusler, E. Richter, N.P.E. Vermeulen, and D.D. Breimer, *J. Chromatogr.*, 1981, 226, 403.
210 S. Floberg, B. Lindström, and G. Lonnerholm, *J. Chromatogr.*, 1980, 221, 166.
211 K.-Y. Tserng, K.C. King, and F.N. Takieddine, *Clin. Pharmacol. Ther.*, 1981, 29, 594.
212 J.L. Brazier, B. Ribon, M. Desage, and B. Salle, *Biomed. Mass Spectrom.*, 1980, 7, 189.
213 J.L. Brazier, B. Salle, B. Ribon, M. Desage, and H. Renaud, *Dev. Pharmacol. Ther.*, 1981, 2, 137.
214 J.L. Brazier, B. Ribon, M. Desage, and B. Salle, *Anal. Chem. Symp. Ser.*, 1981, 7, 27.
215 E. Houghton, *Biomed. Mass Spectrom.*, 1982, 9, 103.
216 M.J. Arnaud, A. Thelin-Doerner, E. Ravussin, and K.J. Acheson, *Biomed. Mass Spectrom.*, 1980, 7, 521.
217 J.J. Rafter and L. Nilsson, *Xenobiotica*, 1981, 11, 771.
218 E. Tarrus, I. Garcia, and J. Segura, *J. Pharm. Sci.*, 1981, 70, 542.
219 K. Selvig and K.S. Bjerve, *Drug Metab. Dispos.*, 1980, 8, 456.
220 S.K. Saha and W. Pfleiderer, *Indian J. Chem., Sect. B*, 1980, 19, 325.
221 B.J. Gudzinowicz, M.J. Gudzinowicz, J. Hologgitas, and J.L. Driscoll, *Adv. Chromatogr.*, 1980, 18, 197.
222 D.J. Harvey, *Biomed. Mass Spectrom.*, 1981, 8, 575.
223 D.J. Harvey, *Biomed. Mass Spectrom.*, 1981, 8, 579.
224 D.J. Harvey, *Biomed. Mass Spectrom.*, 1981, 8, 366.

225 S. Björkman, *J. Chromatogr.*, 1982, 237, 389.
226 C.E. Turner, O.J. Bouwsma, S. Billets, and M.A. Elsohly, *Biomed. Mass Spectrom.*, 1980, 7, 247.
227 M.E. Wall and M. Perez-Reyes, *J. Clin. Pharmacol.*, 1981, 21, 178S.
228 D.J. Harvey, B.R. Martin, and W.D.M. Paton, *J. Pharm. Pharmacol.*, 1980, 32, 267.
229 Z. Ben-Zvi, *Xenobiotica*, 1980, 10, 805.
230 S. Burstein and T.S. Shoupe, *Drug Metab. Dispos.*, 1981, 9, 94.
231 M. Binder and A. Popp, *Helv. Chim. Acta*, 1980, 63, 2515.
232 D.J. Harvey, J.T.A. Leuschner, and W.D.M. Paton, *Res. Commun. Substance Abuse*, 1980, 1, 159.
233 D.J. Harvey and W.D.M. Paton, *Res. Commun. Substance Abuse*, 1981, 2, 193.
234 D.J. Harvey, E.W. Gill, M. Slater, and W.D.M. Paton, *Drug Metab. Dispos.*, 1980, 8, 439.
235 M. Binder and U. Barlage, *Helv. Chim. Acta*, 1980, 63, 255.
236 R.E. Billings, G.W. Whitaker, and R.E. McMahon, *Xenobiotica*, 1980, 10, 33.
237 H. Ishii, T. Niwaguchi, Y. Nakahara, and M. Hayashi, *J. Chem. Soc., Perkin Trans. 1*, 1980, 902.
238 T. Inoue, T. Niwaguchi, and T. Murata, *Xenobiotica*, 1980, 10, 343.
239 T. Toda and N. Oshino, *Drug Metab. Dispos.*, 1981, 9, 108.
240 F.I. Carroll, G.A. Brine, K.G. Bolt, E.J. Cone, D. Yousefnejad, D.B. Vaupel, and W.F. Buchwald, *J. Med. Chem.*, 1981, 24, 1047.
241 J.K. Baker, J.G. Wohlford, B.J. Bradbury, and P.W. Wirth, *J. Med. Chem.*, 1981, 24, 666.
242 J. Ambre, G. Smith, and K. Smith, *Clin. Pharmacol. Ther.*, 1981, 29, 231.
243 K. Uba, N. Miyata, K. Watanabe, and M. Hirobe, *Chem. Pharm. Bull.*, 1980, 28, 356.
244 R.M. Rodgers, S.M. Taylor, R.G. Dikinson, A.M. Ilias, R.K. Lynn, and N. Gerber, *Drug Metab. Dispos.*, 1980, 8, 390.
245 S.M. Taylor, R.M. Rodgers, R.K. Lynn, and N. Gerber, *J. Pharmacol. Exp. Ther.*, 1980, 213, 289.
246 M.E. Wall, D.R. Brine, and M. Perez-Reyes, *Drug Metab. Dispos.*, 1981, 9, 369.
247 R.C. Gaver, M. Vasiljev, H. Wong, I. Monkovic, J.E. Swigor, D.R. Van Harken, and R.D. Smyth, *Drug Metab. Dispos.*, 1980, 8, 230.
248 J.V. Evans, J.L. Leeling, and R.J. Helms, *Biomed. Mass Spectrom.*, 1982, 9, 191.
249 H. Watanabe, J.A. Menzies, N. Jordan, and J.C.K. Loo, *Res. Commun. Chem. Pathol. Pharmacol.*, 1981, 31, 435.
250 H. Peter, R. Jung, H.M. Bolt, and F. Oesch, *J. Steroid Biochem.*, 1981, 14, 83.
251 C.J. Lewis, C.W. Vose, P.N. Spalton, G.C. Ford, N.J. Haskins, and R.F. Palmer, *Xenobiotica*, 1980, 10, 705.
252 E. Houghton, M.C. Dumasia, and J.K. Wellby, *Biomed. Mass Spectrom.*, 1981, 8, 558.
253 H.W. Dürbeck and I. Büker, *Biomed. Mass Spectrom.*, 1980, 7, 437.
254 E. Houghton and P. Teale, *Biomed. Mass Spectrom.*, 1981, 8, 358.
255 L. Tökés, D. Chu, M.L. Maddox, M.D. Chaplin, and N.I. Chu, *Drug Metab. Dispos.*, 1981, 9, 485.
256 P.J. Teitelbaum, N.I. Chu, D. Cho, L. Tökes, J.W. Patterson, P.J. Wagner, and M.D. Chaplin, *J. Pharmacol. Exp. Ther.*, 1981, 218, 16.

257 J.H. Sherry, J.P. O'Donnell, and H.D. Colby, *Life Sci.*, 1981, 29, 2727.
258 J.M.P. Holly, D.J.H. Trafford, J.W. Sear, and H.L.J. Makin, *J. Pharm. Pharmacol.*, 1981, 33, 427.
259 Y. Mori, M. Tsuboi, and M. Suzuki, *Chem. Pharm. Bull.*, 1981, 29, 2478.
260 Y. Mori, M. Tsuboi, M. Suzuki, A. Saito, and H. Ohnishi, *Chem. Pharm. Bull.*, 1981, 29, 2646.
261 J.W. Steele, L.J. Boux, and R.C.S. Audette, *Xenobiotica*, 1981, 11, 117.
262 E. Houghton and M.C. Dumasia, *Xenobiotica*, 1980, 10, 381.
263 J.W. Sear, J.M.P. Holly, D.J.H. Trafford, and N.L.J. Makin, *J. Pharm. Pharmacol.*, 1980, 32, 349.
264 B.A. Clement and A.H. Beckett, *Xenobiotica*, 1981, 11, 609.
265 C.J. Reddrop, W. Riess, and T.F. Slater, *J. Chromatogr.*, 1980, 192, 375.
266 E.A. Dewey, G.A. Maylin, J.G. Ebel, and J.D. Henion, *Drug Metab. Dispos.*, 1981, 9, 30.
267 I.G. Taulov, J. Tamás, J. Hegedüs-Vajda, and D. Simov, *Acta Chim. Acad. Sci. Hung.*, 1980, 105, 109.
268 I.G. Taulov, J. Tamás, J. Hegedüs-Vajda, and D. Simov, *Acta Chim. Acad. Sci. Hung.*, 1980, 105, 117.
269 H. Johnsen and S.G. Dahl, *Drug Metab. Dispos.*, 1982, 10, 63.
270 B.C. Mayo, D.R. Hawkins, M. Hümpel, L.F. Chasseaud, and R. Kirkin, *Xenobiotica*, 1980, 10, 413.
271 H. Nau and C. Liddiard, *Biochem. Pharmacol.*, 1980, 29, 447.
272 M. Fujimoto, Y. Tsukinoki, K. Hirose, K. Kuruma, R. Konaka, and T. Okabayashi, *Chem. Pharm. Bull.*, 1980, 28, 1378.
273 J.P. Cano, Y.C. Sumirtapura, W. Cautreels, and Y. Sales, *J. Chromatogr.*, 1981, 226, 413.
274 T.C. Orton and C. Lowery, *J. Pharmacol. Exp. Ther.*, 1981, 219, 207.
275 V.T. Vu and F.P. Abramson, *Drug Metab. Dispos.*, 1980, 8, 300.
276 B.K. Wasson, J. Scheigetz, C.S. Rooney, R.A. Hall, N.N. Share, W.J.A. VandenHeuvel, B.H. Arison, O.D. Hensens, R.L. Ellsworth, and D.J. Tocco, *J. Med. Chem.*, 1980, 23, 1178,
277 A. Barrow, R.D. Brownsill, P.N. Spalton, C.M. Walls, Y. Gunn, N.J. Haskins, D.A. Rose, and R.F. Palmer, *Xenobiotica*, 1980, 10, 219.
278 A.R. Waller, L.F. Chasseaud, R. Bonn, T. Taylor, A. Darragh, R. Girkin, W.H. Down, and E. Doyle, *Drug Metab. Dispos.*, 1982, 10, 51.
279 K.K. Midha, J.K. Cooper, K. Bailey, and J.W. Hubbard, *Xenobiotica*, 1981, 11, 137.
280 J.W. Hubbard, K. Bailey, K.K. Midha, and J.K. Cooper, *Drug Metab. Dispos.*, 1981, 9, 250.
281 R.T. Coutts and G.R. Jones, *Res. Commun. Chem. Pathol. Pharmacol.*, 1982, 36, 173.
282 R.M. Matsumoto and A.K. Cho, *Biochem. Pharmacol.*, 1982, 31, 105.
283 R.C. Kammerer, E. Distefano, J. Jonsson, and A.K. Cho, *Biochem. Pharmacol.*, 1981, 30, 2257.
284 P. Peura, I. Kari, and M.M. Airaksinen, *Biomed. Mass Spectrom.*, 1980, 7, 553.
285 I. Kari, P. Peura, and M.M. Airaksinen, *Biomed. Mass Spectrom.*, 1980, 7, 549.
286 B. Sjöquist and E. Magnuson, *J. Chromatogr.*, 1980, 183, 17.
287 S.B. Sjöquist and E.T. Magnuson, *Adv. Mass Spectrom.*, 1980, 8, 1204.
288 K.L. Rominger and G. Kitzenberger, *Int. J. Clin. Pharmacol. Ther. Toxicol.*, 1980, 18, 150.

289 R.T. Coutts, B.C. Foster, and F.M. Pasutto, *Life Sci.*, 1981, 29, 1951.
290 J.A. Timbrell, S.J. Harland, and V. Facchini, *Clin. Pharmacol. Ther.*, 1980, 28, 350.
291 V. Facchini and J.A. Timbrell, *Br. J. Clin. Pharmacol.*, 1981, 11, 345.
292 T. Inaba, S.V. Otton, and W. Kalow, *Clin. Pharmacol. Ther.*, 1980, 27, 547.
293 D.S. Davies, G.C. Kahn, S. Murray, M.J. Brodie, and A.R. Boobis, *Br. J. Clin. Pharmacol.*, 1981, 11, 89.
294 T. Talseth, K.D. Haegele, and J.L. McNay, *Drug Metab. Dispos.*, 1980, 8, 73.
295 K.D. Haegele, T. Talseth, H.B. Skrdlant, A.M.M. Shepherd, and S.L. Huff, *Arzneim.-Forsch.*, 1981, 31, 357.
296 H.M. Ali and A.H. Beckett, *J. Chromatogr.*, 1981, 210, 350.
297 H.M. Ali and A.H. Beckett, *J. Chromatogr.*, 1980, 202, 287.
298 R.M. Gorsen, A.J. Weiss, and R.W. Manthei, *J. Chromatogr.*, 1980, 221, 309.
299 S.H. Kuttab, S. Tanglertpaibul, and P. Vouros, *Biomed. Mass Spectrom.*, 1982, 9, 78.
300 P. Toffel-Nadolny and W. Gielsdorf, *Arzneim.-Forsch.*, 1981, 31, 719.
301 H. Tachizawa, K. Sudo, H. Sasano, and M. Sano, *Drug Metab. Dispos.*, 1981, 9, 442.
302 J. Segura, O.M. Bakke, G. Huizing, and A.H. Beckett, *Drug Metab. Dispos.*, 1980, 8, 87.
303 G. Huizing, J. Segura, and A.H. Beckett, *J. Pharm. Pharmacol.*, 1980, 32, 650.
304 G. Huizing, A.H. Beckett, J. Segura, and O.M. Bakke, *Xenobiotica*, 1980, 10, 211.
305 H.H. Van Rooy, N.P.E. Vermeulen, and J.G. Bovill, *J. Chromatogr.*, 1981, 223, 85.
306 J.M. Frincke and G.L. Henderson, *Drug Metab. Dispos.*, 1980, 8, 425.
307 H. Davi, P. DuPont, J.P. Jeanniot, R. Roncucci, and W. Cautreels, *Xenobiotica*, 1981, 11, 735.
308 I. Midgley, S.R. Biggs, D.R. Hawkins, L.F. Chasseaud, A. Darragh, R.R. Brodie, and L.M. Walmsley, *Xenobiotica*, 1981, 11, 595.
309 C. Mas-Chamberlin, G. Gillet, J. André, R. Gomeni, G. Dring, and P.L. Morselli, *Drug Metab. Dispos.*, 1981, 9, 150.
310 D. Greenslade, M.E. Havler, M.J. Humphrey, B.J. Jordan, C.J. Lewis, and M.J. Rance, *Xenobiotica*, 1981, 11, 89.
311 R. Venkataramanan, F.S. Abbott, and J.E. Axelson, *J. Pharm. Sci.*, 1982, 71, 491.
312 A. McLean, D. Newell, G. Baker, and T. Connors, *Biochem. Pharmacol.*, 1980, 29, 2039.
313 S.Y. Chang, B.J. Larcom, D.S. Alberts, B. Larsen, P.D. Walson, and I.G. Sipes, *J. Pharm. Sci.*, 1980, 69, 80.
314 C. Liddle, G.G. Graham, R.K. Christopher, S. Bhuwapathanapun, and A.M. Duffield, *Xenobiotica*, 1981, 11, 81.
315 J.A. Slack, A.W. Ford-Hutchinson, W. Cautreels, and R. Roncucci, *Adv. Mass Spectrom.*, 1980, 8, 1280.
316 G. Tybring, C. Von Bahr, L. Bertilsson, H. Collste, H. Glaumann, and M. Solbrand, *Drug Metab. Dispos.*, 1981, 9, 561.
317 K. Lertratanangkoon and M.G. Horning, *Drug Metab. Dispos.*, 1982, 10, 1.
318 J.P. Uetrecht, R.L. Woosley, R.W. Freeman, B.J. Sweetman, and J.A. Oates, *Drug Metab. Dispos.*, 1981, 9, 183.
319 Y. Yasuda, T. Shindo, N. Mitani, N. Ishida, F. Oono, and T. Kageyama, *J. Pharm. Sci.*, 1982, 71, 565.

320 M.S. Miller and A.J. Gandolfi, *Life Sci.*, 1980, 27, 1465.
321 B.E. Pape, *J. Anal. Toxicol.*, 1981, 5, 113.
322 W. Lintz, S. Erlaçin, E. Frankus, and H. Uragg, *Arzneim.-Forsch.*, 1981, 31, 1932.
323 J. Imbs, A. Leibenguth, J.C. Koffel, and L. Jung, *Arzneim.-Forsch.*, 1982, 32, 503.
324 G. Remberg, M. Ende, M. Eichelbaum, and M. Schomerus, *Arzneim.-Forsch.*, 1980, 30, 398.
325 D.P. Benziger, A.K. Fritz, S.D. Clemans, and J. Edelson, *Drug Metab. Dispos.*, 1981, 9, 424.
326 J. Wal, J. Peleran, and G.F. Bories, *FEBS Lett.*, 1980, 119, 38.
327 J.L. Martin, B.J. Gross, P. Morris, and L.R. Pohl, *Drug Metab. Dispos.*, 1980, 8, 371.
328 D. Greenslade, M.E. Havler, M.J. Humphrey, B.J. Jordan, and M.J. Rance, *Xenobiotica*, 1980, 10, 753.
329 T. Kuninaka, K. Sugai, T. Saito, N. Mori, R. Kimura, and T. Murata, *Yakugaku Zasshi*, 1981, 101, 232.
330 S.B. Pedersen, B. Alhede, O. Buchardt, J. Møller, and K. Bock, *Arzneim.-Forsch.*, 1981, 31, 1944.
331 R. Tonani, E. Pianezzola, and E. Pella, *Adv. Mass Spectrom.*, 1980, 8, 1274.
332 F. Rubio, S. Seawall, R. Pocelinko, B. DeBarbieri, W. Benz, L. Berger, L. Morgan, J. Pao, T.H. Williams, and B. Koechlin, *J. Pharm. Sci.*, 1980, 69, 1245.
333 J.E. Ray and D.N. Wade, *Biopharm. Drug Dispos.*, 1982, 3, 29.
334 F. Nielsen-Kudsk, P. Jakobsen, and I. Magnussen, *Acta Pharmacol. Toxicol.*, 1980, 47, 11.
335 C. Feniou, B. Neau, G. Prat, C. Cheze, F. Fauran, and J. Roquebert, *J. Pharm. Pharmacol.*, 1980, 32, 104.
336 K. Krohg and E. Jellum, *Anesthesiology*, 1981, 54, 329.
337 L.T. Wong and G. Solomonraj, *Xenobiotica*, 1980, 10, 201.
338 G. Paeme, W. Sonck, D. Tourwé, R. Grimée, and A. Vercruysse, *Drug Metab. Dispos.*, 1980, 8, 115.
339 F.-J. Leinweber, A.J. Szuna, T.H. Williams, G.J. Sasso, and A. Debarbieri, *Drug Metab. Dispos.*, 1981, 9, 284.
340 P. Ventura and A. Selva, *Biomed. Mass Spectrom.*, 1982, 9, 18.
341 J.P. Horwitz, W. Brukwinski, J. Treisman, D. Andrzejewski, E.B. Hills, H.L. Chung, and C.Y. Wang, *Drug Metab. Dispos.*, 1980, 8, 77.
342 K.M. Williams, A.M. Duffield, R.K. Christopher, and P.J. Finlayson, *Biomed. Mass Spectrom.*, 1981, 8, 179.
343 C.F. Gelijkens, P. Sandra, F. Belpaire, and A.P. De Leenheer, *Drug Metab. Dispos.*, 1980, 8, 363.
344 T. Marunaka, Y. Minami, Y. Umeno, A. Yasuda, T. Sato, and S. Fujii, *Chem. Pharm. Bull.*, 1980, 28, 1795.
345 R. Kaul, G. Kiefer, S. Erhardt, and B. Hempel, *J. Pharm. Sci.*, 1980, 69, 531.
346 T. Kobari, Y. Iguro, A. Ujiie, and H. Namekawa, *Xenobiotica*, 1981, 11, 57.
347 M.-C. Cosyns-Duyck, A.A.M. Cruyl, A.P. De Leenheer, A. De Schryver, J.V. Huys, and F.M. Belpaire, *Biomed. Mass Spectrom.*, 1980, 7, 61.
348 H. Isomura, S. Higuchi, and S. Kawamura, *J. Chromatogr.*, 1981, 224, 423.
349 T. Marunaka, Y. Umeno, K. Yoshida, M. Nagamachi, Y. Minami, and S. Fujii, *J. Pharm. Sci.*, 1980, 69, 1296.
350 T. Marunaka, Y. Umeno, Y. Minami, and T. Shibata, *Biomed. Mass Spectrom.*, 1980, 7, 331.
351 T. Marunaka, *Biomed. Mass Spectrom.*, 1981, 8, 105.
352 T. Inaba, H. Uchino, and W. Kalow, *Res. Commun. Chem. Pathol. Pharmacol.*, 1981, 33, 3.

353 H. Blecher, R. Blecher, W. Wegst, J. Eberspaecher, and F. Lingens, *Xenobiotica*, 1981, 11, 749.
354 M. Tateishi, C. Koitabashi, and S. Ichihara, *Biochem. Pharmacol.*, 1980, 29, 2705.
355 F.S. Chiccarelli, H.J. Eisner, and G.E. Van Lear, *Arzneim.-Forsch.*, 1980, 30, 707.
356 F.S. Chiccarelli, H.J. Eisner, and G.E. Van Lear, *Arzneim.-Forsch.*, 1980, 30, 728.
357 J.E. Bakke, J.J. Rafter, P. Lindeskog, V.J. Feil, J.-Å. Gustafsson, and B.E. Gustafsson, *Biochem. Pharmacol.*, 1981, 30, 1839.
358 J.F. Ancher, A. Donath, A. Malnoë, J.P. Morizur, and M.S. Benedetti, *Xenobiotica*, 1981, 11, 519.
359 G. Paulson and C. Struble, *Life Sci.*, 1980, 27, 1811.
360 C.T. Gombar, W.P. Tong, and D.B. Ludlum, *Biochem. Pharmacol.*, 1980, 29, 2639.
361 S.J. Kohlhepp, H.E. May, and D.J. Reed, *Drug Metab. Dispos.*, 1981, 9, 135.
362 A.B. Foster, L.J. Griggs, M. Jarman, J.M.S. Van Maanen, and H.-R. Schulten, *Biochem. Pharmacol.*, 1980, 29, 1977.
363 J. Lundström, T. Högberg, T. Gosztonyi, and T. de Paulis, *Arzneim.-Forsch.*, 1981, 31, 486.
364 D.P. Benziger, A. Fritz, and J. Edelson, *Drug Metab. Dispos.*, 1981, 9, 493.
365 G. Kauert, L.V. Meyer, and G. Drasch, *Arch. Toxicol.*, 1980, 45, 149.
366 D.C. Hobbs and T.M. Twomey, *Drug Metab. Dispos.*, 1981, 9, 114.
367 R. Achari and A.H. Beckett, *J. Pharm. Pharmacol.*, 1980, 33, 747.
368 P.C. Ruenitz, *Drug Metab. Dispos.*, 1981, 9, 456.
369 D. Kirkpatrick, R. Girkin, L.F. Chasseaud, D.R. Hawkins, and B. Conway, *Xenobiotica*, 1980, 10, 797.
370 K.J. Kripalani, J. Dreyfuss, J. Nemec, A.I. Cohen, F. Meeker, and P. Egli, *Xenobiotica*, 1981, 11, 481.
371 T. Horie, T. Ohno, and K. Kinoshita, *Xenobiotica*, 1981, 11, 197.
372 N. Kurobe, S. Nakamura, and M. Shimizu, *Xenobiotica*, 1980, 10, 37.
373 S. Goenechea, G. Eckhardt, and W. Fahr, *Arzneim.-Forsch.*, 1980, 30, 1580.
374 H. Egger, F. Bartlett, R. Dreyfuss, and J. Karliner, *Drug Metab. Dispos.*, 1981, 9, 415.
375 M. Masada, Y, Kurodu, T. Nakagawa, and T. Uno, *Chem. Pharm. Bull.*, 1980, 28, 3527.
376 Y. Murai, T. Nakagawa, and T. Uno, *Chem. Pharm. Bull.*, 1980, 28, 362,
377 J.L. Gower, C. Beaugrand, and C. Sallot, *Biomed. Mass Spectrom.*, 1981, 8, 36.
378 J. Haginaka, T. Nakagawa, and T. Uno, *J. Antibiot.*, 1980, 33, 236.
379 T. Shomura, S. Someya, S. Murata, K. Umemura, and M. Nishio, *Chem. Pharm. Bull.*, 1981, 29, 2413.
380 C. Fraire, P. Lecointe, and C. Paoletti, *Drug Metab. Dispos.*, 1981, 9, 156.
381 B. Gaugain, C. Fraire, P. Lecointe, C. Paoletti, and B.P. Roques, *FEBS Lett.*, 1981, 129, 70.
382 L.R. Pohl, R.V. Branchflower, R.J. Highet, J.L. Martin, D.S. Nunn, T.J. Monks, J.W. George, and J.A. Hinson, *Drug Metab. Dispos.*, 1981, 9, 334.
383 M. Przybylski, S.D. Nelson, R.L. Cysyk, and D. Shoemaker, *Adv. Mass Spectrom.*, 1980, 8, 1186.

384 M. Przybylski, R.L. Cysyk, D. Shoemaker, and R.H. Adamson, *Biomed. Mass Spectrom.*, 1981, 8, 485.
385 P.C.C. Feng, C.C. Fenselau, and R.R. Jacobson, *Drug Metab. Dispos.*, 1981, 9, 521.
386 F. Borchers and G. Achtert, *Anal. Chem. Symp. Ser.*, 1981, 7, 39.
387 H.G. Jonen, F. Oesch, and K.L. Platt, *Drug Metab. Dispos.*, 1980, 8, 446.
388 J.A. Ingelfinger, G. Isakson, D. Shine, C.E. Costello, and P. Goldman, *Clin. Pharmacol. Ther.*, 1981, 29, 570.
389 H. Maurer and K. Pfleger, *J. Chromatogr.*, 1981, 222, 409.
390 R.M. Kok and D.B. Faber, *J. Chromatogr.*, 1981, 222, 389.
391 M.J. Whitehouse and L.V. Jones, *Anal. Chem. Symp. Ser.*, 1980, 4, 303.
392 E.P. Yeager, U. Goebelsmann, J.R. Soares, J.D. Grant, and S.J. Gross, *J. Anal. Toxicol.*, 1981, 5, 81.
393 R.A. Bergman, T. Lukaszewski, and S.Y.S. Wang, *J. Anal. Toxicol.*, 1981, 5, 85.
394 J.E. O'Connor and T.A. Rejent, *J. Anal. Toxicol.*, 1981, 5, 168.
395 M. Prez-Reyes, S. Di Guiseppi, K.H. Davis, V.H. Schindler, and D.E. Cook, *Clin. Pharmacol. Ther.*, 1982, 31, 617.
396 J. Fenton, M. Schaffer, N. Wu Chen, and E.W. Bermes, jun., *J. Forensic Sci.*, 1980, 25, 314.
397 K. Fotherby, J.O. Akpoviroro, L. Siekmann, and H. Breuer, *J. Steroid Biochem.*, 1981, 14, 499,
398 H. Tanaka, S. Komatsu, M. Shibuya, M. Maeda, and A. Tsuji, *Yakugaku Zasshi*, 1981, 101, 521.
399 K.K. Midha, C. Charette, J.K. Cooper, and I.J. McGilveray, *J. Anal. Toxicol.*, 1980, 4, 237.
400 J. Vasiliades, *J. Chromatogr.*, 1982, 228, 195.
401 D.A. Breutzmann and L.D. Bowers, *Clin. Chem.*, 1981, 27, 1907.
402 A.J. Malanoski, C.J. Barnes, and T. Fazid, *J. Assoc. Off. Anal. Chem.*, 1981, 64, 1386.
403 R. Jochemsen and D.D. Breimer, *J. Chromatogr.*, 1981, 223, 438.
404 E. Bailey and E.J. Barron, *J. Chromatogr.*, 1980, 183, 25.
405 A. Ranise, E. Benassi, and G. Besio, *J. Chromatogr.*, 1981, 222, 120.
406 F. Mari and E. Bertol, *J. Pharm. Pharmacol.*, 1981, 33, 814.
407 L.M. Walmsley and L.F. Chasseaud, *J. Chromatogr.*, 1981, 226, 155.
408 M.M. Moulin, F. Albessard, J. Lacotte, and P. Camsonne, *J. Chromatogr.*, 1981, 226, 250.
409 T. Marunaka, T. Shibata, Y. Minami, and Y. Umeno, *J. Chromatogr.*, 1980, 183, 331.
410 M.G. Horning and K. Lertratanangkoon, *J. Chromatogr.*, 1980, 181, 59.
411 Y. Bergqvist and S. Eckerbom, *J. Chromatogr.*, 1981, 226, 91.
412 D.G. Calverley, G.B. Baker, R.T. Coutts, and W.G. Dewhurst, *Biochem. Pharmacol.*, 1981, 30, 861.
413 A.K. Mitra, C.L. Baustian, and T.J. Mikkelson, *J. Pharm. Sci.*, 1980, 69, 257.
414 R. Riva, F. Albani, and A. Baruzzi, *J. Chromatogr.*, 1980, 221, 75.
415 J.L. Woolley, jun., C.W. Sigel, and C.M. Wels, *Life Sci.*, 1980, 27, 1819.
416 J. Vink, H.J.M. Van Hal, and P.C.J.M. Koppens, *Adv. Mass Spectrom.*, 1980, 8, 1251.
417 F.C. Falkner, *Biomed. Mass Spectrom.*, 1981, 8, 43.
418 A.P. De Leenheer and H.J.C.F. Nelis, *Analyst (London)*, 1981, 106, 1025.

419 B.N. Colby, A.E. Rosecrance, and M.E. Colby, *Anal. Chem.*, 1981, 53, 1907.
420 E.D. Bush and W.F. Trager, *Biomed. Mass Spectrom.*, 1981, 8, 211.
421 D.A. Schoeller, *Biomed. Mass Spectrom.*, 1980, 7, 457.
422 C. Van Peteghem, J. Demeter, and A. Heyndrickx, *Anal. Chem. Symp. Ser.*, 1980, 4, 497.
423 G. Tomson, M. Garle, B. Thalme, H. Nisell, L. Nylund, and A. Rane, *Br. J. Clin. Pharmacol.*, 1982, 13, 653.
424 S. Murray, *Biomed. Mass Spectrom.*, 1980, 7, 179.
425 M. Tetsuo, M. Axelson, and J. Sjövall, *J. Steroid Biochem.*, 1980, 13, 847.
426 W. Stüber, E. Mutschler, and D. Steinbach, *J. Chromatogr.*, 1982, 227, 193.
427 M. Koyama, M. Hashimoto, N. Asakawa, M. Ishibashi, and H. Miyazaki, *Biomed. Mass Spectrom.*, 1980, 7, 372.
428 C. Aubert, C. Luccioni, P. Coassolo, P.J. Sommadossi, and J.P. Cano, *Arzneim.-Forsch.*, 1981, 31, 2048.
429 P. Daniel, S.J. Gaskell, H. Bishop, C. Campbell, and R.I. Nicholson, *Eur. J. Cancer Clin. Oncology*, 1981, 17, 1183.
430 F.A. Wolheim, P. Stenberg, B. Nilsson, and G. Mellbin, *Eur. J. Clin. Pharmacol.*, 1981, 20, 423.
431 D. Hazelby and K.T. Taylor, *Ann. Chim. (Rome)*, 1980, 70, 201.
432 S.J. Gaskell, *Postgrad. Med. J.*, 1980, 56 (Suppl. 1), 90.
433 J. Taskinen, N. Vahvelainen, and P. Nore, *Biomed. Mass Spectrom.*, 1980, 7, 556.
434 E. Sturm and K. Zech, *Anal. Chem. Symp. Ser.*, 1982, 11, 241.
435 N.K. Lee, P.A. Gnad, P.R. Trebilco, G.J. Wright, and L.K. Pannell, *Biomed. Mass Spectrom.*, 1980, 7, 418.
436 D.J. Harvey, J.T.A. Leuschner, and W.D.M. Paton, *J. Chromatogr.*, 1980, 202, 83.
437 D.J. Harvey, J.T.A. Leuschner, and W.D.M. Paton, *Br. J. Pharmacol.*, 1981, 74, 771P.
438 D.J. Harvey, J.T.A. Leuschner, and W.D.M. Paton, *J. Chromatogr.*, 1982, 239, 243.
439 D.A. Durden, *Anal. Chem.*, 1982, 54, 666.
440 J.A. Thompson and F.H. Leffert, *J. Pharm. Sci.*, 1980, 69, 707.
441 J.A. Thompson, D.C. Bloedow, and F.H. Leffert, *J. Pharm. Sci.*, 1981, 70, 1284.
442 R. Whelpton, S.H. Curry, and G.M. Watkins, *J. Chromatogr.*, 1982, 228, 321.
443 K.K. Midha, R.M.H. Roscoe, K. Hall, E.M. Hawes, J.K. Cooper, G. McKay, and H.U. Shetty, *Biomed. Mass Spectrom.*, 1982, 9, 186.
444 W. Ihn, W. Schade, B. Müller, and G. Peiker, *Pharmazie*, 1980, 35, 598.
445 S.P. Jindal, T. Lutz, and P. Vestergaard, *J. Anal. Toxicol.*, 1982, 6, 34.
446 S.P. Jindal, T. Lutz, C. Hallstrom, and P. Vestergaard, *Clin. Chim. Acta*, 1981, 112, 267.
447 M. Ervik, I. Kylberg-Hanssen, and P.-O. Lagerström, *J. Chromatogr.*, 1982, 229, 87
448 G. Van Peteghem, A. Heyndrickx, and W. Van Zele, *J. Pharm. Sci.*, 1980, 69, 118.
449 M. Eckert and P.H. Hinderling, *Agents Actions*, 1981, 11, 520.
450 L. Palmér, J. Edgar, G. Lundgren, B. Karlén, and J. Hermansson, *Acta Pharmacol. Toxicol.*, 1981, 49, 72.
451 A. Ohlsson, J.-E. Lindgren, A. Wahlen, S. Agurell, L.E. Hollister, and H.K. Gillespie, *Clin. Pharmacol. Ther.*, 1980, 28, 409.

452 S. Agurell, S. Carlsson, J.-E. Lindgren, A. Ohlsson, H. Gillespie, and L. Hollister, *Experientia,* 1981, 37, 1090.

453 L.E. Hollister, H.K. Gillespie, A. Ohlsson, J.-E. Lindgren, A. Wahlen, and S. Agurell, *J. Clin. Pharmacol.*, 1981, 21, 171S.

454 J.-E. Lindgren, A. Ohlsson, S. Agurell, L. Hollister, and H. Gillespie, *Psychopharmacology,* 1981, 74, 208.

455 J.A.A. Jonckheere, L.M.R. Thienpont, A.P. De Leenheer, P. De Backer, M. Debackere, and F.M. Belpaire, *Biomed. Mass Spectrom.*, 1980, 7, 582.

456 N.-E. Larsen and K. Marinelli, *J. Chromatogr.*, 1981, 222, 482.

457 J. Rosenfeld, A. Phatak, T.L. Ting, and W. Lawrence, *Anal. Lett.*, 1980, 13, 1373.

458 I. Björkhem, O. Lantto, and A. Löf, *J. Steroid Biochem.*, 1980, 13, 169.

459 K. Kawahara, M. Matsumura, and K. Kimura, *J. Chromatogr.*, 1981, 223, 202.

460 S. Baba, T. Goromaru, K. Yamazaki, and Y. Kasuya, *J. Pharm. Sci.*, 1980, 69, 1300.

461 H. Ehrsson, S. Eksborg, I. Wallin, Y. Mårde, and B. Joansson, *J. Pharm. Sci.*, 1980, 69, 710.

462 K. Ohya, S. Shintani, and M. Sano, *J. Chromatogr.*, 1980, 221, 293.

463 M.R. Holdiness, Z.H. Israili, and J.B. Justice, *J. Chromatogr.*, 1981, 224, 415.

464 R. Endele, M. Senn, and U. Abshagen, *J. Chromatogr.*, 1982, 227, 187.

465 U. Abshagen, G. Betzien, B. Kaufmann, and G. Endele, *Eur. J. Clin. Pharmacol.*, 1982, 21, 293.

466 D. Alkalay, S. Carlsen, and W.E. Wagner, jun., *Anal. Lett.*, 1981, 14, 1745.

467 E.L. Crampton, R.C. Glass, B. Marchant, and J.A. Rees, *J. Chromatogr.*, 1980, 183, 141.

468 S.-N. Lin, T.-P.F. Wang, R.M. Caprioli, and B.P.N. Mo, *J. Pharm. Sci.*, 1981, 70, 1276.

469 S.P. Pallante, C. Fenselau, R.G. Mennel, R.B. Brundrett, M. Appler, N.B. Rosenshein, and M. Colvin, *Cancer Res.*, 1980, 40, 2268.

470 S.-E. Jacobsson, S. Jönsson, C. Lindberg, and L.-Å. Svensson, *Biomed. Mass Spectrom.*, 1980, 7, 265.

471 S. Murray, K.A. Waddel, and D.S. Davies, *Biomed. Mass Spectrom.*, 1981, 8, 500.

472 K.P. Maguire, T.R. Norman, G.D. Burrows, and B.A. Scoggins, *J. Chromatogr.*, 1981, 222, 399.

473 K.P. Maguire, G.D. Burrows, T.R. Norman, and B.A. Scoggins, *Br. J. Clin. Pharmacol.*, 1981, 12, 405.

474 K.P. Maguire, T.R. Norman, G.D. Burrows, and B.A. Scoggins, *Eur. J. Clin. Pharmacol.*, 1982, 21, 517,

475 S.P. Jindal, T. Lutz, and P. Vestergaard, *J. Pharm. Sci.*, 1980, 69, 684.

476 M. Ervik, K.-J. Hoffmann, and K. Kylberg-Hanssen, *Biomed. Mass Spectrom.*, 1981, 8, 322.

477 C.-G. Regardh, L. Jordö, M. Ervik, P. Lundborg, R. Olsson, and O. Rönn, *Clin. Pharmacokinet.*, 1981, 6, 375.

478 K.-U. Seiler, K.J. Schuster, G.-J. Meyer, W. Niedermayer, and O. Wassermann, *Clin. Pharmacokinet.*, 1980, 5, 192.

479 L. Jordö, P.O. Attman, M. Aurell, L. Johansson, G. Johnsson, and C.-G. Regardh, *Clin. Pharmacokinet.*, 1980, 5, 169.

480 B. Vandel, M. Sandoz, S. Vandel, G. Allers, and R. Volmat, *Eur. J. Clin. Pharmacol.*, 1982, 22, 239.

481 D.M. Chinn, T.A. Jennison, D.J. Crouch, M.A. Peat, and G.W. Thatcher, *Clin. Chem.*, 1980, 26, 1201.
482 A. Lapin, *Eur. J. Mass Spectrom.*, 1980, 1, 121.
483 A. Lapin and M. Karobath, *Biomed. Mass Spectrom.*, 1980, 7, 588.
484 J.G. Leferink, J. Dankers, and R.A.A. Maes, *J. Chromatogr.*, 1982, 229, 217.
485 D.M. Chinn, D.J. Crouch, M.A. Peat, B.S. Finkle, and T.A. Jennison, *J. Anal. Toxicol.*, 1980, 4, 37.
486 C. Lartigue-Mattei. J.L. Chabard, H. Bargnoux, J. Petit, and J.A. Berger, *J. Chromatogr.*, 1982, 229, 211.
487 T. Marunaka and Y. Umeno, *J. Chromatogr.*, 1980, 221, 382.
488 I.H. Patel, R.H. Levy, J.M. Neal, and W.F. Trager, *J. Pharm. Sci.*, 1980, 69, 1218.
489 I.H. Patel, R.H. Levy, and R.E. Cutler, *Clin. Pharmacol. Ther.*, 1980, 27, 515.
490 E. Tomori and I. Elekes, *J. Chromatogr.*, 1981, 204, 355.
491 Y. Matsuki, T. Ito, K. Fijuhara, T. Nakamura, M. Kimura, and H. Ono, *J. Chromatogr.*, 1982, 239, 585.
492 D.E. Schwartz and U.B. Ranalder, *Biomed. Mass Spectrom.*, 1981, 8, 589.
493 B. Lindeke, H. Broetell, B. Karlen, G. Rietz, and A. Vietorisz, *Acta Pharm. Suecica*, 1981, 18, 25.
494 B.R. Kuhnert, P.M. Kuhnert, and A.L.P. Reese, *J. Chromatogr.*, 1981, 224, 488.
495 K. Matsumura, O. Kubo, T. Sakashita, Y. Adachi, and H. Kato, *J. Chromatogr.*, 1981, 222, 53.
496 C.F. Gelijkens, A.P. De Leenheer, and P. Sandra, *Biomed. Mass Spectrom.*, 1980, 7, 572.
497 R.G. Smith and L.K. Cheung, *J. Chromatogr.*, 1982, 229, 464.
498 Y.M. Chan, S.J. Soldin, J.M. Swanson, C.M. Deber, J.J. Thiessen, and S. MacLeod, *Clin. Biochem.*, 1980, 13, 266.
499 W.F. Bayne, F.T. Tao, G. Rogers, L.C. Chu, and F. Theeuwes, *J. Pharm. Sci.*, 1981, 70, 75.
500 H. Watanabe, S. Nakano, N. Ogawa, and T. Suzuki, *Biomed. Mass Spectrom.*, 1980, 7, 160.
501 H. Watanabe, S. Nakano, and N. Ogawa, *J. Chromatogr.*, 1982, 229, 95.
502 H. Nau and W. Wittfoht, *J. Chromatogr.*, 1981, 226, 69.
503 H. Nau and W. Loscher, *J. Pharmacol. Exp. Ther.*, 1982, 220, 654.
504 J. Schmid and F.-W. Koss, *J. Chromatogr.*, 1982, 227, 71.
505 H. Hakusui, W. Suzuki, and M. Sano, *J. Chromatogr.*, 1980, 182, 47.
506 K. Kishida, R. Manabe, K. Bando, and Y. Miwa, *Anal. Lett.*, 1981, 14, 335.
507 N. Hashimoto, T. Aoyama, and T. Shioiri, *Chem. Pharm. Bull.*, 1981, 29, 1478.
508 J.B. Fourtillan, M.A. Lefebvre, J. Girault, and Ph. Courtois, *J. Pharm. Sci.*, 1981, 70, 573.
509 H. Studel, A. Steudel, and G.E. Vonunruh, *J. Clin. Chem. Clin. Biochem.*, 1982, 20, 267.
510 E.J. Cone, W. Buchwald, and D. Yousefnejad, *J. Chromatogr.*, 1981, 223, 331.
511 T.A. Robert, A.N. Hagardorn, E.A. Diagneault, and R.D. Brown, *J. Chromatogr.*, 1981, 224, 116.
512 W. Cautreels and J.P. Jeanniot, *Biomed. Mass Spectrom.*, 1980, 7, 565.
513 P. Jakobsen and A.K. Pedersen, *J. Pharm. Pharmacol.*, 1981, 33, 89.
514 A.K. Pedersen and P. Jakobsen, *Br. J. Clin. Pharmacol.*, 1981, 11, 597.

515 W.A. Ritschel, G. Bykadi, K.A. Hoffmann, E.J. Norman, P.W. Lucker, and W. Rindt, *Arzneim.-Forsch.*, 1982, 32, 64.
516 W.A. Ritschel, G. Bykadi, E.J. Norman, and P.W. Lucker, *Arzneim.-Forsch.*, 1980, 30, 1535.
517 G. Idzu, M. Ishibashi, and H. Miyazaki, *J. Chromatogr.*, 1982, 229, 327.
518 M. Miyazaki, M. Ishibashi, Y. Hashimoto, G. Idzu, and Y. Furuta, *J. Chromatogr.*, 1982, 239, 277.
519 J.D. Gilbert, R.I. Aylott, G.H. Draffan, and H.H. Sögtrop, *Arzneim.-Forsch.*, 1982, 32, 571.
520 H.-J. Egger, W. Wittfoht, H. Nau, G. Dielmann, and U. Rapp, *Adv. Mass Spectrom.*, 1980, 8, 1219.
521 J. Pao and J.A.F. de Silva, *J. Chromatogr.*, 1980, 221, 97.
522 B.J. Miwa, W.A. Garland, and P. Blumenthal, *Anal. Chem.*, 1981, 53, 793.
523 L.E. Martin, J. Oxford, and R.J.N. Tanner, *Xenobiotica*, 1981, 11, 831.
524 L.E. Martin, J. Oxford, and R.J.N. Tanner, *J. Chromatogr.*, 1982, 251, 215.
525 S.L. Abidi, *J. Chromatogr.*, 1982, 234, 187.
526 Y. Nakagawa, Y. Ikenishi, K. Sugeno, H. Kambara, and I. Kanomata, *Adv. Mass Spectrom.*, 1980, 8, 1261.
527 U. Bahr, H.-R. Schulten, O.R. Hommes, and F. Aerts, *Clin. Chim. Acta*, 1980, 103, 183.
528 U. Bahr and H.-R. Schulten, *Biomed. Mass Spectrom.*, 1981, 8, 553.
529 U. Bahr and H.-R. Schulten, *J. Labelled Comp. Radiopharm.*, 1981, 18, 571.
530 W.D. Lehmann and H.-R. Schulten, *Anal. Chem. Symp. Ser.*, 1982, 11, 635.
531 M. Przybylski, J. Preiss, R. Dennebaum, and J. Fischer, *Biomed. Mass Spectrom.*, 1982, 9, 22.
532 G.M. Schier, I.E.T. Gan, B. Halpern, and J. Korth, *Clin. Chem. (Winston-Salem, N.C.)*, 1980, 26, 147.
533 M.J. Kreek, F.A. Bencsath, and F.H. Field, *Biomed. Mass Spectrom.*, 1980, 7, 385.
534 D.A. Shiba and R.J. Weinkam, *J. Chromatogr.*, 1982, 229, 397.
535 W.J. Stone and T. Walle, *Clin. Pharmacol. Ther.*, 1980, 28, 449.
536 D.A. Riopel and T. Walle, *Clin. Pharmacol. Ther.*, 1980, 28, 743.
537 T.C. Fagan, T. Walle, U.K. Walle, E.C. Conradi, G. Harmon, and T.E. Gaffney, *Br. J. Clin. Pharmacol.*, 1982, 13, 571.
538 T. Walle, T.C. Fagan, U.K. Walle, M.-J. Oexmann, E.C. Conradi, and T.E. Gaffney, *Clin. Pharmacol. Ther.*, 1981, 30, 790.
539 T. Walle, E.C. Conradi, U.K. Walle, T.C. Fagan, and T.E. Gaffney, *Clin. Pharmacol. Ther.*, 1980, 27, 22.
540 C. Graffner, K.-J. Hoffmann, G. Johnson, P. Lundborg, and O. Rönn, *Eur. J. Clin. Pharmacol.*, 1981, 20, 91.
541 H. Liedholm, A. Melander, P.-O. Bitzén, G. Helm, G. Lönnerholm, I. Mattiasson, B. Nilsson, and E. Wahlin-Boll., *Eur. J. Clin. Pharmacol.*, 1981, 20, 229.
542 W. Kirch, H. Köhler, W. Braum, and Ch.V. Gizycki, *Clin. Pharmacokinet.*, 1980, 5, 476.
543 J.L. Reid, C. Zamboulis, and C.A. Hamilton, *Br. J. Clin. Pharmacol.*, 1980, 10, 183S.
544 C. Zamboulis and J.L. Reid, *Eur. J. Clin. Pharmacol.*, 1981, 19, 19.
545 W. Kirch, H. Köhler, and W. Braun, *Br. J. Clin. Pharmacol.*, 1980, 10, 33S.

546 M.V. Rudorfer and R.C. Young, *Commun. Psychopharmacol.*, 1980, 4, 185.
547 M.V. Rudorfer and R.C. Youhg, *Clin. Pharmacol. Ther.*, 1980, 28, 703.
548 C. Lindberg, U. Bondesson, and P. Hartvig, *Biomed. Mass Spectrom.*, 1980, 7, 88.
549 R.K. Verbeeck, R.A. Branch, and G.R. Wilkinson, *Clin. Pharmacol. Ther.*, 1981, 30, 619.
550 U. Bondesson, S. Arnér, P. Anderson, L.O. Boréus, and P. Hartvig, *Eur. J. Clin. Pharmacol.*, 1980, 17, 45.
551 P. Anderson, S. Arnér, L.O. Boréus, U. Bondesson, and P. Hartvig, *Eur. J. Clin. Pharmacol.*, 1981, 19, 217.
552 U. Bondesson, P. Hartvig, and B. Danielsson, *Drug Metab. Dispos.*, 1981, 9, 376.
553 B. Mellström, L. Bertilsson, J. Säwe, H.-U. Schulz, and F. Sjöqvist, *Clin. Pharmacol. Ther.*, 1981, 30, 189.
554 L. Bertilsson, M. Eichelbaum, B. Mellström, J. Sawe, H.-U. Schulz, and F. Sjöqvist, *Life Sci.*, 1980, 27, 1673.
555 G. Lönnerholm and B. Lindström, *Br. J. Clin. Pharmacol.*, 1982, 13, 729.
556 R.G. Devlin and P.M. Fleiss, *J. Clin. Pharmacol.*, 1981, 21, 110.
557 P.-O. Bitzén, B. Gustafsson, K.G. Jostell, A. Melander, and E. Wåhlin-Boll, *Eur. J. Clin. Pharmacol.*, 1981, 20, 123.
558 L.F. Gram, J. Schou, J. Heltberg, and W.L. Way, *Acta Pharmacol. Toxicol.*, 1980, 46, 395.
559 K.-C. Khoo, J. Mendels, M. Rothbart, W.A. Garland, W.A. Colburn, B.N. Min, R. Lucek, J.J. Carbone, H.G. Boxenbaum, and S.A. Kaplan, *Clin. Pharmacol. Ther.*, 1980, 28, 368.
560 J.L. Cunningham and D.A. Price-Evans, *Br. J. Clin. Pharmacol.*, 1981, 11, 591.
561 W.E. Dodson, *Neurology*, 1980, 30, 196.
562 J.B. Whitlam, K.F. Brown, M.J. Crooks, and G.F.W. Room, *Clin. Pharmacol. Ther.*, 1981, 29, 487.
563 W.A. Ritschel, G. Bykadi, E.J. Norman, and R.J. Cluxton, *J. Clin. Pharmacol.*, 1981, 21, 461.
564 P. Kragh-Sorensen and N.-E. Larsen, *Clin. Pharmacol. Ther.*, 1980, 28, 796.
565 D.E. Schwartz, G. Eckert, D. Hartmann, B. Weber, D. Richard-Lenoble, J.M.K. Ekue, and M. Gentilini, *Chemotherapy*, 1982, 28, 70.
566 J. Greaves, D.A.P. Evans, H.M. Gilles, K.A. Fletcher, D. Bunnag, and T. Harinasuta, *Br. J. Clin. Pharmacol.*, 1980, 10, 399.
567 J. Greaves, D.A.P. Evans, and K.A. Fletcher, *Br. J. Clin. Pharmacol.*, 1980, 10, 293.
568 R. Jauch, E. Griesser, and G. Oesterhelt, *Arzneim.-Forsch.*, 1980, 30, 60.
569 S.-M. Aquilonius, S.-Å. Eckernäs, P. Hartvig, B. Lindström, and P.O. Osterman, *Eur. J. Clin. Pharmacol.*, 1980, 18, 423.
570 D.N. Bateman, C. Kahn, and D.S. Davies, *Br. J. Clin. Pharmacol.*, 1980, 9, 371,
571 J.G. Leferink, W. Van Den Berg, I. Wagemaker-Engels, J. Kreukniet, and R.A.A. Maes, *Arzneim.-Forsch.*, 1982, 32, 159.
572 P.M. Kuhnert, B.R. Kuhnert, J.M. Stitts, and T.L. Gross, *Anesthesiology*, 1981, 55, 611.
573 B.A. Scoggins, K.P. Maguire, T.R. Norman, and G.D. Burrows, *Clin. Chem. (Winston-Salem, N.C.)*, 1980, 26, 5.
574 B.A. Scoggins, K.P. Maguire, T.R. Norman, and G.D. Burrows, *Clin. Chem. (Winston-Salem, N.C.)*, 1980, 26, 805.
575 N. Narasimhachari, *J. Chromatogr.*, 1981, 225, 189.

576 M. Eichelbaum, A. Somogyi, C.E. Von Unruh, and H.J. Dengler, *Eur. J. Clin. Pharmacol.*, 1981, 19, 133.
577 A. Somogyi, M. Albrecht, G. Kliems, K. Schäfer, and M. Eichelbaum, *Br. J. Clin. Pharmacol.*, 1981, 12, 51.
578 R.K. Verbeck, R.C. James, D.F. Taber, B.J. Sweetman, and G.R. Wilkinson, *Biomed. Mass Spectrom.*, 1980, 7, 58.
579 U. Meresaar, M.I. Nilsson, J. Holmstrand, and E. Angaard, *Eur. J. Clin. Pharmacol.*, 1981, 20, 473,
580 J.R. Carlin, R.W. Walker, R.O. Davies, R.T. Ferguson, and W.J.A. VandenHeuvel, *J. Pharm. Sci.*, 1980, 69, 1111.
581 N.J. Haskins, G.C. Ford, R.F. Palmer, and K.A. Waddell, *Biomed. Mass Spectrom.*, 1980, 7, 74.
582 N.J. Haskins, K.A. Waddell, G.C. Ford, P.N. Spalton, C.M. Walls, T.J. Forrest, and R.F. Palmer, *Biomed. Mass Spectrom.*, 1980, 7, 80.
583 N. Haskins, C. Ford, P.N. Spalton, and K. Waddell, *Adv. Mass Spectrom.*, 1980, 8, 286.
584 W.A. Colburn, I. Bekersky, B.H. Min, B.J. Hodshon, and W.A. Garland, *Res. Commun. Chem. Pathol. Pharmacol.*, 1980, 27, 73.
585 F. Varin, C. Marchand, P. Larochelle, and K.K. Midha, *J. Pharm. Sci.*, 1980, 69, 640.
586 M. Eichelbaum, H.J. Dengler, A. Somogyi, and E.E. Von Unruh, *Eur. J. Clin. Pharmacol.*, 1981, 19, 127.
587 D. Alkalay, W.E. Wagner, jun., S. Carlsen, L. Khemani, J. Volk, M.F. Bartlett, and A. LeSher, *Clin. Pharmacol. Ther.*, 1980, 27, 697.
588 H. De Bree, D.J.K. Van Der Stel, J.M.A. Kaal, and J.B. Van Der Schoot, *Anal. Chem. Symp. Ser.*, 1982, 11, 203.
589 J. Schmid, A. Prox, H. Zipp, and F.W. Koss, *Biomed. Mass Spectrom.*, 1980, 7, 560.
590 S. Higuchi and Y. Shiobara, *Biomed. Mass Spectrom.*, 1980, 7, 339.
591 M.L. Powell, R.R. Wagoner, C.-H. Chen, and W.L. Nelson, *Res. Commun. Chem. Pathol. Pharmacol.*, 1980, 30, 387.
592 T.R. Burke, jun., W.N. Howald, and W.L. Nelson, *Res. Commun. Chem. Pathol. Pharmacol.*, 1980, 28, 399.
593 U.K. Walle, M.J. Wilson, and T. Walle, *Biomed. Mass Spectrom.*, 1981, 8, 78.
594 C. Hignite, J. Uetrecht, C. Tschanz, and D. Azarnoff, *Clin. Pharmacol. Ther.*, 1980, 28, 99.
595 R.A. Orielly, W.F. Trager, C.H. Motley, and W. Howald, *Clin. Pharmacol. Ther.*, 1980, 28, 187.
596 J.A. Thompson and G.W. Shioshita, *Drug Metab. Dispos.*, 1981, 9, 5.
597 D.L. Hachey, K. Nakamura, M.J. Kreek, and P.D. Klein, *Anal. Chem. Symp. Ser.*, 1982, 11, 235.
598 K. Nakamura, D.L. Hachey, M.J. Kreek, C.S. Irving, and P.D. Klein, *J. Pharm. Sci.*, 1982, 71, 40.
599 A. Van Langenhove, C.E. Costello, J.E. Biller, K. Biemann, and T.R. Browne, *Biomed. Mass Spectrom.*, 1980, 7, 576.
600 A. Van Langenhove, C.E. Costello, J.E. Biller, K. Biemann, and T.R. Browne, *Clin. Chim. Acta*, 1981, 115, 263.
601 T.R. Browne, A. Van Langenhove, C.E. Costello, K. Biemann, and D.J. Greenblatt, *Clin. Pharmacol. Ther.*, 1981, 29, 511.
602 A. Van Langenhove, J.E. Biller, K. Biemann, and T.R. Browne, *Biomed. Mass Spectrom.*, 1982, 9, 201.
603 I.M. Kapetanovic and H.J. Kupferberg, *Biomed. Mass Spectrom.*, 1980, 7, 47.
604 I.M. Kupetanovic, H.J. Kupferberg, R.J. Porter, W. Theodore, E. Schulman, and J.K. Penry, *Clin. Pharmacol. Ther.*, 1981, 29, 480.

605 G.E. Von Unruh, B.Ch. Jancik, and F. Hoffmann, *Biomed. Mass Spectrom.*, 1980, 7, 164.
606 F. Hoffmann, G.E. Von Unruh, and B.C. Jancik, *Eur. J. Clin. Pharmacol.*, 1981, 19, 383.
607 W. Kochen, B. Tauscher, M. Klemens, and E. Depene, *Anal. Chem. Symp. Ser.*, 1982, 11, 271.
608 W. Wittfoht, H. Hau, D. Rating, and H. Helge, *Anal. Chem. Symp. Ser.*, 1982, 11, 265.
609 L. Bertilsson, B. Hojer, G. Tybring, J. Osterloh, and A. Rane, *Clin. Pharmacol. Ther.*, 1980, 27, 83.
610 B. Mellstrom and C. von Bahr, *Drug Metab. Dispos.*, 1981, 9, 565.
611 L.A. Bauer, T. Brown, M. Gibaldi, L. Hudson, S. Nelson, V. Raisys, and J.P. Shea, *Clin. Pharmacol. Ther.*, 1982, 31, 433.
612 A. Ohlsson, J.lE. Lindgren, A. Wahlen, S. Agurell, L.E. Hollister, and H.K. Gillespie, *Biomed. Mass Spectrom.*, 1982, 9, 6.

9

Natural Products

BY D. E. GAMES

1 Introduction

In addition to other relevant chapters in this volume, the recent biannual review on mass spectrometry in *Anal. Chem.*[1] and a volume that collates a number of workers' experiences with soft-ionization techniques in biological chemistry[2] provide a valuable background to this chapter.

Since the previous report,[3] two major developments relevant to the mass-spectrometric study of natural products have occurred. One is the advent of a new technique, fast atom bombardment (f.a.b.) mass spectrometry for obtaining mass-spectral data from polar molecules of low volatility.[4-8] F.a.b. differs from secondary-ion mass spectrometry (s.i.m.s.) in that a primary-neutral-atom bombarding beam is used, as opposed to an ionic beam. Also the sample is dissolved in a thin film of a viscous matrix for loading on the target. F.a.b. mass spectrometry has the advantage that it can be readily adapted to high-resolution double-focusing instruments, because the incoming neutral beam is not affected by the high source potential. Hence, coupled with the second development, which is the availability of new, extended mass-range options on medium-resolution mass spectrometers, which enable a mass range of 2500 dalton to be achieved at maximum accelerating voltage, a whole range of molecules that previously could be studied only in specialized laboratories has become amenable to the general practitioner of mass spectrometry. For those fortunate enough to have access to unlimited funds, even larger molecules can be studied, since two high-performance double-focusing sector instruments are commercially available with mass ranges of 3000 dalton at 8 kV accelerating potential. The power of f.a.b. is illustrated by the recently reported identification of the molecular species of the large oligopeptides melittin, glucagon, and the B chain of bovine insulin.[9] In addition to providing relative-molecular-mass data readily from mass-spectrometrically difficult molecules, the technique, in many cases, also provides an abundance of structurally useful fragment ions. However, recent studies of compounds of relative molecular mass greater than 3000 dalton highlight the need for further instrumental development to enable even higher mass ranges to be achieved with better instrumental sensitivity. Whether this development will come more readily from conventional mass spectrometers or from Fourier-transform mass spectrometers[1,10] is difficult to assess.

Although controversy surrounds the development of f.a.b. mass spectrometry,[11] the UMIST group was the first to show its relevance in the solution of real problems, and the rapidity with which the technique has appeared in laboratories throughout the world testifies to its importance.

So where do other techniques now stand? Because of the use of time-of-flight mass analysis, ^{252}Cf plasma desorption has the advantage that it can be effectively used for high-mass molecules, and this is illustrated by the detection of a $(2M + Na)^+$ species at m/z 12637 for a fully protected oligonucleotide.[13] F.a.b. suffers from a number of problems including the high background due to the ionization of the matrix and difficulties in handling mixtures (because of preferential or suppressed ionization of some components of the mixture). If relative-molecular-mass profiling is being undertaken, the presence of fragment ions, the possibility of chemical reaction taking place with the matrix, and selection of a suitable matrix for non-polar or more difficult compounds are additional complications. In a number of these respects, laser desorption,[14,15] field desorption (f.d.),[16] and s.i.m.s. currently all have advantages for certain types of problem. Of particular interest in the surface-ionization area have been recent studies indicating the viability of high-performance liquid chromatography-mass spectrometry (l.c.-m.s.) with laser desorption,[17] s.i.m.s.,[18,19] and ^{252}Cf plasma desorption.[20] These ionization methods show potential for the development of systems for handling more difficult compounds by l.c.-m.s. since the necessity of having a solvent matrix is avoided. However, it may well be that with the continual sample supply of moving-belt interfaces for f.a.b.

the need for a matrix can be avoided or the presence of small amounts of solvent remaining on the belt may be sufficient to effect efficient ionization.

A number of reviews[21-24] have appeared collating the types of information obtainable from molecules of relative molecular mass in excess of 1000 dalton with high-magnetic-field instruments and some of the ionization techniques referred to earlier. A symposium report[25] on a meeting concerned with fast-atom and ion-induced mass spectrometry of involatile organics is also of interest in that it describes the different approaches that can be used for studies in this area.

Developments in other forms of ionization that can be readily incorporated into current instrumentation continue. The approach using desorption of compounds from extended probes has been reviewed.[26] Recent developments in this area, which have been illustrated with examples of mass-spectrometrically difficult natural products, include the coating of wires with polyimide for direct chemical ionization, which reduces thermal decomposition,[27] the use of a specially constructed gold support,[28] and a recent modification in which the gold wire is at the same potential as the ion source,[29] for e.i. studies. The use of volatile amines, which by adduct formation preclude water loss from compounds containing labile hydroxy functions, has been advocated for this type of c.i. study.[30]

Other relevant advances in ionization methodology include the development of a new ionization technique as a result of studies of l.c.-m.s.[31-33] An oxy-hydrogen flame or electrical heating is used to vaporize rapidly the total effluent from a l.c. column. A neutral beam is formed and produces ions upon impinging on a nickel surface (heated to 250 °C). The system shows considerable promise for enabling l.c.- m.s. to be carried out on more difficult compounds. Developments whereby polar organic molecules have been successfully handled by atmospheric-pressure ionization[34] and use of an inductively coupled plasma are of interest in the context of future l.c.-m.s. developments.[35] In the context of ionization techniques for l.c.-m.s., two recent reviews[36,37] have critically surveyed methods that may be of use in this area, and a strong case has been put for not removing solvent in the liquid chromatograph-mass spectrometer interface.[37]

Attention to the method of sample preparation can have considerable advantages in the quality of spectra obtained from desorption-ionization techniques[38] and is illustrated in the use of *p*-toluenesulphonic acid as an additive to give improved $(M + 1)^+$ ion sensitivity in f.d. mass spectrometry for zwitterionic compounds such as phospholipids and amino-acids.[39] In the case of f.d. mass spectrometry, methods for improving the quality of samples and hence the spectra obtained have also been described.[40] Negative-ion f.d. appears to be useful for obtaining mass-spectral data from anionic species.[41-43]

Polyamide thin-layer plates have been shown to give excellent spectra for a range of natural products when portions of the plates are introduced into the mass-spectrometer ion source by means of the direct-insertion probe.[44]

The use of gas-phase ion/molecule isotope-exchange reactions, whereby hydrogen atoms in specific organic structural environments can be located,[45] would appear to have considerable utility for the characterization of new natural products when only very small sample amounts, insuffient for n.m.r. study, are available.

A major strength of mass spectrometry for studies in the natural-product area remains its ability through combined techniques of chromatography-mass spectrometry and mass spectrometry-mass spectrometry to study complex mixtures. The introduction of on-column capillary g.c. techniques together with bonded-phase fused-silica capillary columns has further extended the range of compounds amenable to capillary g.c.-m.s. There are many examples in the literature of the use of direct liquid introduction and moving-belt l.c.-m.s. systems in the natural-product area. The advantage of the chromatographic approach is that isomeric species can usually be resolved and, in the case of l.c.-m.s., the l.c. conditions can be used to separate new compounds of interest for further structural study, since apart from, for example, studies of peptides mass spectrometry is still unable to characterize a new compound completely. Mass spectrometry-mass spectrometry (m.s.-m.s.) has the advantage of speed of sample analysis, but a recent comparative study of l.c.-m.s. and m.s.-m.s.[46] for the investigation of a crude extract of ergot illustrates the problems that can be encountered if new compounds are sought. Even the more sophisticated instrumentation currently on offer with double sectors and double quadrupole and with triple quadrupole and four sectors is unable to differentiate between isomeric species when they are present as mixtures. It is of interest to note that the recent identification of the quaternary alkaloids choline and muscarine in mushrooms by thin-layer chromatography-s.i.m.s. concludes that this is a more specific approach than m.s.-m.s.[47] The former approach was also concluded to be better than l.c.-m.s. with a moving-belt interface and c.i. mass spectrometry for this type of compound, although it would appear that no on-line experiments were performed.

Apart from the developments in ionization methods relevant to l.c.-m.s. referred to earlier, another important development has been the use of microbore l.c.[48] Although early studies in this area suffered from poor chromatography, recent studies[49] in our laboratories using glass-lined stainless-steel microbore columns have effected considerable improvements in terms of chromatographic efficiencies so that performance is now comparable with conventional columns, but improved sensitivity is obtainable. For example, with conventional l.c.-m.s., 3 μg of gibberellin A_3 requires to be injected on-column for a full spectrum, whereas 20 ng gave a spectrum of similar quality by microbore l.c.-m.s. In addition, mobile phases containing high percentages of water are readily handled with moving-belt interfaces using this approach.

Methods for the structure elucidation of natural products by computer-assisted interpretation of high-resolution mass-spectral data have been described.[50] A number of general reviews of methods for sequence determination of biological macromolecules have also appeared.[51-53]

2 Alkaloids

E.i. mass-spectral fragmentations of porphine,[54] tropine ester,[55] and pyrrolizidine alkaloids[56] have been studied. A detailed study[57] of the complex indole alkaloids the penitrems has been undertaken using linked-scan techniques. Using model compounds, the mechanism of formation of doubly charged fragments from bis-benzyltetrahydroisoquinolines has been investigated.[58]

F.d. mass spectrometry has been shown to be suitable for the characterization of unstable indole alkaloid immonium salts[59] and was used together with other techniques for the characterization of the spermine alkaloid kukoamine A.[60] This technique, together with e.i. and m.i.k.e.s. with collisional activation, has been used to investigate the mass-spectral behaviour of *N*-6-substituted ergoline derivatives,[61] and c.i. mass spectrometry has been found to be useful for the characterization of ergot cyclol alkaloids.[62] Use of a magnetic-sector instrument with high-resolution s.i.m.s. has been demonstrated for a number of compounds including quinidine, which shows an $(M + 1)^+$ ion as its base peak and some fragment ions of low intensity.[63]

M.s.-m.s. techniques have been used for chemotaxonomic studies of tetrahydroisoquinoline alkaloids in Mexican cacti, where the technique was able to show, by comparison of data from ethanol extracts and from plant material itself, that an *N*-Me to 1-Me isomerization had occurred during the extraction procedure.[64] The same technique has been applied in the analysis of alkaloids present in cocoa[65] and mushrooms.[47] This approach is probably useful if a full range of spectra of possible isomeric species is available or if isomer identification is not important. A more useful application of this type of instrumentation is in the differentiation of isomers that cannot be differentiated by their conventional spectra. This is illustrated in recent studies[46] of ergot alkaloids where B/E linked scans, with or without collisional activation, enabled pure isomers to be identified readily; cross-correlation between series of compounds with and without *N*-Me groups showed the possibilities of obtaining stereochemical information about unknowns by this technique. In the same paper, the power of l.c.-m.s. for taxonomic studies and for the location of new structural types of alkaloids was demonstrated.

By mass-spectral study of the alkaloid products formed from feeding appropriately isotope-labelled precursors, direct hydroxylation in the α-position of α-amino-acids was found to be the favoured pathway in their conversion into α-hydroxy-α-amino-acids, a key step in ergot peptide alkaloid biosynthesis.[66]

3 Aromatic Compounds and Oxygen Heterocycles

The power of modern mass-spectral techniques in structural studies is illustrated by the characterization of the host recognition substance for parasitic angiosperms xenognosin (1).[67] Deuterium-exchange c.i. mass spectrometry[45] showed the presence of two exchangeable hydrogen atoms, which was confirmed by derivatization. Evidence for the oxygenation pattern on the two aromatic rings was obtained from the mass-spectral fragmentation pattern, and the position of the methoxy substituent was assigned by comparison of the collisional-activation spectra obtained from the ion at *m/z* 137 under methane c.i. conditions with model benzyl alcohols.

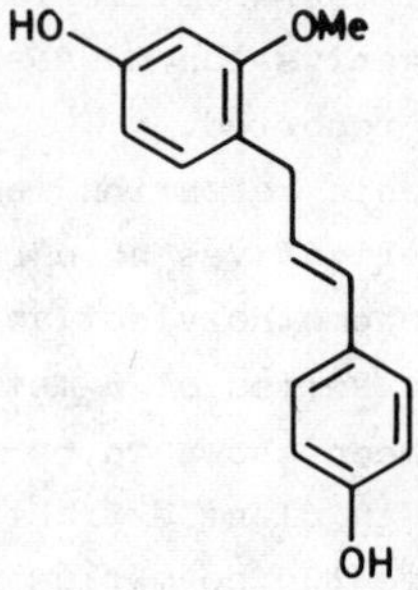

HO OMe OH

(1)

A detailed investigation of the mass-spectral behaviour of the trimethylsilyl derivatives of the ginger constituents zingerone, the gingerols, shogaols, gingerdiols, gingerdiones, and hexahydrocurcumins has been reported.[68] The mass-spectral behaviour of some hydroxymethylanthraquinones has been investigated.[69]

The mass-spectral behaviour of cannabinoids has been reviewed.[70] ^{2}H-Labelling has been used in elucidating the fragmentation behaviour of trimethyl derivatives of Δ^{1}- and Δ^{6}-tetrahydrocannabinol[71] and their hydroxylated and carboxylated metabolites;[72] similar studies have been reported for *cis*- and *trans*-hexahydrocannabinol and their hydroxy and acid analogues.[73] A detailed study of the utility of electron-voltage selected-ion monitoring[74] has been undertaken.[75] The mass spectra of a large number of cannabinoids were recorded at various electron energies. Plots of electron energy (eV) against relative abundance (electron-voltage selected-ion recordings) were found to be useful for structural analysis. Homologues were recognized and positional and double-bond isomers differentiated.

Catechol[76] and dihydroxyphenylbenzodioxin[77] derivatives isolated from insect cuticle, and hence providing information about the nature of cross-linking occurring, have been studied by a variety of mass-spectral techniques. E.i. was found to be best for characterization of the catechols and a combination of e.i. and f.d. mass spectrometry for the benzodioxins. Problems were encountered with the use of c.i. mass spectrometry owing to the formation of dimers in the ion source.

Mass spectrometry, together with other spectroscopic techniques, has been used to identify lignans in human and animal urine.[78]

Unlike those of plant origin, they have phenolic hydroxy groups only at the *meta* position of the aromatic ring. The electron-attachment positive- and negative-ion mass spectra of usnic acid and related compounds have been reported.[79]

Use of gas-chromatographic retention indices and the mass spectra of trimethylsilyl derivatives have been used to distinguish between isomeric dihydroxymonomethoxyisoflavones,[80] and the e.i. mass-spectral behaviour of a series of 6-methoxyaurones has been reported.[81] L.c.-m.s. has been shown to be useful for the identification of rotenoids in plant extracts.[82,83] This technique has been used to characterize the coumarins present in extracts of the roots of *Imperatoria ostruthium*.[84] G.c.-m.s. has been used for similar studies of coumarins in the roots of Citrus and Poncirus species.[85]

4 Isoprenoids

A collection of mass spectra of monoterpenes is available,[86] and the e.i. mass-spectral behaviour of a series of naturally occurring bornyl, 4-terpinenyl, and α-terpinyl esters[87] and the isotetronic acids[88] and their *O*-methyl ethers[89] have been reported. The electron-attachment mass-spectral behaviour of plumieride and its penta-acetyl derivative has been studied,[90] and the e.i. mass spectra of zoapatanol and montanol and related compounds, novel diterpenes from the Mexican plant zoapatle, have been reported.[91] Characteristic mass-spectral fragmentations of a number of classes of sesquiterpenes have been described.[92-94] Identification of thapsigargin (2), the major skin-irritating constituent of the root of *Thaspia garganica*, was assisted by the complementary use of e.i. and f.d. mass spectrometry.[95]

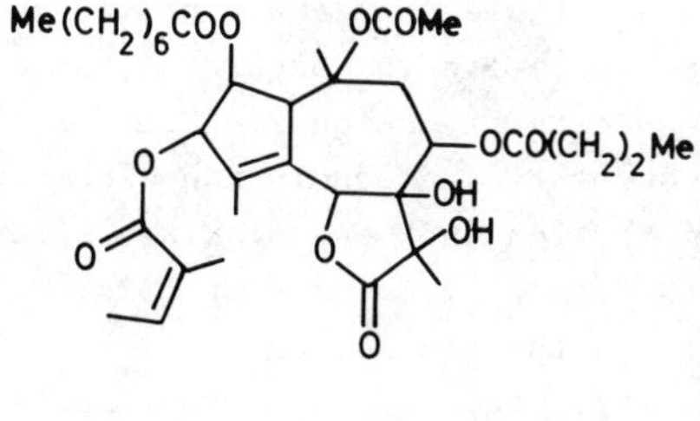

(2)

Mass-spectral studies of diterpenoids include those of anhydrocinnzeylanin and its derivatives[96] and ring-A-saturated gibberellins.[97] G.c.-m.s. of permethylated derivatives has been used to identify endogenous gibberellins and their conjugates.[98] Recent studies indicate that l.c.-m.s. may also be useful in this context.[99] Mixtures of standard compounds have been studied and, by means of a combination of e.i. and positive- and negative-ion c.i. with l.c.-m.s., constituent compounds were characterized. With microbore l.c. columns, full mass spectra in the low ng range were obtainable.[99]

The mass-spectral behaviour of bruceolides under e.i., c.i., and f.d. conditions has been reported and the use of these techniques illustrated in a determination of the structure of dihydrobruceine.[100] Other triterpenes studied by mass spectrometry include fusidic acid and its derivatives,[101] 18α,19βH-ursane derivatives,[102] bryonolic acid and its derivatives,[103] and A-norallobetulines.[104] The mass-spectral behaviour of a large number of pentacyclic triterpenoids has been collated, and base peaks characteristic of specific stereoskeletons have been identified, assisting in their structural elucidation.[105]

Characteristics of the e.i. mass-spectral behaviour of ecdysteroids with different numbers of OR groups have been described.[106] By means of a combination of n.m.r. techniques and f.a.b. mass spectrometry, ecdysone-22-phosphate (3a) and 2-deoxyecdysone-22-phosphate (3b) have been identified in the eggs of the desert locust *Schistoceria gregaria*.[107]

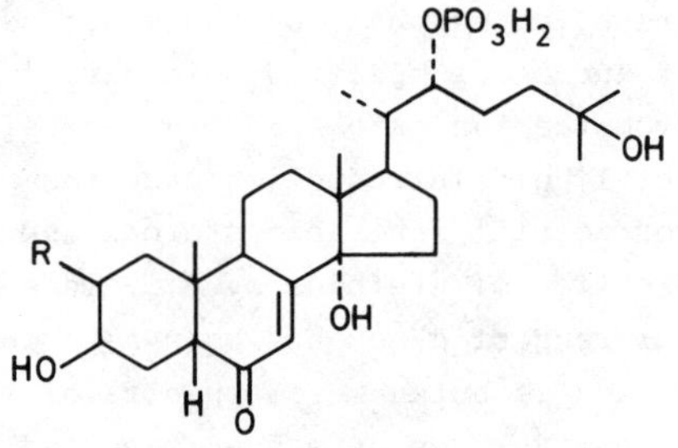

OPO_3H_2 OH R OH HO H O

(3) a; R = OH
b; R = H

A detailed investigation of the linked-scan spectra of carotenoid isomers has shown that this technique enables ready differentiation of isomeric species, which is not possible if their e.i. mass spectra are used.[108]

5 Steroids, Sterols, and Bile Acids

A number of reviews covering the types of mass-spectrometric techniques used in this area have appeared.[109-112] The e.i. mass-spectral fragmentation of androstane-7,17-diones,[113] ring-D-hydroxylated 3-methoxy-1,3,5(10)-estratrienes,[114] steroidal cyclopropylketones,[115] and dioxygenated C_{27}, C_{28}, and C_{29} steroids[116] has been reported. Studies of the effect of additional double bonds on the mass-spectrometric fragmentations of Δ^4-3-keto steroids have been undertaken.[117]

A computer-aided approach to the interpretation of mass spectra has been applied to a series of marine sterols.[118] The method consists of the construction of biosynthetically acceptable candidate structures whose major mass-spectral fragmentation pathways are predicted by use of rules derived from a training set of known compounds. A system of ranking of candidate structures, based on the comparison of predicted and observed spectra, is applied. In order to utilize this approach, sufficient biosynthetic and analytical data must be available to delimit the range of possible candidate structures. The fragmentation behaviour of the structural types must also be available and expressed in the form of programmes of fragmentation rules.

A large number of new marine sterols has been identified by mass spectrometry and other spectroscopic techniques.[119] Halistanol sulphate, an antimicrobial substance isolated from marine sponge, was assigned the structure 24,25-dimethylcholestane-2β,3α,6α-triyl trisodium sulphate on the basis of e.i. and f.d. mass spectrometry and other spectroscopic data.[120] Studies[121] of epidoxysterols from the tunicate *Ascida nigra* using l.c.-m.s. with a system of direct liquid introduction show the potential of this technique for identifying known sterols and locating new ones.

The identification of steroids by c.i. mass spectrometry using a variety of reagent gases has been described,[122] and the ammonia, methane, and isobutane c.i. spectra of twenty monohydroxylated cholesterol derivatives have been reported.[123] Ammonia c.i. is recommended as the preferred reagent gas for studies of bile acid methyl ester acetates,[124] and analysis of conjugated bile salts in human duodenal bile has been described by the use of f.d. and f.a.b. mass spectrometry.[125]

Measurements of the m.i.k.e. spectra of steroid epimers differing in their stereochemistry at the A/B and C/D ring junctions have been reported. Reproducible differences are observed that are sufficiently large to enable the stereochemistry of the ring junctions to be determined.[126,127] Mass-spectral studies of the structure and stereochemistry of vitamin D_3 and related compounds have been reviewed.[128]

6 Antibiotics, Toxins, and Anti-tumour Agents

Detailed reports of the e.i. mass-spectral fragmentation of maleimycin[129] and hexahydropyrrolobenzodiazepines isolated from lower fungi and the bacterium *Klebsiella pneumoniae*[130] have appeared. C.i. mass spectrometry with a variety of reagent gases has been shown to be useful for the characterization of the macrolide antibiotics platenomycin[131] and M-4365[132] and related compounds. Desorption c.i.[133,134] and s.i.m.s.[134,135] spectra of some kanamycins have been reported. Resonance electron-capture negative-ion c.i. of aflatoxins and related mycotoxins has been studied,[136] very simple spectra being produced. Studies of the positive- and negative-ion desorption c.i. spectra,[137] f.d. spectra,[138] and f.a.b. spectra[139] of penicillins indicate that f.a.b. is the method of choice for studies in this area, particularly if alkali-metal salts are to be characterized.

The previously derived structure of the peptide antibiotic trichotoxin A-40 has been verified by means of both e.i. and c.i. mass spectrometry,[140] and mass-spectral methods for the localization of the amide function in its sequence have been described.[141] A combination of f.d., e.i. mass spectrometry, and g.c.-m.s. was used in the characterization of lipopeptine A, an antifungal peptide antibiotic.[142] F.d. mass spectrometry using lithium attachment was used in the identification of six new and two known antibiotics of the quinomycine class (4; R^1, R^2 = heteroaryl).[143] The streptovirudins, tunicamycin-related antibiotics, have also been characterized partially by the use of f.d. mass spectrometry,[144] and the technique has been applied in studies of the degradation products of the marine toxin polytoxin.[145] In-beam e.i. and f.d. mass spectrometry has been used in studies of the peptide antibiotics trichopolyns I and II.[146] The glycopeptide antibiotics the bleomycins have been the subject of f.d.,[147] f.a.b.,[148] and a comparative f.d. and f.a.b. study,[149] and the f.d. and f.a.b. spectra of hydroxamate-containing siderophores have been reported.[150]

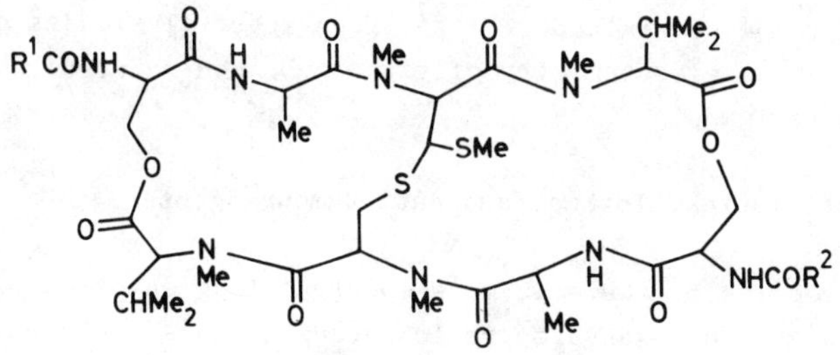

(4)

The impact of f.a.b. mass spectrometry in this area is well illustrated in studies of eleven zervamicin and two emerimicin peptide antibiotics.[151] F.a.b. together with g.c. and g.c.-m.s. was used to assign structures to zervamicins IA, IB, IB^1, IC (acidic) and IIA, IIB, II-1 to II-5 (basic); the emerimicins IIA and IIB were found to be identical to zervamicins IIA and IIB. The major acidic zervamicin IC was found to have the structure Ac-Trp-Ile-Glu-Iva-Ile-Thr-Aib-Leu-Aib-Hyp-Gln-Aib-Hyb-Aib-Hyb-Aib-Pro-Phol, whereas the major neutral antibiotic zervamicin IIB has the same structure but with Glu replaced by Gln. Other zervamicins differ from these by isovaline/aminoisobutyrate, isoleucine/valine, and leucine/valine exchanges.

^{252}Cf plasma desorption has been of considerable utility in structural studies in this area.[12,152] Examples of its use include the characterizations of Sch 18640, a new thiostrepton type of antibiotic,[153] (β-lysyloxy)myoinositol guanidino-glycoside antibiotics from a *Nocardia* species,[154] adenomycin (C_{19-97}) substance, a nucleoside antibiotic,[155] and kijanimicin, a new class of tetronic acid-containing antibiotic.[156]

7 Nucleic Acid Components and Cytokinins

The role of mass spectrometry in the identification of modified nucleosides in tRNA has been briefly reviewed.[157] Pattern-recognition techniques have been applied to the interpretation of nucleoside mass spectra by means of a set of 125 nucleoside mass spectra.[158] A detailed e.i. mass-spectrometric study of trialkylsilyl and acyl derivatives of 2'-deoxynucleosides has been reported.[159] With ammonia as the reagent gas, desorption c.i. spectra of pyridinium nucleoside salts and purine and pyrimidine nucleosides and nucleotides have been obtained.[160] F.d. mass-spectral studies of some cyclic nucleotides have been undertaken,[161] and the technique has been used in combination with high-performance liquid-chromatographic separation to identify protected deoxyribonucleotides.[162] In a comparative study of the pulsed laser and ^{252}Cf plasma desorption mass spectra of a series of nucleosides and nucleotides and their bases, similar spectra were obtained and provided relative-molecular-mass information together with structurally significant fragment ions.[163] Similar results were obtained in s.i.m.s. studies,[164] and the technique has been used for the identification of alkylated nucleosides[165] and, in combination with n.m.r., in the study of the extent of methylation of polycytidylic acid.[166] S.i.m.s., using time-of-flight mass analysis and Cs^+ ions, has been used to study a series of protected diribonucleoside monophosphates.[167] The technique provided relative molecular mass, and the bases present, together with the base sequence, could be identified. Studies with a protected triribonucleoside diphosphate showed that studies of this type could be extended to molecules with a mass of 2100 dalton.

The success of ^{252}Cf plasma desorption in this area was referred to in the introduction, and the technique has been applied to development of a method for sequencing fully protected oligonucleotides.[168] Positive- and negative-ion spectra were found to be complementary and to enable the purity and base sequences of the oligonucleotides to be determined unambiguously.

Pyrolysis m.s.-m.s. has been used to identify modified bases in DNA or RNA.[3] Pyrolysis experiments[169] have confirmed the possibility that misidentification can occur using this approach owing to artefact formation. This finding was confirmed in pyrolysis studies[165] of poly(dA.dT).poly(dA.dT) alternating copolymer, where 1-methyladenosine, an artefact, was found.

G.c.-m.s. analysis of trimethylsilyl derivatives of a mixture of ribonucleosides derived from RNase T_2 digestion of *E. coli* $tRNA^{Leu}_{UUR}$ revealed the presence of a novel nucleoside containing unmodified ribose and a base equal in mass to acetyladenine.[170] On the basis of e.i. mass-spectral and n.m.r. studies, the structure 6-[(*o*-hydroxybenzyl)amino]-9-β-D-ribofuranosylpurine has been assigned to the spathe regreening factor in *Zantedeschia aethiopica*.[171] High-resolution e.i. mass spectra of its trimethylsilyl derivative were used to characterize the substrate for enzyme transfer RNA, guanine transglycosylase, as 7-(4,5-*cis*-dihydroxy-1-cyclopenten-3-ylaminomethyl)-7-deazaguanine.[172] A combination of e.i., c.i., and f.d. mass spectrometry together with ^{13}C-n.m.r., degradative studies, and synthesis enabled structure (5) to be assigned to a pharmacologically active agent

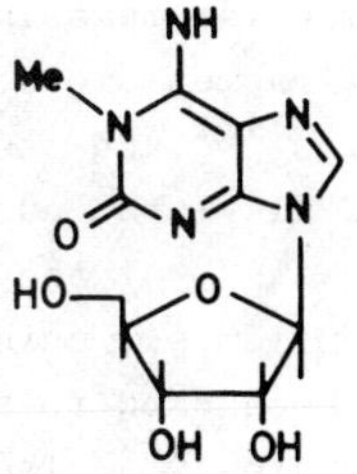

(5) 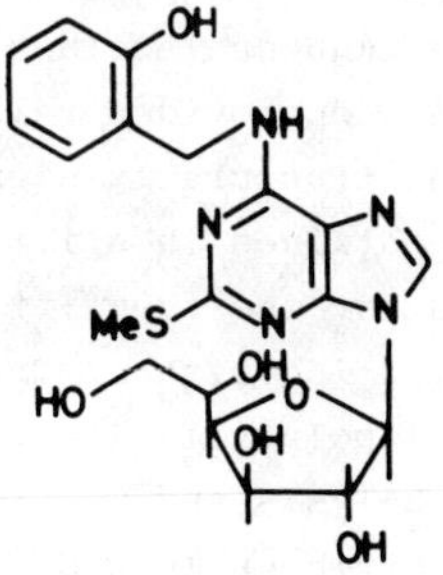 (6)

from marine sponge.[173] A new cytokinin (6) from cuckoo-pint fruits was characterized on the basis of its e.i. mass spectrum, n.m.r., the e.i. mass spectrum of its trimethylsilyl derivative, and chemical degradation.[174]

8 Pyrrole Pigments

Methane-enhanced negative c.i. spectra of some metalloporphyrins have been reported.[175] This class of compound has been shown to undergo cation exchange between lithium halide or transition-metal halide surfaces in f.d. investigations.[176] Hydrogen c.i. mass spectrometry shows promise for structural studies in this area.[177] Investigation using several standard porphyrins showed that fragmentation proceeds by hydrogenation of the porphyrin to porphyrinogen, which cleaves at the *meso* positions to yield mono-, di-, and tri-pyrrolic ions. Secondary fragmentation by cleavage of substituent groups attached to the pyrrole ring may then occur. In some cases, the data were sufficient to distinguish between porphyrin isomers, and, in addition, information on substituents at *meso* positions may be obtained. By means of a combination of preparative h.p.l.c. and mass spectrometry, porphyrins present in deep-sea sediments, bitumens, shales, and petroleums have been characterized.[178]

F.a.b.[179] and laser-desorption[180] mass spectra of cobolamines have been reported. In both cases, relative molecular mass and structurally significant fragment ions were observed. F.d. mass spectrometry has provided evidence for the depletion of carbon-13 in the biosynthesis of vitamin B_{12}.[181]

Fast desorption from a gold support has yielded e.i. and c.i. mass spectra of chlorophylls, pheophytins, and pheophorbides comparable with those obtained by f.d. or desorption c.i.[182] In-beam e.i. mass spectrometry, together with ^{1}H-n.m.r., was used to characterize a bacteriochlorophyll allomer (7).[183]

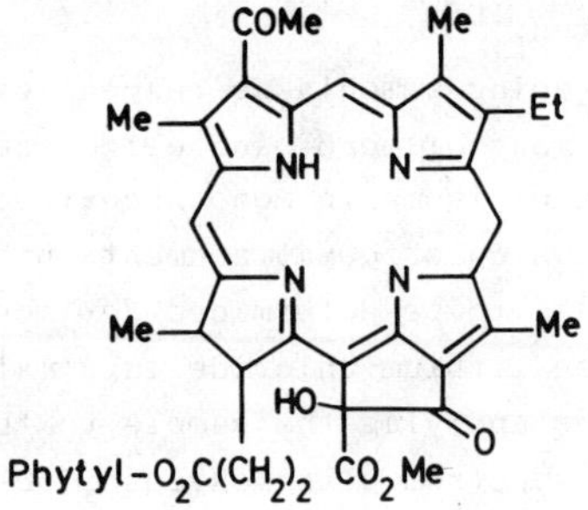

COMe Me Me Et NH N N N Me Me HO O Phytyl-$O_2C(CH_2)_2$ CO_2Me

(7)

The self-assembly and oligmer formation of chlorophyll *a* have been studied by ^{252}Cf plasma desorption.[184] Laser-assisted f.d.[185] of chlorophyll *a* showed doubly charged dimers $(2M + 2H)^{2+}$ confirming these findings. ^{252}Cf plasma desorption of chlorophyll *a* and its fully deuteriated analogue enabled elemental compositions to be assigned to the various fragment ions.[186]

9 Carbohydrates and Glycosides

Mass-spectrometric behaviour of carbohydrates has been reviewed.[187] Reports on the e.i. mass-spectral behaviour of variously derivatized carbohydrates include those on dideoxy derivatives of 1,6-anhydro-β-D-hexopyranoses,[188] methylated amino-sugars,[189] partially methylated and acetylated methylglycosides of galactose, mannose, glucose, and *N*-acetylglucosamine,[190] xylobioses and xylotrioses,[191] and 6-deoxyhexapyranoses.[192] Studies of the ammonia and isobutane c.i. mass spectra of a series of dimethyl acetals and dimethyl dithioacetals of aldopentoses and aldohexoses showed significant differences in ion abundances as a function of the stereochemistry of the compounds.[193] The c.i. spectra of the methyl ethers of methylglycoside[194] and e.i. and c.i. mass spectra of peracetylated oxime derivatives of ketoses[195] have been reported, and c.i. is reported as giving improved sensitivities for the analysis of partially methylated aldetol acetates.[196] Methane and isobutane c.i. mass spectra of permethylated disaccharides show characteristic differences that enable glycosidic linkages to be differentiated.[197] Ammonia c.i. mass spectra were not as useful in this context. The isobutane c.i. mass spectra of peracetates of aldabiuronic acids and peracetate and permethyl derivatives of dialdose dianhydrides have also been reported.[198]

Desorption c.i. using ammonia as reagent gas gave $(M + NH_4)^+$ and $(M + NH_4 - H_2O)^+$ ions of mono- and oligo-saccharides,[199] and f.d. mass spectra of isomeric monosaccharides are reported.[200] Negative-ion f.d. shows chloride attachment anions, enabling relative molecular mass to be determined for mono-, di-, and tetra-saccharides when lithium chloride in a matrix of polyethylene oxide 4000 is used for applying the sample to the emitter.[201] The laser-desorption spectrum of stachycose, obtained by means of a repetitive pulsing laser, showed mainly cationized species.[202] Studies of s.i.m.s.[203] and matrix-assisted s.i.m.s.[204] spectra of sugars have been reported. The spectra are similar to those obtained by other desorption techniques.

E.i. mass spectrometry has been extensively used to characterize a wide variety of sugars. Examples include the use of alditol acetate derivatives in the identification of a novel 2-aminotetrose, 2-acetamido-2-deoxytetrose, in a lipid-soluble fraction of rat liver[205] and a new acidic amino sugar from a *P. aeruginosa* lipopolysaccharide,[206] permethylated reduced di- and tri-saccharides from oligosaccharides isolated from human meconium,[207] and oligosaccharides from patients with α-mannosidosis[208,209] and from the kidney of a goat with β-mannosidosis.[210] New fucose-[211] and mannose-containing[212] oligosaccharides have been identified, and oligosaccharides isolated from A^+, H^+, and A-H-hog submaxillary gland mucin glycoproteins have been characterized by a combination of 360 mHz n.m.r., permethylation analysis, and mass spectrometry.[213]

The saccharides and oligosaccharides formed from starch by degradation in the mass-spectrometer ionization chamber have been studied by a combination of positive and negative d.c.i.[214] F.d.

mass spectrometry has been used successfully to analyse mycobacterial methylmannose polysaccharides of general structure Man_xMeMan_y-OMe. Man_1MeMan_{13}-OMe was the largest homologue to be studied and gave the expected signal for a cationized $(M + Na)^+$ ion at *m/z* 2506.[215] Complete sequence information was obtained for the polysaccharides up to Man_1MeMan_9-OMe. The same technique has been used to characterize a series of human-milk sugars that were purified by h.p.l.c. as their alditol acetates.[216] Polysaccharides such as xylon, agarase, and alginic acid, which contain monomer units of different elemental composition, are readily differentiated by the f.i. mass spectra of their Curie-point pyrolysates.[217] For polysaccharides with hexosyl subunits such as cellulose, galactan, laminarian, and mannan, differentiation can be effected by differences in the relative heights of the f.i. peaks.

The potential of direct-liquid-introduction l.c.-m.s. for characterizing polysaccharides is illustrated in studies of permethylated, partially hydrolysed, and ethylated lichenan and starch.[218] This approach has been incorporated into structural studies of the acidic polysaccharide secreted by *Rhizobium meliloti* strain 1021.[219] In earlier studies,[220] a combination of reversed-phase h.p.l.c. and then g.c.-m.s. was used to identify peralkylated oligosaccharide alditols obtained by hydrolysis of the polysaccharides; however, not all the products were amenable to g.c.-m.s. studies, and l.c.-m.s. has extended the range of compounds that can be sequenced.

The e.i. mass-spectral fragmentation of permethylated 6-*C*-glycosylflavones has been studied[221] and used, together with other spectroscopic techniques, for the characterization of xylostosidine, a member of a new class of monoterpene alkaloid glycosides,[222] veproside (a new iridoid glycoside),[223] the cardiac glycoside 3'-*O*-methylevomonoside,[224] and protogenkwanin 4'-glucoside (a new type of flavanoid).[225] E.i. mass-spectral studies of permethylated derivatives have assisted in structural assignments to a novel steroidal cyclic glycoside isolated from star fish[226] and wedeloside, a toxic diterpenoid amino-glycoside.[227]

Ammonia c.i. has been used for the identification of steryl glucosides found in yeast cells.[228] The in-beam e.i. and positive and negative c.i. spectra of 21 flavanoid glycosides,[229] the d.c.i. mass spectra of a variety of different types of naturally occurring glycosides,[230] and the f.d. mass spectra of flavanoid glycosides[231] have been reported. Analogies between the sugar cleavages observed by acid hydrolysis of oligoglycosidic saponins and their f.d. mass-spectral behaviour have been described.[232] F.d. mass spectrometry has assisted in the structural characterization of ginsenoside saponins,[233] triterpenoid saponins,[234] and lignan glycosides.[235]

Both e.i.[236] and c.i.,[237] with a variety of reagent gases, fail to provide relative-molecular-mass data for glucosinolates (mustard oil glycosides) and desulphoglucosinolates, although both techniques provide diagnostically important fragment ions. For the disulphoglucosinolates the problem can be overcome by derivatization.[238]

L.c.-m.s. using an interface of the direct-liquid-introduction type has been shown to be useful for characterizing a flavanoid and other types of glycosides.

10 Amino-acids and Peptides

The isobutane c.i. mass spectra of L-methionine and some analogues have been reported,[240] and the same ionization method has been used to analyse phenyl thiohydantoin and 2-anilino-5-thiazolinone amino-acids obtained by Edman degradation of proteins and peptides.[241] Amino-acids and small peptides have been studied by laser-induced mass spectrometry,[242] s.i.m.s.,[243] and f.a.b. mass spectrometry;[244] the spectra obtained using s.i.m.s. have been compared with those obtained from other ionization methods.[245]

High-resolution e.i. mass spectrometry of a trimethylsilyl derivative of a new amino-acid intermediate in aromatic biosynthesis [L-(85)-arogenate] unambiguously confirmed its structure.[246] E.i. mass spectrometry was also used in the characterization of cyclopentenylglycine, a novel non-proteingenic amino-acid in *Flacourtiaceae*.[247] A combination of e.i. and c.i. g.c.-m.s. of *N*-acetyl/methyl esters and *N*-TFA/methyl esters was used to characterize new sulphur-containing amino-acids in man.[248] Laser-microprobe mass-analyser mass spectrometry assisted in the identification of aminocitric acid in ribonucleoproteins.[249] F.d. mass spectrometry has assisted in the identification of clithionenine, a new amino-acid betaine, isolated from a poisonous mushroom,[250] *NNN*-trimethylalanine as the *N*-terminal amino-acid of the ribosomal protein L11 from *E. coli*,[251] and tyrosine sulphate in dogfish protamine.[252]

The potassium-cationized f.d. mass spectra of some penta- and hexa-peptides derived from substance P have been reported,[253] and the use of f.d. with h.p.l.c. for the characterization of hypothalmic oligopeptides has been described.[254] Other peptides studied by f.d. include oligopeptidic hormones and peptide conjugates,[255] tryptic peptides in human haemoglobin chains,[256] and peptides from thermolysin and staphylococcal protease-treated human globins.[257] The effects of emitter-current temperature on the f.d. spectra of a number of derivatized synthetic peptides have been described,[258] and studies of the f.d. mass spectra of hexapeptides containing adjacent arginyl residues indicated that there are preferential cleavages at higher emitter-current temperatures between the arginyl residues.[259] Sequencing of peptides by use of a combination of Edman degradation and f.d. mass spectrometry has been described[260] and used together with a computer program[261] to sequence the *N*-terminal cyanogen bromide fragment of *Streptomyces erythraeus* lysozyme.[262] Specific enzymic digestion in combination with f.d. mass spectrometry has also been used for peptide sequencing.[263] Using model peptides, collision-induced dissociation of $(M + H)^+$ ions formed by f.d. has been shown to provide abundant fragments containing sequence information.[264]

S.i.m.s. of oligopeptides yields cationized and protonated molecular ions, together with structurally useful fragment ions.[265] Xenon is reported[266] as being better than argon for studies of this type. A pulsed beam of Cs^+ ions has been used to obtain s.i.m.s. spectra of the neuropeptides Leu-enkephalin and Met-enkephalin.[267] The spectra showed ions characteristic of the intact molecule and diagnostic of the amino-acid sequence.

The most important advance in this area has been the availability of f.a.b. mass spectrometry, since in many cases the technique provides relative-molecular-mass and considerable sequence information from underivatized peptides with large numbers of amino-acid residues (see introduction). A wide variety of peptides has been studied, including two isomeric tripeptides,[268] bradykinin and related peptides,[269] angiotensin peptides,[270] and enkephalins.[271] The potential of the technique for peptide sequencing has been described,[272] and it has been used in combination with enzymic cleavage, Edman degradation, and g.c.-m.s. in the sequencing of the anti-tumour proteins macromomycin and neocarcinostatin.[273] A similar approach[274] used the technique with carboxypeptidase digestion and Edman degradation, and it is evident that in most cases full sequence information can only be obtained by a combined approach. Other studies have used a combination of f.a.b. and e.i. mass spectrometry to sequence two blocked *N*-terminal peptides from Paracoccus cytochrome C-550.[275] The utility of f.a.b. mass spectrometry in structural studies of peptides has been reviewed.[276]

In addition to the f.d. collision-induced dissociation studies of peptide sequencing mentioned earlier, a number of other ionization techniques has been used in this type of approach. Secondary-ion mass spectrometry with Xe ions or atoms and collision-induced dissociation in a triple quadrupole has been shown to yield fragmentations characteristic of amino-acid sequence.[277] M.i.k.e.s. and B/E linked-scan studies, with or without collision-induced dissociation, have been reported on a variety of peptides,[278,279] and m.i.k.e.s., together with negative-ion c.i., has been shown to be useful for sequencing underivatized di- and tri-peptides.[280] At present, it appears that a triple-quadrupole instrument is most suitable for studies of this type because it is fully computerized and can readily handle mixtures. Using isobutane c.i., mixtures of $(^{2}H_{0})$,$(^{2}H_{3})$-*N*-acetylated and *NO*-permethylated peptides have been sequenced by means of collision-induced dissociation in a triple-quadrupole instrument, and, by means of electron-capture negative c.i., neuropeptides can be sequenced as their *N*-acetyl-*O*-pentafluorobenzyl esters at the picomole level.[281]

The cyclodepsipeptides of *Beauveria bassiaria* have been further characterized, and structures for beauverolides J_aK_a and A-F have been assigned on the basis of high-resolution e.i. mass-spectral data and amino-acid analysis.[282]

A mass-spectrometric method for the determination of *N*-terminal tryptophan and *N*-terminal histidine in peptides, using derivatization with aldehydes and subsequent mass-spectral examination of the derivatized peptides, has been described.[283] A similar approach using *ortho*-phthalaldehyde has been described as a potential method for peptide sequencing. Peptides, after derivatization, are isolated by reversed-phase h.p.l.c. The isolated derivatized peptides are permethylated and studied by mass spectrometry. The spectra obtained were found to be similar to those from *N*-acyl/permethyl derivatives.[284]

Further improvements and applications in gas-chromatographic mass-spectrometric protein sequencing have been described,[285] and mass-spectrometric methods for structure determination of neuropeptides, proteins, and glycoproteins have been reviewed.[286] The alkylamine-reactive site in human α_2-macroglobin has been determined by a combination of chemical sequencing procedures and mass-spectral identification of α-glutamylmethylamide as its PTH and *N*-benzyl-*O*-methyl ester derivatives.[287] A combination of e.i. mass spectrometry and protein chemical techniques was used to establish the structure of an aldosterone-stimulating component of pituitary gland as the *N*- and *C*-terminally blocked 13-residue peptide α-melanocyte-stimulating hormone (α-MSH).[288] Analogous procedures together with f.a.b. mass spectrometry were used to characterize a second active compound as bisacetyl-Ser-α-MSH.[289] Low-resolution e.i. mass spectrometry was a major tool in sequencing chloramphenicol transferase,[290] a superoxide dismulase from *Bacillus stearothermophilus*,[291] eleven tryptic peptides from papaya mosaic virus protein,[292] and the factor XIIIa acceptor site in bovine plasma fibronectin.[293]

Combined l.c.-m.s. has considerable potential for sequence studies of peptide mixtures, and this has been illustrated in studies of *N*-acetyl-*NOS*-permethylated peptides using a moving-belt system.[294]

11 Fatty Acids and Lipids

Hydrogen and methane c.i. studies of a series of *cis*- and *trans*-dicarboxylic acids and their esters show that they can be differentiated on the basis of the extent of water loss for acids from protonated molecular ions and of alcohol loss for the esters.[295] The laser-desorption mass-spectrometric behaviour of some organic acids has been described.[296]

Location of the double-bond position in polyunsaturated fatty acids continues to excite interest. A c.i. method has been described,[297] and improvements in the mixtures of reagent gases when vinyl methyl ether is used have been described, whereby better spectra are obtained.[298,299] In addition to enabling double-bond positions to be located, cyclopropanoid and monenoic fatty-acid methyl esters can be differentiated.[300] Use of nitric oxide as the c.i. reagent gas has been shown to be useful in characterizing homoconjugated triene and tetraene units in esters,[301] and the use of negative ions for double-bond location in esters has been described.[302] In a detailed study, the e.i.

mass spectra 3-picolinyl esters of n-, iso-, anteiso-, and unsaturated long-chain fatty acids have been shown to provide data enabling the position of methyl branching and the double bond to be located.[303] Capillary g.c.-m.s. with positive and negative c.i. has been shown to be effective in detecting hydroxy fatty acids in biological samples.[304]

A review of the utility of l.c.-m.s. in the lipid area has appeared.[305] In addition, further details of an interface specifically designed for analysis of complex lipid samples are described, and a method for quantification of the 19 most common fatty acids occurring in major amounts in animal and plant tissue is described.[306]

Detailed studies of e.i. mass-spectral behaviour of some prostaglandins of the F series are reported.[307] The f.a.b. mass spectra of leukotriene-C_4 of natural and synthetic derivation have been compared.[308] The major difference in the spectra was in the abundances of cationized species, probably due to differences in salt contact of the two samples.

Ammonia d.c.i. is a convenient and rapid technique for obtaining information regarding carbon numbers and extent of unsaturation in triglycerides present in fat samples.[309] F.d. mass spectrometry has been shown to yield almost exclusively relative-molecular-mass profiles of natural waxes.[310]

Methylation techniques for the structural analysis of glycolipids have been reviewed.[311] Mass-spectral methods developed for the fingerprinting of glycolipids by the study of the e.i. mass spectra of permethylated and permethylated/reduced derivatives[312] continue to yield impressive results in studies of glycosphingolipid fractions from a variety of sources.[313] To date, the largest molecule investigated by the group is a ceramide dodecasaccharide from rat urine.[314] The e.i. mass spectra of the permethylated and permethylated/reduced glycolipid enabled the overall size, sugar sequence, and nature of the ceramide to be determined.

Low-eV e.i. mass spectrometry and n.m.r. have been used to characterize permethylated glycosphingolipids from human plasma,[315] and f.d.[316-319] and d.c.i.[320] have been shown to be of assistance in determining the structures of glycosphingolipids. The f.d. technique is particularly useful for high-molecular-weight derivatized compounds.[316,319] An alternative approach has used electron-impact desorption mass spectrometry for glycolipids containing up to five sugar residues,[321] and the technique has been applied in a number of other studies.[322] Further studies have used a combination of f.d., c.i., and e.i. methods together with n.m.r. in studies of glycoconjugates.[323] The ammonia c.i. mass spectra of a series of permethylated cerebrosides and ceramide oligosaccharides provide relative-molecular-mass data and sequence information.[324] The general applicability of c.i. mass spectrometry in phospholipid structure determination has been discussed and the technique used to characterize the *O*-deacetyl platelet-activity factor from leukocytes.[325]

(References begin overleaf)

References

1 A.L. Burlingame, A. Dell, and D.H. Russel, *Anal. Chem.*, 1982, 54, 363R.
2 'Soft Ionization Biological Mass Spectrometry', ed. H.R. Morris, Heyden, London, 1981.
3 D.E. Games in 'Mass Spectrometry', ed. R.A.W. Johnstone (Specialist Periodical Reports), The Royal Society of Chemistry, London, 1981, Vol. 6, p. 241.
4 M. Barber, R.S. Bordoli, R.D. Sedgwick, and A.N. Tyler, *J. Chem. Soc., Chem. Commun.*, 1981, 325.
5 M. Barber, R.S. Bordoli, R.D. Sedgwick, and A.N. Tyler, *Nature (London)*, 1981, 293, 270.
6 M. Barber, R.S. Bordoli, and R.D. Sedgwick in ref. 1, p. 137.
7 M. Barber, R.S. Bordoli, G.J. Elliott, R.D. Sedgwick, and A.N. Tyler, *Anal. Chem.*, 1982, 54, 645A.
8 D.H. Williams, C. Bradley, G. Bojesen, S. Santikarn, and L.C.E. Taylor, *J. Am. Chem. Soc.*, 1981, 103, 5700.
9 M. Barber, R.S. Bordoli, R.D. Sedgwick, A.N. Tyler, G.V. Garner, D.B. Gordon, L.W. Tetler, and B.C. Hider, *Biomed. Mass Spectrom.*, 1982, 9, 265.
10 C.L. Wilkins and M.L. Gross, *Anal. Chem.*, 1981, 53, 1661A.
11 F.M. Devienne and J.-C. Roustan, *Org. Mass Spectrom.*, 1982, 17, 173.
12 R.D. Macfarlane in ref. 1, p. 449.
13 C.J. McNeal and R.D. Macfarlane, *J. Am. Chem. Soc.*, 1981, 103, 1609.
14 P.G. Kistemaker, G.J.Q. van der Peyl, and J. Haverkamp in ref. 1, p. 120.
15 E. Denoyer, R. Van Grieken, F. Adams, and D.F.S. Natusch, *Anal. Chem.*, 1982, 54, 26A; D.M. Hercules, R.J. Day, K. Balasanmugam, T.A. Dang, and C.P. Li, *Anal. Chem.*, 1982, 54, 280A.
16 H.-R. Schulten in ref. 1, p. 6.
17 E.D. Hardin and M.L. Vestal, *Anal. Chem.*, 1981, 53, 1492.
18 A. Benninghoven, A. Eicke, M. Junack, W. Sichtermann, J. Krizek, and H. Peters, *Org. Mass Spectrom.*, 1980, 15, 459.
19 R.D. Smith, J.E. Burger, and A.L. Johnson, *Anal. Chem.*, 1981, 53, 1603.
20 H. Jungclas, H. Danigel, and L. Schmidt, *Org. Mass Spectrom.*, 1982, 17, 86.
21 C. Fenselau, R. Cotter, G. Hansen, T. Chen, and D. Heller, *J. Chromatogr.*, 1981, 218, 21.
22 H.R. Morris, A. Dell, and R.A. McDowell, *Biomed. Mass Spectrom.*, 1981, 8, 463.
23 A. Dell, R.A. McDowell, H.R. Morris, and G.W. Taylor in ref. 1, p. 69.
24 C. Fenselau, *Anal. Chem.*, 1982, 54, 105A.
25 C.J. McNeal, *Anal. Chem.*, 1982, 54, 43A.
26 R.J. Cotter, *Anal. Chem.*, 1980, 52, 1589A.
27 V.N. Reinhold and S.A. Carr, *Anal. Chem.*, 1982, 54, 499.
28 E. Constantin, Y. Nakatani, G. Ourisson, R. Hueber, and G. Teller, *Tetrahedron Lett.*, 1980, 4745.
29 P. Traldi, *Org. Mass Spectrom.*, 1982, 17, 245.
30 D.I. Carroll, J.G. Nowlin, R.N. Stillwell, and E.C. Horning, *Anal. Chem.*, 1981, 53, 2007.
31 C.R. Blakley, J.J. Carmody, and M.L. Vestal, *J. Am. Chem. Soc.*, 1980, 102, 5931.
32 C.R. Blakley, J.J. Carmody, and M.L. Vestal, *Anal. Chem.*, 1980, 52, 1636.

33 C.R. Blakley, J.C. Carmody, and M.L. Vestal, *Clin. Chem.*, 1980, 26, 1467.
34 H. Kambara, *Anal. Chem.*, 1982, 54, 143.
35 R.S. Hook, V.A. Fassel, and H.J. Svec, *Org. Mass Spectrom.*, 1982, 17, 240.
36 N.M.M. Nibbering, *J. Chromatogr.*, 1982, 251, 93.
37 P.J. Arpino and C. Guiochon, *J. Chromatogr.*, 1982, 251, 153.
38 K.L. Busch, S.E. Unger, A. Vincze, R.C. Cooks, and T. Keough, *J. Am. Chem. Soc.*, 1982, 104, 1507.
39 T. Keough and A.S. DeStefano, *Anal. Chem.*, 1981, 53, 25.
40 W.D. Lehmann, *Anal. Chem.*, 1982, 54, 299.
41 T. Higuchi, E. Kubota, F. Kunihiro, and Y. Itagaki, *Adv. Mass Spectrom.*, 1980, 8A, 1061.
42 K.H. Ott, F.W. Röllgen, J.W. Zwinselmann, R.H. Fokkens, and N.M.M. Nibbering, *Angew. Chem., Int. Ed. Engl.*, 1981, 20, 111.
43 J.J. Zwinselmann, R.H. Fokkens, N.M.M. Nibbering, K.H. Ott, and F.W. Röllgen, *Biomed. Mass Spectrom.*, 1981, 8, 312.
44 R. Kraft, A. Otto, A. Makower, and G. Etzold, *Anal. Biochem.*, 1981, 113, 193.
45 D.F. Hunt and S.K. Sethi, *J. Am. Chem. Soc.*, 1980, 102, 6953.
46 C. Eckers, D.E. Games, D.N.B. Mallen, and B.P. Swann, *Biomed. Mass Spectrom.*, 1982, 9, 162.
47 S.E. Unger, A. Vincze, R.G. Cooks, R. Chrisman, and L.D. Rothman, *Anal. Chem.*, 1981, 53, 976.
48 D.E. Games, M.S. Lant, S.A. Westwood, M.J. Cocksedge, N. Evans, J. Williamson, and B.J. Woodhall, *Biomed. Mass Spectrom.*, 1982, 9, 215, and references therein.
49 N.J. Alcock, L. Corbelli, D.E. Games, M.S. Lant, and S.A. Westwood, *Biomed. Mass Spectrom.*, in press.
50 N.A.B. Gray, A. Buchs, and D.H. Smith, *Anal. Chem. Symp. Ser.*, 1981, 7, 295.
51 H.R. Morris, *Nature (London)*, 1980, 286, 447.
52 A. Molorni, G. Marino, and A. Parente, *Ann. Chim.*, 1981, 71, 59.
53 H.R. Morris, *Ann. Chim.*, 1981, 71, 45.
54 I.A. Israilov, S.V. Karimova, M.S. Yunusov, and S. Yu. Yunusov, *Khim. Prir. Soedin.*, 1980, 16, 279.
55 P.-Z. Cong, *Hua Hsueh Hsueh Pao*, 1981, 39, 75.
56 Y.-W. Fang, Z.L. Bian, G.-H. Wang, and W.G. Chai, *Hua Hsueh Hsueh Pao*, 1981, 39, 139.
57 P.A. Fellows, N. Kyriakidis, P.G. Mantle, and E.S. Waight, *Org. Mass Spectrom.*, 1981, 16, 403.
58 P. Bel and A. Mandelbaum, *Org. Mass Spectrom.*, 1980, 15, 568.
59 J. Respondek, *Org. Mass Spectrom.*, 1980, 15, 544.
60 S. Funayama, K. Yoshida, C. Konno, and H. Hikino, *Tetrahedron Lett.*, 1980, 1355.
61 J. Schmidt, K. Seifert, S. Haertling, S. Johne, and H.J. Veith, *Biomed. Mass Spectrom.*, 1981, 8, 13.
62 J.K. Porter and D. Betowski, *J. Agric. Food Chem.*, 1981, 29, 650.
63 K.D. Klöppel and G. von Bünau, *Int. J. Mass Spectrom. Ion Phys.*, 1981, 39, 85.
64 S.E. Unger, R.G. Cooks, R. Mata, and J.L. McLaughlin, *J. Nat. Prod.*, 1980, 43, 288.
65 R.G. Cooks, R.W. Kondrat, M. Youssefi, and J.L. McLaughlin, *J. Ethnopharmacol.*, 1981, 3, 299.
66 C.M. Belzecki, F.R. Quigley, H.G. Floss, N.C. Perellino, and A. Guicciardi, *J. Org. Chem.*, 1980, 45, 2215; F.R. Quigley and H.G. Floss, *J. Org. Chem.*, 1981, 46, 464.
67 D.G. Lynn, J.C. Steffens, V.S. Kamat, D.W. Graden, J. Shabanowitz, and J.L. Riopel, *J. Am. Chem. Soc.*, 1981, 103, 1868.

68 D.J. Harvey, *Biomed. Mass Spectrom.*, 1981, 8, 546.
69 S.A.M. Metwally, M.S.K. Youssef, and M.I. Younes, *Indian J. Chem., Sect. B*, 1980, 19, 984.
70 B.J. Gudzinowicz and M.J. Gudzinowicz, 'Analysis of Drugs and Metabolites by GC/MS', Marcel Dekker, New York, 1980, Vol. 7.
71 D.J. Harvey, *Biomed. Mass Spectrom.*, 1981, 8, 575.
72 D.J. Harvey, *Biomed. Mass Spectrom.*, 1981, 8, 579.
73 D.J. Harvey, *Biomed. Mass Spectrom.*, 1981, 8, 366.
74 C.A.L. Brecht, R.J.H.Ch. Lousberg, E.J.E.M. Kuppers, C.A. Salemink, T.B. Vree, and J.M. Van Rossum, *J. Chromatogr.*, 1973, 81, 163; F.S. El Feraly, M.A. Elsohly, and C.E. Turner, *Acta Pharm.*, 1977, 27, 43.
75 C.E. Turner, O.J. Bouwsma, S. Billets, and M.A. Elsohly, *Biomed. Mass Spectrom.*, 1980, 7, 247.
76 P. Roepstorff and S.O. Andersen, *Biomed. Mass Spectrom.*, 1980, 7, 317.
77 P. Roepstorff and S.O. Andersen, *Biomed. Mass Spectrom.*, 1981, 8, 174.
78 K.D.R. Setchell, A.M. Lawson, F.L. Mitchell, H. Adlercreutz, D.N. Kirk, and M. Axelson, *Nature (London)*, 1980, 287, 740; K.D.R. Setchell, A.M. Lawson, E. Conway, N.F. Taylor, D.N. Kirk, G. Cooley, R.D. Farrant, S. Wynn, and M. Axelson, *Biochem. J.*, 1981, 197, 447.
79 S. Huneck and J. Schmidt, *Biomed. Mass Spectrom.*, 1980, 7, 301; J. Schmidt, S. Huneck, and P. Franke, *Biomed. Mass Spectrom.*, 1981, 8, 293.
80 M.D. Woodward, *Phytochemistry*, 1981, 20, 532.
81 R.J. Goldsack and J.S. Shannon, *Org. Mass Spectrom.*, 1980, 15, 545.
82 D.E. Games, P. Hirter, W. Kuhnz, E. Lewis, N.C.A. Weerasinghe, and S.A. Westwood, *J. Chromatogr.*, 1981, 203, 131.
83 D.E. Games, *Biomed. Mass Spectrom.*, 1981, 8, 454.
84 C. Eckers, D.E. Games, M.L. Games, W. Kuhnz, E. Lewis, N.C.A. Weerasinghe, and S.A. Westwood in 'Recent Developments in Mass Spectrometry in Biochemistry, Medicine and Environmental Research', ed. A. Frigerio, Elsevier, Amsterdam, 1981, Vol. 7, p. 169.
85 H.E. Nordby and S. Nagy, *J. Chromatogr.*, 1981, 207, 21.
86 A.A. Swigar and R.M. Silverstein, 'Monoterpenes: I.R., Mass Proton, and C-13 N.M.R. Spectra and Kovats Indexes', Aldrich Chem. Co., Milwaukee, U.S.A., 1981.
87 A.M. Bambagrotti, S.A. Coran, V. Giannellini, G. Moneti, F.F. Vincieri, A. Selva, and P. Traldi, *Biomed. Mass Spectrom.*, 1981, 8, 343.
88 C.C. Bonini, C. Iavarone, C. Trogolo, and G.A. Poulton, *Org. Mass Spectrom.*, 1980, 15, 516.
89 C.C. Bonini, C. Iavarone, C. Trogolo, and C.A. Poulton, *Org. Mass Spectrom.*, 1981, 16, 68.
90 D. Voigt, G. Adam, and W. Schade, *Org. Mass Spectrom.*, 1981, 16, 85.
91 C.J. Shaw, R. Chen, M.L. Cotter, R.M. Kanojia, and M.P. Wachter, *Org. Mass Spectrom.*, 1981, 16, 281.
92 U.A. Abdullaev, Ya. V. Rashkes, V.A. Tarasov, and S.Z. Kasymov, *Khim. Prir. Soedin.*, 1980, 636.
93 U.A. Abdullaev, Ya. V. Rashkes, M.I. Yusupov, and Sh. Z. Kasymov, *Khim. Prir. Soedin.*, 1980, 796.
94 A.K. Borg-Karlson, T. Norin, and A. Tolvitie, *Tetrahedron*, 1981, 37, 425.
95 C.B. Christensen, U. Rasmussen, and C. Christophersen, *Tetrahedron Lett.*, 1980, 3829.

96 K. Nakano, T. Nohara, T. Tomimatsu, and I. Nishioka, *Yakugaku Zasshi*, 1981, 101, 1052.
97 D. Voight, G. Adam, and P. Franke, *Org. Mass Spectrom.*, 1980, 15, 587.
98 L. Rivier, P. Gaskin, K.-Y.S. Albone, and J. MacMillan, *Phytochemistry*, 1981, 20, 687.
99 D.E. Games, M.S. Lant, and B.J. Woodhall, unpublished work.
100 M.A. Baldwin, D.M. Carter, F.A. Darwish, and J.D. Phillipson, *Biomed. Mass Spectrom.*, 1981, 8, 362.
101 N.A. Klyuev, G.S. Sichkova, G.B. Lokshin, and A.D. Kuzovkov, *Khim. Prir. Soedin.*, 1981, 228.
102 J. Protiva, E. Klinotova, H. Skorkovska, and A. Vystreil, *Collect. Czech. Chem. Commun.*, 1981, 46, 1023.
103 A.G. Panosyan, G.M. Avetisyan, V.A. Mnatsakanyan, and B.V. Rozynov, *Bioorg. Khim.*, 1980, 6, 1094.
104 J. Schmidt and S. Huneck, *J. Prakt. Chem.*, 1981, 323, 654.
105 L. Ogunkoya, *Phytochemistry*, 1981, 20, 121.
106 Ya.V. Rashkes and N.K. Abubakirov, *Khim. Prir. Soedin.*, 1980, 518.
107 R.E. Isaac, M.E. Rose, H.H. Rees, and T.W. Goodwin, *J. Chem. Soc., Chem. Commun.*, 1982, 249.
108 M.E. Rose in 'Carotenoid Chemistry and Biochemistry', ed. C. Britton and T.W. Goodwin, Pergamon Press, Oxford, 1982, 167.
109 H. Adlercreutz, *Adv. Mass Spectrom.*, 1980, 8B, 1165.
110 C.J.W. Brooks, W.J. Cole, H.B. McIntyre, and A.G. Smith, *Lipids*, 1980, 15, 745.
111 M.J. Thompson, G.W. Patterson, S.R. Dutky, J.A. Svoboda, and J.N. Kaplanis, *Lipids*, 1980, 15, 719.
112 A. Kukis and P. Child, *Lipids*, 1980, 15, 770.
113 J.E. Gurst and A.K. Schrock, *J. Org. Chem.*, 1980, 45, 4062.
114 W. Schade, W. Ihn, J. Vokoun, B. Schonecker, and A. Schubert, *Steroids*, 1980, 36, 547.
115 A. Kasal and A. Trka, *Collect. Czech. Chem. Commun.*, 1980, 45, 2985.
116 L. Aringer and L. Nordström, *Biomed. Mass Spectrom.*, 1981, 8, 183.
117 F.J. Brown and C. Djerassi, *J. Org. Chem.*, 1981, 46, 954.
118 A. Lavanchy, T. Varkony, D.H. Smith, N.A.B. Gray, W.C. White, R.E. Carhart, B.G. Buchanan, and C. Djerassi, *Org. Mass Spectrom.*, 1980, 15, 355; A.A.B. Gray, A. Buchs, D.H. Smith, and C. Djerassi, *Helv. Chim. Acta*, 1981, 64, 458.
119 M.W. Khalil, C. Djerassi, and D. Sica, *Steroids*, 1980, 35, 707; S. Carmely and Y. Kashman, *Tetrahedron*, 1981, 37, 2397; R.J. Stonard, J.C. Petrovich, and R.J. Andersen, *Steroids*, 1981, 36, 81; M. Guyot and M. Durgeat, *Tetrahedron Lett.*, 1981, 22, 1391; M. Alam, T.B. Sarsing, J.R. Guerra, and A.D. Harmon, *Steroids*, 1981, 38, 375; S.I. Teshima, G.W. Patterson, and S.R. Dutky, *Lipids*, 1980, 15, 1004; U. Sjöstrand, L. Bohlin, L. Fischer, M. Colin, and C. Djerassi, *Steroids*, 1981, 38, 347; M. Rohmer, W.C.M.C. Kokke, W. Fenical, and C. Djerassi, *Steroids*, 1980, 35, 219; A.A.L. Gunatilaka, Y. Gopichand, F.J. Schmitz, and C. Djerassi, *J. Org. Chem.*, 1981, 46, 3860; L.N. Li, U. Sjöstrand, and C. Djerassi, *J. Org. Chem.*, 1981, 46, 3860; L.N. Li, U. Sjöstrand, and C. Djerassi, *J. Org. Chem.*, 1981, 46, 3867; L.N. Li and C. Djerassi, *J. Am. Chem. Soc.*, 1981, 103, 3606.
120 N. Fusetani, S. Matsunaga, and S. Konosu, *Tetrahedron Lett.*, 1981, 22, 1985.
121 F.R. Sugnaux and C. Djerassi, *J. Chromatogr.*, 1982, 251, 189.
122 Y.Y. Lin, *Lipids*, 1980, 15, 756.

123 Y.Y. Lin, C.E. Low, and L.L. Smith, *J. Steroid Biochem.*, 1981, 14, 563.
124 B.R. DeMark and P.D. Klein, *J. Lipid Res.*, 1981, 22, 166.
125 J. Whitney, S. Lewis, K.M. Straub, M.M. Thaler, and A.L. Burlingame, *Koenshu-Iyo Masu Kenkyukai*, 1981, 6, 33.
126 E.A. Larka, I. Howe, J.H. Beynon, and Z.V.I. Zaretskii, *Tetrahedron*, 1981, 37, 2625.
127 E.A. Larka, I. Howe, J.H. Beynon, and Z.V.I. Zaretskii, *Org. Mass Spectrom.*, 1981, 16, 465.
128 Z.V.I. Zaretskii in 'Recent Developments in Mass Spectrometry in Biochemistry, Medicine and Environmental Research', ed. A. Frigerio, Elsevier, Amsterdam, 1981, Vol. 7, p. 227.
129 D. Ghiringhelli, A. Griffini, and P. Traldi, *Biomed. Mass Spectrom.*, 1981, 8, 155.
130 H. Budzikiewicz and N. Mohr, *Org. Mass Spectrom.*, 1981, 16, 329.
131 M. Suzuki, K. Harada, N. Takeda, and A. Tatematsu, *Heterocycles*, 1981, 15, 1123.
132 M. Suzuki, K. Harada, N. Takeda, and A. Tatematsu, *Biomed. Mass Spectrom.*, 1981, 8, 332.
133 N. Takeda, M. Umemura, K. Harada, M. Suzuki, and A. Tatematsu, *J. Antibiot.*, 1981, 34, 617.
134 M. Suzuki, K. Harada, N. Takeda, A. Tatematsu, and H. Kambara, *Tennen Yuki Kagobutsu Toronkai Koen Yoshisv*, 1981, 24, 215.
135 K. Harada, M. Suzuki, N. Takeda, A. Tatematsu, and H. Kambara, *J. Antibiot.*, 1982, 35, 102.
136 W.C. Brumley, S. Nesheim, M. Trucksess, E.W. Trucksess, P.A. Dreifuss, J.A. Roach, D. Andrzejewski, R.M. Eppley, and A.E. Pohland, *Anal. Chem.*, 1981, 53, 2003.
137 J.L. Gower, G. Beaugrand, and C. Sallot, *Biomed. Mass Spectrom.*, 1981, 8, 36.
138 M. Shimizu, H. Matsumoto, A. Koda, T. Ichimura, and Y. Koboyashi, *Yamanouchi Seiyaku Kenkyu Hokoku*, 1980, 4, 112.
139 M. Barber, R.S. Bordoli, R.D. Sedgwick, A.N. Tyler, B.N. Green, V.C. Parr, and J.L. Gower, *Biomed. Mass Spectrom.*, 1982, 9, 11.
140 W.A. König, H. Krohn, M. Greiner, H. Brückner, and G. Jung, *Adv. Mass Spectrom.*, 1980, 8B, 1109.
141 M. Aydin, D.H. Bloss, W.A. König, H. Brückner, and G. Jung, *Biomed. Mass Spectrom.*, 1982, 9, 39.
142 M. Nishii, T. Kihara, K. Isono, T. Higashijima, T. Miyazawa, S.K. Sethi, and J.A. McCloskey, *Tetrahedron Lett.*, 1980, 4627.
143 G. Bojesen, D. Gauvreau, D.H. Williams, and M.J. Waring, *J. Chem. Soc., Chem. Commun.*, 1981, 46.
144 A.D. Elbein, J.L. Occolowitz, R.L. Hamill, and K. Eckardt, *Biochemistry*, 1981, 20, 4210.
145 R.E. More and G. Bartolini, *J. Am. Chem. Soc.*, 1981, 103, 2491.
146 T. Fujita, Y. Takaishi, A. Okamura, E. Fujita, and K. Fuji, *J. Chem. Soc., Chem. Commun.*, 1981, 585.
147 A. Dell, H.R. Morris, S.M. Hecht, and M.D. Levin, *Biochem. Biophys. Res. Commun.*, 1980, 97, 977.
148 M. Barber, R.S. Bordoli, R.D. Sedgwick, A.N. Tyler, and B.W. Bycroft, *Biochem. Biophys. Res. Commun.*, 1981, 101, 632.
149 A. Dell, H.R. Morris, M.D. Levin, and S.M. Hecht, *Biochem. Biophys. Res. Commun.*, 1981, 102, 730.
150 A. Dell, R.C. Hider, M. Barber, R.S. Bordoli, R.D. Sedgwick, A.N. Tyler, and J.B. Neilands, *Biomed. Mass Spectrom.*, 1982, 9, 158.

151 K.L. Rinehart, jun., L.A. Gaudioso, M.L. Moore, R.C. Pardey, J.C. Cook, jun., M. Barber, R.D. Sedgwick, R.S. Bordoli, A.N. Tyler, and B.N. Green, *J. Am. Chem. Soc.*, 1981, 103, 6517.
152 R.D. Macfarlane, *Biomed. Mass Spectrom.*, 1981, 8, 449.
153 M.S. Puar, A.K. Ganguly, A. Afonso, R. Brambilla, P. Mangiaracina, O. Sarre, and R.D. Macfarlane, *J. Am. Chem. Soc.*, 1981, 103, 5231.
154 W.J. McGarhen, B.A. Hardy, G.O. Morton, F.M. Lovell, N.A. Perkinson, R.T. Hargreaves, D.B. Borders, and G.A. Ellestad, *J. Org. Chem.*, 1981, 46, 792.
155 N. Otake, T. Ogita, Y. Miyazaki, H. Yonehara, R.D. Macfarlane, and C.J. McNeal, *J. Antibiot.*, 1981, 34, 130.
156 A.K. Mallams, M.S. Puar, R.R. Rossman, A.T. McPhail, and R.D. Macfarlane, *J. Am. Chem. Soc.*, 1981, 103, 3940.
157 P.F. Crain, H. Yamomoto, J.A. McCloskey, Z. Yamaizumi, S. Nishimura, K. Limbury, M. Raba, and H.J. Gross, *Adv. Mass Spectrom.*, 1980, 8B, 1135.
158 R.G.A.R. Maclagan and M.J. Mitchell, *Aust. J. Chem.*, 1980, 33, 1401.
159 M.A. Quilliam, K.K. Ogilvie, and J.B. Westmore, *Org. Mass Spectrom.*, 1981, 16, 129.
160 E.L. Esmans, E.J. Freyne, J.H. Vanbroeckhoven, and F.C. Alderweireldt, *Biomed. Mass Spectrom.*, 1980, 7, 377.
161 C.-H. Wang, Y.-X. Wang, K.-X. Cao, L.-H. Chang, L.-T. Ma, and X. Wang, *Hua Hsueh Tung Pao*, 1981, 11.
162 H.M. Schiebel and H.-R. Schulten, *Z. Naturforsch., Teil B*, 1981, 36, 967.
163 B. Schueler and F.R. Krueger, *Org. Mass Spectrom.*, 1980, 15, 295.
164 A. Eicke, W. Sichtermann, and A. Benninghoven, *Org. Mass Spectrom.*, 1980, 15, 289.
165 S.E. Unger, A.E. Schoen, R.G. Cooks, D.J. Ashworth, J.D. Gomes, and C.-J. Chang, *J. Org. Chem.*, 1981, 46, 4765.
166 D.J. Ashworth, C.-J. Chang, S.E. Unger, and R.G. Cooks, *J. Org. Chem.*, 1981, 46, 4770.
167 E. Werner, K.G. Standing, J.B. Westmore, K.K. Ogilvie, and M.J. Nemer, *Anal. Chem.*, 1982, 54, 960.
168 C.J. McNeal, K.K. Ogilvie, N.Y. Theriault, and M.J. Nemer, *J. Am. Chem. Soc.*, 1982, 104, 972, 976, 981.
169 M. Jarman, *J. Anal. Appl. Pyrolysis*, 1980, 2, 217.
170 Y. Yamaizumi, Y. Kuchino, F. Harada, S. Nishimura, and J.A. McCloskey, *J. Biol. Chem.*, 1980, 255, 2220
171 H.J. Chavas das Neves and M.S.S. Pais, *Biochem. Biophys. Res. Commun.*, 1980, 95, 1387.
172 P.F. Crain, S.K. Sethi, M.R. Katze, and J.A. McCloskey, *J. Biol. Chem.*, 1980, 207, 8405.
173 A.F. Cook, R.T. Bartlett, R.P. Gregson, and D.J. Quinn, *J. Org. Chem.*, 1980, 45, 4020.
174 H.J. Chavas das Neves and M.S.S. Pais, *Tetrahedron Lett.*, 1980, 4387.
175 N.B.H. Henis, K.L. Busch, and M.M. Bursey, *Inorg. Chim. Acta*, 1981, 53, L31.
176 N.B.H. Henis, T.L. Youngless, and M.M. Bursey, *J. Chem. Soc., Dalton Trans.*, 1980, 1416.
177 C.J. Shaw, G. Eglinton, and J.M.E. Quirke, *Anal. Chem.*, 1981, 53, 2014.
178 S.K. Hajibrahim, J.M.E. Quirke, G. Eglinton, S.E. Palmer, and E.W. Baker, *Chem. Geol.*, 1982, 35, 69.
179 M. Barber, R.S. Bordoli, R.D. Sedgwick, and A.N. Tyler, *Biomed. Mass Spectrom.*, 1981, 8, 492.

180 S.W. Graham, P. Dowd, and D.M. Hercules, *Anal. Chem.*, 1982, 54, 649.
181 H.M. Schiebel and H.-R. Schulten, *Naturwissenschaften*, 1980, 67, 256.
182 E. Constantin, Y. Nakatani, G. Teller, R. Heieber, and G. Ourisson, *Bull. Soc. Chim. Fr.*, 1981, 303.
183 R.G. Brereton, B. Rajananda, T.J. Blake, J.K.M. Sanders, and D.H. Williams, *Tetrahedron Lett.*, 1980, 1671.
184 J.E. Hunt, R.D. Macfarlane, T.J. Kutz, and R.C. Dougherty, *Proc. Natl. Acad. Sci. U.S.A.*, 1980, 77, 1745.
185 H.-R. Schulten, P.B. Monkhouse, and R. Müller, *Anal. Chem.*, 1982, 54, 654.
186 J.E. Hunt, R.D. Macfarlane, J.J. Katz, and R.C. Dougherty, *J. Am. Chem. Soc.*, 1981, 103, 6775.
187 D.C. DeJongh in 'Physical Methods for Structural Analysis. II Mass Spectrometry. Carbohydrate, Chemistry and Biochemistry', 2nd Edn., ed. W.W. Pigman and D. Horton, Academic Press, New York, 1981, Vol. IB, p. 1327.
188 F. Turecek, T. Trnka, and M. Cerny, *Collect. Czech. Chem. Commun.*, 1981, 46, 2390.
189 G.C. Wong, S.S. Sung, and C.C. Sweeley in 'Methods Carbohydrate Chemistry', ed. R.L. Whistler and J.N. Bemiller, Academic Press, New York, 1980, Vol. 8, p. 55.
190 B. Fournet, G. Strecker, Y. Leroy, and J. Montreuil, *Anal. Biochem.*, 1981, 116, 489.
191 V. Kovacik, V. Mihalov, and P. Kovac, *Carbohydr. Res.*, 1981, 88, 189.
192 V. Kovacik, P. Kovac, and A. Liptak, *Carbohydr. Res.*, 1981, 98, 242.
193 M. Blanc-Muesser, J. Defaye, R.L. Foltz, and D. Horton, *Org. Mass Spectrom.*, 1980, 15, 317.
194 V.I. Kadentsev, A.G. Kaimarazov, and O.S. Chishov, *Izv. Akad. Nauk SSSR, Ser. Khim.*, 1980, 330.
195 F.R. Seymour, E.C.M. Chen, and J.E. Stouffer, *Carbohydr. Res.*, 1980, 83, 201.
196 R.A. Laine, *Anal. Biochem.*, 1981, 116, 383.
197 E.G. De Jong, W. Heerma, and G. Dijkstra, *Adv. Mass Spectrom.*, 1980, 8B, 1314.
198 T. Fujiwara and K. Arai, *Carbohydr. Res.*, 1982, 101, 287.
199 T. Murata, S. Takahashi, T. Takeda, S. Ohnishi, and K. Miseki, *Shimadzu Hyoron*, 1980, 37, 49.
200 J. Deutsch, *Org. Mass Spectrom.*, 1980, 15, 240.
201 J.J. Zwinselman, R.H. Fokkens, N.M.M. Nibbering, K.H. Ott, and F.W. Röllgen, *Biomed. Mass Spectrom.*, 1981, 8, 312; K.H. Ott, F.W. Röllgen, J.J. Zwinselman, R.H. Fokkens, and N.M.M. Nibbering, *Org. Mass Spectrom.*, 1980, 15, 419.
202 F. Heresch, E.R. Schmid, and J.F.K. Huber in 'Recent Developments in Mass Spectrometry in Biochemistry, Medicine and Environmental Research', ed. A. Frigerio, Elsevier, Amsterdam, 1981, Vol. 7, p. 351.
203 H. Kambara and S. Hishida, *Org. Mass Spectrom.*, 1981, 16, 167.
204 L.K. Liu, K.L. Busch, and R.G. Cooks, *Anal. Chem.*, 1981, 53, 109.
205 D.A. Cummings, C.G. Hellerquist, and O.J. Touster, *J. Biol. Chem.*, 1981, 256, 7723.
206 Y.A. Knirel, N.A. Kocharova, A.S. Shaskov, B.A. Dimitriev, and N.K. Kochetkov, *Carbohydr. Res.*, 1981, 93, C12.
207 M.C. Herlant-Peers, J. Montreuil, G. Strecker, L. Dorland, H. van Halbeek, G.A. Veldink, and J.F.G. Vliegenthart, *Eur. J. Biochem.*, 1981, 117, 291.

208 F. Matsuura, A.H. Nunez, G.A. Grabowski, and C.C. Sweeley, *Arch. Biochem. Biophys.*, 1981, 207, 337.
209 H. Egge, J.C. Michalski, and G. Strecker, *Arch. Biochem. Biophys.*, 1982, 213, 318.
210 F. Matsuura, R.A. Laine, and M.Z. Jones, *Arch. Biochem. Biophys.*, 1981, 211, 485.
211 H. Egge and V. Von Nicolai, *Hoppe-Seyler's Z. Physiol. Chem.*, 1981, 362, 195.
212 H. Egge, G. Strecker, and J.-C. Michalski, *Hoppe-Seyler's Z. Physiol. Chem.*, 1981, 362, 196.
213 H. Van Halbeek, L. Dorland, J. Haverkamp, G.A. Veldnik, J.F.G. Vliegenthart, B. Fournet, G. Ricart, J. Montreuil, W.D. Gathmann, and D. Aminoff, *Eur. J. Biochem.*, 1981, 118, 487.
214 J. Metzger, *Fresenius' Z. Anal. Chem.*, 1981, 308, 29.
215 M. Linscheid, J. D'Angona, A.L. Burlingame, A. Dell, and C.E. Ballou, *Proc. Natl. Acad. Sci. U.S.A.*, 1981, 78, 1471.
216 H. Egge, J. Dabrowski, P. Hanfland, H.V. Nicolai, U. Dabrowski, and A. Dell in 'Glycoconjugates', ed. T. Yamatawa, T. Osawa, and S. Handa, Japan Scientific Society Press, Tokyo, 1981, p. 318.
217 H.-R. Schulten, U. Bahr, and W. Goertz, *J. Anal. Appl. Pyrolysis*, 1982, 3, 229.
218 C.N. Kenyon, A. Melera, and F. Erni, *J. Anal. Toxicol.*, 1981, 5, 216.
219 P. Aman, M. McNeil, L.E. Franzen, A.G. Darvill, and P. Albersheim, *Carbohydr. Res.*, 1981, 95, 263.
220 B.S. Valent, A.G. Darvill, M. McNeil, B.K. Robertson, and P. Albersheim, *Carbohydr. Res.*, 1980, 79, 165; B.K. Robertson, P. Aman, A.G. Darvill, M. McNeil, and P. Albersheim, *Plant Physiol.*, 1981, 67, 389; M. McNeil, A.G. Darvill, P. Aman, L.E. Franzen, and P. Albersheim, *Methods Enzymol.*, 1982, 83, 3.
221 M.L. Bouillant, A. Besset, J. Favre-Bonvin, and J. Chopin, *Phytochemistry*, 1980, 19, 1755.
222 R.K. Chaudhuri, O. Sticher, and T. Winkler, *Helv. Chim. Acta*, 1980, 63, 1045.
223 F.U. Afifi-Yazar and O. Sticher, *Helv. Chim. Acta*, 1980, 63, 1905.
224 S.D. Jolad, J.J. Hoffman, J.R. Cole, M.S. Tempesta, and R.B. Bates, *J. Org. Chem.*, 1981, 46, 1946.
225 M. Hauteville, J. Chopin, H. Geiger, and L. Schüler, *Tetrahedron Lett.*, 1980, 1227.
226 F. De Simone, A. Dini, E. Finamore, L. Minale, C. Pizza, R. Riccio, and F. Zollo, *J. Chem. Soc., Perkin Trans. 1*, 1981, 1855.
227 J.V. Eichholzer, I.A.S. Lewis, J.K. MacLeod, and P.B. Oelrichs, *Tetrahedron*, 1981, 37, 1881; I.A.S. Lewis, J.K. MacLeod, and P.B. Oelrichs, *Tetrahedron*, 1981, 37, 4305.
228 T. Kastelic-Suhadok, *Eur. J. Mass Spectrom. Biochem. Med. Environ. Res.*, 1980, 1, 193.
229 H. Ikokawa, Y. Oshida, A. Ikuta, and Y. Shida, *Chem. Lett.*, 1982, 49.
230 K. Hostettmann, J. Doumas, and M. Hardy, *Helv. Chim Acta*, 1981, 64, 297.
231 H. Geiger and G. Schwinger, *Phytochemistry*, 1980, 19, 897.
232 T. Komori, I. Maetani, N. Okamura, T. Kawasaki, T. Nohara, and H.-R. Schulten, *Liebigs Ann. Chem.*, 1981, 683.
233 H.-R. Schulten and F. Soldati, *J. Chromatogr.*, 1981, 212, 37.
234 K. Hostettmann, *Helv. Chim. Acta*, 1980, 63, 606.
235 K. Kudo, T. Nohara, T. Komori, T. Kawasaki, and H.-R. Schulten, *Planta Med.*, 1980, 40, 250.

236 G.R. Fenwick, J. Eagles, R. Gmelin, and D. Rakow, *Biomed. Mass Spectrom.*, 1980, 7, 410.
237 J. Eagles, G.R. Fenwick, R. Gmelin, and D. Rakow, *Biomed. Mass Spectrom.*, 1981, 8, 265.
238 B.W. Christensen, A. Kjaer, J.O. Madsen, C.E. Olsen, O. Olsen, and H. Sørensen, *Tetrahedron*, 1982, 38, 353.
239 R. Schuster, *Chromatographia*, 1980, 13, 379.
240 A.J.L. Cooper, O.W. Griffith, A. Meister, and F.H. Field, *Biomed. Mass Spectrom.*, 1981, 8, 95.
241 T. Fairwell and H.R. Brewer, jun., *Anal. Biochem.*, 1980, 107, 140.
242 C. Schiller, K.D. Kupka, and F. Hillenkamp, *Fresenius' Z. Anal. Chem.*, 1981, 308, 304; in 'Recent Developments in Mass Spectrometry in Biochemistry, Medicine and Environmental Research', ed. A. Frigerio, Elsevier, Amsterdam, 1981, Vol. 7, p. 287.
243 A. Benninghoven and W. Sichtermann, *Int. J. Mass Spectrom. Ion Phys.*, 1981, 38, 351; *ibid.*, 1981, 40, 177.
244 D.J. Surman and J.C. Vickerman, *J. Chem. Res. (S)*, 1981, 70.
245 K.D. Kloeppel in 'Recent Developments in Mass Spectrometry in Biochemistry, Medicine and Environmental Research', ed. A. Frigerio, Elsevier, Amsterdam, 1981, Vol. 7, p. 283.
246 L.O. Zamir, R.A. Jensen, B.H. Arison, A.W. Douglas, G. Albers-Schonbery, and J.R. Bowen, *J. Am. Chem. Soc.*, 1980, 102, 4499.
247 U. Cramer, A.G. Rehfeldt, and F. Spener, *Biochemistry*, 1980, 19, 3074.
248 R.J.W. Truscott, D. Malegan, E. McCairns, B. Halpern, J. Hammond, R.G.H. Cotton, J.F.B. Mercer, S. Hunt, J.G. Rogers, and D.M. Danks, *Biomed. Mass Spectrom.*, 1981, 8, 99.
249 G. Wilhelm and K.-D. Kupka, *FEBS Lett.*, 1981, 123, 141.
250 K. Konno, H. Shirahama, and T. Matsumoto, *Tetrahedron Lett.*, 1981, 1617.
251 M.J. Dognin and B. Wittman-Liebold, *Hoppe-Seyler's Z. Physiol. Chem.*, 1980, 361, 1697.
252 P. Sautière, G. Briand, M. Gusse, and P. Chevallier, *Eur. J. Biochem.*, 1981, 119, 251.
253 J. Deutsch, C. Gilon, and M. Chorev, *Int. J. Pept. Protein Res.*, 1981, 18, 203.
254 D.M. Desiderio, J.L. Stein, M.D. Cunningham, and J.Z. Sabbatani, *J. Chromatogr.*, 1980, 195, 369.
255 M. Przybylski and W. Volter, *Struct. Act. Nat. Pept. Proc. Fall Meet. Ges. Biol. Chem.*, *1979*, 1981, 141.
256 T. Matsuo, H. Matsuda, I. Katakuse, Y. Wada, T. Fujita, and A. Hayashi, *Biomed. Mass Spectrom.*, 1981, 8, 25; Y. Wada, A. Hayashi, T. Fujita, T. Matsuo, I. Katakuse, and H. Matsuda, *Biochem. Biophys. Acta*, 1981, 667, 233.
257 T. Matsuo, I. Katakuse, H. Matsuda, Y. Wada, T. Fujita, and A. Hayashi, *Koenshu-Iyo Masu Kenkyukai*, 1981, 6, 107.
258 V.L. Sadovskaya, T.M. Andronova, V.G. Merimson, and B.V. Rosynov, *Org. Mass Spectrom.*, 1980, 15, 473.
259 B. Calas, J. Mery, J. Parello, J.C. Prome, J. Roussel, and D. Patouraux, *Biomed. Mass Spectrom.*, 1980, 7, 288.
260 Y. Shimonishi, Y.-M. Hong, T. Kitagishi, T. Matsuo, H. Matsuda, and I. Katakuse, *Eur. J. Biochem.*, 1980, 112, 251.
261 T. Matsuo, H. Matsuda, and I. Katakuse, *Biomed. Mass Spectrom.*, 1981, 8, 137.
262 Y. Shimonishi, Y.-M. Hong, I. Katakuse, and S. Aara, *Bull. Chem. Soc. Jpn.*, 1981, 54, 3069; I. Katakuse, T. Matsuo, H. Matsuda, Y. Shimonishi, Y.-M. Hong, and Y. Izumi, *Biomed. Mass Spectrom.*, 1982, 9, 64.

263 A. Tsugita, R. Van den Broek, and M. Przybylski, *FEBS Lett.*, 1982, 137, 19; Y. Shimonishi, Y.-M. Hong, T. Takao, S. Aimoto, H. Matsuda, and Y. Izumi, *Proc. Jpn. Acad., Ser. B*, 1981, 57, 304.

264 R. Weber and K.L. Levsen, *Biomed. Mass Spectrom.*, 1980, 7, 314; D.M. Desiderio and J.Z. Sabbatini, *Biomed. Mass Spectrom.*, 1980, 8, 565; T. Matsuo, H. Matsuda, I. Katakuse, Y. Shimonishi, Y. Maruyama, T. Higuchi, and E. Kubota, *Anal. Chem.*, 1981, 53, 416.

265 H. Kambara, S. Hishida, and H. Naganawa, *Org. Mass Spectrom.*, 1982, 17, 67.

266 H. Kambara, *Org. Mass Spectrom.*, 1982, 17, 29.

267 J.B. Westmore, W. Ens, and K.G. Standing, *Biomed. Mass Spectrom.*, 1982, 9, 119.

268 M. Barber, R.S. Bordoli, R.D. Sedgwick, and L.W. Tetler, *Org. Mass Spectrom.*, 1981, 16, 256.

269 M. Barber, R.S. Bordoli, R.D. Sedgwick, A.N. Tyler, and E.T. Whalley, *Biomed. Mass Spectrom.*, 1981, 8, 337.

270 M. Barber, R.S. Bordoli, R.D. Sedgwick, and A.N. Tyler, *Biomed. Mass Spectrom.*, 1982, 9, 208.

271 M. Barber, R.S. Bordoli, G.V. Garner, D.B. Gordon, R.D. Sedgwick, L.W. Tetler, and A.N. Tyler, *Biochem. J.*, 1981, 197, 401.

272 H.R. Morris, M. Panico, M. Barber, R.S. Bordoli, R.D. Sedgwick, and A. Tyler, *Biochem. Biophys. Res. Commun.*, 1981, 101, 623.

273 K. Biemann, *Koenshu-Iyo Masu Kenkyukai*, 1981, 6, 21.

274 C.V. Bradley, D.H. Williams, and M.R. Hanley, *Biochem. Biophys. Res. Commun.*, 1982, 104, 1223.

275 D.H. Williams, G. Bojesen, A.D. Auffret, and L.C.E. Taylor, *FEBS Lett.*, 1981, 128, 37.

276 H.R. Morris, A. Dell, A.T. Etienne, M. Judkins, R.A. McDowell, M. Panico, and A.W. Taylor, *Pure Appl. Chem.*, 1982, 54, 267.

277 D.F. Hunt, W.M. Bone, J. Shabanowitz, J. Rhodes, and J.M. Ballard, *Anal. Chem.*, 1981, 53, 1704.

278 R. Steinauer, H. Walther, and U.P. Schlunegger, *Helv. Chim. Acta*, 1980, 63, 610; R. Steinauer and U.P. Schlunegger, *Biomed. Mass Spectrom.*, 1982, 9, 153.

279 H.W. Ronge and H.F. Gruetzmacher in 'Recent Developments in Mass Spectrometry in Biochemistry, Medicine and Environmental Research', ed. A. Frigerio, Elsevier, Amsterdam, 1981, Vol. 7, p. 255.

280 C.V. Bradley, I. Howe, and J.H. Beynon, *Biomed. Mass Spectrom.*, 1981, 8, 85.

281 D.F. Hunt, A.M. Buko, J.M. Ballard, J. Shabanowitz, and A.B. Giordani, *Biomed. Mass Spectrom.*, 1981, 8, 397; in ref. 2, p. 85.

282 J.F.J. Grove, *J. Chem. Soc., Perkin Trans. 1*, 1980, 2878; J.F. Elsworth and J.F.J. Grove, *J. Chem. Soc., Perkin Trans. 1*, 1980, 1795.

283 K. Jayasimhulu and R.A. Day, *Biomed. Mass Spectrom.*, 1980, 7, 321.

284 B.R. Larson, *Life Sci.*, 1982, 30, 1003.

285 S.A. Carr, W.C. Herlihy, and K. Biemann, *Biomed. Mass Spectrom.*, 1981, 8, 51; W.C. Herlihy, R.J. Anderegg, and K. Biemann, *Biomed. Mass Spectrom.*, 1981, 8, 62; W.C. Herlihy and K. Biemann, *Biomed. Mass Spectrom.*, 1981, 8, 70; W.C. Herlihy, N.J. Royal, K. Biemann, S.D. Putney, and P.R. Schimmel, *Proc. Natl. Sci. U.S.A.*, 1980, 77, 6531.

286 H.R. Morris, A. Dell, T. Etienne, and G.W. Taylor, *Dev. Biochem.*, 1980, 10, 193.

287 R.P. Swenson and J.B. Howard, *J. Biol. Chem.*, 1980, 255, 8087.
288 G.P. Vinson, B.J. Whitehouse, A. Dell, T. Etienne, and H.R. Morris, *Nature (London)*, 1980, 284, 464.
289 H.R. Morris, A. Dell, M. Judkins, R.A. McDowell, M. Panico, and G.W. Taylor in 'Proceedings of the 7th American Peptide Symposium', ed. D.H. Rich and E. Gross, Pierce Chemical Co., Rockford, IL, U.S.A., 1981, p. 745.
290 A. Dell and H.R. Morris, *Biomed. Mass Spectrom.*, 1981, 8, 128.
291 A.D. Auffret, T.J. Blake, and D.H. Williams, *Eur. J. Biochem.*, 1981, 113, 333.
292 A. Parente, M.N. Short, R. Self, and K.R. Parsley, *Biomed. Mass Spectrom.*, 1982, 9, 141.
293 R.P. McDonagh, J. McDonagh, T.E. Petersen, H.C. Thøgersen, K. Skorstengaard, L. Sottrup-Jensen, S. Magnusson, A. Dell, and H.R. Morris, *FEBS Lett.*, 1981, 127, 174.
294 T.J. Yu, H. Schwartz, R.W. Gise, B.L. Karger, and P. Vouros, *J. Chromatogr.*, 1981, 218, 519.
295 A.G. Harrison and R.K.M.R. Kallury, *Org. Mass Spectrom.*, 1980, 15, 277.
296 R.J. Day, A.L. Forbes, and D.M. Hercules, *Spectrosc. Lett.*, 1981, 14, 703.
297 M. Suzuki, T. Ariga, M. Sekine, E. Araki, and T. Miyatake, *Anal. Chem.*, 1981, 53, 985.
298 R.J. Greathead and K.R. Jennings, *Org. Mass Spectrom.*, 1980, 15, 431.
299 R. Chai and A.G. Harrison, *Anal. Chem.*, 1981, 53, 34.
300 R.K. Christopher and A.M. Duffield, *Biomed. Mass Spectrom.*, 1980, 7, 429.
301 A. Brauner, H. Budzikiewicz, and W. Boland, *Org. Mass Spectrom.*, 1982, 17, 161.
302 V.I. Khvostenko, E.G. Galkin, U.M. Dzhemilev, G.A. Tolstikov, and V.S. Fal'ko, *Izv. Akad. Nauk SSSR, Ser. Khim.*, 1980, 1663.
303 D.J. Harvey, *Biomed. Mass Spectrom.*, 1982, 9, 33.
304 H.-J. Stan and M. Scheutwinkel-Reich, *Lipids*, 1980, 15, 1044.
305 O.S. Privett and W.L. Ardahl, *Methods Enzymol.*, 1981, 72, 56.
306 W.L. Ardahl, W. Beck, C. Jones, D.E. Jarvis, and O.S. Privett, *Lipids*, 1981, 16, 614.
307 G. Horvath, *Adv. Mass Spectrom.*, 1980, 8B, 1334; G. Horvath, *Kem. Kozl.*, 1980, 54, 340.
308 R.C. Murphy, W.R. Mathews, J. Rokach, and C. Fenselau, *Prostaglandins*, 1982, 23, 201.
309 E. Schulte, M. Hoehn, and U. Rapp, *Fresenius' Z. Anal. Chem.*, 1981, 307, 115.
310 K.E. Murray and H.-R. Schulten, *Chem. Phys. Lipids*, 1981, 29, 11.
311 H. Ravvala, J. Finne, T. Krusius, J. Karkkainen, and J. Jarnefelt, *Adv. Carbohydr. Chem. Biochem.*, 1981, 38, 389.
312 K.-A. Karlsson, *Adv. Exp. Med. Biol.*, 1980, 125, 47.
313 M.E. Breimer, G.C. Hansson, K.-A. Karlsson, G. Larson, H. Leffler, L. Pascher, W. Pimlott, and B. Samuelsson, *Adv. Mass Spectrom.*, 1980, 8B, 1097; M.E. Breimer, G.C. Hansson, K.-A. Karlsson, and H. Leffler, *Biochem. Biophys. Res. Commun.*, 1980, 95, 416; *Biochem. Biophys. Acta*, 1980, 617, 85; in 'Cell Surface Glycolipids', ed. C.C. Sweeley, American Chemical Society, Washington, D.C., 1980, p. 79; *FEBS Lett.*, 1980, 114, 51; *J. Biochem.*, 1981, 90, 589; *J. Biol. Chem.*, 1982, 257, 50, 557; G.C. Hansson, K.-A. Karlsson, and J. Thurin, *J. Biochem. Biophys. Acta*, 1980, 620, 270; K.-E. Falk, K.-A. Karlsson, and B.E. Samuelsson, *FEBS Lett.*, 1981, 124, 173; *Chem. Phys. Lipids*, 1980, 27, 9; K.-A. Karlsson and

G. Larson, *J. Biol. Chem.*, 1981, 256, 3512; *FEBS Lett.*, 1981, 128, 71; M.E. Breimer, K.-A. Karlsson, and B.E. Samuelsson, *J. Biol. Chem.*, 1981, 256, 3810; P. Fredman, J.-E. Mansson, L. Svennerholm, B.E. Samuelsson, I. Pascher, W. Pimlott, K.-A. Karlsson, and G.W. Klinghardt, *Eur. J. Biochem.*, 1981, 116, 553.

314 M.E. Breimer, G.C. Hansson, K.-A. Karlsson, H. Leffler, W. Pimlott, and B.E. Samuelsson, *FEBS Lett.*, 1981, 124, 299.

315 H. Egge and P. Hanfland, *Arch. Biochem. Biophys.*, 1981, 210, 396.

316 P. Hanfland, H. Egge, U. Dabrowski, S. Kuhn, D. Roelcke, and J. Dabrowski, *Biochemistry*, 1981, 20, 5310.

317 G. Puzo, H. Aurelle, C. Lacave, and J.C. Prome, *Adv. Mass Spectrom.*, 1980, 8B, 1283.

318 C.E. Costello, B.W. Wilson, K. Biemann, and V.N. Reinhold in 'Cell Surface Glycolipids', ed. C.C. Sweeley, American Chemical Society, Washington, D.C., 1980, p. 35.

319 P. Hanfland, J. Dabrowski, A. Dell, U. Dabrowski, S. Kuhn, and H. Egge in 'Glycosoconjugates', ed. T. Yamakarva, T. Osawa, and W. Hands, Japan Scientific Societies Press, Tokyo, 1981, p. 98.

320 T. Ariga, T. Murata, M. Oshima, M. Maezawa, and T. Miyatake, *J. Lipid Res.*, 1980, 21, 879.

321 B.A. Macher and J.C. Klock in 'Cell Surface Glycolipids', ed. C.C. Sweeley, American Chemical Society, Washington, D.C., 1980, p. 127.

322 B.A. Macher and J.C. Klock, *J. Biol. Chem.*, 1980, 255, 2092; J.C. Klock, J.L. D'Angona, and B.A. Macher, *J. Lipid Res.*, 1981, 22, 1079; W.M.F. Lee, J.C. Klock, and B.A. Macher, *Biochemistry*, 1981, 20, 3810, 6505.

323 C.C. Sweeley, J.R. Moskai, H. Nunez, and F. Matsuura, '27th International Congress of Pure and Applied Chemistry', ed. A. Varmavuori, Pergamon Press, New York, 1980, p. 233.

324 T. Ariga, T. Murata, M. Oshima, M. Maezawa, and T. Miyatake, *J. Lipid Res.*, 1980, 21, 879.

325 J. Polonsky, M. Tence, P. Varenne, B.C. Das, J. Lunel, and J. Benveniste, *Proc. Natl. Acad. Sci. U.S.A.*, 1980, 77, 7019.

10

Organometallic, Co-ordination, and Inorganic Compounds Investigated by Mass Spectrometry

BY R. H. CRAGG

1 Introduction

This chapter, which follows the format of previous volumes, covers the period June 1980 - May 1982. For convenience, those papers concerned with secondary-ion mass spectrometry and Knudsen-cell mass spectrometry are reviewed in separate sections. The following abbreviations have been used in the text: c.i. (chemical ionization), e.i. (electron impact), f.d. (field desorption), f.i. (field ionization), g.c.-m.s. (gas chromatography-mass spectrometry), i.c.r. (ion-cyclotron resonance), m.i.k.e. (metastable-ion kinetic energy), and s.i.m.s. (secondary-ion mass spectrometry).

2 Main-group Organometallics

Group II.- The mass spectra of a series of dinorbornylzinc compounds have been reported.[1] Vacuum-pyrolysis mass spectrometry has been of value in a study[2] of the mechanism of pyrolysis of $RHgCCl_3$ (R = Cl, Ph, or CCl_3); abundances and appearance energies were measured for major ions at various temperatures. The reactions of methyl- and ethyl-mercuric chloride with diphenyl mercury have been studied by g.c.-m.s.[3]

Group III.- The spectra of a series of dimesitylboryl compounds, *e.g.* $(Mes)_2BR$ (R = alkyl, alkoxy, alkylthio, or dialkylamino), have been recorded.[4] All compounds gave a molecular ion in high abundance, except for secondary and tertiary alkyl substituents where the only significant boron-containing fragment was $(Mes_2B)^+$.

Large peaks attributed to doubly charged ions are observed in the spectra of phenyl-substituted borazines when the phenyl groups are co-planar with the borazine ring.[5] It was suggested that the ions are stabilized by exocyclic ring closures as a result of the loss of three radicals. It was observed that fragmentation of compounds (1) involves the loss of all or part of the alkyl substituents and that less ring cleavage occurred than in the fragmentation of oxygen analogues (2).[6] As a result of the analysis of the spectrum of the dioxaborinan (3), comments have been made

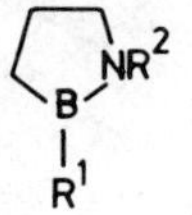

(1) R^1 = H, Pr, Bu, $Me_2CHCH_2CH_2$, or Ph, R^2 = H, Me, Et, Pr, or Bu

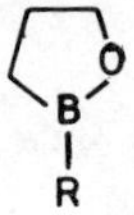

(2) R = Bu, $Me_2CHCH_2CH_2$, or hexyl

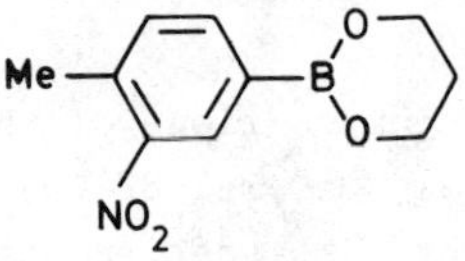

(3)

concerning the problems that arise from the assignment of fragmentation pathways on the basis of metastable-ion data.[7] A review on the use of cyclic boronates for characterization of polar bifunctional organic compounds by g.c.-m.s. has been published.[8] Mass spectrometry has been used to characterize arylboronate derivatives of 2- and 3-hydroxy fatty acids,[9] diols and amino-alcohols,[10] amidoximes (*e.g.* (4)),[11] and acyclic vicinal pentaols (*e.g.* (5) - (7)).[12] Ions corresponding to the molecular ions of boronate esters were observed when a series of diols, in which there were 2 - 6 carbon atoms between the OH groups, were desorbed from a 1-naphthalenylboronic acid surface under f.d. conditions.[13]

In a study of the spectra of the series of dialkylaluminium alkoxides $R^1{}_2AlOR^2$ (R^1 = R^2 = Me, Et, or Me_2CHCH_2; R^1 = Me_2CHCH_2, R^2 = Me_2CH) it was observed that the tendency towards association decreased with increasing size of the substituent.[14] Fragmentation schemes for the compounds studied were proposed. The mass spectra

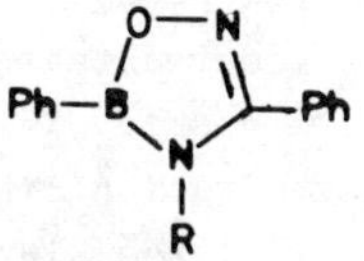

(4) R = H, Me, Et, Pr, or Bu

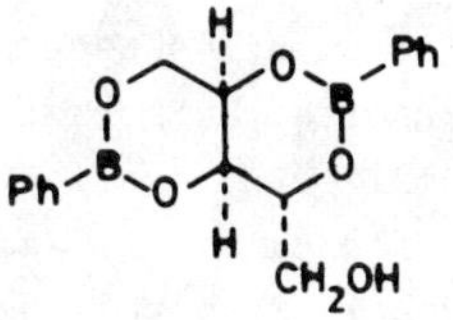

(5)

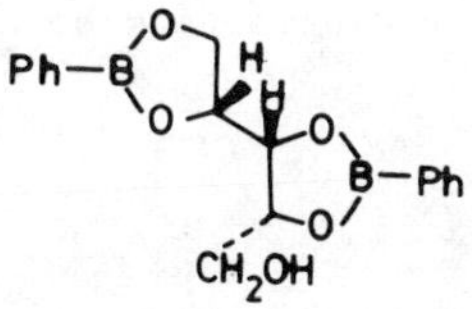

(6)

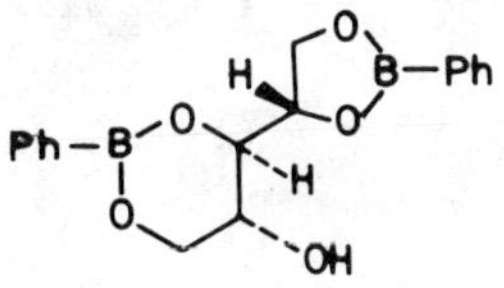

(7)

of a series of dialkylaluminium mercaptides and of trimethylalane, recorded in the presence of oxygen and moisture, have been reported.[15]

Group IV.- Unimolecular decompositions of the tetramethyl derivatives MMe_4 (M = C, Si, Ge, Sn, or Pb) have been studied by means of a neutral-fragment mass spectrometer.[16] Four neutral species were observed, the most abundant being excited $M(Me_4)^*$. The fragmentation pathways for the series of trialkyl hydrides R_3MH (R = Me, Et, Pr, or Bu, M = Si, Ge, or Sn) have been reported.[17] Mass spectra of a series of pentafluorophenyl-alkylmethylsilanes and chlorosilanes have been reported,[18] and p-$RC_6H_4SiMe_2CH_2COMe$ and p-$RC_6H_4SiMe_2OCMe{=}CH_2$ (*e.g.* R = H or Me) have been characterized.[19] In the fragmentation of $Cl_2Si(CH_2CH_2SiClMe_2)_2$ the pseudo-molecular ion $(Cl_2Si{=}SiMe_2)^+$ was observed.[20] In the mass spectra of a series of some silyl and hydroxy silylalkenes, as for example *trans*-$HOCMe_2CH{=}CHSiEt_3$, it was observed that rearrangement involving 1,4-hydroxyl migration from carbon to silicon proceeded *via* hydroxyl transfer to an unsaturated silicon centre accompanied by expulsion of a neutral substituted cyclopropene derivative.[21] The results of a mass-spectral study of 1,2-disilyl derivatives of ethane, ethylene, and acetylene have been reported,[22] and large peaks assigned to species containing the Si=C group have been observed in the spectra of silicon heterocycles of type (8).[23] In a study of some silathietanes (9) by e.i. and photoionization mass spectrometry, the formation of silathione ions $(R_2Si{=}S)^+$ *via* loss of C_2H_4

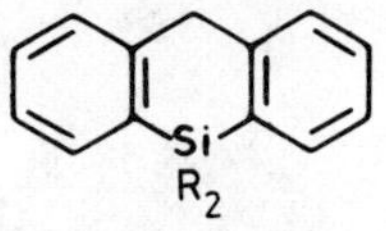

(8) R = Cl, Et, MeO, or F

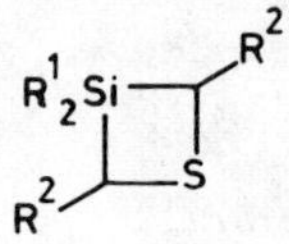

(9) R^1, R^2 = Me, H; Et, Me

from the molecular ions was found to be the main fragmentation process.[24] In addition, silaethylene ions $(R^1{}_2Si{=}CHR^2)^+$ were also observed. The spectra of some phenyl-substituted silacyclobutanes have been reported.[25] The spectra of the heterocycles (10) - (12),[26] (13) - (15),[27] (16),[28] as well as 5β-pregnane-3α,17,20α-triol derivatives of (17)[29] have been published. An examination of the mass spectra of aryl- and vinyl-dichlorosilanes and the heterocycles (18) and (19) suggested that $Cl_2Si{=}$ was eliminated from the molecular ions,[30] and the intermediates $Cl_2Si{=}CH_2$ and $Cl_2Si{=}$ were identified in the thermal degradation of 1,1-dichloro-1-silacyclobutane.[31] Vinyldimethylsilane derivatives of some steroids and cannabinoids have been characterized,[32,33] and full decomposition pathways have been identified in the mass spectra of a series of halovinylsilanes of the types $CH_2{=}CHSiCl_3$, $CCl_2{=}CHSiCl_3$, and $CCl_2{=}CClSiCl_2$.[34] M.i.k.e. data have been determined for $CH_2{=}CBrSiMe_3$.[35] The reaction of the trimethylsilyl cation with alkenyl ethers and alkoxy ketones has been investigated,[36] and the relative stabilities of trimethylsilyl and t-butyl cations have been determined by i.c.r. mass spectrometry.[37] The fragmentation pathways of some allyloxydimethylsilanes, allyloxytrimethylsilanes,[38] and a series of triethylsilyl ethers of enols and their germanium analogues have been reported.[39]

Details of the mass spectra of a number of trimethylsilyl derivatives of organic molecules have been published. These include *cis- and trans*-hexahydrocannabinol, their hydroxy and acid analogues,[40] Δ^1- and Δ^6-tetrahydrocannabinol,[41,42] ginger constituents, *e.g.* zingerone and gingerdione,[43] alkyl porphyrins,[44] C_{21} steroids (pregnane derivatives),[45] testosterone,[46] dihydroxyarenediols,[47] acylglycines,[48] prostaglandins $F_{1\alpha}$, $F_{2\alpha}$, E_1, and E_2,[49] pyridine monoamidoximes,[50] 2-ketoglutarate,[51] and

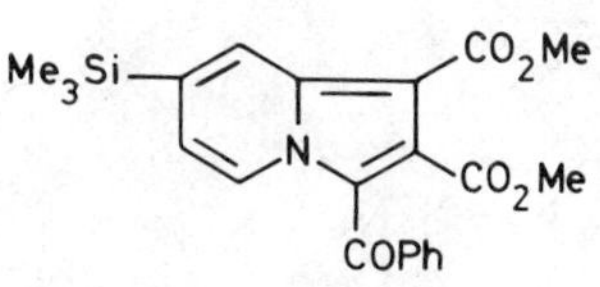

(10)

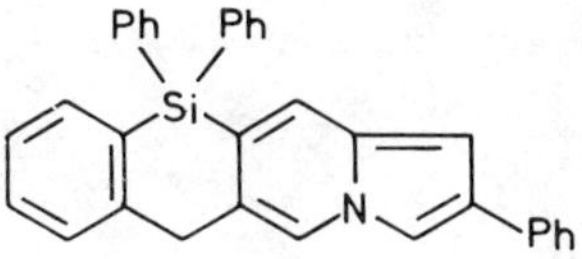

(11)

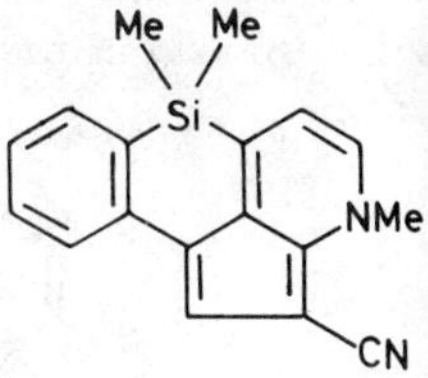

(12)

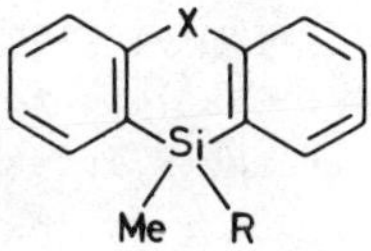

(13) X = S, SO, or SO_2, R = Me or CH_2Cl

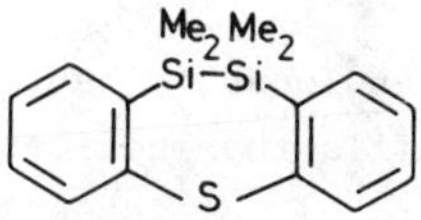

(14)

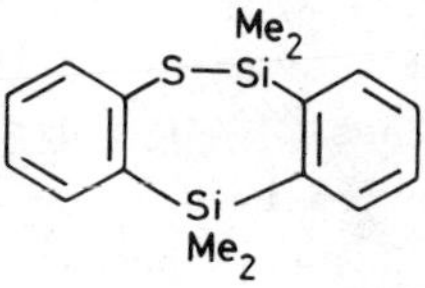

(15)

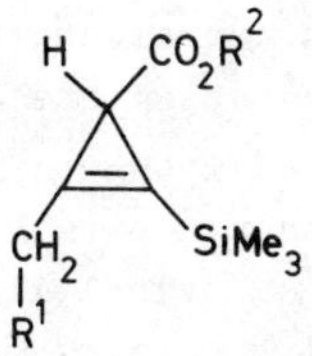

(16) R^1,R^2 = H, Me; Me, Me; H, CD_3; Me, CD_3

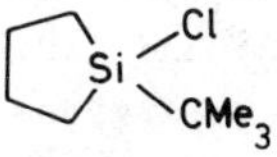

(17)

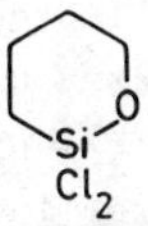

(18)

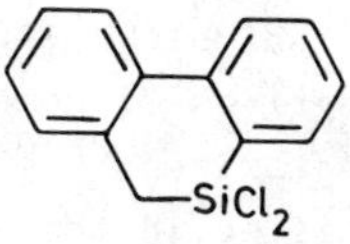

(19)

stereoisomers of 5,7-undecanediol.[52] A review of the application of g.c. and m.s. to the characterization of trimethylsilyl derivatives of plant glycosides has been published.[53] C.i. (isobutane) mass spectrometry has been used to assign the stereochemistry of tricyclononanes and tricyclodecanones (20).[54] A McLafferty rearrangement of the silicon-free alkyl group followed by a McLafferty-type cleavage involving migration of the silyl group to yield a radical ion of the corresponding trimethylsilylenol ether was observed in the fragmentation of $Me_3Si(CH_2)_3$-$CO(CH_2CH_2)_nMe$ (n = 0 - 5).[55] Mass-spectral studies on some dimethyl-t-butylsilyl derivatives of nucleosides,[56] 2'-deoxynucleosides,[57] androstanolones,[58] and organic acids[59] have been reported.

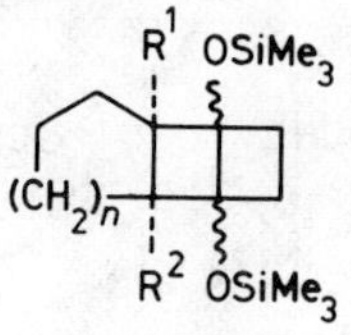

(20) $n = 1$, $R^1, R^2 = H, Me$
$n = 2$, $R^1, R^2 = H, Me, CO_2Et$, or CN

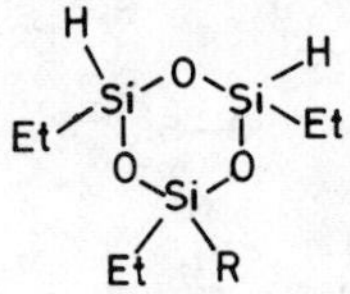

(21) R = H, OH, or OMe

There has been continued interest in the application of mass spectrometry to cyclosiloxanes.[60,61] A fragmentation mechanism for oligoethylhydrocyclosiloxanes, *e.g.* (21), has been proposed,[62] and D-labelling and peak matching were used to support fragmentation mechanisms of aryl-substituted di- and tri-siloxanes.[63] It was observed that fragmentations of cyclosiloxanes $(OSiRMe)_n$ (R = OMe or OEt, n = 3 or 4) were very similar to those for alkoxymethylsilanes.[64] It has been proposed that the large $[M - X]^+$ peak in the spectra of compounds $X_3SiOSiX_3$ (X = Cl or OEt) and of octachlorotrisiloxane was due to formation of four-membered oxonium and chloronium ions. A study of the spectra of vinylsiloxanes (22) - (24) showed that the fragmentation of the cyclosiloxanes was not affected by the number of vinyl groups.[65] Octavinyloctasilsesquioxane,[66] methylethyloctasilsesquioxanes,[67] and 1-subperethylhomoligosilsesquioxanes[68] have been studied. A series of cyclosiloxanes obtained from the pyrolysis (1 s at 980 °C) of dimethylsiloxanes have been identified.[69] Dimethoxymethylsilyl derivatives of hydroxysteroids have been studied by g.c.-m.s.[70] Quadrupole m.s. has been used to determine the chemical species present during radiofrequency discharges in methyltrimethoxysilane.[71] The fragmentation pathways for the heterocycles (25),[72] (26),[73] and (27)[74] have been reported, and (28) was characterized.[75] The fragmentation of silatranes (29) was observed to proceed by two pathways, cleavage of the R–Si bond and rupture of the silatrane skeleton with retention of the R–Si bond.[76] Doubly charged ions were detected in all cases. Transannular N → Si interaction was indicated by a mass-spectral study of compound (30), and this interaction was suggested to be stronger than an analogous N → Ge interaction.[77] It has been suggested that dissociative ionization of $(Me_3Si)_2NOR$ (R = CH_2Ph or alkyl) with cleavage of the O–C bond

(22) $R^1 = R^2 = Me$ or vinyl
R^1 = vinyl, $R^2 = Me$

(23)

(24)

(25) R = F, Cl, MeO, Ph, or Me, X = H
R = Cl, X = $SiCl_3$
R = Me, X = Br

(26) R^1 = H, Me, or Ph
R^2 = H or Me
R^3 = Cl, Me, or EtO

(27) R^1, R^2 = Me, Ph,
R^3 = H, 6-Cl, 6-MeO, or 8-Me,
X = O, S, or NMe

(28) R = Me, Et, Me_2CH, or Ph

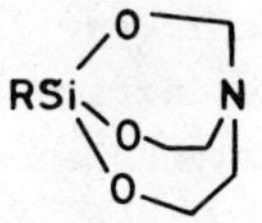

(29) R = Me, Ph, EtO, PhO, 4-ClC_6H_4O, or 4-$O_2NC_6H_4O$

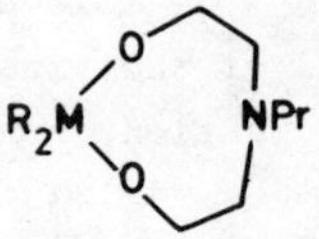

(30) R = Me, M = Si; R = Me or Et, M = Ge

(31)

(32) R = Me_3C or Me_3Si

is anchimerically assisted by migration of the Me_3Si group to the ether oxygen.[78] The loss of both R groups followed by loss of Me_3Si was observed in the spectra of a series of diaminosulphanes, *e.g.* $(Me_3SiNR)_2S$ (R = Bu or Bu^t), and similar observations were made for silanes (31) and (32).[79] Fragmentation pathways for a series of alkoxymethylsilylamines have been reported.[80] A study of the spectrum of the cyclic azasilane (33) (R = *O*-tolyl) suggested that the Si–N bonds are weaker than those in the N–Ph derivatives containing phenyl groups co-planar with the four-membered ring.[81] The spectrum of the thiasilane (34) suggested that the methyl on Si was cleaved off easily and that the S–Si–S moiety had considerable stability.[82] Fragmentation of the tetrathiasilane (35) was observed to proceed *via* successive losses of C_3H_6 and HS.

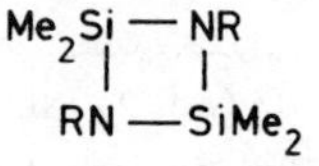

(33) R = Ph or *o*-tolyl

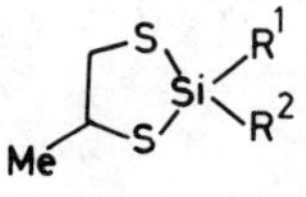

(34) R^1, R^2 = Me, Ph

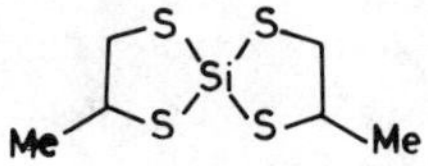

(35)

The mass spectra of a series of trialkyl(2-perfluoroalkyl-1-iodovinyl) derivatives of silicon, germanium, and tin have been reported,[83] compounds Me_3GeO_3SR (R = Me or CF_3) have been characterized,[84] and the spectra of dodecaphenylcyclohexagermane have been reported.[85]

F.i. and f.d. spectra on a series of organotin compounds have been reported.[86] In contrast to e.i., f.i. spectra are dominated by the molecular ion and show little if any fragmentation. The spectra of the tin compounds (36) indicate intramolecular P–Sn co-ordination.[87] The spectra of dibutylstannylenes of

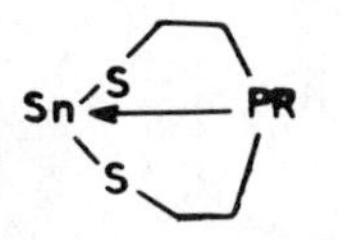

(36) R = Me, Et, Me_3C, or Ph

carbohydrates derived from chiral and other diols have been reported.[88] Studies on a series of alkyl- or aryl-tin(IV) and tin(II) substituted oxinates showed that, with the exception of the tin(II) chelates, the molecular ions were either not observed or were present only in low abundance.[89] The absence of molecular ions or any ions of mass greater than that of the molecular ion or any di- or poly-tin-bearing ions in the spectra of a series of di- and tri-organotin(IV) diphenylphosphate esters demonstrated that there was no association in the gas phase.[90] The mass spectra of some di- and tri-organotin(IV) derivatives of diphenylphosphinyl- and diphenylthiophosphinyl-acetic acid have been reported.[91]

Group V.- The acid-base properties and ion/molecule reactions of a number of organophosphorus compounds in the gas phase have been investigated by i.c.r. spectrometry.[92] The spectra of a series of *p*-substituted triarylphosphines and triarylphosphine oxides have been reported,[93] and bis(di-t-butylphosphinito-*P*) mercury and the products of reaction with halogens and mercuric halides have been characterized.[94] Two main fragmentation pathways were observed for phosphine oxides $Ph_2P(O)CH=CHC_6H_4R$-4 (R = H, NMe_2, Cl, OMe, or CN), namely loss of Ph_2PO and Ph.[95] The mass spectra of $(Ph_2P)_2NR$ (R = Me, Ph, or *p*-C_6H_4Me) have been reported.[96] The fragmentation of the cyclic compound (37) involved the loss of $MeP(O)(OH_2)$ with the formation in the spectra of a peak corresponding to this ionized radical.[97] The mass spectra of a series of 5,5-dimethyl-1,3,2-dioxaphosphorinane 2-oxides and 2-sulphides[98] and some unsymmetrical organobisphosphorus ligands[99] have been discussed in detail. A comparison of the spectra of a series of linear phosphono-acetals, $(EtO)_2P(O)(CH_2)_nCH(OEt)_2$ (n = 1,2, or 3), 2-ethylphosphono-esters, substituted 1,3-dioxolane, and 1,3-dioxane systems (*e.g.* (38)) indicated a direct involvement of the phosphonyl group on the formation of intermediates from the acetal group.[100] The fragmentation pathways of a series of alkyl esters of chloro-substituted phosphoric acids have been reported.[101]

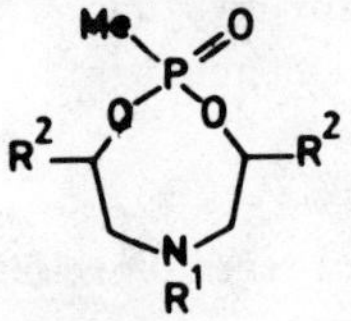

(37) R^1 = Me, Bu, or Ph, R^2 = H
$R^1 = R^2$ = Me

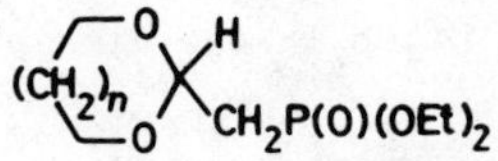

(38) *n* = 0 or 1

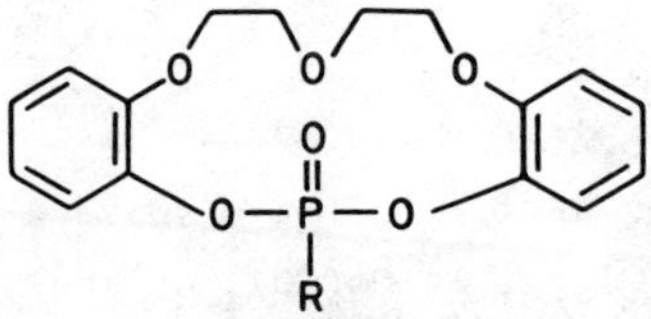

(39) R = Me, Ph, or PhO

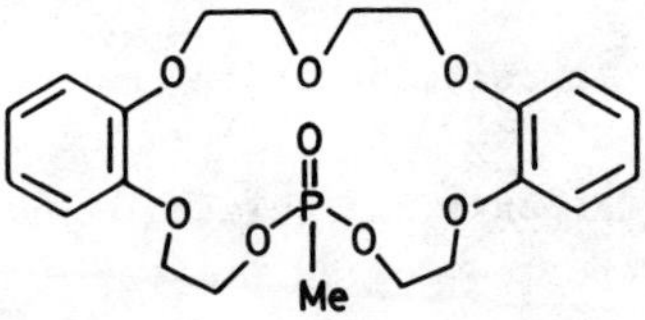

(40)

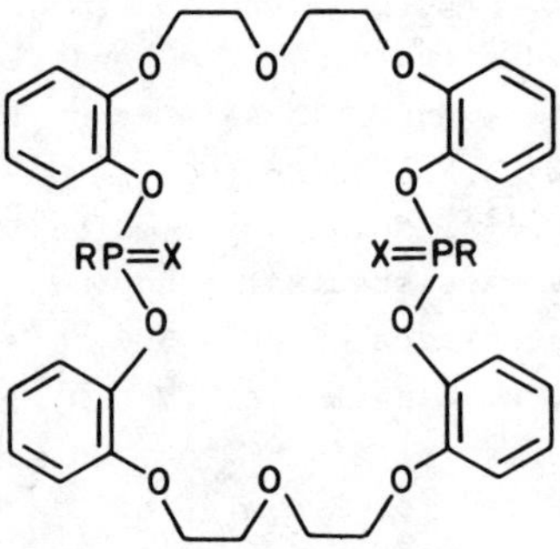

(41) R = Me or PhO, X = O
R = Me, X = S

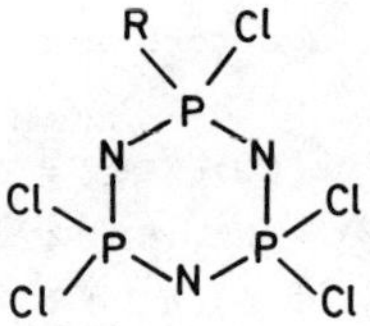

(42) R = Me, Et, Pr^n, Pr^i, Bu^n, or Bu^t

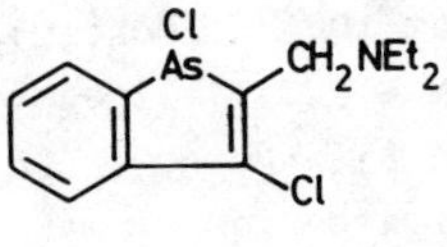

(43)

F.d.m.s. has been used to determine the molecular weights of the cyclic ethers (39) - (41),[102] and it was observed that the compounds had high thermal stability. A series of organophosphorus pesticides has been studied by high-performance liquid chromatograpy-negative c.i. mass spectrometry.[103] In the spectrum of $P_4Me_8N_4$, the ions $(P_4N_4Me_8)^+$ and $(PNH)^+$ were identified, and a fragmentation scheme based on the stability of the cyclic P_nN_n system has been proposed.[104] The spectra of azaphosphines (42) have been reported.[105]

The spectrum of the arsenic compound (43) has been compared with that of PhCCl=CHAsCl_2 and other chloroarsines.[106]

G.c.-m.s. has been used to determine concentrations of Me_4Pb and Et_4Pb in the atmosphere.[107]

Group VI.- The mass spectra of a series of trifluoromethylselenol derivatives CF_3SeR (R = H, Cl, Br, CN, CF_3, or $SeCF_3$) have been reported.[108] In the mass spectra of benzeneselenic acid, $PhSeO_2H$, the ion with the highest mass number appeared in a multiplet and corresponded to the molecular ion of $(PhSe)_2$.[109]

3 Transition-metal Organometallics

Metal Carbonyl and Related Compounds.- F.i./f.d. mass spectra have been reported for $Mn_2(CO)_{10}$, $Cd[Co(CO)_4]_2$, $In[Co(CO)_4]_3$, and the adducts bipy·$Zn[Co(CO)_4]_2$ and bipy·$M[Mn(CO)_5]_2$ (M = Zn or Cd).[110] The spectra were simple, and the small number of easily identified species included the molecular ion in some cases. It was suggested that the failure to observe spectra for $Fe_2(CO)_9$ and $Co_2(CO)_8$ was possibly due to their carbonyl-bridged structure. The negative-ion spectra for $Re(CO)_5X$ (X = Cl, Br, or I), $Mn(CO)_5X$ (X = Br or I), $Re_2(CO)_8X_2$ (X = Cl, Br, or I), $Mn_2(CO)_8Br_2$, and $Rh_2(CO)_4X_2$ (X = Cl, Br, or I) have been reported.[111] No negative molecular ions were observed. The results of a mass-spectral study of $Mn(CO)_5Cl$ at 70 eV and for inlet temperatures of 20 - 500 °C indicated that $Mn_2(CO)_{10}$ is formed by combination of $[Mn(CO)_5]$ radicals at 100 °C; $Mn_2(CO)_8Cl_2$ is formed at 250 °C.[112] Molecular-ion and $[M - (CO)_n]^{+\cdot}$ peaks were observed for f.d.m.s. of $Pt_3(CO)_3(Ph_3P)_4$, $Pt_4(CO)_5(Ph_3P)_4$, and $Pt_5(CO)_6(Et_3P)_4$.[113] The mass spectra of a series of manganese and rhenium 1-3-η-alkyl tetracarbonyl complexes have been reported,[114] the fragmentations of β-alcoholic tricarbonyl (diene)iron(0) complexes have been investigated,[115] and iron tetracarbonyl complexes with thioacetamides as ligands have been characterized.[116] Mass spectrometry has been used to study the gas-phase protonation reactions between η^4-dienetricarbonyliron, η^4-polyenetricarbonyliron complexes, and Brönsted acid reagent ions (*e.g.* H_3^+ and Me_3C^+).[117] **F.d.m.s.** has been a valuable method for the characterization of involatile and thermally labile metal carbonyl 1,4-diazabutadiene complexes.[118] In most cases, only the molecular ions were observed, with the $[M - CO]^+$ fragment ions observed in a few cases. A comparison of the e.i. and c.i.m.s. of some arenetricarbonyl chromium complexes has been reported.[119]

The fragmentation of iron complexes (44) has been suggested to involve three successive decarbonylations followed by the loss of RH; the structures of $[C_5H_5SiRFe]^+$, $[C_5H_5Si]^+$, and $[C_3H_3Si]^+$ were discussed.[120] It was observed that the fragmentation of the chromium complex (45) is followed by a number of H-transfer processes, and these have been studied by D-labelling of the propyl

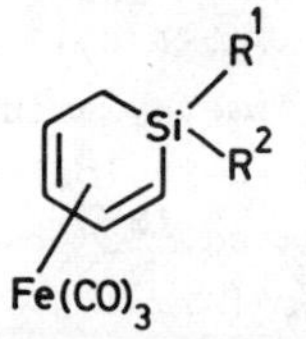

(44) $R^1 = R^2$ = H, Cl, Me, Ph, OH, or EtO
R^1 = Me, R^2 = Cl

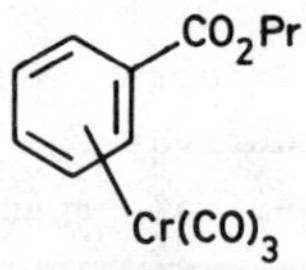

(45)

group in the 1-, 2-, or 3-position.[121] F.d.m.s. has been of considerable value in the characterization of neutral π-hydrocarbon metal complexes.[122] For example, in the case of the relatively volatile species $(\eta^6-C_7H_8)Met(CO)_3$ (Met = Cr, Mo, or W) the only observed peaks were the molecular and $[M - 1]^+$ and $[M + 2]^+$ ions.[123] In contrast, the f.d. spectra of the less volatile neutral complexes $(\eta^6-C_6H_6)RuX_2L$ (X = Cl or Br, L = DMSO or Ph_3P) and $[(\eta^6-C_6H_6)RuX_2]_2$ (X = Cl, Br, or I) were observed to be more complex. The mass spectra of negative ions from η^6-$RPhCr(CO)_3$ (R = H, Cl, F, I, OMe, or NMe_2),[123] η^6-$LCr(CO)_3$ (L = benzene, thiophene, indene, fluorene, or 4-azafluorene),[124] and (η^6-arene)$Cr(CO)_3$ [arene = *p*-$RC_6H_4CO_2Me$ (R = H, NH_2, OH, OMe, Me, Cl, or CO_2Me), benzene, mesitylene, or 2,2-dimethylindan-1,3-dione][125] have been reported. The spectra of some arene chromium tricarbonyl complexes[126] and of $[CF_3EMn(CO)_4]$ and $[CF_3EFe(CO)_3]_2$ (E = S or Se) have been reported,[127] and $Os_6(CO)_{20}$ has been characterized.[128]

Hydrocarbon Metal π-Complexes.- A study has been made of the effect of Et, Pr^n, and Pr^i substituents in the cyclopentadiene ring on the spectra of $(RC_5H_4)_3Sc$.[129] It was observed that the substituents decrease the stability of the compounds to electron ionization. The mass spectra of a series of deuteriated metallocenes $(C_5D_5)_2M$ (M = V, Cr, Mn, Fe, Co, or Ni) have been reported,[130] and the interaction of py/D_2O with substituted ferrocenes showed that the distribution of incorporated deuterium was not statistical.[131] Ion/molecule reactions resulting in the formation of bimetallic ions were suggested to occur as a result of ionization of vapours of ferrocene and its derivatives.[132] The influence of the structure of R (R = Et, Pr, Me_2CH, Bu, or Me_3C) on the fragmentation of $(RC_5H_4)_2Ni$ has been investigated.[133] A series of dimethylaminomethylferrocene adducts of organotin halides[134] and $4[(\eta^5\text{-}C_5H_5)Fe(\eta^5\text{-}C_5H_4)]\text{-}2,3,7,8\text{-}Me_4C_4B_8H_7$[135] and *N*-substituted ferrocenylmethyladenines[136] [*e.g.* (46)] have been characterized. Dimethylaminomethyl and α-(dimethylamino)ethyl derivatives of ferrocene and cymantrene and their iodine methylates and deuterio derivatives have been investigated.[137] F.d.m.s. showed that neutral organometallic complexes [*e.g.* $(\eta^5\text{-}C_5H_5)Mn(CO)_3$] were dominated by molecular ions in contrast to ionic organometallic complexes {*e.g.* $[(\eta^5\text{-}C_6H_7)Fe(CO)_3]^+BF_4^-$}, which were dominated by the monocations.[138] It has been suggested that the triflate complex $(\eta^5\text{-}C_5H_5)Fe(CO)_2(O_3SCF_3)$ has covalent bonding. Protonated (deuteriated) molecular ions and fragment ions were observed in the spectra of a series of π-complexes of transition metals [*e.g.* (47)], which were prepared by simultaneous evaporation with organic molecules.[139] The influence of substituents on the fragmentation of manganese compounds (48)

NHCH$_2$Fc

(46) Fc = ferrocenyl

Mn(CO)$_3$

(47)

R — Mn(CO)$_3$

(48) R = H, halogen, or MeCO$_2$

(49) Q = Mn, R = H, Cl, Br, or Ac, n = 3
Q = Re, R = H, n = 3
Q = V, R = H, n = 4

(50) L = aromatic or heterocyclic compound

has been investigated.[140] It has been observed that ion/molecule reactions occur in the ionization chamber of a mass spectrometer during the combined vaporization of $ArCr(CO)_3$ (Ar = C_6H_6, C_6H_5Cl, or $PhNMe_2$) or complex (49) with aromatic or heterocyclic compounds (L) to give secondary corresponding sandwich ions of type $ArCrL^+$ or (50) containing new metal-ligand bonds.[141] I.c.r. techniques have been used to study the positive-ion chemistry of $(\eta^5\text{-}C_5H_5)Co(CO)_2$.[142] It was observed that the relative ligand-binding energies for $(\eta^5\text{-}C_5H_5)Co^+$ are benzene > MeCN ≈ NH_3 > THF ≈ Me_2CO > propylene ≈ MeCHO ≈ Me_2O ≈ NO > MeOH > CO. The suggestion has been made that the main process of successive decarbonylation of $\eta^5\text{-}RC_5H_4Re(CO)_3$ (R = H, Me, I, CH_2OH, CHO, CO_2H, CN, COMe, CO_2Me, CH_2Cl, or CH═NPh) competes only with decomposition reactions of molecular ions that give stable 18-electron cations.[143] Complexes of the type $(\eta^5\text{-}C_5H_5)W(CO)_2PMe_3C{\equiv}CH$ have been characterized.[144] The spectra of a series of compounds containing bonds between main-Group IV elements and transition metals, as for example $(\eta^5\text{-}C_5H_5)_2Mo(L)SnMe_3$ (L = H, Cl, Br, or I), $(\eta^5\text{-}C_5H_5)_2TaH_2\text{-}SnMe_3$, and $Me_3SnMn(CO)_5$, have been reported.[145] A study has been made of the related series $(MeC_5H_4)MnCO_2L$ and $(MeC_5H_4)MnCo(CS)L$ (L = Ph_3P, Ph_3As, or Ph_3Sb), and ionization energies of all manganese-containing ions with intact ligands were determined.[146] In the fragment ions, the relative bond strengths of the ligands followed the order CO < CS < L. The appearance energies of manganese-containing ions with intact ligands have been determined for $RC_5H_4MnCS(NO)I$ (R = H or Me).[147] In the fragment ions, the manganese-ligand bond strengths lie in the order I > RC_5H_4 > CS >> NO. The first fragmentation pathway of $(C_5H_5)_2VSR$ (R = Me, Me_2CH, Ph, or $PhCH_2$) has been characterized by the formation of $Cp_2V^{+\cdot}$, $Cp_2VH^{+\cdot}$, $Cp_2VS^{+\cdot}$, and $CpSR^{+\cdot}$ ions.[148] The results of a f.i. and m.i.k.e. study of some alkyne-substituted heterometallic cluster complexes have been reported,[149] and a study has been made of the collisional-activation spectra of $C_6H_6M^+$ and $C_7H_7M^+$ (M = V, Cr, or Fe) fragment ions arising from transition-metal π-complexes.[150] The spectra of cyclopentadienyl, cyclo-octatetraenyl, and methyl-cyclo-octatetraenyl complexes of uranium have been studied and appearance energies and dissociation energies determined.[151] The compound $[(\eta^5\text{-}C_5H_5)_3U]_2\cdot C_4H_4N_2$ has been characterized.[152] Thermal-desorption mass spectrometry has been used to study the chemisorption of acetylene and ethylene on the (111) surface of rhodium[153] and cyclopropane on the (110) surface of iridium.[154]

Metal Complexes.- F.d.m.s. of crown-ether complexes with Cu, Ag, Co, Fe, Mg, and Sr indicated that the transition metals exchanged with Co emitter surfaces at medium and high heating currents, but not with Ni emitters.[155] The same technique has been used to characterize perchlorate complexes of Ni^{II} and Cu^{II}.[156] The values of the technique are (i) simple spectra and (ii) ease of analysis of perchlorate complexes. Similarly, f.d.m.s. has been used to characterize involatile monocationic co-ordination complexes.[157] With octahedral complexes, as for example *cis*-$Co(NH_2CH_2CH_2NH_2)_2(NO_2)_2I$, no fragmentation was observed and the base peak corresponded to the molecular ion for the cation. Molecular ions have been observed[158] for the complexes ML_2 ($M = Ni^{II}$, Pd^{II}, or Cu^{II}, $L = HOC_6H_4CH{=}NR$, R = H, Me, Et, or Pr^i). The low-energy electron-attachment reactions of complexes $M(HL)_2$ (M = Ni, Pd, or Pt, HL = dimethylglyoxime) have been discussed in terms of their negative-ion spectra and the results compared with the mass spectrum of the complex (51).[159] The complexes (52),[160] (53),[160] and (54)[161] have been characterized. The stability of molecular ions in the negative-ion mass spectra of nickel complexes (55) was suggested to be related to the nature of the donor atoms, increasing in the order ($X^1 = X^2 = O$) $\leqslant$ ($X^1 = O$, $X^2 = S$) < ($X^1 = X^2 = S$) for most combinations of R^1 and R^2.[162] A study of the mass spectra of a series of europium(III) fluorinated β-diketones indicated two main fragmentation pathways, one involving the breaking of the $C{-}CF_3$ bond and the other the splitting of the alkyl chain.[163] The spectra of a series of Nd and Ni β-diketones have been reported.[164] The 1:1 complexes of $Nd(acac)_3$ with *O*-phenanthroline have been studied by mass spectrometry,[165] and complexes of anions 1,1,1,5,5,5,-hexafluoro-2,4-pentanedione with Co^{III}, Fe^{III}, Cr^{III}, Al^{III}, Zn^{II}, Ni^{II}, Mn^{II}, and Cu^{I} have been examined.[166] In many cases, peaks corresponding to the rearrangement of fluorine to metal were observed. Fluorinated nickel dialkyldithiocarbamates NiS_2CNR (*e.g.* $R = C_3H_2F_5$) have been characterized,[167] and fragmentation pathways of some dialkyldithiophosphate complexes of nickel(II)[168] and some rare-earth complexes of thiopheneformamide trifluoroacetone[169] have been reported.

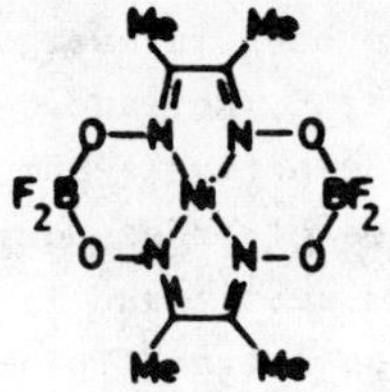

(51)

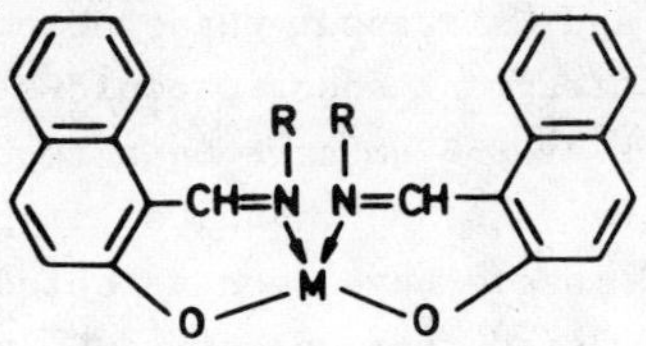

(52) R = OH or Ph, M = Ni or Co

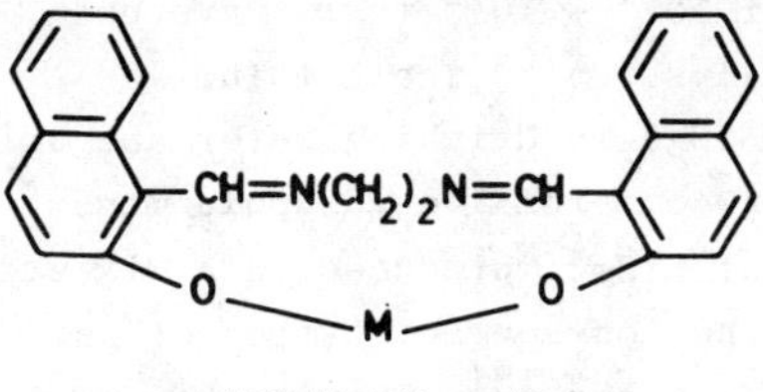

(53) M = Co or Ni

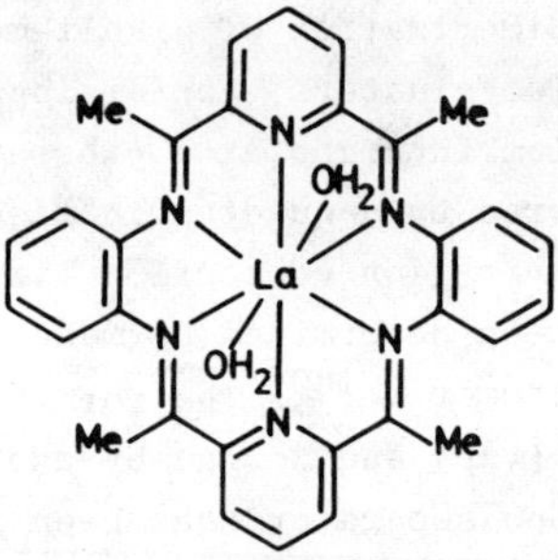

(54)

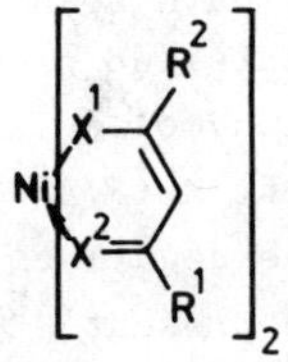

(55) $X^1 = X^2 = O$ or S, R^1, R^2 = Me, CF_3, Me_3C, OEt, or Ph

4 Inorganic Compounds

Group I.- The dissociations of $MetMnF_3$ and $MetMnF_4$ (Met = alkali metal) have been studied.[170] The heats of dissociation by e.i. of $MetAlF_4$ into AlF_3 and MetF (Met = alkali metal) and differences in appearance energies of M^+ ions from dissociative ionization have been determined.[171] The values for the enthalpies of formation of a number of polyatomic negative ions including CsI_2^-, $Cs_2I_2^-$, KF_2^-, and $KAlF_5^-$ have been reported.[172] F.d.m.s. has given spectra of a series of alkali-metal (Li^+, Na^+, K^+) cryptates; the cryptate cations were observed to be significant.[173] The KF/UF_4 and NaF/AlF_3

systems and CsI vapour-phase composition have been investigated,[174] and photofragment spectroscopic studies of Na, K, Rb, and Cs chlorides at 266 nm have been reported.[175] Selective accumulations of Li^+, Na^+, K^+, Rb^+, and Cs^+ at protein sites of freeze-dried embedded muscle have been detected by l.a.m.m.a.,[176] and it has been suggested that selective K^+ adsorption on intracellular proteins is the main cause of selective accumulation of K^+ in living cells.[177] Mass spectrometry has been used to investigate the dimerization of alkali-metal chlorides and bromides.[178] Lithium clusters, containing up to 15 atoms, prepared by evaporating lithium in argon at a pressure of 10^{-4} - 10^{-2} torr, have been measured by a quadrupole mass analyser.[179] The lithium concentration (detection limit ≈ 5pg) in drinking water and urine has been determined by means of a commercial quadrupole mass spectrometer.[180] The formation of $LiHe^+$, produced in a mixture of LiCl (vap) and He (g) by use of the ion-source chamber of an isotope separator, has been investigated.[181] The ion had an appearance energy of 19.3 ± 0.7 eV. Gaseous lithium compounds, as for example LiMg and Li_2Mg_2, have been studied by mass spectrometry[182] and sodium acetate and propionate by f.d.m.s.[183] The results of a study of the ion/molecule equilibrium in the saturated vapour of two-component MetF/AlF_3 systems (Met = Li, Na, K, Rb, or Cs) suggested that the negative ions F^-, AlF_4^-, $Al_2F_7^-$, $MetAl_2F_8^-$, and KF_2 were present.[184] The heat of formation of KF_2^- of -806.3 ± 11.3 kcal mol^{-1} has been determined,[185] and the ion-current ratio method has been used to determine activities of the KCl/NaCl system in both solid and liquid solutions.[186]

The spectra of a series of silver salts RCO_2Ag (R = alkyl, fluoroalkyl, aryl, or fluoroaryl) have been reported.[187] It was observed that the compounds vaporized as dimers and that the most abundant silver-containing ion in most cases was $\{Ag_2(O_2CR)\}^+$. It has been observed that silver-ion affinities for alcohols increase with the number of carbon atoms and with the degree of branching for structurally isomeric alcohols.[188] Using a sequential ionization process, threshold ionization energies of Au^{2+}, Au^{3+}, and Au^{4+} have been determined.[189]

Group II.- Mass spectrometry has been used to study the vaporization of BeO, MgO, and CaO, heated over a tungsten filament, as a function of temperature.[190] The fragmentation pathways of a series of $MetL_2$ complexes (L = 1,1,1-trifluoro-5,5-dimethylhexane-2,4-dione,

Met = Be, Zn, Cu, or Cd) have been reported,[191] and the molecular ion was observed to be the base peak. Cluster ions formed by heating $Mg(OAc)_2$ and $Zn(OAc)_2$ on a tungsten filament have been investigated by mass spectrometry.[192] In-beam electron-ionization mass spectrometry has been used to characterize the bacteriochlorophyll allomer (56).[193] The dissociation energy of magnesium

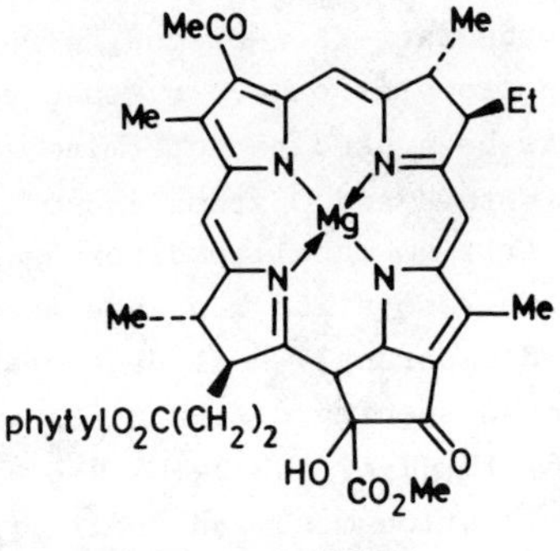

(56)

hydroxide (73.5 ± 5 kcal mol^{-1}) has been measured in a high-temperature mass spectrometer,[194] and the results of a study of the vaporization of magnesium oxide from magnesium aluminate spinel have been reported.[195] The use of mass spectrometry for the determination of calcium isotopes in blood plasma has been evaluated.[196] Cationization with Ca^{2+} ions of polyethylene glycols prior to f.d.m.s. resulted in the formation of abundant $(F + Ca)^{2+}$ ions (F = fragment ion) in approximate proportion to the species present.[197] On treating oligosaccharides with Ba^{2+}, doubly charged ions, including $(M + Ba)^{2+}$, were observed. The axial and radial distributions of trace elements (25 observed) in commercial calcium fluoride single crystals have been determined by spark-source mass spectrometry.[198] The heats of formation of $BaMoO_4$, Ba_2MoO_4, Ba_2MoO_5, and $Ba_2Mo_2O_8$[199] and bond energies in Ba_2O_2 and Ba_2O[200] have been calculated. The main fragmentation pathway, involving the elimination of $Zn(OAc)_2$ or Ac_2O, has been proposed for π_4-oxohexa-μ-acetatotetrazinc.[201] The mass spectrum of mercury(II)dithiocarbamate has been reported[202] as well as the results of a study concerned with the stability of the gaseous complex $MetAl_2Cl_8$ (Met = Be, Zn, or Cd).[203]

Group III.- The dissociation energies of the gaseous hydrides AlH,

GaH, and InH (67.8 ± 2, 62.7 ± 2, and 53.7 ± 2 kcal mol^{-1}, respectively) have been determined.[204] The results of a mass-spectral study of the thermal decomposition of $Met(ReO_4)_3$ (Met = Al, Ga, or In) in a vacuum suggested that the relative thermal stability decreased in the order Al > Ga > In.[205] A value of 8.84 ± 0.10 eV has been obtained for the ionization energy of the BF_2 radical.[206] The reactions of BCl_3 with H, O, and OH radicals have been studied by the fast-flow-discharge reactor technique coupled to a mass spectrometer or e.p.r. spectrometer;[207] the same technique has been used to determine rate constants for the reaction of BBr_3 with H and O atoms and OH radicals.[208] The specific activity of tritium-labelled diborane has been determined,[209] and high-resolution m.s. has been used to characterize di(*nido*-decaboranyl)-*nido*-decaborane, $B_{10}H_{13}B_{10}H_{12}B_{10}H_{13}$.[210] Anisotopic mass spectra of B_4H_{10}, B_5H_9, B_5H_{11}, $B_{10}H_{14}$, $2\text{-}ClB_{10}H_{13}$, $6\text{-}ClB_{10}H_{18}$, and $2\text{-}IB_{10}H_{13}$ have been determined by high-resolution m.s. and their fragmentation pathways elucidated.[211] The $(M - OCHCF_2)^+$ ion was observed to be the base peak in the spectra of *C*-(trifluoromethyl)-substituted boratranes (57).[212] The mass spectra of tris{bis(methanethio)boryl}amine and the heterocycles (58) - (60) have been reported; fragmentation leading to preferential retention of B–N bonding was observed.[213] The molecular ion of the azathiaboron compound (58) was observed, with the loss of $CH_2{=}CH_2$ in each of three successive fragmentations, to form the ion (61). G.c.-m.s. properties of borate esters of

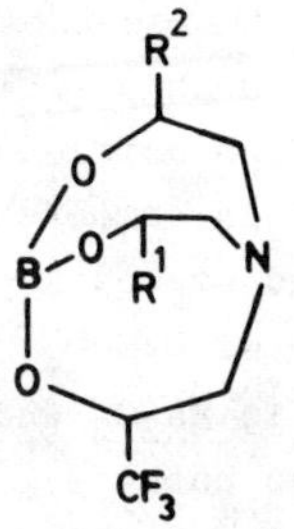

(57) $R^1 = H, R^2 = CF_3$
$R^1 = R^2 = CF_3$

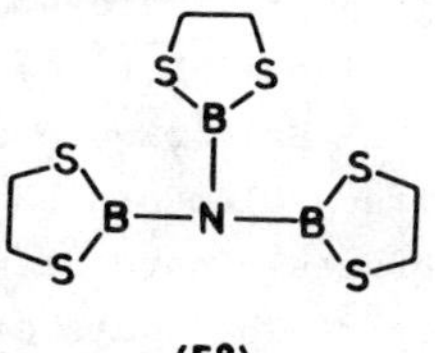

(58)

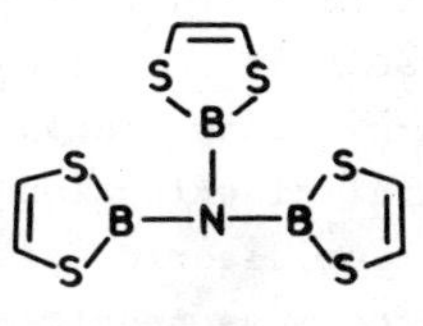

(59)

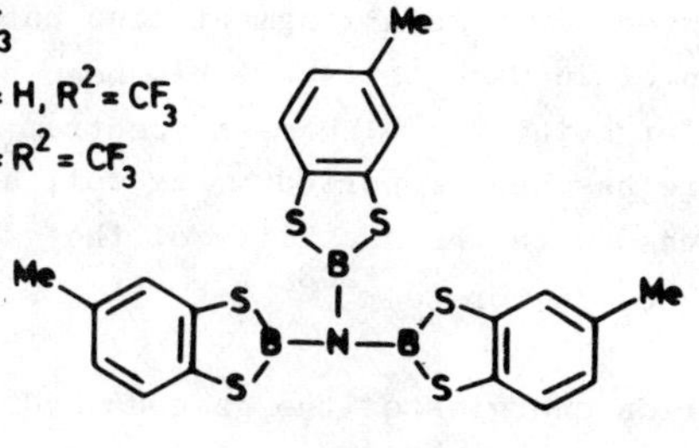

(60)

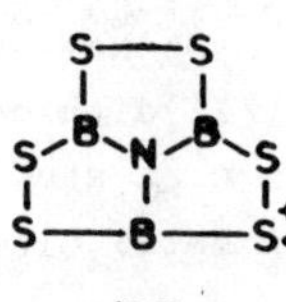

(61)

24*R*,25-dihydroxycholecalaferol[214] and the results of a f.d./f.i. investigation of borate ester formation, from alcohols adsorbed on boric acid surfaces, have been reported.[215] A mechanism has been proposed for the e.i.-induced fragmentation of a series of trialkylamine-borane complexes R_3NBX_3 (*e.g.* R = Et or Bu, X = Cl, Br, or H), the studies being at 100 °C and 250 °C.[216] Deuterio-tetrakis(1-imidazolyl)borate has been identified and characterized by mass spectrometry,[217] and the fragmentation pathways of a series of aromatic azo compounds containing boron have been reported.[218]

The MetF/AlF_3 system (Met = alkali metal) has been studied by mass spectrometry over the temperature range 780 - 930 K, and the AlF_4^- ion was observed.[219] The systems $Al_2Cl_6/CoCl_2$[220] and Al_2Cl_6/VCl_3 and Al_2Cl_6/VCl_2[221] have been investigated, and a number of oxide (and hydroxide)/chloride complexes of aluminium, formed by the interaction of gaseous Al_2Cl_6 and the quartz of a glass wall, have been identified by m.s.[222] It has been observed that AlO_2 is a minor species in the gaseous species evaporating from a high-purity Al_2O_3 crystal at 2200 - 2318 K.[223] Ion-probe mass analysis has been used to study the diffusion of ^{29}Mg and ^{6}Li in aluminium at 431 - 540 °C and 455 - 522 °C.[224] The results of a study of the dissociative ionization process, by e.i., on TlI, TlBr, InBr, and InI have been reported,[225] and a series of indium porphyrin complexes has been characterized.[226] The TlCl/TlBr system has been investigated,[227,228] and the spectra of the vapours, over the temperature range 540 - 600 K, were observed to contain Tl^+, $TlCl^+$, $TlBr^+$, Tl_2^+, Tl_2Cl^+, and Tl_2Br^+ ions as well as trace amounts of $Tl_2Cl_2^+$, $Tl_2Br_2^+$, and Tl_2ClBr^+. A study of the fragmentation of a series of monovalent thallium carboxylates TlO_2CR suggested that the formation of $TlCO_2^+$ was dependent on the nature of R and on the nature and position of substituents on the aromatic ring.[229]

Group IV.- The fragmentation pathways of Group IV metal tetranitrates have been reported, and it was observed that the heaviest fragment ions were formed by the loss of one NO_3 group from the molecular ion.[230] Systematic measurements of ion kinetic energies using an electrostatic-energy analyser have been used to investigate the dissociative ionizations of GeS, SnS, and PbS.[231]

Oxygen exchange between molecular oxygen and carbon monoxide has been studied,[232] and the results of a g.c.-m.s. study of $C^{18}O$ in air have been reported.[233] The reactions of carbon dioxide **cluster** ions have been investigated.[234] The ions were observed to undergo the reaction $(CO_2)_n^+ \rightarrow (CO_2)_{n-1}^+ + CO_2$ with only one CO_2 loss at a time. The average kinetic energy released during the unimolecular decomposition of CO_2 positive-ion clusters, in the field-free region of a mass spectrometer, has been measured.[235] Molecular-beam photoionization mass spectrometry has been used to study CS_2, CS_2 dimers, and CS_2 clusters.[236] The half-life of ^{32}Si, 108 ± 18 years,[237] 101 ± 18 years,[238] has been determined by tandem-accelerator mass spectrometry. Spark-source mass spectrometry has been used in testing and controlling the purification steps in the manufacture of solar-grade silicon.[239] The results of an effusion mass-spectral study of the reactions of $SiBr_4$ (g) with Si (g) over the temperature range 1054 - 1603 K have been reported, and the standard heats of formation for SiBr (g), $SiBr_2$ (g), and $SiBr_3$ (g) of 433 ± 10.5, -43.9 ± 4, and -200.8 ± 2.1 kJ mol^{-1}, respectively, have been determined.[240] M.s. isotope-dilution analysis has been used to detect traces of chloride in silicon rocks,[241] and the high-temperature gaseous equilibrium of Si_2O_2 (g) and LiSiO (g), over Li_2O-Si (g) mixtures, has been determined by mass spectrometry.[242] The heats and free energies of formation, at 298 K, of $SiAl_2O_2N_2$ and $SiAl_4O_2N_4$ have been determined,[243] and quadrupole mass spectrometry has been used to determine the hydrogen content in plasma-activated silicon nitride films.[244] The spectra of germanium compounds (63) and (64)[245] have been reported, and the ion (62) was observed.

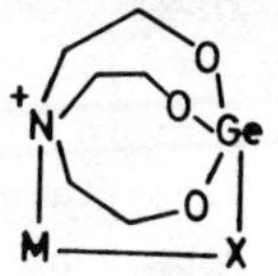

(62) M = Ge or Si, X = O or CH_2 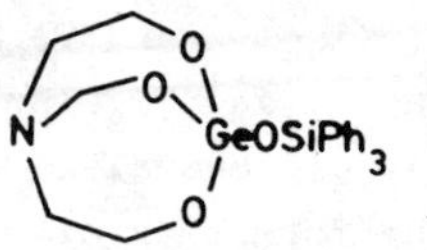 (63) 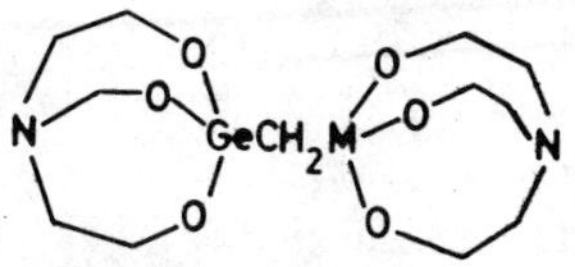 (64) M = Si or Ge

The ionization energy for SnI_2 (9.3 ± 0.5 eV) and appearance energy of SnI^+ (10.1 ± 0.5 eV) have been determined.[246] The thermal decomposition of Sn_2S_3 (g)[247] and tetrakis(*NN*-diethyl-dithiocarbamato)tin(IV)[248] has been studied by mass spectrometry, and the fragmentation pathways of a series of tin(II) bis(dithio-phosphate) esters have been proposed.[249]

The results of a mass-spectral study concerning the properties of lead clusters, generated in a helium atmosphere, have been reported.[250]

Group V.- The thermal-desorption mass spectra of a series of ammonium and phosphonium salts indicated that the formation of positive ions, *e.g.* Me_4N^+, resulted from a thermal process.[251] The results of an e.i. fragmentation study of thirty-four cyanamides, cyanophosphines, cyanoarsines, and their vinylogs have been published,[252] and $Met(SCMe_3)_3$ (Met = As, Sb, or Bi) have been characterized.[253] The kinetic-energy distribution of N^+ ions, produced by dissociative ionization of N_2, has been measured by high-resolution measurement of drift times in a time-of-flight mass spectrometer.[254] A drift-chamber mass spectrometer has been used to investigate the reactions of N_2^+ and H_2O^+ ions with N_2O, CO_2, SO_2, C_2H_2, NH_3, and NO.[255] It was observed that the first two reactions gave displacement products whereas the remainder proceeded mainly by charge transfer. Total and partial cross-sections have been measured for e.i. ionization, at 25 - 100 eV, for ammonia, and the ion currents of NH_3^+, NH_2^+, NH^+, N^+, H_2^+, H^+, and NH_3^{2+} have been recorded.[256] Ammonia cluster ions, $(NH_3)_{n-2}NH_4^+$ (n = 2 - 8), have been detected by multi-photon ionization mass spectrometry.[257] A field mass-spectrometric study of a catalyst surface used for the catalysis of ammonia production, by FeO or Fe_3O_4-Al_2O_3 with H_2-N_2 mixtures at 100 - 500 °C, showed that the catalytic surfaces contained adsorbed NH_3, NH_4^+, and H_2O molecules.[258] The results of an e.i.m.s. study of a series of tetra-alkylammonium halide salts have been reported,[259] and it was observed that the spectra of the tetra-alkylammonium ions are dominated by ions resulting from the homolytic cleavage of the N–C bond and loss of one substituent accompanied by hydrogen transfer.[260] High-pressure mass spectrometry has been used to study the formation of NO_3-$(HNO_3)_n$ clusters (n < 6) in mixtures of HNO_3 and H_2 or HNO_3 and CH_4,[261] and a charge-exchange mass spectrometer has been used to detect the first five vibrational levels in the electronic ground-state NO^+ ion.[262]

The products and intermediates of the reaction of PCl_3 with H, O, and OH radicals have been analysed,[263] and $(Ph_2N)_2PCl$ and $(Ph_2N)_2PNEt_2$ have been characterized.[264] The mass spectra of a series of bromofluorocyclotriphosphazenes ($N_3P_3Br_nF_{6-n}$) (n = 2 - 5),[265] phenoxycyclotriphosphazenes $N_3P_3Cl_{6-n}OPh_n$ (n = 1-6),[266] and the anti-tumour cyclophosphazene (65)[267] have been reported, and mass-spectral data for compound (66) have been tabulated.[268] In addition, the thermal-decomposition products of some poly(organophosphazenes)[269] have been characterized. The results of an i.c.r. study of the gas-phase ion/molecule reactions of trimethyl phosphite, trimethyl phosphate, triethyl phosphate, and trimethylphosphorothionate have been reported, and it has been

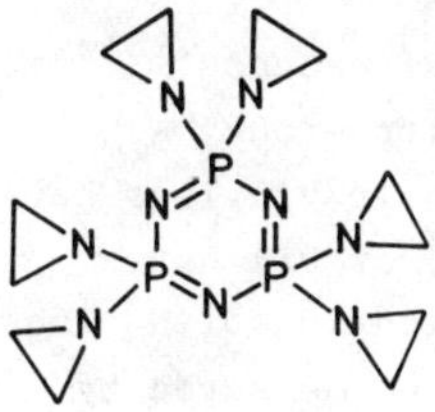

(65)

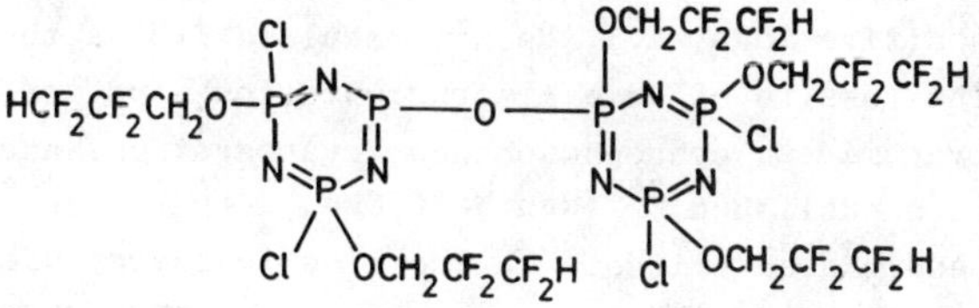

(66)

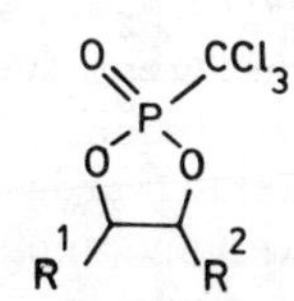

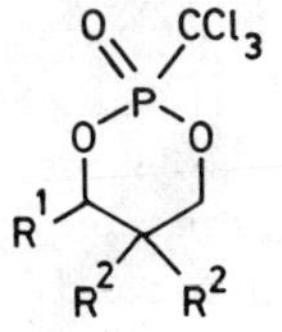

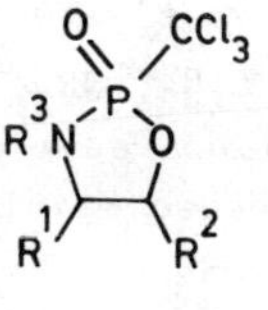

(67) R^1 = H, R^2 = Me, $ClCH_2$, or Ph
R^1 = R^2 = Me

(68) R^1 = Me, R^2 = H
R^1 = H, R^2 = Et

(69) R^1 = R^2 = H, R^3 = Me, Et, Ph, or $PhCH_2$
R^1 = Me, R^2 = H, R^3 = H
R^1 = H, R^2 = Me, R^3 = H

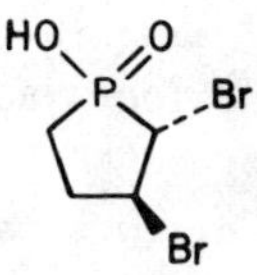

(70)

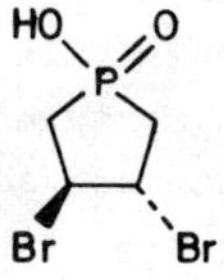

(71)

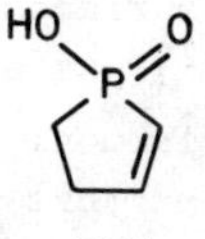

(72)

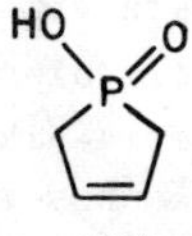

(73)

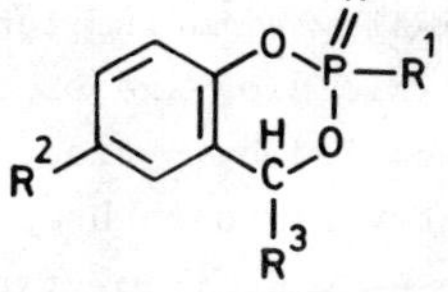

(74) R^1 = MeO, PhO, OH, 4-(MeO)C_6H_4O, SMe, Me, or Ph,
R^2 = H or Cl, R^3 = H or Me,
X = O or S

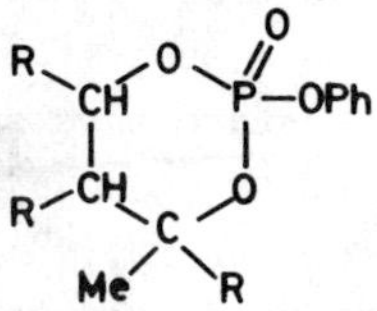

(75) R = H or Me

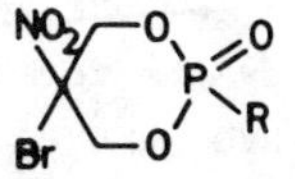

(76) R = Cl, p-$O_2NC_6H_4O$, MeO, or aziridinyl

suggested that the first-ionization energy of $(MeO)_3P$ should be assigned to the P lone pair.[270] Negative-ion mass spectra have been reported for tris(dichloropropyl)phosphate[271] and some organophosphorus pesticides.[272] The fragmentation pathways of the heterocycles (67) - (69),[273] (70) - (73),[274] (74) and (75),[275] (76),[276] (77),[277] (78),[278] and (79)[279] have been reported. In the spectra of a series of dioxaphosphorepines (80), it was observed that P–R bond cleavage was not a common process.[280] A Finnigan 4000 mass spectrometer has been used to measure the $^{18}O/^{16}O$ ratio in the isomeric compounds (81).[281] F.d.m.s. has been used to detect the heterocycle (82), taken orally, in samples of urine, serum and cerebrospinal fluid of multiple-sclerosis patients,[282,283] and the spectra of a series of acetoinyl phosphomono-, di-, and tri-esters have been reported.[284]

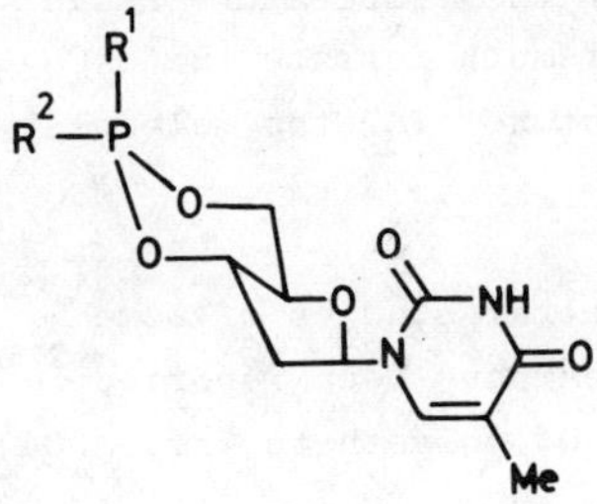

(77) $R^1 = S{=}, R^2 = NHPh$
$R^1 = NHPh, R^2 = S{=}$

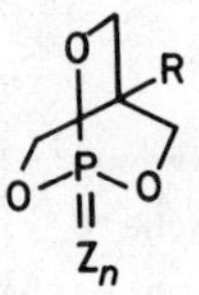

(78) R = Me, Et, or Me_2CH, n = 0 or 1
Z = O, S, or Se

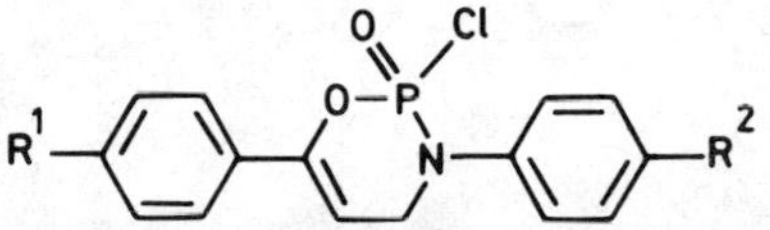

(79) R^1 = H, MeO, or NO_2,
R^2 = H, Me, MeO, or Cl

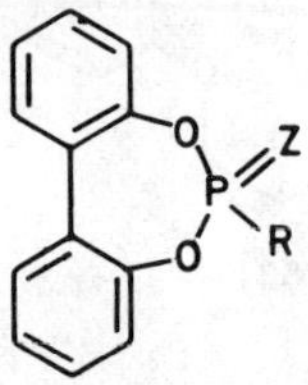

(80) Z = S, R = OH, OMe, SMe, Cl, or SH
Z = Se, R = Cl; Z = O, R = OH, OMe, SMe, or Cl
Z = lone pair, R = Cl

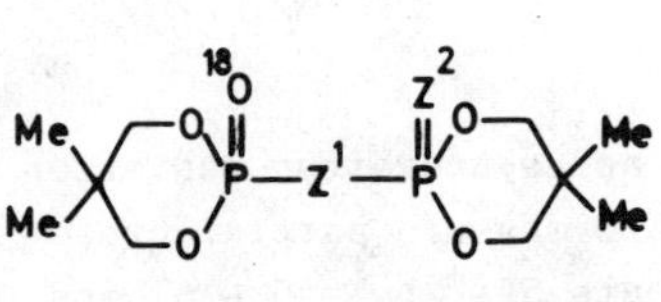

(81) $Z^1 = S, Z^2 = {}^{18}O; Z^1 = {}^{18}O, Z^2 = S$

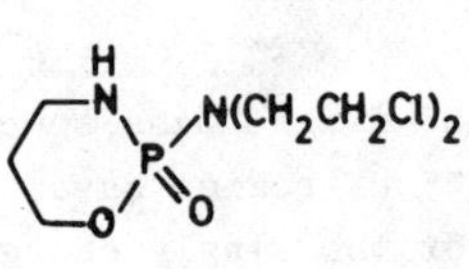

(82)

The mass spectra of $As_2P_2S_7$ and $(As_2P_2S_8)_n$ have been reported,[285] and i.c.r. spectroscopy has been used to study the nucleophilic addition-elimination reactions of weak bases with the trifluoroarsonium ion in the gas phase.[286]

The mass spectra of $Sb(OAc)_3$, $Sb(SAc)_3$,[287] $Sb(OEt)_3$, $ClSb(OEt)_2$, and $Cl_2Sb(OEt)$ have been reported,[288] and a series of antimony(III) complexes with tridentate Schiff-base ligands have been characterized.[289] The use of time-of-flight mass spectrometry has been of value in studying microclusters of metal atoms of antimony and bismuth.[290,291,292] The volatility of Bi_2O_3-GeO_2 glasses has been studied over the temperature range 1000 - 1520 K,[293] and the spectrum of bismuth formate, $Bi(O_2CH)_3$, contained peaks for the $BiO(O_2CH)$ ion but not for the molecular ion.[294]

Group VI.- The results of f.d. mass-spectral studies of some Group I crown-ether (*e.g.* (83)) complexes have been reported,[295,296] and fragmentation pathways for a number of polyethers (*e.g.* (84) and (85)) have been proposed.[297]

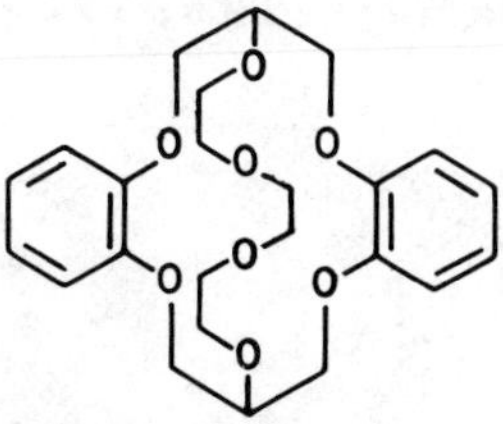

(83)

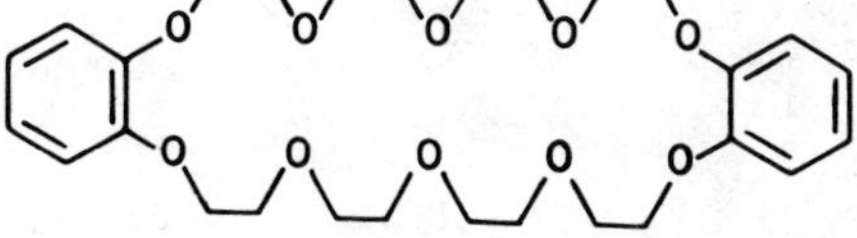

(84)

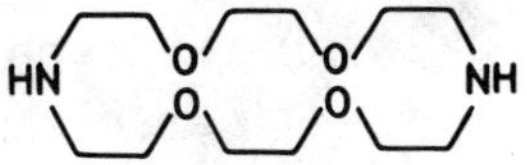

HN NH

(85)

Time-of-flight mass spectrometry and synchrotron radiation in the 75 - 125 eV region have been used to measure partial cross-sections for the singly charged fragments SF_m^+ (m = 0 - 5) and F^+ and the doubly charged fragments SF_n^{2+} (n = 0 - 4) in SF_6.[298] Mass spectrometry has been used to investigate plasma etching of silicon by sulphur hexafluoride.[299] The results of a photo-ionization mass-spectral study of sulphur dichloride, disulphur di-chloride, and disulphur dibromide have been reported, and the heats of formation, bond energies, and ionization energies for a number of molecular fragments have been determined.[300] The reactions and energy distribution in dissociative electron-capture processes in the sulphuryl halides SO_2F_2, SO_2Cl_2, and SO_2ClF have been investigated.[301] Fourteen volatile sulphur compounds, mainly mono-, di-, tri-, and tetra-sulphides and esters of MeSH, have been identified by g.c.-m.s. of six North Sea fish oils.[302] The interaction of sulphur dioxide gas and surfaces of Na_2O-CaO-SiO_2 glasses has been investigated.[303] In an e.i.m.s. study of a series of thiazoles and selenazoles, no major differences were observed between the sulphur- and selenium-containing compounds.[304] Stable-isotope-dilution g.c.-m.s. has been used to determine selenium quantitatively, in p.p.b., in biological materials.[305]

Gaseous products, for example $TeOCl_2$, from the thermal decomposition of tellurium oxychloride ($Te_6O_{11}Cl_{12}$) have been studied by mass spectrometry.[306]

Group VII.- The thermal decompositions of compounds of graphite with BrF_3 and ClF_3 have been studied,[307] and collision-induced dissociation m.i.k.e.s. have been recorded for the cubane (86).[308]

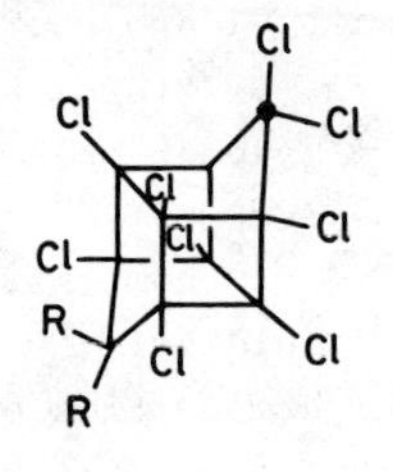

(86) R_2 = O; R = Cl

Traces of Br^- and Cl^- have been determined in environmental samples[309] and inorganic matrixes[310] by mass-spectral isotope-dilution techniques. The results of an e.i.m.s. study of ClO_2 and ClO_2F have been reported,[311] and the perfluoro-n-propylbromine(V) tetrafluoride has been characterized.[312]

Group VIII.- Trace impurities (detection limit 10^{-7} - 10^{-4} vol. %) of He, Ar, Xe, and Kr in high-purity inert gases have been determined by mass spectrometry.[313] The results of an e.i.m.s. study of Ar_2, Kr_2, Xe_2, ArKr, HeXe, NeXe, and ArXe have been reported, and the existence of the doubly charged dimer molecules $NeXe^{2+}$ and $ArXe^{2+}$ has been confirmed.[314] An injected-ion drift-tube mass spectrometer has been used to investigate the mobilities of Ne^+ and Ar^+ ions in helium gas at 82 K.[315] Microclusters, Xe_n, formed in the gas phase, have been studied by mass spectrometry.[316] The observed values for *n* (13, 55, and 147) coincide with the numbers of hard spheres needed for complete-shell icosahedra. The ionization energies for Xe^{2+} (33.11 ± 0.04 eV) and Xe^{3+} (64.35 ± 0.10 eV) have been determined.[317] The existence of KrF^- ion has been confirmed by studying the mass spectrum of KrF_2 in the KRF region.[318]

5 Knudsen-cell Mass Spectrometry

There have been several publications that report the application of Knudsen-cell mass-spectrometric techniques.

Reports on the following studies have appeared: a determination of the partial pressures of Li (g), LiO (g), Li_2O (g), Li_3O (g), Li_2O_2 (g), and O_2 (g) over Li_2O (c) over the temperature range 1225 - 1507 K;[319] partial pressures of Li (g), LiO (g), Li_2O (g), and Li_2SiO_3 (g) over Li_2SiO_3 (c);[320] composition of the gas phase over lithium alloys with Bi and Pb;[321] KF + ZrF_4 and C_3F + ZrF_4 systems in which $MetZrF_5$, $MetZr_2F_9$, and Met_2ZrF_6 were observed (Met = Cs or K);[322] vapour-phase composition over solid and fused caesium antimonide (Cs_3Sb);[323] studies on caesium aluminosilicate compounds;[324,325] composition of the vapour in equilibrium with solid and liquid AgI;[326] MgHg and $MgHg_2$ amalgams;[327] CaHg;[328] vapour composition of CdO;[329] activities of CaO and Al_2O_3 in lime-alumina melts;[330] dissociation energies of gaseous CuSi and AgSi;[331] thermodynamic activities and excess Gibbs energies for the liquid CuSi system;[332] $CuReO_4$-CuI and $CuReO_4$-CuBr systems;[333] vapour-phase composition above KF-AlF_3 system in which AlF_4^-, $Al_2F_7^-$, KF_2^-, $K_2F_3^-$, $KAlF_5^-$, and $KAl_5F_8^-$ were observed;[334] activities of indium and bismuth in Bi-In melts;[335] determination of the dissociation energies of TlSb (g) and $TlSb_2$ (g);[336] thermal dissociation of Tl_2S;[337] vapour-phase composition over GeO_2-GeO with GeO_2, Ge_2O_2, and Ge_3O_3 being observed;[338] dissociation energy of PbSb (g);[339] dissociation energies of $CrPO_2$ (g) and CrO (g);[340] heats of formation of FeF_3^- and FeF_4^-;[341] dissociation energy of FeOH;[342] vapour-phase compositions over $MoSF_3$-$(MoSF_3)_2$ and $(MoSF_3)_3$ were observed;[343] heats of formation of $MoSF_3$ and $MoSF_4$;344 stabilities of gaseous carbides of V, Nb, and Mo;[345] stabilities of Pt_2 and PtY;[346] heats of formation of gaseous and solid WSF_4;[347] identification and atomization energies of ScC_2, ScC_3, ScC_4, and ScC_6;[348] LaC_7 and LaC_8;[349] atomization energies and heats of formation of LaC_2, LaC_3, LaC_4, LaC_5, and LaC_6;[350] heat of formation of LaF_4^-;[351] heats of atomization and formation of YC_2, YC_3, YC_4, YC_5, and YC_6;[352] heat of formation of UF_5^- and electron affinity of UF_5;[353] heats of formation of UF_5^-, UF_6^-, and $UF_2F_9^-$ and electron affinities of UF_5^- and UF_6^-;[354] heat of formation of UR_3;[355] observation of ThC, ThC_2, ThC_3, ThC_4, ThC_5, and ThC_6 in the vapour phase above a Th-U-Rh-graphite system.[356]

6 Secondary-ion Mass Spectrometry

A number of papers have been published over the last two years concerning the application of secondary-ion mass spectrometry. It is felt that, rather than refer to s.i.m.s. throughout the review, it is of more value to have a separate section.

A bibliography of s.i.m.s. 1976 - 1980 has been published,[357] and s.i.m.s. has been used to investigate the spacial and depth distributions of halides and dopants, *e.g.* Pb, In, and Ir, in silver halides[358] and the stability of alkali iodide clusters.[359-361]

The magnesium-isotopic abundance in Allende carbonaceous chondrite has been determined,[362,363] and trace elements have been detected and localized in areas of human eyes by s.i.m.s.[364] For example, calcium was localized in the melanin-containing cell associated with strontium and barium, indicating the affinity of melanin for metal ions.

Impurities, at levels of 10^{14} atom cm^{-3}, of boron-doped silicon have been determined,[365] boron and lithium impurities in a range of nuclear materials,[366] surface impurities of aluminium sheets,[367] and Mg, Si, Mn, Fe, and Cu impurities, in the range 3 - 6700 $\mu g\ g^{-1}$, have been detected in aluminium.[368] S.i.m.s. has been of value in analysis of AlAs-GaAs solid solutions,[369] GaAs,[370,371] InSb,[371] dopant concentrations in GaAs,[372,373] and diffusion of gallium in silicon.[374] Detection of C and O, limit 10^{17} atom cm^{-3}, has been measured in semiconductor materials such as Si and GaAs,[375] adsorption of oxygen on silicon surfaces,[376] impurities in amorphous silicon,[377] analysis of silanes covalently attached to glassy-carbon or tin oxide substrates,[378] analysis of silicate glasses,[379] fluorine adsorption on silicon,[380] determination of lead in oils,[381] s.i.m.s. of N_2O, NO, N_2O_3, and N_2O_4,[382] and β-diketone complexes of transition metals, *e.g.* $Met(acac)_3$ (Met = **Fe, Cr, Mn,** or Co),[383] have been reported. The interactions of ethylene and deuterium on nickel,[384] ethylene and acetylene with oxygen on nickel,[385] ethylene and acetylene with hydrogen on nickel,[386] hydrogen and water with nickel oxide surfaces,[387] and oxygen with nickel surfaces[388] have been investigated by s.i.m.s.

References

1 K.H. Thiele, V. Dimitrov, J. Thieleman, W. Brueser, and A. Zschunke, *Z. Anorg. Allg. Chem.*, 1981, 483, 145.
2 K. Ujszaszy, J. Tamas, N.D. Kagramanov, A.K. Maltsev, and O. M. Nefedov, *J. Anal. Appl. Pyrolysis*, 1980, 2, 231.
3 A. Touabet, K. Abdeddaim, and M.H. Guermouche, *J. High Resolut. Chromatogr. Chromatogr. Commun.*, 1981, 4, 525.
4 F. Davidson and J.W. Wilson, *Org. Mass Spectrom.*, 1981, 16, 467.
5 A. Meller and G. Beer, *Z. Anorg. Allg. Chem.*, 1980, 460, 169.
6 A.S. Bleshinskaya, B.M. Zolotarev, O.S. Chizhov, V.A. Dorokhov, O.G. Boldyreva, and B.M. Mikhailov, *Izv. Akad. Nauk. SSSR, Ser. Khim.*, 1980, 1566.
7 M.E. Rose, *Org. Mass Spectrom.*, 1981, 16, 323.
8 W.C. Kossa, *Chem. Deriv. Anal. Chem.*, 1981, 1, 99.
9 S.G. Batrakov, B.V. Rozynov, and A.N. Ushakov, *Khim. Prir. Soedin.*, 1981, 283.
10 S. Singhawangcha, C.F. Poole, and A. Zlatkis, *Org. Mass Spectrom.*, 1980, 15, 505.
11 R.H. Cragg and T.J. Miller, *J. Chem. Soc., Dalton Trans.*, 1982, 907.
12 C.J. Griffiths and H. Weigel, *Carbohydr. Res.*, 1980, 81, 7.
13 T.L. Youngless and M.M. Bursey, *Int. J. Mass Spectrom. Ion Phys.*, 1980, 34, 9.
14 V.N. Bochkarev, A.I. Belokon, N.N. Korneev, A.V. Mikhalev, and O.G. Levina, *Zh. Obshch. Khim.*, 1980, 50, 1547.
15 V.N. Bochkarev, A.I. Belokon, A.V. Kuchin, and S.I. Smogorzhevskii, *Zh. Obshch. Khim.*, 1981, 51, 1235.
16 G.D. Flesch and H.J. Svec, *Int. J. Mass Spectrom. Ion Phys.*, 1981, 38, 361.
17 K. Hottmann, *J. Prakt. Chem.*, 1981, 323, 399.
18 C.F. Poole, W.F. Size, S. Singhawangcha, and A. Zlatkis, *Org. Mass Spectrom.*, 1980, 15, 486.
19 A. Oliva and G.L. Larson, *Rev. Latinoam. Quim.*, 1981, 12, 70.
20 V.N. Bochkarev, A.N. Polivanov, T.F. Slyusarenko, V.I. Zhun, S.D. Vlasenko, and V.D. Sheludyakov, *Zh. Obshch. Khim.*, 1981, 51, 246.
21 C.A. Tsipis, *Chem. Chron.*, 1980, 9, 111.
22 V.N. Bochkarev, T.F. Slyusarenko, A.N. Polivanov, V.D. Sheludyakov, V.I. Zhun, S.D. Vlasenko, and V.G. Lakhtin, *Zh. Obshch. Khim.*, 1981, 51, 1364.
23 V.N. Bochkarev, A.N. Polivanov, T.F. Slyusarenko, and A.A. Bernadskii, *Zh. Obshch. Khim.*, 1980, 50, 1783.
24 L.E. Gusel'nikov, V.V. Volkova, V.G. Zaikin, A.N. Tarasenko, A.A. Tishenkov, N.S. Nametkin, M.G. Voronkov, and S.V. Kirpichenko, *J. Organomet. Chem.*, 1981, 215, 9.
25 N. Auner and J. Grobe, *J. Organomet. Chem.*, 1980, 197, 147.
26 N.S. Prostakov, N. Saxena, P.I. Zakharov, and A.V. Fresenko, *J. Organomet. Chem.*, 1982, 228, 37.
27 M. Yu. Eismont, G.V. Zaikin, V.F. Traven, and B.I. Stepanov, *Izv. Akad. Nauk SSSR, Ser. Khim.*, 1981, 1619.
28 K. Vebey, J. Tamas, G.P. Okonnishnikova, I.E. Dolgy, and O.M. Nefedov, *Org. Mass Spectrom.*, 1981, 16, 209.

29 M.A. Quilliam and J.B. Westmore, *Eur. J. Mass Spectrom. Biochem., Med. Environ. Res.*, 1980, 1, 53.
30 V.N. Bochkarev, A.N. Polivanov, T.F. Slyusarenko, A.A. Bernadskii, N.N. Sukina, and B.N. Klimentov, *Zh. Obshch. Khim.*, 1981, 51, 824.
31 L.E. Gusel'nikov, V.M. Sokolova, E.A. Volnina, Z.A. Kerzina, N.S. Nametkin, N.G. Komalenkova, S.A. Bashkirova, and E.A. Chernyshev, *Dokl. Akad. Nauk SSSR*, 1981, 260, 348.
32 D.J. Harvey, *Biomed. Mass Spectrom.*, 1980, 7, 211.
33 D.J. Harvey, *Anal. Chem. Symp. Ser.*, 1981, 7, 315.
34 V.N. Bochkarev, T.F. Slyusarenko, V.I. Zhun, G.V. Lakhtin, G.P. Sheludyakov, and A.N. Polivanov, *Zh. Obshch. Khim.*, 1981, 51, 2271.
35 J.A. Sonderquist and A. Hassner, *J. Organomet. Chem.*, 1981, 217, 151.
36 M.F. Dottore, C.V. Craige, D.J.M. Stone, and J.H. Bowie, *Org. Mass Spectrom.*, 1981, 16, 339.
37 J.R. Eyler, G. Silverman, and M.A. Battiste, *Organometallics*, 1982, 1, 477.
38 K.P. Steele and W.P. Weber, *Org. Mass Spectrom.*, 1982, 17, 222,
39 V. Yu. Vitkovskii, D.V. Gendin, O.A. Kruglaya, and N.S. Vyazankin, *Zh. Obshch. Khim.*, 1981, 51, 1771.
40 D.J. Harvey, *Biomed. Mass Spectrom.*, 1981, 8, 366.
41 D.J. Harvey, *Biomed. Mass Spectrom.*, 1981, 8, 575.
42 D.J. Harvey, *Biomed. Mass Spectrom.*, 1981, 8, 579.
43 D.J. Harvey, *Biomed. Mass Spectrom.*, 1981, 8, 546.
44 R. Alexander, G. Eglinton, J.P. Gill, and J.K. Volkman, *J. High Resolut. Chromatogr. Commun.*, 1980, 3, 521.
45 P. Vouros and D.H. Harvey, *Biomed. Mass Spectrom.*, 1980, 7, 217
46 M. Donike and J. Zimmermann, *J. Chromatogr.*, 1980, 202, 483.
47 C.J.W. Brooks, W.J. Cole, J.H. Borthwick, and G.M. Brown, *J. Chromatogr.*, 1982, 239, 191.
48 P.V. Fennessey and S.S. Tjva, *Org. Mass Spectrom.*, 1980, 15, 202.
49 T. Murata, S. Takahashi, T. Takada, and S. Ohnishi, *Shimadzu Hyoron*, 1980, 37, 43.
50 G.A. Pearse and S. Jacobsson, *Org. Mass Spectrom.*, 1980, 15, 331.
51 F. Rocchiccioli, J.P. Leroux, and P. Cartier, *J. Chromatogr.*, 1981, 226, 325.
52 R. Wolfschuetz, H. Schwarz, W. Blum, and W.J. Richter, *Org. Mass Spectrom.*, 1981, 16, 37.
53 E.M. Martinelli, *Eur. J. Mass Spectrom. Biomed., Med. Environ. Res.*, 1980, 1, 33.
54 M. Van Audenhove, D. DeKeukeleire, and M. Vandewalle, *Bull. Soc. Chim. Belg.*, 1980, 89, 371.
55 G. Sonnek, K.G. Baumgarten, and D. Habisch, *J. Prakt. Chem.*, 1980, 322, 94.
56 M.A. Quilliam, K.K. Ogilvie, K.L. Sadana, and J.B. Westmore, *Org. Mass Spectrom.*, 1980, 15, 207.
57 M.A. Quilliam, K.K. Ogilvie, and J.B. Westmore, *Org. Mass Spectrom.*, 1981, 16, 129.
58 S.J. Gaskell and A.W. Pike, *Biomed. Mass Spectrom.*, 1981, 8, 125.
59 A.P.J.M. De Jong, J. Elema, and B.J.T. VandeBerg, *Biomed. Mass Spectrom.*, 1980, 7, 359.
60 V.N. Bochkarev, A.N. Polivanov, A.A. Bernadskii, T.F. Slyusarenko, N.N. Silkina, and B.N. Klimentov, *Fiz.-Khim. Metody Issled. Elementoorgan. Soedin., M*, 1980, 63.
61 V.N. Bochkarev, A.N. Polivanov, A.A. Bernadskii, and N.E. Rodzevich, *Zh. Obshch. Khim.*, 1980, 50, 1074.
62 V.I. Lavrent'ev and V.G. Kostrovskii, *Zh. Obshch. Khim.*,

1980, 50, 1337.

63 R.E. Swaim, W.P. Weber, H.G. Beottger, M. Evans, and F.M. Bockhoff, *Org. Mass Spectrom.*, 1980, 15, 304.

64 A.N. Polivanov, A.A. Bernadskii, N.N. Silkina, B.N. Klimentov, and V.N. Bochkarev, *Zh. Obshch. Khim.*, 1980, 50, 1780.

65 A.N. Polivanov, A.A. Bernardskii, V.N. Bochkarev, V.I. Zhun, and V.D. Sheludyakov, *Zh. Obshch. Khim.*, 1980, 50, 614.

66 M.G. Voronkov, T.N. Martynova, V.P. Korchkov, P.P. Semyannikov, V.M. Grankin, and R.G. Mirskov, *Izv. Sib. Otd. Akad. Nauk SSSR, Ser. Khim. Nauk*, 1981, 125.

67 V.I. Lavrent'ev, V.M. Kovrigin, and G.G. Treer, *Zh. Obshch. Khim.*, 1981, 51, 124.

68 V.I. Lavrent'ev, M.G. Voronkov, and V.M. Kovrigin, *Zh. Obshch. Khim.*, 1980, 50, 382.

69 J.C. Kleinert and C.J. Weschler, *Anal. Chem.*, 1980, 52, 1745.

70 D.J. Harvey, *J. Chromatogr.*, 1980, 196, 156.

71 A.K. Hays, *Thin Solid Films*, 1981, 84, 401.

72 V.N. Bochkarev, A.N. Polivanov, N.N. Silkina, T.L. Krasnova, V.V. Stepanov, and A.A. Chernyshev, *Zh. Obshch. Khim.*, 1981, 51, 2274.

73 V.N. Bochkarev, A.N. Polivanov, T.L. Krasnova, M.O. Labartkava, N.N. Silkina, and E.A. Chernyshev, *Zh. Obshch. Khim.*, 1981, 51, 119.

74 R.H. Cragg and R.D. Lane, *J. Organomet. Chem.*, 1981, 212, 301.

75 M.G. Voronkov, V. Yu. Vitkovskii, N.M. Kudyakov, and R.K. Valetdinov, *Zh. Obshch. Khim.*, 1981, 51, 2176.

76 T. Mueller, P. Hencsei, and L. Bihatsi, *Period. Polytech. Chem. Eng.*, 1981, 25, 181.

77 V.N. Bochkarev, T.F. Slyusarenko, N.N. Silkina, A.N. Polivanov, T.K. Gar, and N. Yu. Khromova, *Zh. Obshch. Khim.*, 1980, 50, 1080.

78 B. Ciommer, H. Schwarz, A. Maaroufi, M.T. Reetz, and K. Levsen, *Z. Naturforsch., Teil B*, 1981, 36, 771.

79 R. Neidlein, A. Hotzel, and W. Lehr, *Arch. Pharm.*, 1981, 314, 138.

80 J. Pikies, W. Wojnowski, and A. Meller, *Z. Anorg. Allg. Chem.*, 1981, 473, 215.

81 L. Bihatsi, P. Hencsei, and L. Parkanyi, *J. Organomet. Chem.*, 1981, 219, 145.

82 T. Mueller, M. El-Kersh, K. Becker-Palossy, and J. Balla, *Acta Chem. Acad. Sci. Hung.*, 1980, 103, 9.

83 V. Yu. Vitkovskii, V.I. Rakhlin, R.G. Mirskov, A.L. Kuznetsov, S.K.L. Khangazheev, and M.G. Voronkov, *Zh. Obshch. Khim.*, 1981, 51, 1776.

84 J.E. Drake, L.N. Khasrou, and A. Majid, *J. Inorg. Nucl. Chem.*, 1981, 43, 1473.

85 M. Draeger, L. Ross, and D. Simon, *Z. Anorg. Allg. Chem.*, 1980, 466, 145.

86 R. Weber, F. Visel, and K. Levsen, *Anal. Chem.*, 1980, 52, 2299.

87 A. Tzschlach and W. Uhlig, *Z. Anorg. Allg. Chem.*, 1981, 475, 251.

88 S. David, A. Thieffy, and A. Forchioni, *Tetrahedron Lett.*, 1981, 22, 2647.

89 P. Umapathy, S.N. Bhide, K.D. Ghuge, and D.N. Sen, *J. Indian Chem. Soc.*, 1981, 58, 33.

90 K.C. Molloy, F.A.K. Nasser, and J.J. Zuckerman, *Inorg. Chem.*, 1982, 21, 1711.

91 S.W. Ng and J.J. Zuckermann, *Organometallics*, 1982, 1, 714.

92 J.L. Beauchamp, *Gov. Rep. Announce. Index (U.S.)*, 1981, 81, 2184.

93 G. Marshall, S. Franks, and F.R. Hartley, *Org. Mass Spectrom.*, 1981, 16, 272.

94 P. Peringer and J. Eichbichler, *J. Inorg. Nucl. Chem.*, 1981, 43, 2033.
95 K.G. Berndt, D. Gloyna, and H.G. Henning, *J. Prakt. Chem.*, 1981, 323, 445.
96 J. Ellermann and L. Mader, *Spectrochim. Acta, Part A*, 1981, 37, 449.
97 V. Yu. Vitkovskii, M.G. Voronkov, and G.A. Kuznetsova, *Zh. Obshch. Khim.*, 1981, 51, 1769.
98 R.S. Edmundson, *Phosphorus Sulfur*, 1981, 9, 307.
99 J.C. Briggs, C.A. McAuliffe, W.E. Hill, D.M. Minahan, and G. Dyer, *J. Chem. Soc., Perkin Trans. 2*, 1982, 321.
100 S. Yanai, *Phosphorus Sulfur*, 1982, 12, 369.
101 N.S. Ovchinnikova, L.T. Zhuravlev, K. Ya. Shengeliya, M.P. Glazunov, and Ya. A. Lozovoi, *Izv. Akad. Nauk SSSR, Ser. Khim.*, 1981, 1814.
102 V.G. Golovalyi and E.N. Korol, *Teor. Eksp. Khim.*, 1981, 17, 849.
103 C.E. Parker, C.A. Haney, and J.R. Hass, *J. Chromatogr.*, 1982, 237, 233.
104 A. Soto and R. Cea, *Rev. Roum. Chim.*, 1980, 25, 529.
105 P.J. Harris, R.D. Minard, and H.R. Allcock, *Org. Mass Spectrom.*, 1982, 17, 351.
106 D.W. Allen, *J. Heterocycl. Chem.*, 1980, 17, 1341.
107 T. Nielsen, H. Egsgaard, E. Larsen, and G. Schroll, *Anal. Chim. Acta*, 1981, 124, 1.
108 W. Gormbler and H.U. Weile, *J. Fluorine Chem.*, 1980, 15, 279.
109 A. Beneditti, C. Preti, G. Tosi, and P. Zannini, *J. Chem. Soc., Dalton Trans.*, 1980, 1467.
110 D.G. Tuck, G.W. Wood, and S.Z. Landhive, *Can. J. Chem.*, 1980, 58, 833.
111 P.M. Lausarot, G.A. Vaglio, M. Valle, and P. Volpe, *J. Organomet. Chem.*, 1980, 201, 459.
112 M.K. Chaudhuri, H.S. Dasgupta, N. Roy, and D.T. Khathing, *Org. Mass Spectrom.*, 1981, 16, 303.
113 T. Wuerminghausen, H.J. Reinecke, and P. Braunstein, *Org. Mass Spectrom.*, 1980, 15, 38.
114 B.J. Bridson, D.A. Edwards, J.W. White, and M.G.B. Drew, *J. Chem. Soc., Dalton Trans.*, 1980, 2129.
115 B.F.G. Johnson, J. Lewis, D.G. Parker, and G.R. Stephenson, *J. Organomet. Chem.*, 1980, 197, 77.
116 W. Petz, *Z. Naturforsch., Teil B, 1980*, 35, 860.
117 M.R. Blake, J.L. Garnett, I.K. Gregor, and D. Nelson, *J. Organomet. Chem.*, 1980, 188, 203.
118 L.H. Staal, G. Van Koten, R.H. Fokkens, and N.M.M. Nibbering, *Inorg. Chim. Acta*, 1981, 50, 205.
119 W.J.A. Vanden Leuvel, R.W. Walker, S.B. Nagelberg, and B.R. Willeford, *J. Organomet. Chem.*, 1980, 190, 73.
120 Yu. S. Nekrasov, O.B. Afanasova, Yu. N. Sukharev, G.A. Nurgalieva, N.G. Komalenkova, and E.A. Chernyshev, *Izv. Akad. Nauk SSSR, Ser. Khim.*, 1981, 2243.
121 J. Mueller, G. Krebs, F. Leudemann, and E. Baumgartner, *J. Organomet. Chem.*, 1981, 218, 61.
122 D.E. Games, J.L. Gower, M. Gower, and L.A.P. Kane-Maguire, *J. Organomet. Chem.*, 1980, 193, 229.
123 Yu. S. Nekrasov, V.I. Khvostenko, I.I. Furlei, N.I. Vasyukova, V.K. Mavrodiev, A.Sh. Sultanov, and G.A. Tolstikov, *J. Organomet. Chem.*, 1981, 212, 373.
124 V.I. Khvostenko, Yu. S. Nekrasov, I.I. Furlei, N.I. Vasyukova, and G.A. Tolskikov, *J. Organomet. Chem.*, 1981, 212, 369.
125 M.R. Blake, J.L. Garrett, I.K. Gregor, and S.B. Wild, *Org. Mass Spectrom.*, 1980, 15, 369.
126 N.I. Vasyukova, Yu. S. Nekrasov, V.V. Krivykh, and M.I. **Rybinskaya**, *J. Organomet. Chem.*, 1980, 201, 283.

127 M.K. Chaudhuri, H.S. Dasgupta, N. Roy, and D.T. Khathing, *Org. Mass Spectrom.*, 1981, 16, 534.
128 D.H. Farrar, B.F.G. Johnson, J. Lewis, N.J. Nicholls, P.R. Raithby, and M.J. Rosales, *J. Chem. Soc., Chem. Commun.*, 1981, 273.
129 G.G. Devyatykh, P.E. Gaivoronskii, E.M. Gavrishchuk, N.P. Chernyaev, and Yu. B. Zuerev, *Zh. Neorg. Khim.*, 1980, 25, 2558.
130 V.P. Mar'in, O.N. Druzhkov, Yu. A. Andrianov, T.I. Aren'eva, and I.I. Grinval'd, *Zh. Obshch. Khim.*, 1980, 50, 1830.
131 D.W. Slocum, D.L. Beach, C.R. Ernst, R. Fellows, M. Moronski, B. Conway, J. Bencini, and A. Siegel, *J. Chem. Soc., Chem. Commun.*, 1980, 1043.
132 D.V. Zagorevskii, Yu. S. Nekrasov, and G.A. Nurgalieva, *J. Organomet. Chem.*, 1980, 194, 77.
133 Yu. A. Andrianov, O.N. Druzhkov, A.S. Smirnov, and V.A. Dodonov, *Zh. Obshch. Khim.*, 1981, 51, 2508.
134 B. Bajpai and K.K. Bajpai, *J. Inorg. Nucl. Chem.*, 1981, 43, 1938.
135 R.N. Grimes, N.M. Maxwell, R.B. Maynard, and E. Sin, *Inorg. Chem.*, 1980, 19, 2981.
136 S.-C. Chen, *J. Organomet. Chem.*, 1980, 202, 183.
137 D.V. Zagorevskii, N.M. Loim, Yu. S. Nekrasov, V.F. Sizoi, and Yu. N. Sukharev, *J. Organomet. Chem.*, 1980, 202, 201.
138 N.B. Henis, W. Lamanna, M.B. Humphrey, M.M. Bursey, and M.S. Brookhart, *Inorg. Chim. Acta*, 1981, 54, L11.
139 G.A. Nurgalieva, Yu. S. Nekrasov, D.V. Zagorevskii, and D.N. Kursanov, *J. Organomet. Chem.*, 1980, 202, 77.
140 V.F. Sizoi and Yu. S. Nekrasov, *Izv. Akad. Nauk SSSR, Ser. Khim.*, 1982, 285.
141 Yu. S. Nekrasov, N.I. Vasynkova, D.V. Zagorevskii, V.F. Sizoi, G.A. Nurgalleva, and L.I. Dyubina, *J. Organomet. Chem.*, 1980, 201, 433.
142 R.W. Jones and R.H. Staley, *Int. J. Mass Spectrom. Ion Phys.*, 1981, 39, 35.
143 V.F. Sizoi, Yu. S. Nekrasov, Yu. N. Sukharev, and N.E. Kolobova, *J. Organomet. Chem.*, 1981, 210, 97.
144 K. Eberl, W. Uedelhoven, M. Wolfgruber, and F.R. Kreissl, *Chem. Ber.*, 1982, 115, 504.
145 D.H. Harris and T.R. Spalding, *Inorg. Chim. Acta*, 1980, 39, 187.
146 E. Efraty, D. Liebman, M.H.A. Huang, and C.A. Weston, *Inorg. Chim. Acta*, 1980, 39, 105.
147 A. Efraty and M.H.A. Huang, *Inorg. Chem.*, 1980, 19, 2296.
148 G.S. Claude, R. Tabacchi, G. Facchinetti, and C. Floriani, *Bull. Soc. Chim. Belg.*, 1980, 89, 875.
149 A. Marinetti and E. Sappa, *Ann. Chim. (Rome)*, 1981, 71, 707.
150 J. Mueller and F. Luedemann, *Z. Naturforsch., Teil B*, 1981, 36, 74.
151 T.L. Agafonov, N.F. Shushunov, P.E. Gaivoronskii, T.L. Spirina, M.R. Leonov, V.A. Il'yushchenkov, G.V. Solov'eva, and N.I. Gramoteev, *Radiokhimiya*, 1981, 23, 62.
152 C.W. Eigenbrot, *INIS Atomindex*, 1981, 12, Abstr. No. 621354.
153 L.H. Dubois, D.G. Castner, and G.A. Somorjai, *J. Chem. Phys.*, 1980, 72, 5234.
154 T.S. Wittrig, P.D. Szuromi, and W.H. Weinberg, *J. Chem. Phys.*, 1982, 76, 716.
155 N.B. Henis, D.F. Fraley, and M.M. Bursey, *Inorg. Nucl. Chem. Lett.*, 1981, 17, 121.

156 S.P. Roe, P.G. Cullis, P.J. Derrick, and J.O. Hill, *Aust. J. Chem.*, 1982, 35, 287.
157 D.E. Games, J.L. Gower, and L.A.P. Kane-Maguire, *J. Chem. Soc., Dalton Trans.*, 1981, 1994.
158 Zl.Yu. Vaisbein, *Izv. Akad. Nauk Mold. SSR, Ser. Biol., Khim. Nauk*, 1980, 69.
159 P.L. Beaumont, J.L. Garnett, and I.K. Gregor, *Inorg. Chim. Acta*, 1980, 45, 99.
160 J.R. Majer and A.S.P. Azzouz, *J. Inorg. Nucl. Chem.*, 1981, 43, 1793.
161 W. Radecka-Paryzek, *Inorg. Chim. Acta*, 1981, 54, L251.
162 D.R. Dakternieks, I.W. Fraser, J.L. Garnett, I.K. Gregor, and M. Guilhaus, *Org. Mass Spectrom.*, 1980, 15, 556.
163 O.R. Cea, J. Gomez-Lara, and E. Cortes, *J. Inorg. Nucl. Chem.*, 1980, 42, 1530.
164 V.S. Khomenko, T.A. Rasshinina, V.P. Suboch, and N.E. Akimova, *Vestsi Akad. Navuk BSSR, Ser. Khim. Navuk*, 1981, 54.
165 N.G. Dyubenko, E.M. Gavrishchuk, L.I. Martyneako, N.P. Chernyaev, Yu. B. Zuerev, and V.I. Spitsyn, *Dokl. Akad. Nauk SSSR*, 1980, 253, 353.
166 M.L. Morris and R.D. Koob, *Inorg. Chem.*, 1981, 20, 2737.
167 J. Magner, H. Meierer, and R. Neeb, *Fresenius' Z. Anal. Chem.*, 1982, 311, 249.
168 W. Szczepaniak and A.S. Plaziak, *Pol. J. Chem.*, 1981, 55, 353.
169 G.-Z. Xu, C. Wang, J. Sun, and Y. Tang, *Kexue Tongbao*, 1981, 26, 1278.
170 L.N. Sidorov and V.C. Gubarevich, *Koord. Khim.*, 1982, 8, 463.
171 M.I. Niktin and L.N. Sidorov, *Int. J. Mass Spectrom. Ion Phys.*, 1980, 35, 101.
172 A.V. Gusarov, L.N. Gorokhov, A.T. Pyatenko, and I.V. Sidorova, *Adv. Mass Spectrom.*, 1980, 8A, 262.
173 G.W. Wood, M.-K. Au, N. Mak, and P.-Y. Lau, *Can. J. Chem.*, 1980, 58, 681.
174 L.N. Sidorov, *Int. J. Mass Spectrom. Ion Phys.*, 1981, 38, 49.
175 T.-M. R. Su and S.J. Riley, *J. Chem. Phys.*, 1980, 72, 6632.
176 L. Edelmann, *Fresenius' Z. Anal. Chem.*, 1981, 303, 218.
177 L. Edelmann, *Physiol. Chem. Phys.*, 1980, 12, 509.
178 H.H. Emons, W. Horlbeck, and D. Kiessling, *Z. Chem.*, 1981, 21, 416.
179 K. Kimoto, I. Nishida, H. Takahashi, and H. Kato, *Jpn. J. Appl. Phys.*, 1980, 19, 1821.
180 J.R. Lloyd and F.H. Field, *Biomed. Mass Spectrom.*, 1981, 8, 19.
181 Y. Murano, G. Izawa, K. Yoshihara, M. Takahashi, M. Kishimoto, and S. Suzuki, *Int. J. Mass Spectrom. Ion Phys.*, 1982, 41, 179.
182 C.H. Wu and H.R. Ihle, *Adv. Mass Spectrom.*, 1980, 8A, 374.
183 G.W. Wood and W.F. Sun, *Can. J. Chem.*, 1981, 59, 2218.
184 L.N. Sidorov, M.I. Nikitin, E.V. Skokan, and I.D. Sorokin, *Int. J. Mass Spectrom. Ion Phys.*, 1980, 35, 203.
185 M.I. Nikitin, L.N. Sidorov, E.V. Skokan, and I.D. Sorokin, *Zh. Fiz. Khim.*, 1981, 55, 1944.
186 M. Itoh, T. Sasamoto, and T. Sata, *Bull. Chem. Soc. Jpn.*, 1981, 54, 3391.
187 G.D. Roberts and E.V. White, *Org. Mass Spectrom.*, 1981, 16, 546.
188 S.A. McLuckey, A.E. Schoen, and R.G. Cooks, *J. Am. Chem. Soc.*, 1982, 104, 848.
189 A.V. Korgaonkar, C.P. Gopolarman, and V.K. Rohatgi, *Int. J. Mass Spectrom. Ion Phys.*, 1981, 40, 127.
190 S.V. Ghaisas, S.M. Chaudhari, M.R. Bhiday, and V.N. Bhoraskar, *J. Phys. Soc. Jpn.*, 1980, 48, 1795.

191 V.P. Suboch, V.S. Khomenko, and T.A. Rasshinma, *Vestsi Akad. Navuk BSSR, Ser. Khim. Navuk*, 1981, 96.
192 K. Matsumoto and Y. Kosugi, *Org. Mass Spectrom.*, 1981, 16, 249.
193 R.G. Brereton, V. Rajananda, T.J. Blake, J.K.M. Sanders, and D.H. Williams, *Tetrahedron Lett.*, 1980, 21, 1671.
194 E. Murad, *Chem. Phys. Lett.*, 1980, 72, 295.
195 T. Sasamoto, H. Hara, and T. Sata, *Bull. Chem. Soc. Jpn.*, 1981, 54, 3327.
196 R. Kownatski, R. Peters, and G.H. Reil, *Fresenius' Z. Anal. Chem.*, 1980, 301, 179.
197 G.W. Wood and W.F. Sun, *Biomed. Mass Spectrom.*, 1980, 7, 399.
198 H.J. Dietze and H. Zahn, *ZfI-Mitt.*, 1981, 46, 60.
199 L.S. Kudin, G. Balducci, G. Gigli, and M. Guido, *Izv. Vyssh. Uchebn. Zavid. Khim. Tekhnol.*, 1982, 25, 259.
200 L.S. Kudin, *Izv. Vyssh. Uchebn. Zavid. Khim. Tekhnol.*, 1981, 24, 837.
201 V.A. Sipachev, L.N. Reshetova, Y.S. Nekrasov, and S. Yu. Sil'vestrova, *Org. Mass Spectrom.*, 1980, 15, 192.
202 C. Chieh and S.K. Cheung, *Can. J. Chem.*, 1981, 59, 2746.
203 H. Schaefer and U. Fluerke, *Z. Anorg. Allg. Chem.*, 1981, 478, 57.
204 A. Kant and K.A. Moon, *High Temp. Sci.*, 1981, 14, 23.
205 K.V. Ovchinnikov, E.N. Nikolaev, and G.A. Semenov, *Zh. Obshch. Khim.*, 1981, 51, 261.
206 K.H. Lau and D.L. Hildenbrand, *J. Chem. Phys.*, 1980, 72, 4928.
207 J.L. Jourdain, G. Laverdet, G. Le Bras, and J. Combourieu, *J. Chim. Phys. Phys.-Chim. Biol.*, 1981, 78, 253.
208 J.L. Jourdain, G. Le Bras, and J. Combourieu, *J. Phys. Chem.*, 1981, 85, 655.
209 Y. Murano, G. Izawa, and T. Shiokawa, *Radiochem. Radioanal. Lett.*, 1980, 44, 315.
210 S.K. Boocock, N.N. Greenwood, and J.D. Kennedy, *J. Chem. Res. (S)*, 1981, 50.
211 N.N. Greenwood, T.R. Spalding, and D. Taylorson, *J. Inorg. Nucl. Chem.*, 1980, 42, 317.
212 M.G. Voronkov, V. Yu. Vitkovskii, and V.P. Baryshok, *Zh. Obshch. Khim.*, 1981, 51, 1773.
213 H. Nöth, R. Staudigl, and W. Storch, *Chem. Ber.*, 1981, 114, 3024.
214 J.M. Halket, I. Ganschow, and B.P. Lisboa, *J. Chromatogr.*, 1980, 192, 434.
215 T.L. Youngless and M.M. Bursey, *Int. J. Mass Spectrom. Ion Phys.*, 1980, 34, 1.
216 S.U. Sheikh and A. Ghafoor, *Pak. J. Sci. Ind. Res.*, 1980, 23, 23.
217 D.C. Hockman, C.E. Moore, J.L. Huston, and S. Chao, *Anal. Lett.*, 1981, 14, 719.
218 E. Hohaus, W. Riepe, and K. Wessendorf, *Z. Naturforsch., Teil B*, 1980, 35, 316.
219 L.N. Sidorov, M.I. Nikitin, E.V. Skokan, and I.D. Gorokin, *Adv. Mass Spectrom.*, 1980, 8A, 462.
220 H. Schaefer and U. Floerke, *Z. Anorg. Allg. Chem.*, 1980, 469, 172.
221 H. Schaefer, U. Floerke, and M. Trenkel, *Z. Anorg. Allg. Chem.*, 1981, 478, 191.
222 H. Schaefer, M. Binnewies, and U. Floerke, *Z. Anorg. Allg. Chem.*, 1981, 477, 31.
223 P. Ho and R.P. Burns, *High Temp. Sci.*, 1980, 12, 31.
224 J. Verlinden and R. Gijbels, *Adv. Mass Spectrom.*, 1980, 8A, 485.

225 A. Brunot, M. Cottin, M.H. Donnart, and S.C. Muller, *J. Mass Spectrom. Ion Phys.*, 1980, 33, 417.
226 P. Cocolios, P. Fournari, R. Guilard, C. Lecomte, J. Protas, and J.C. Boubel, *J. Chem. Soc., Dalton Trans.*, 1980, 2081.
227 J. Kapala and K. Skudlarski, *Commun. Czech.-Pol. Colloq. Chem. Thermodyn. Phys. Org. Chem., 2nd,* 1980, 142.
228 J. Kapala and K. Skudlarski, *Int. J. Mass Spectrom. Ion Phys.*, 1981, 40, 255.
229 Yu. S. Nekrasov, S. Yu. Sil'vestrova, T.V. Lysyak, Z.M. Alaudinova, Yu. Ya. Kharitovnov, and I.S. Kolomnikov, *Koord. Khim.*, 1981, 7, 1270.
230 Yu. S. Nekrasov, V.A. Sipachev, and N.I. Tuseev, *J. Inorg. Nucl. Chem.*, 1980, 42, 1677.
231 A. Brunot, M. Cottin, P. Gotchiguian, and J.C. Muller, *J. Mass Spectrom. Ion Phys.*, 1981, 41, 31.
232 S. Yokoyama, K. Tanaka, K. Miyahara, I. Takakuwa, and M. Seisho, *Radioisotopes*, 1981, 30, 679.
233 A.R. Swanson and M.W. Aaders, *J. Chromatogr.*, 1982, 234, 268.
234 A.J. Stace and A.K. Shukla, *Int. J. Mass Spectrom. Ion Phys.*, 1980, 36, 119.
235 A.J. Stace and A.K. Shukla, *Chem. Phys. Lett.*, 1982, 85, 157.
236 Y. Ono, S.H. Linn, H.F. Prest, M.E. Gress, and C.Y. Ng, *J. Chem. Phys.*, 1980, 73, 2523.
237 D. Elmore, N. Anantaraman, H.W. Fulbright, H.E. Grove, H.S. Hans, K. Nishizumi, M.T. Murrell, and M. Honda, *Phys. Rev. Lett.*, 1980, 45, 589.
238 W. Kutschera, W. Henning, M. Paul, R.K. Smither, E.J. Stephenson, J.L. Yatema, D.E. Alburger, J.B. Cumming, and G. Harbottle, *Phys. Rev. Lett.*, 1980, 45, 592.
239 A. Hurrle and J. Dietl, *Fresenius' Z. Anal. Chem.*, 1981, 309, 277.
240 M. Farber and R.D. Srivastava, *High Temp. Sci.*, 1980, 12, 21.
241 K.G. Heumann, F. Beer, and R. Kifmann, *Talanta*, 1980, 27, 567.
242 C.H. Wu, H.R. Ihle, and K. Zmbov, *J. Chem. Soc., Faraday Trans. 2*, 1980, 76, 447.
243 K.A. Moon, A. Kant, and W.J. Croft, *J. Am. Ceram. Soc.*, 1980, 63, 698.
244 T. Yoshimi, H. Sakai, and K. Tanaka, *J. Electrochem. Soc.*, 1980, 127, 1853.
245 V.N. Bochkarev, T.F. Slyusarenko, A.N. Polivaaov, N.N. Silkina, T.K. Gar, N. Yu. Khromova, and B.M. Zolotarev, *Zh. Obshch. Khim.*, 1980, 50, 2145.
246 C. Hirayama and R.L. Kleinosky, *Thermochim. Acta*, 1981, 47, 355.
247 H. Wiedemeir and F.J. Csillag, *Z. Anorg. Allg. Chem.*, 1980, 469, 197.
248 G.K. Bratspies, J.F. Smith, and J.O. Hill, *J. Anal. Appl. Pyrolysis*, 1980, 2, 35.
249 J.L. Lefferts, K.C. Molloy, M.B. Hossain, D. Van der Helm, and J.J. Zuckerman, *Inorg. Chem.*, 1982, 21, 1410.
250 J. Mushlbach, E. Rechnagel, and K. Sattler, *Vide. Couches. Minces*, 1980, 201.
251 R. Stoll and F.W. Roellgen, *J. Chem. Soc., Chem. Commun.*, 1980, 789.
252 R.G. Kostyanovsky, A.P. Pleshkova, V.N. Voznesensky, and Yu. I. El'natanov, *Org. Mass Spectrom.*, 1980, 15, 397.
253 A.F. Janzen, O.C. Vaidya, and C.J. Willis, *J. Inorg. Nucl. Chem.*, 1981, 43, 1469.
254 M. Armenante, A. Brancaccio, E. Burattini, V. Santoro, N. Spinelli, and F. Vanoli, *J. Mass Spectrom. Ion Phys.*, 1980, 36, 213.

255 L.J. McCrumb and P. Warneck, *Int. J. Mass Spectrom. Ion Phys.*, 1981, 37, 127.
256 K. Bederski, L. Wojcik, and B. Adamczyk, *Int. J. Mass Spectrom. Ion Phys.*, 1980, 35, 171.
257 H. Shinohara and N. Nishi, *Chem. Phys. Lett.*, 1982, 87, 561.
258 Yu. N. Artyukh, V.G. Golovatyi, and E.N. Korol, *Teor. Eksp. Khim.*, 1980, 16, 714.
259 T.D. Lee, W.R. Anderson, and G.D. Daves, *Anal. Chem.*, 1981, 53, 304.
260 H.J. Veith, *Adv. Mass Spectrom.*, 1980, 8A, 768.
261 S. Wlodek, Z. Luczynski, and H. Wincel, *Int. J. Mass Spectrom. Ion Phys.*, 1980, 35, 39.
262 J.M. Tedder and P.H. Vidand, *Chem. Phys. Lett.*, 1980, 76, 380.
263 J.L. Jourdain, G. Laverdet, G. Le Bras, and J. Combourieu, *J. Chim. Phys. Phys.-Chim. Biol.*, 1980, 77, 809.
264 M.J. Babin, *Z. Anorg. Allg. Chem.*, 1980, 467, 218.
265 P. Clare and D.B. Sowerby, *J. Inorg. Nucl. Chem.*, 1981, 43, 467.
266 W. Sulkowski, A.A. Volodin, K. Brandt, V.V. Kireev, and V.V. Korshak, *Zh. Obshch. Khim.*, 1981, 51, 1221.
267 B. Monsarrat, J.C. Prome, J.F. Labarre, F. Sournies, and J.C. Van de Grampel, *Biomed. Mass Spectrom.*, 1980, 7, 405.
268 G.S. Gol'din, S.G. Fedorov, E.V. Kotova, V.N. Bochkarev, and T.F. Slyusarenko, *Zh. Obshch. Khim.*, 1980, 50, 21.
269 A. Ballistreri, S. Foti, G. Montaudo, S. Lora, and G. Pezzin, *Makromol. Chem.*, 1981, 182, 1319.
270 R.V. Hodges, T.J. McDonnell, and J.L. Beauchamp, *J. Am. Chem. Soc.*, 1980, 102, 1327.
271 T. Hudec, J. Thean, D. Kuehl, and R.C. Dougherty, *Science*, 1981, 211, 951.
272 R.C. Dougherty and J.D. Wander, *Biomed. Mass Spectrom.*, 1980, 7, 401.
273 B.M. Kwon and D.Y. Oh, *Phosphorus Sulfur*, 1981, 11, 177.
274 K. Moedritzer and R.E. Miller, *Phosphorus Sulfur*, 1981, 10, 279.
275 N. Fukuhara and M. Eto, *Nippon Noyaku Gakkaishi*, 1980, 5, 63.
276 V. Radu and N. Ion, *Rev. Chim. (Bucharest)*, 1980, 31, 964.
277 Z.J. Lesnikowski, W.J. Stec, and B. Zielinska, *Org. Mass Spectrom.*, 1980, 15, 454.
278 H. Kenttamaa and J. Enqvist, *Org. Mass Spectrom.*, 1980, 15, 520.
279 V.N. Gogte, P.S. Kulkarni, A.S. Modak, and B.D. Tilak, *Org. Mass Spectrom.*, 1981, 16, 515.
280 H. Keck, W. Kuchen, and H.F. Mahler, *Org. Mass Spectrom.*, 1980, 15, 591.
281 W. Reimschussel and P. Paneth, *Org. Mass Spectrom.*, 1980, 15, 302.
282 U. Bahr, H.R. Schulten, O.R. Hommes, and F. Aerts, *Clin. Chim. Acta*, 1980, 103, 183.
283 U. Bahr and H.R. Schulten, *Biomed. Mass Spectrom.*, 1981, 8, 553.
284 S. Meyerson, E.S. Kukn, F. Ramirez, J.F. Marecek, and H. Okazaki, *Adv. Mass Spectrom.*, 1980, 8A, 577.
285 C. Wibbelmann and W. Brockner, *Z. Naturforsch., Teil A*, 1981, 36, 836.
286 C.E. Doiron and T.B. McMahon, *Inorg. Chem.*, 1980, 19, 3037.
287 M. Hall and D.B. Sowerby, *J. Chem. Soc., Dalton Trans.*, 1980, 1292.
288 G.E. Binder, W. Schwarz, W. Rozdzinski, and A. Schmidt, *Z. Anorg. Allg. Chem.*, 1980, 471, 121.

289 F. Di Bianca, N. Bertazzi, G. Alonzo, G. Ruisi, and T.C. Gibb, *Inorg. Chim. Acta,* 1981, 50, 235.
290 K. Sattler, J. Muehlbach, and E. Recknagel, *Phys. Rev. Lett.,* 1980, 45, 821.
291 K. Sattler, J. Muehlbach, and A. Reyes-Flotte, *J. Phys. E,* 1980, 13, 673.
292 J. Muehlbach, E. Recknagel, and K. Sattler, *Surf. Sci.,* 1981, 106, 188.
293 E.N. Nikolaev, G.A. Semenov, K.E. Frantseva, S.N. Sharov, and L.F. Yurkov, *Fitz. Khim. Stekla,* 1981, 7, 606.
294 G. Gattow and K. Sarter, *Z. Anorg. Allg. Chem.,* 1980, 463, 163.
295 D.E. Games, L.A.P. Kane-Maguire, D.C. Parsons, and M.R. Truter, *Inorg. Chim. Acta,* 1981, 49, 213.
296 D.G. Parsons, M.R. Truter, P.W. Brookes, and K. Hall, *Inorg. Chim. Acta,* 1982, 59, 15.
297 E. Blasius, H. Lauder, and M. Keller, *Fresenius' Z. Anal. Chem.,* 1980, 304, 10.
298 T. Masuoka and J.A.R. Lincon, *J. Chem. Phys.,* 1981, 75, 4946.
299 J.J. Wagner and W.W. Brandt, *Plasma Chem. Plasma Process,* 1981, 1, 201.
300 R. Kaufel, G. Vahl, R. Minkwitz, and H. Baumgaertel, *Z. Anorg. Allg. Chem.,* 1981, 481, 207.
301 J.-S. Wang and J.L. Franklin, *Int. J. Mass Spectrom. Ion Phys.,* 1980, 36, 233.
302 B.W. Christensen, A. Kjaet, and J.O. Madsen, *JOACS, J. Am. Oil Chem. Soc.,* 1981, 58, 1053.
303 J. Greim, H. Knoedler, and H.A. Schaeffer, *Verres Refract.,* 1981, 35, 315.
304 G.N. Jham, R.N. Hanson, R.W. Giese, and P. Vouros, *J. Heterocycl. Chem.,* 1981, 18, 1335.
305 D.C. Reamer and C. Veillon, *Anal. Chem.,* 1981, 53, 2166.
306 G.H. Westphal, F. Rosenberger, P.R. Cunningham, and L.L. Ames, *J. Chem. Phys.,* 1980, 72, 5192.
307 Yu. I. Nikonorov and V.V. Marusin, *Zh. Neorg. Khim.,* 1981, 26, 2662.
308 T.A. Lehman, D.J. Harvan, and J.R. Hass, *Org. Mass Spectrom.,* 1980, 15, 437.
309 K.G. Henmann, R. Kifmann, W. Schindlmeier, and M. Unger, *Int. J. Environ. Anal. Chem.,* 1981, 10, 39.
310 K.G. Hermann, K. Baier, F. Beer, R. Kifmann, and W. Schindlmeier, *Adv. Mass Spectrom.,* 1980, 8A, 318.
311 A.V. Balnev, Z.K. Nikitina, L.I. Fedorova, and V. Ya. Rosolovskii, *Izv. Akad. Nauk SSSR, Ser. Khim.,* 1980, 1963.
312 M.H. Habibi and L.C. Sams, *J. Fluorine Chem.,* 1981, 18, 277.
313 I.L. Agafonov and A.I. Kuz'michev, *Zh. Anal. Khim.,* 1980, 35, 940.
314 K. Stephan, H. Helm, and T.D. Maerk, *Contrib. Symp. At. Surf. Phys.,* 1980, 123.
315 T. Koizumi, N. Kobayashi, and Y. Kaneko, *J. Phys. Soc. Jpn.,* 1980, 48, 1678.
316 O. Echt, K. Sattler, and E. Recknagel, *Phys. Rev. Lett.,* 1981, 47, 1121.
317 R. Dutil and P. Marmet, *J. Mass Spectrom. Ion Phys.,* 1980, 35, 371.
318 E. Minehara and S. Abe, *Nucl. Instrum. Methods Phys. Res.,* 1981, 190, 215.
319 H. Kimura, M. Asano, and K. Kubo, *J. Nucl. Mater.,* 1980, 92, 221.
320 H. Nakagawa, M. Asano, and K. Kubo, *J. Nucl. Mater.,* 1981, 102, 292.

321 A. Neubert, H.R. Ihle, and K.A. Gingerich, *J. Chem. Phys.*, 1980, 73, 1406.
322 L.N. Sidorov, N.M. Karasev, and Yu. M. Korenev, *J. Chem. Thermodyn.*, 1981, 13, 915.
323 B. Busse and K.G. Weil, *Ber. Bunsenges. Phys. Chem.*, 1982, 86, 93.
324 R. Odoj and K. Hilpert, *High Temp. High Pressures*, 1980, 12, 93.
325 R. Odog and K. Hilpert, *Z. Naturforsch., Teil A*, 1980, 35, 9.
326 U. Pittermann and K.G. Weil, *Ber. Bunsenges. Phys. Chem.*, 1980, 84, 542.
327 K. Hilpert, *Ber. Bunsenges. Phys. Chem.*, 1980, 84, 494.
328 K. Hilpert, *Adv. Mass Spectrom.*, 1980, 8A, 382.
329 R.G. Behrens and C.F.V. Mason, *J. Less Common Met.*, 1981, 77, 169.
330 M. Allibert, C. Chatillon, K.T. Jacob, and R. Lourtau, *J. Am. Ceram. Soc.*, 1981, 64, 307.
331 G. Riebert, P. Lamparter, and S. Steeb, *Z. Metallkd.*, 1981, 72, 765.
332 G. Riebert, P. Lamparter, and S. Steeb, *Z. Naturforsch., Teil A*, 1981, 36, 447.
333 K. Skudlarski, J. Kapala, and M. Kowalska, *J. Less Common Met.*, 1982, 83, L39.
334 A.V. Gusarov, A.T. Pyatenko, and L.N. Gorokhov, *Teplofiz. Vys. Temp.*, 1980, 18, 961.
335 P. Lamparter, D.L. Cocke, and S. Steeb, *Z. Matallkd.*, 1982, 73, 149.
336 G. Balducci, D. Ferro, and V. Piacente, *High Temp. Sci.*, 1981, 14, 207.
337 R.A. Isokova, S.M. Kozhakmetov, A.I. Red'ko, and V.A. Spitsyn, *Kompleksn. Ispol'z. Miner. Syr'ya*, 1981, 41.
338 T. Sasamoto, M. Kobayashi, and T. Sata, *Shitsuryo Bunseki*, 1981, 29, 249.
339 K.F. Zmbov, A. Neubert, and H.R. Ihle, *Z. Naturforsch., Teil A*, 1981, 36, 914.
340 G. Balducci, G. Gigli, and M. Guido, *J. Chem. Soc., Faraday Trans. 2*, 1981, 77.
341 I.D. Sorokin, L.N. Sidorov, M.I. Nikitin, and E.V. Skokan, *J. Mass Spectrom. Ion Phys.*, 1981, 41, 45.
342 E. Murad, *J. Chem. Phys.*, 1980, 73, 1381.
343 I.P. Malkerova, A.S. Alikhanyan, V.S. Perrov, V.D. Butskii, and V.I. Gorgoraki, *Zh. Neorg. Khim.*, 1980, 25, 2067.
344 I.P. Malkerova, A.S. Alikhanyan, and V.I. Gorgoraki, *Zh. Neorg. Khim.*, 1980, 25, 3181.
345 S.K. Gupta and K.A. Gingerich, *J. Chem. Phys.*, 1981, 74, 3584.
346 S.K. Gupta, B.M. Nappi, and K.A. Gingerich, *Inorg. Chem.*, 1981, 20, 966.
347 I.P. Malkerova, A.S. Alikhanyan, V.D. Butskii, V.S. Perovov, and V.I. Gorkoraki, *Zh. Neorg. Khim.*, 1981, 26, 1955.
348 R. Haque and K.A. Gingerich, *J. Chem. Phys.*, 1981, 74, 6407.
349 K.A. Gingerich, R. Haque, and M. Pelino, *J. Chem. Soc., Faraday Trans. 1*, 1982, 78, 341.
350 K.A. Gingerich, M. Pelino, and R. Haque, *High Temp. Sci.*, 1981, 14, 137.
351 A.T. Pyatenko, A.V. Gusarev, and L.N. Gorokhov, *Teplofiz. Vys. Temp.*, 1981, 19, 329.
352 K.A. Gingerich and R. Haque, *J. Chem. Soc., Faraday Trans. 2*, 1980, 76, 101.
353 L.N. Sidorov, E.V. Skokan, M.I. Nikitin, and I.D. Sorokin, *Int. J. Mass Spectrom. Ion Phys.*, 1980, 35, 215.

354 A.T. Pyatenko, A.V. Gusarov, and L.N. Gorokhov, *Teplofiz. Vys. Temp.*, 1980, 18, 1154.
355 J.G. Edwards, J.S. Starzynski, and D.E. Peterson, *J. Chem. Phys.*, 1980, 73, 908.
356 S.K. Gupta and K.A. Gingerich, *J. Chem. Phys.*, 1980, 72, 2795.
357 S.Y. Yin, *Microbeam Anal.*, 1981, 16th, 342.
358 B.K. Furman, G.H. Morrison, V.I. Saunders, and Y.T. Tan, *Photogr. Sci. Eng.*, 1981, 25, 121.
359 T.M. Barlak, J.E. Campana, R.J. Colton, J.J. De Corpo, and J.R. Wyatt, *J. Phys. Chem.*, 1981, 85, 3840.
360 T.M. Barlak, J.R. Wyatt, R.J. Colton, J.J. De Corpo, and J. E. Campana, *J. Am. Chem. Soc.*, 1982, 104, 1212.
361 J.E. Campana, T.M. Barlak, R.J. Colton, J.J. De Corpo, J.R. Wyatt, and B.I. Dunlap, *Phys. Rev. Lett.*, 1981, 47, 1046.
362 J. Okano, *Adv. Mass Spectrom.*, 1980, 8A, 513.
363 H. Nishimura and J. Okano, *Mem. Natl. Inst. Polar Res., Spec. Issue*, 1981, 20, 229.
364 C. Chassard-Bouchaud and P. Galle, *Electron Microsc. Proc. Eur. Congr., 7th*, 1980, 3, 222.
365 Y. Ochiai, Y. Nannichi, and K. Masuda, *Jpn. J. Appl. Phys.*, 1980, 19, L777.
366 W.H. Christie, R.E. Eby, R.J. Warmack, and L. Landau, *Anal. Chem.*, 1981, 53, 13.
367 E. Janssen, *Mater. Sci. Eng.*, 1980, 42, 309.
368 F. Degreve, *J. Microsc. Spectrosc. Electron.*, 1981, 6, 223.
369 A.G. Koval, V.N. Mel'nikov, and S.A. Bondar, *Zh. Tekh. Fiz.*, 1981, 51, 148.
370 L.L. Kazmerski, P.J. Ireland, S.S. Chu, and Y.T. Lee, *J. Vac. Sci. Technol.*, 1980, 17, 521.
371 G.F. Romanova, P.I. Didenko, and R.I. Marchenko, *Pis'ma Zh. Tech. Fiz.*, 1981, 7, 404.
372 A.E. Morgan and J.B. Clegg, *Spectrochim. Acta, Part B*, 1980, 35, 281.
373 F. Simondet, P. Dolizy, and M. Decaesteker, *Vide, Couches Minces*, 1980, 201, 275.
374 A. Rupert, S. Haridoso, and F. Beniere, *Analusis*, 1980, 8, 142.
375 V.R. Deline, R.J. Blattner, and C.A. Evans, *Proc. Annu. Conf. Microbeam Anal. Soc.*, 1980, 239.
376 D. Marton, *Appl. Surf. Sci.*, 1980, 5, 65.
377 C. Magee and D.E. Carlson, *Sol. Cells*, 1980, 2, 365.
378 M.R. Ross and J.F. Evans, *Midl. Macromol. Monogr.*, 1980, 7, 99.
379 Y. Tsuhawaki and N. Iwamoto, *Adv. Mass Spectrom.*, 1980, 8A, 527.
380 J.W. Coburn, A.E. Knabbe, and E. Kay, *J. Vac. Sci. Technol.*, 1982, 20, 480.
381 P.A. Bertrand, R. Bauer, and P.D. Fleischauer, *Anal. Chem.*, 1980, 52, 1279.
382 R.G. Orth, H.T. Jonkman, and J. Michl, *J. Am. Chem. Soc.*, 1982, 104, 1834.
383 J.L. Pierce, K.L. Busch, R.G. Cooks, and R.A. Walton, *Inorg. Chem.*, 1982, 21, 2597.
384 D.J. Surman, J.C. Vickerman, and J. Wolsteholme, *Vide, Couches Minces*, 1980, 201, 525.
385 A. Benninghoven, P. Beckmann, and M. Schemmer, *Vide, Couches Minces*, 1980, 201, 1299.
386 A. Benninghoven, P. Beckmann, D. Greifendorg, and M. Schemmer, *Appl. Surf. Sci.*, 1980, 6, 288.
387 S. Bourgeois and M. Perdereau, *J. Microsc. Spectrosc. Electron.*, 1981, 6, 335.
388 S. Bourgeois and M. Perdereau, *J. Microsc. Spectrosc. Electron.*, 1981, 6, 565.